Biocommunication of Ciliates

Guenther Witzany · Mariusz Nowacki
Editors

Biocommunication of Ciliates

Springer

Editors
Guenther Witzany
Telos-Philosophische Praxis
Buermoos
Austria

Mariusz Nowacki
Institut für Zellbiologie
Universität Bern
Bern
Switzerland

ISBN 978-3-319-32209-4 ISBN 978-3-319-32211-7 (eBook)
DOI 10.1007/978-3-319-32211-7

Library of Congress Control Number: 2016935955

Printed on acid-free paper

This Springer imprint is published by Springer Nature
The registered company is Springer International Publishing AG Switzerland

Preface

Whenever living cells, whether prokaryotic or eukaryotic, coordinate their behavior, communication processes are necessary to reach coordinative goals. Each activity on all levels of biocommunication (intraorganismic, interorganismic, transorganismic, and generating response behavior to abiotic influences) is usually achieved by means of specialized signaling. If these signaling processes are disturbed, damaged, or incomplete, the activity remains incomplete or rudimentary or is deleted.

Communication means interactions that are mediated by signals in contrast to purely physico-chemical interactions where no signals are present. Additionally these sign-mediated interactions need rules on how the signals may be combined to transport more complex informational content. Last but not least, communicative interactions mediated by signs depend essentially on living agents that are able to follow such rules of sign use. Additionally it must be mentioned that sign-mediated rule-governed interactions are a kind of social interaction, i.e., communication processes are social events, involving groups of interacting agents that share the rules on how to use signals, and this means that group identity is essential in living nature.

The prerequisites of communication are: (1) its social character, (2) its dependence on the use of signals according to three levels of rules, (3) the primacy of context (pragmatics) which determines the meaning/function (semantics) of the used signs, and (4) the inherent capability of sign-using agents to change these rules of sign use according to environmental or adaptational needs, none of which is found in inanimate nature. For example, no signs, no semiotic rules, and no socially interacting living agents are present when water freezes to ice.

The change of sign-using rules, which gives signals and sign sequences new meanings that never existed before and are not the result of a recombination of former ones, is an inherent feature of living agents competent to use a natural language or a natural code. The generation of new sign sequences is essentially not the result of the selection of beneficial mutations out of an abundance of defective variants. In contrast to this passive derivation narrative of positive selective forces the generation of new sequences is an active process in which natural code-using agents produce new sequences in vivo.

This short description of the essentials of biocommunication contradicts former opinions of communication in natural sciences, information theory, systems theory, mechanistic and other reductionistic approaches which rely on a mathematical theory of language, i.e., the analysis of a quantifiable set of signs. The history of the philosophy of science clearly demonstrates a variety of such approaches in which signals are molecules which are subject to quantitative investigations and comparisons usually based on investigations of the molecular syntax structure of the natural codes. This means that molecular syntax as a result of chance mutations (error replications) and selection represents the material reality of the physico-chemical world which can best be represented by mathematical equations. In this perspective material reality is the only reality because it can be objectivized, measured and empirically investigated. The molecular syntax of natural codes is therefore the information-bearing content out of which its functions and its meaning can be distilled.

Yet this fundamental paradigm was falsified in the 1980s by pragmatic philosophy and sociology which empirically tested that meaning was not represented by the syntax structure of natural codes, but by the context in which sign sequences are used by in vivo interactions of living agents. In contrast to the former narrative, pragmatics (context) determines the meaning of sign sequences, with the consequence that identical sequences of signs may transport different meanings, even contradictory ones. This makes sense in the light of energy costs: it is not necessary to represent an ontological entity or event by unequivocal representations. One sign sequence can designate multiple meanings according to contextual needs.

A sign- and rule-sharing population only needs a limited number of signals and a limited number of rules to produce multiple variant communications; even de novo generation, although rare, is possible in principle.

This means it is not the syntax which is the relevant information for extracting the meaning of signals used to coordinate and organize behavior. It is the context in which social interactions occur, i.e., that in which signs and sign sequences are used. This means the sociological aspect is essential for deciphering the meaning of natural codes.

Biocommunication processes have been documented meanwhile on the whole area of living nature, i.e., plants, animals, fungi, prokaryotes, viruses, and even RNA consortia. The missing publication on communication of single-celled protozoa is presented now. The published works on biocommunication are not the end but the start of a coherent process of investigations and data mining regarding communicative acts within cells, between cells and between non-related organisms during the whole biosphere, and could lead to a better understanding of the principles governing living nature and a better picture of life on this planet to restructure the behavioral motifs of humankind in its relation to non-human living nature.

Communication between ciliates or between ciliates and non-ciliates is not the only subject of this book. Of equal interest was to find the communication pathways within the ciliates how the information transfer between nuclei is organized, i.e., how RNA-mediated genome arrangements and rearrangements are coordinated or conserved or remain an important source for genome plasticity as indicated between parental and offspring ciliates.

The benefit of this new type of research which integrates empirically derived knowledge about ciliate physics and chemistry with pragmatic action theory is its more coherent explanatory power. It complements the current knowledge about the physiology of ciliates and motifs on each level of ciliate life with the available habitats and contexts in which ciliate species live. In contrast to pure reductionistic biology it can integrate the basic motifs of ciliate signaling within varying contexts with the knowledge about all physiological interactions. In contrast to mechanistic biology—it only recalls on the outdated narrative of "information" transfer—the biocommunication approach focuses on the real-life situations in which signaling directs the various forms of interaction. In this context communication is not restricted to information transfer but predominantly acts as a kind of social interaction.

Finally, the editors hope that *Biocommunication of Ciliates* will integrate a diversity of research goals on the function, taxonomy, and genetics of ciliates, representing their main principles of life, evolution, and developmental stages. Understanding the full range of ciliate life will have repercussions for the understanding of life and its evolution in general.

Buermoos, Austria — Guenther Witzany
Bern, Switzerland — Mariusz Nowacki
May 2016

Contents

Chapter 1
Introduction: Keylevels of Biocommunication of Ciliates

Guenther Witzany

Abstract Similarly to other organisms of all domains of life from bacteria to complex animals ciliates are sensitive organisms that assess their surroundings, estimate how much energy they need for particular goals and then realize the optimum variant. They take measures to control certain environmental resources. They perceive themselves and can distinguish between self and non-self. They process and evaluate information and then modify their behavior accordingly. These competences are made possible by sign-mediated communication processes within the ciliate body (intraorganismic), between the same, related and different ciliate species (interorganismic) and between ciliates and non-ciliate organisms (transorganismic). In order to generate an appropriate response behavior ciliates must be able not only to sense but also to interpret and memorize important indices from the abiotic environment and adapt to them accordingly. This is decisive in coordinating growth and development, mating, shape and dynamics. However, these communication, interpretation and memory processes can also fail. In such cases the overall consequences can mean disease or even death for the ciliate. In this respect biocommunication method applied to ciliates could integrate the various levels of ciliate signaling processes.

1.1 Introduction

If we look at recent advances in biocommunication research we find that the investigations of sign-mediated interactions between cell tissues, organs and organisms in all kingdoms have increased in the last decade. It is common knowledge that if such biological entities want to coordinate their behavior or in more specialized cases the cells of organs which share a genetically conserved identity it needs not only the generation of biomolecules that serve as signaling

---

G. Witzany (✉)
Telos-Philosophische Praxis, Vogelsangstr. 18c, 5111 Buermoos, Austria
e-mail: witzany@sbg.at

G. Witzany and M. Nowacki (eds.), *Biocommunication of Ciliates*,
DOI 10.1007/978-3-319-32211-7_1

tools to transport important messages and their use in different contextual circumstances but the signals to act within interactional motifs to reach certain individual or common goals. Communication is essential to reach such goals and the investigations on biocommunication tell us whether communication functions as the ultimate pre-condition to reach such goals or it does not function for one or several reasons and the message is deformed or damaged so common goals cannot be reached. In the case of cell coordination or organization of organs or group behavior in organisms this might lead to disease or similar problems.

Ciliates are well investigated unicellular eukaryotes and together with other phyla of alveolates have a broad variety of lifeforms and lifecycles (Lynn 2008). Of additional interest in this context of identifying sign-mediated interactions of ciliates is that these single-celled eukaryotes seem to represent the most evolved and complex protozoans. As they are very old in evolutionary terms and date back to the origins of unicellular eukaryotes they can be found on the whole planet except in dry land ecospheres. This tells us that they are excellently adapted and have successfully colonized most ecospheres on this planet and in several cases are interaction partners of symbiotic lifestyles.

Ciliates in most cases are free-living, few of them are parasitic, and some may cause diseases in fish aquacultures. Others are stabilizing bacteria in the gut of hoofed mammals, which is a kind of symbiotic lifestyle. Most ciliates feed on bacteria, other ciliates (cannibalism), fungi or algae.

Of special interest is the parallel representation of two nuclei in most ciliates: a micronucleus which serves as a competence center for reproduction and a macronucleus which serves as a competence center for metabolism and development (Schoeberl and Mochizuki 2011). Ciliates reproduce asexually by division. The micronucleus undergoes mitosis whereas the macronucleus splits into two. Yet ciliates can also reproduce sexually in that two partners of opposite mating types conjugate via a cytoplasmatic bridge. Then the micronuclei divide by meiosis, the macronuclei disintegrate and the conjugating cells exchange haploid micronuclei. Then they segregate, reform the macronuclei from their micronuclei and divide.

1.2 The Concept of Biocommunication

When we consider the biocommunication of ciliates, we must first become familiar with the current terms of communication (and with the signaling system, which we would term language or code, the essential tool for communication).

When we speak about language and communication we usually think of humans that talk to each other and communicate to organize common goals and to coordinate common behavior. However, since Karl von Frisch received the Nobel Prize for detection and investigation of bee languages and dialects it is evident that even non-human social animals may communicate to construct complex behavioral patterns. The Nobel Laureate Manfred Eigen insisted nearly at the same time that when we speak about the genetic code we are speaking about a real language, not just a metaphor (Witzany 1995). Concerning these fundamental insights I developed the biocommunicative approach which investigates both communication (1) and language (2) as universal requirements for life. The first such investigation was published in 1993 (Witzany 1993) followed by an updated version in 2000 (Witzany 2000) as a draft on the theory of communicative nature. The results in virology and the role of viruses in evolution and developmental processes in particular exemplified a variety of the proposed pre-assumptions therein (Witzany 2009; Villarreal and Witzany 2010, 2013a, b, 2015). The theory of biocommunication was outlined first as a program in 2007 updated in 2010 (Witzany 2010) concerning bees, corals, plants, fungi, bacteria, viruses and subviral RNAs. Several more detailed exemplifications followed in a series of books I edited between 2011 and 2014 with leading researchers in their field (Witzany 2011, 2012a, b, 2014; Witzany and Baluska 2012a, b).

1.2.1 The Mechanistic Narrative Used to Explain Communication Is Outdated

The method for analyzing any part of a machine in detail to get a picture of the whole functional blueprint, which can then be used to reproduce or manipulate it, or to produce an even more perfect example, to take artificial genetic engineering as an example, is still useful if we are dealing with machines.

In contrast, communication between cells, cellular components, tissues, organs and organisms is far from being a procedure that can be reduced to mechanistic input/output or cause/reaction descriptions. It is evident that communication processes between living organisms include a variety of non-mechanistic circumstances and competences that must be satisfied in parallel if communicative acts are to have successful consequences; for example, innovating common coordination to adapt to new environmental conditions. Machines cannot create new programs out of a functional blueprint, which is in contrast to the abilities of living organisms that are able to communicate with each other. The universal Turing machine and the self-reproducing machines of von Neumann still remain at the conceptual stage. However, no single self-reproducing machine has ever been produced within the last 80 years. There are good reasons for this, because it is, in principle, impossible that an artificial machine could reproduce itself (Witzany and Baluska 2012b). In contrast to the artificial machines which cannot reproduce themselves, living cells and organisms can reproduce themselves and additionally generate an abundance of behavioral motifs for which no algorithm can be constructed, such as de novo generation of coherent nucleotide sequences.

1.2.2 Communication Is Interaction Between Living Agents Mediated by Sign(al)s

Coherently with current knowledge about natural communication processes communication is defined as the sign-mediated interaction between at least two living agents which share a repertoire of signs (which represent a kind of natural language) that are combined (according to syntactic rules) in varying contexts (according to pragmatic rules) to transport content (according to semantic rules). This means monological concepts such as the sender-receiver narrative, in which the sender codes information and the receiver decodes it according to inherited programs, cannot explain the emergence of commonly shared meanings.

Contrary to former concepts the importance of this result is that these three levels of semiotic rules (*semeion* = sign) are complementary parts of any natural language or code. If one level is missing, according to Charles Morris, we cannot seriously speak of language- or signal- mediated communication. Therefore, the most recent definition of communication is: sign-mediated and rule-governed interactions, i.e. interactions that depend on a shared repertoire of signs and rules (Witzany 2010). However, these features are lacking in abiotic interactions; no semiotic rules are necessary if water freezes to ice.

Additionally, we know that mathematical and mechanistic theories of language are less helpful in investigating natural language and real-life communication processes, because such theories cannot explain typical features of communication between living entities, which are not formal (i.e. for which no algorithm is available), such as (1) the de novo generation of coherent, sentences or sequences or (2) different and even contradictory meanings of identical syntactic sequences/sentences. This means that no natural language or code speaks, or codes, itself but requires living agents that are competent in such languages or codes (Witzany 2010).

In the biology of the twentieth century, the physiology of all manner of cells, tissues, organs and organisms was the mainstream direction of biological research and experiments. In the 1970s, an increasing use of "communication" as a metaphor also arose in biology. During the last decade of the twentieth century, interest in communication (no longer used as a metaphor) within, and between, organisms overtook that of the purely physiological understanding of organisms. This was owed to concrete communication processes designating varying contexts in real-life circumstances. Cell-to-cell communication now dominates contemporary cell biology, including knowledge of a great variety of signaling pathways, serving for both organization and coordination of production, release, uptake, interpretation and processing of context-dependent "information" (content) within and between cells. Context dependency determines the crucial fact that it is not the syntax (grammar) of a sequence of signs (information) which determines the meaning (semantics), but the context (pragmatics) in which the concrete use of the sequence occurs.

In parallel, the use of "language" as a metaphor has increased since the middle of the twentieth century with the improved knowledge of the genetic code. Most of the processes that evolve, constitute, conserve and rearrange the genetic storage medium (DNA) are terms that were originally used in linguistics. For example: nucleic acid language, genetic code, "code without commas" (F. Crick), coding, copying, translation, transcription, "genetic text" (F. Jacob), sequence homology, etc. Meanwhile, the linguistic approach has also lost its metaphorical character, and the similarity between natural languages and codes and the genetic storage medium of DNA have not only been accepted but adopted in epigenetics, comparative genomics, bioinformatics, biolinguistics, biocommunication theory and biosemiotics.

1.2.3 Communicative and Linguistic Competencies and the Primacy of Pragmatics

First, no single organism is able to communicate as an emerging property; it must be part of a community, society or swarm of organisms that share an identity and have the ability to sense whether others are part of this identity or not (self/non-self differentiation competence), even if this competence is solely shared genetically. For communication it is necessary for organisms to have assets that serve as signs, signals or symbols, such as chemical molecules, either produced directly by the organism or as secondary metabolites, or even molecules in the surroundings, which can be manipulated according to the organism's needs. In the case of animals, especially complex ones, visible and audible sign repertoires have evolved.

Second, organisms must also share the ability to use these signs in a coherent manner, which means using them in a strictly temporal and spatial context. In most cases it is not just one sign but several that are combined in a specific manner to transport messages or information. This represents a common feature of sign use in communication processes, and is termed the correct combination or syntax.

Third, organisms are part of the natural habitat in which they live, together with similar organisms of the same or related species, but usually also with an abundance of unrelated organisms. This historically developed context fully represents the natural history of swarms or communities and the way in which they have evolved certain abilities and are able to mount appropriate response behaviors to enable their survival. These competencies, which include sensing, monitoring, learning and memory, are preconditions for faster adaptation.

Finally, signaling molecules, which serve as signs, transport messages with meanings (semantics). The informational (semantic) content which is transported triggers certain response behaviors in the same, related, or even unrelated, organisms. Interestingly, the signal sequence or content does not necessarily depict a strict meaning, i.e. a function, but can vary according to different situational contexts. This means that identical signs can transport a variety of different messages according to different contextual needs. The different uses of identical signs, or sequences, enable the generation of dialects within the same species that can transport messages which are microecosphere-specific. This includes very sensitive self/non-self recognition between slightly differently adapted populations of the same species in the same ecological habitat (Witzany 2000).

Although sign-mediated interactions (i.e. communication processes) are very reliable in most cases, they do not function mechanistically in a strict sense. Syntax (combination), pragmatics (context) and semantics (content) must function in parallel to ensure and optimize the coordination and thus survival of group members.

These three levels of semiotic rules (syntax, pragmatics and semantics) do not function mechanistically but can be varied, deleted, or, in certain circumstances, and in contrast to the capabilities of machines, generated de novo. Additionally, semiotic rules do not function by themselves but need semiotic subjects, i.e. living organisms that utilize such rules. If no living organism is present, semiotic rules, signs and communication are absent. Although highly conserved semiotic rules are modifiable, environmental circumstances, such as stress, can trigger adaptive responses. In such cases, signals may transport new messages which previously did not exist, broadening the communicative competences of organisms, i.e. broadening their evolutionary capabilities.

Natural communication assembles the full range of signal-mediated interactions that are necessary in order to organize all evolutionary and developmental coordination within, and between, cells, tissues, organs and organisms. To identify biocommunication of organisms we therefore have to look at the interaction motifs in the real lifeworld context of the organisms (Witzany 2010).

1.3 In Vitro Analyses Lack Context-Dependent Behaviors of Real-Life Habitats

In vitro investigations focus on ecological setups, which do not represent the entire interactional context in which an organism is involved in vivo. The evolution and development of each organism depends on the in vivo habitat with its inter-, intra-, and transorganismic triggers of genetic reading patterns, which are absent from in vitro setups. Therefore, it is likely that isolated organisms in laboratory setups lack a variety of features which would be triggered in in vivo habitats by natural phenomena such as symbiotic and parasitic microorganisms. This may lead to limited conclusions regarding their intra- and interorganismic biocommunicative capabilities.

1.4 The Biocommunication Method Applied to Ciliates

The advantage of a methodical adaptation of communication and linguistic terminology is that it provides appropriate tools for differentiation at specific levels, which are otherwise difficult to describe in non-reductive terms with pure

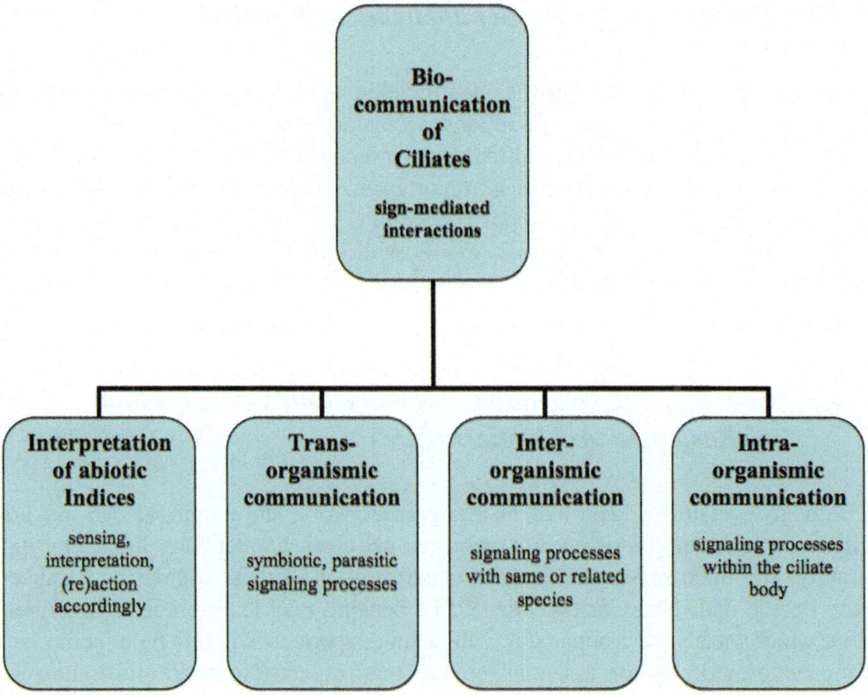

Fig. 1.1 The several levels of biocommunication of ciliates

physiology. The result of this is that language-like structures and communication processes occur at the simplest levels of nature. Language and communication are not the evolutionary inventions of humans, nor are they anthropomorphic adaptations for describing non-human entities. It is an empirical fact that all coordination and organization within and between cells, tissues, organs and organisms needs signs, i.e. molecules that serve as signals or symbols in messages, or serve as vital indicators of environmental conditions. Because no natural code can encode itself, in the way that no natural language can speak itself, these signs must be sensed and interpreted in the correct way by biological agents, i.e. there must be subjects of sign production and sign interpretation. The consequence of this is that sensing, as well as interpretation and memory storage, may fail. This can result in inappropriate behavior of, or even have fatal consequences for, cells, tissues, organs, and organisms.

The biocommunication approach can integrate empirical data into a holistic perspective of an organism and its interactional patterns at various levels (Fig. 1.1).

1.4.1 Designation of the Semiochemical Vocabulary

Investigations may start with the identification of the used semiochemicals, i.e. chemicals such as hormones, secretions, secondary metabolites that serve as signs within interactions of various motifs, such as reproduction, mating, feeding, attack or defense, etc. In this respect it is important to interconnect the semiochemicals with the concrete interactional motif (Luporini et al. 1995, 2006). Identical signals may be used in several interaction motifs with quite different meanings and this indicates the context dependence of signals.

1.4.2 Sensing, Interpretation, Memory and Coordination of Response to Abiotic Indices

The next level to be identified in this perspective is the empirical data on how ciliates sense their abiotic environment, i.e. pH level, temperature, light, moisture, gravity, etc. Such sensing is crucial for the survival of these organisms because it may decide their behavior (Weisse 2014). Sensing may lead to correct interpretation, which means the comparison with lasting experiences in this field memorized. Epigenetic markings are essential tools in this respect (Nowacki and Landweber 2009). The benefit of correct interpretation is that it may lead to faster and more appropriate reactions.

1.4.3 Intraorganismic Communication in Ciliates

The various behavioral motifs as appropriate response behavior to environmental sensing, interpretation and memory leads us to various signaling pathways within ciliate organisms (Prescott 1994, 1999; Zufall et al. 2006). This will focus on signalling within the cell nuclei and between cell nuclei and surrounding cell bodies to generate available resources for the generation of signaling of the ciliates to other organisms (Swart et al. 2014). In ciliates we see the first evolutionary separation of the germline and the soma (Chalker 2008). The special kind of signaling which is needed for that is of exceptional interest and leads us to the interactions between ciliates and persistent viruses.

1.4.3.1 The Role of Persistent Viruses

Viruses have long been accepted only as disease-causing, epidemic phenomena with lytic and therefore dangerous consequences for infected organisms. However, new research has corrected this picture. Viruses are part of the living world, in most cases integrated in the cytoplasm or the nucleoplasma of cells without harming the host. Viruses are on their way to representing the best examples of symbiotic relationships, because there is no living being since the start of life that has not been colonized by them, most often in the form of multiple colonizations (Villarreal 2005).

As mentioned in recent years, the lytic consequences of viral infection are a special case if viruses are not able to develop a sessile lifestyle without harming the host. In most cases, viruses living within organisms help to ward off competing parasites from the host and become part of its evolutionary history. Persistent viruses are decisive for species diversity and host genome editing (Villarreal 2009). The persistent status is most often reached by a competing genetic parasite that counterregulates their competences, now co-opted by the host organism for regulating all replication-relevant pathways and intron excision (Villarreal 2005, 2015).

In ciliates the dual nuclei merit special attention in this respect. The genome-wide DNA rearrangements in the macronucleus are different from most other organisms. In the macronucleus the sequences are precisely excised and degraded. The excision is controlled epigenetically and during development thousands of these sequences are excised. Some thousand of deletion events occur during the formation of the somatic macronucleus which include sequences containing long terminal repeats (LTRs), which resemble transposon-like elements. Such LTRs are indicators of genetic parasites and represent repeat sequences that reached a persistent status in the ciliate genome by counteregulation of other genetic parasites. The remaining LTR sequences clearly resemble regulative functions co-adapted to the ciliate genome (Villarreal 2005).

The benefit for host organisms colonized successfully by persistent viruses that remain as co-adapted regulatory tools is the immune function against similar genetic colonizers (Villarreal 2011; Sharma et al. 2016; Seligman and Raoult 2016). This may explain why there are so few DNA viruses in ciliates. As mentioned, the presence of LTR transposons protects the replicons of the ciliates from DNA degradation and prevents the ciliate organisms from lytic infection by any DNA viruses.

Interestingly there are some indicators that both nuclei derived from large doublestranded DNA virus infections that reached a persistent lifestyle within the ciliates. In particular the silent status of the micronucleus is quite similar to latent genomes of DNA viruses. Also, the silent nucleus involved in germine transmission possesses the general ability of latent viruses to be associated with sexual reproduction. In parallel the transcriptional activation and overreplication of the macronucleus during sexual reproduction represent a virus-like behavior such as the lytic activation of a latent DNA virus characterized by DNA amplification outside cell cycle control and consequent degradation of non-amplified DNA sequences (Villarreal 2005). The ciliate's dual-nucleus identity represents a very interesting example of how infective agents reached a persistent status in host genomes and serve as competent regulatory tools for the host organism (Swart and Nowacki 2015).

1.4.4 Interorganismic Communication in Ciliates

Sign-mediated interactions between ciliates can be investigated by comparing signaling and the behavioral motifs (Schtickzelle et al. 2009; Jacob et al. 2015). In mating ciliates produce and exchange a variety of signals that are sensed and interpreted. Additionally the coordinated behavior between two ciliates depends on correct uptake of the signals by the receptors and transport to the intracellular domain of the ciliate for further information processing (Wood and Rosenbaum 2015).

1.4.5 Transorganismic Communication in Ciliates

If ciliates interact with non-ciliate organisms we can identify several signaling pathways which coordinate this interaction either in a variety of symbiotic or parasitic lifestyles (Nakajima et al. 2013; Lobban and Schefter 2014). As parasitic ciliates can cause serious damage to the aquaculture of fish it is important to identify such parasitic lifestyles to prevent further damage.

1.5 Summary

An overview of all significant levels of ciliate communication shows that identification of signal-mediated processes in signaling pathways are context-dependent both within and among ciliates as well as between ciliates and other organisms. Depending on the context, semiochemicals (molecular components) are integrated into unique signaling pathways where they are used to transport certain meanings. Such meanings are subject to change, i.e. they rely on various behavioral contexts, which differ under altered conditions. The interactional context determines the semantic relationship, i.e. its meaning and the function of the chemical components, and forms a signal-mediated communication pattern in ciliates.

After recognizing how versatile ciliate communication competences really are we can see that one main principle is followed throughout all these signaling processes: ciliates coordinate all their behavioral patterns with a core set of chemical molecules. The interactional context and the different modes of coordinating appropriate response behavior in e.g. development, growth, mating, attack, defense, feeding, etc. determine the combinations of signals that generate the appropriate meaning function, i.e. the informational content of messages. These generating processes normally function in a very conservative way but under certain circumstances may fail, or selective pressure may lead to changes that can be a driving force in ciliate evolution. Additionally it must be recognized that the persistent lifestyle of viruses is a driving force in ciliate evolution in that they are the main resource for immunity and genetic identity.

References

Chalker DL (2008) Dynamic nuclear reorganization during genome remodeling of Tetrahymena. Biochim Biophys Acta 1783:2130–2136

Jacob S, Chaine AS, Schtickzelle N, Huet M, Clobert J (2015) Social information from immigrants: multiple immigrant-based sources of information for dispersal decisions in a ciliate. J Anim Ecol 84:1373–1383

Lobban CS, Schefter M (2014) Cluster dynamics in Maristentor dinoferus, a gregarious benthic ciliate with zooxanthellae and a hypericin-like pigment, in relation to biofilm grazing by the fish Ctenochaetus striatus. Symbiosis 63:137–147

Luporini P, Vallesi A, Miceli C, Bradshaw RA (1995) Chemical signaling in ciliates. J Eukaryot Microbiol 42:208–212

Luporini P, Vallesi A, Alimenti C, Ortenzi C (2006) The cell type-specific signal proteins (pheromones) of protozoan ciliates. Curr Pharm Des 12:3015–3024

Lynn D (2008) The ciliated protozoa, 3rd edn. Springer, Dordrecht

Nakajima T, Matsubara T, Ohta Y, Miyake D (2013) Exploitation or cooperation? Evolution of a host (ciliate)-benefiting alga in a long-term experimental microcosm culture. Biosystems 113:127–139

Nowacki M, Landweber LF (2009) Epigenetic inheritance in ciliates. Curr Opin Microbiol 12:638–643

Prescott DM (1994) The DNA of ciliated protozoa. Microbiol Rev 58:233–267

Prescott DM (1999) The evolutionary scrambling and developmental unscrambling of germline genes in hypotrichous ciliates. Nucleic Acids Res 27:1243–1250

Schoeberl UE, Mochizuki K (2011) Keeping the soma free of transposons: programmed DNA elimination in ciliates. J Biol Chem 286:37045–37052

Schtikzelle N, Fjerdingstad EJ, Chaine A, Clobert J (2009) Cooperative social clusters are not destroyed by dispersal in a ciliate. BMC Evol Biol 9:251

Seligmann H, Raoult D (2016) Unifying view of stem–loop hairpin RNA as origin of current and ancient parasitic and non-parasitic RNAs, including in giant viruses. Curr Opin Microbiol 31:1–8

Sharma V, Colson P, Pontarotti P, Raoult D (2016) Mimivirus inaugurated in the 21st century the beginning of a reclassification of viruses. Curr Opin Microbiol 31:16–24

Swart EC, Nowacki M (2015) The eukaryotic way to defend and edit genomes by sRNA-targeted DNA deletion. Ann NY Acad Sci 1341:106–114

Swart EC, Wilkes CD, Sandoval PY, Arambasic M, Sperling L, Nowacki M (2014) Genome-wide analysis of genetic and epigenetic control of programmed DNA deletion. Nucleic Acids Res 42:8970–8983

Villarreal LP (2005) Viruses and the evolution of life. ASM Press, Washington

Villarreal LP (2009) Origin of group identity: viruses, addiction, and cooperation. Springer, New York

Villarreal LP (2011) Viral ancestors of antiviral systems. Viruses 3:1933–1958

Villarreal LP (2015) Force for ancient and recent life: viral and stem-loop RNA consortia promote life. Ann NY Acad Sci 1341:25–34

Villarreal LP, Witzany G (2010) Viruses are essential agents within the roots and stem of the tree of life. J Theor Biol 262:698–710

Villarreal LP, Witzany G (2013a) The DNA habitat and its RNA inhabitants: at the dawn of RNA sociology. Genomics Insights 6:1–12

Villarreal LP, Witzany G (2013b) Rethinking quasispecies theory: from fittest type to cooperative consortia. World J Biol Chem 4:71–82

Villarreal LP, Witzany G (2015) When competing viruses unify: evolution, conservation, and plasticity of genetic identities. J Mol Evol 80:305–318

Weisse T (2014) Ciliates and the rare biosphere—community ecology and population dynamics. J Eukaryot Microbiol 61:419–433

Witzany G (1993) Natur der Sprache – Sprache der Natur. Sprachpragmatische Philosophie der Biologie. Koenigshausen & Neumann, Würzburg
Witzany G (1995) From the "logic of the molecular syntax" to molecularpragmatism. Explanatory deficits in Manfred Eigen's concept of language and communication. Evol Cogn 1:148–168
Witzany G (2000) Life: the communicative structure. A New Philosophy of Biology, BoD, Norderstedt
Witzany G (2009) Noncoding RNAs: persistent viral agents as modular tools for cellular needs. Ann NY Acad Sci 1178:244–267
Witzany G (2010) Biocommunication and natural genome editing. Springer, Dordrecht
Witzany G (ed) (2011) Biocommunication in soil microorganisms. Springer, Heidelberg
Witzany G (ed) (2012a) Biocommunication of fungi. Springer, Dordrecht
Witzany G (ed) (2012b) Viruses: essential agents of life. Springer, Dordrecht
Witzany G (ed) (2014) Biocommunication of animals. Springer, Dordrecht
Witzany G, Baluska F (eds) (2012a) Biocommunication of plants. Springer, Dordrecht
Witzany G, Baluska F (2012b) Life's code script does not code itself. The machine metaphor for living organisms is outdated. EMBO Rep 13:1054–1056
Wood CR, Rosenbaum JL (2015) Ciliary ectosomes: transmissions from the cell's antenna. Trends Cell Biol 25:276–285
Zufall RA, McGrath CL, Muse SV, Katz LA (2006) Genome architecture drives protein evolution in ciliates. Mol Biol Evol 23:1681–1687

Chapter 2
Principles of Intracellular Signaling in Ciliated Protozoa—A Brief Outline

Helmut Plattner

Abstract Ciliates have available most of the intracellular signaling mechanisms known from metazoans. Long-range signals are represented by firmly installed microtubules serving as gliding rails aiming at specific targets. Many components are distinctly arranged to guarantee locally restricted effects. Short-range signals include Ca^{2+}, provided from different sources, and proteins for membrane recognition and fusion, such as SNAREs, GTPases and high affinity Ca^{2+}-binding proteins (still to be defined). A battery of ion conductances serves for electric coupling from the outside medium to specific responses, notably ciliary activity, which also underlies gravitaxis responses. Eventually cyclic nucleotides are involved, e.g. in ciliary signaling. Furthermore, an elaborate system of protein kinases and phosphatases exerts signaling mechanisms in widely different processes.

2.1 Introduction—Basic Aspects of Signaling in Ciliates

As for every eukaryotic cell one may ask also for ciliates which cellular processes require signaling, how signaling is executed and over which distances, whether principles are shared with metazoans and plants, whether mechanisms are maintained during evolution, abolished or newly invented. Together with *Dictyostelium*, the ciliates *Paramecium* and *Tetrahymena* represent the protozoa which, at this time, are best analyzed with regard to signaling. It is useful to differentiate between long- and short-range signaling, e.g. by microtubules or electrical signals and by molecular interactions or spatially restricted Ca^{2+} signals (Plattner and Klauke 2001), respectively.

H. Plattner (✉)
Department of Biology, University of Konstanz, P.O. Box M625, 78457 Konstanz, Germany
e-mail: helmut.plattner@uni-konstanz.de

G. Witzany and M. Nowacki (eds.), *Biocommunication of Ciliates*,
DOI 10.1007/978-3-319-32211-7_2

2.1.1 Basic Phenomena Applicable to Ciliates

Signaling pertinent to ciliary activity in ciliated protozoa is as elaborate, or even more than in metazoan (Machemer 1988a) as these cells are highly mobile and capable of reacting to various environmental stimuli (Machemer 1988b; Bell et al. 2007). To achieve this, mechanical, electrical, biochemical and molecular signals, i.e. long range and short-range signals, can be combined in some variation to the basic theme.

Ciliates have at their disposal a highly sophisticated vesicle trafficking system, as illustrated (http://www5.pbrc.hawaii.edu/allen/) and summarized (Allen and Fok 2000) for *Paramecium* and *Tetrahymena* (Frankel 2000). The routes have to be addressed here. (i) Endoplasmic reticulum (ER) → Golgi apparatus → lysosomes + dense core-secretory organelles (trichocysts in *Paramecium* and mucocysts in *Tetrahymena*). (ii) Constitutive exocytosis of surface coat materials (Flötenmeyer et al. 1999) and dense core-secretory organelle exocytosis (Plattner et al. 1985; Plattner and Kissmehl 2003a). (iii) Phagocytosis, from cytostome → phagosome → endosomal and lysosomal input → phagolysosome (called "food [digesting] vacuole") → discharge of spent vacuoles at the cytoproct. (iv) Endocytosis via early endosomes → links to phagosomes + lysosomes. (v) Vesicle recycling from the cytoproct to the nascent phagosome. (vi) In addition, the contractile vacuole complex impresses not only by its dynamic activity (Allen and Naitoh 2002) in the context of ongoing osmoregulation (Allen et al. 2009), but it also represents a site endowed with the machinery typical of vesicle trafficking (Plattner 2015b) although vesicle trafficking within the organelle is less obvious. Steps (iii) to (v) have been documented in detail for *Paramecium* (Allen and Fok 2000) as well as for *Tetrahymena* (Frankel 2000). Beyond short-range signaling, steps (i), (iii) and (v) include long-range signaling. All these pathways serve for proper delivery and positioning of signaling elements so that they can execute their signaling function at distinct sites of the cell.

2.1.2 Molecular Key Players

Recent availability of a macronuclear genome database for the most frequently used species, *P. tetraurelia* and *T. thermophila*, has enabled the identification, localization and assessment of the functional relevance of key players. In *Paramecium* such work has included mainly SNARE (*s*oluble *N*-ethylmaleimide sensitive factor [NSF] *a*ttachment protein *re*ceptors) proteins, actin and H^+-ATPase, as summarized previously (Plattner 2010) as well as Ca^{2+}-release channels (CRC) of the type inositol 1,4,5-trisphosphate receptors ($InsP_3R$) and ryanodine receptor-like proteins (RyR-LP) (Ladenburger and Plattner 2011; Ladenburger et al. 2006, 2009; Plattner 2015a), as summarized recently (Plattner and Verkhratsky 2015). This is complemented by monomeric GTP (guanosine trisphosphate) binding proteins

(G-proteins), the GTPases, not only in higher eukaryotes (Zhen and Stenmark 2015) but also in ciliates (Bright et al. 2010). Isoforms, i.e. paralogs or ohnologs (in case of diversification following whole genome duplications, particularly in *P. tetraurelia*), can be assigned to different steps and routes of vesicle trafficking and, thus, mirror the high complexity of the ciliate cell.

2.1.3 Long- and Short-Range Signals

The distinction between short-range and long-range signals has been extensively elaborated elsewhere (Plattner 2016a). A typical long-range signal is the docking of trichocysts (Aufderheide 1978) along microtubules which emanate from ciliary basal bodies and, thus, serve as transport rails (Plattner et al. 1982). This has to be complemented by short-range signals. For instance SNAREs and G-proteins are important for vesicle docking and finally membrane fusion. Local Ca^{2+} increase is another signal which has to arise from a nearby source since Ca^{2+} signals decay rapidly (Neher 1998). This also guarantees selective activation of distinct sites and also avoids cytotoxicity (Plattner and Verkhratsky 2015). Local restriction of Ca^{2+} signaling is most obvious, for instance, by the assignment of different CRCs types to different trafficking organelles, from the cell surface to deep inside, in *Paramecium* (Plattner 2015a). Moreover, ciliates fascinate particularly by their highly regular design that predetermines their vesicle trafficking routes and signaling sites based on epigenetic phenomena (Frankel 2000; Beisson 2008). Accordingly, cilia and secretory organelles are arranged in a strikingly regular surface pattern.

2.2 Overview of Trafficking Regulation Along Different Signaling Pathways

Basic trafficking pathways in ciliates are outlined in Fig. 2.1a. Box 1 outlines different kinds of cytoplasmic signaling operating in ciliates. Despite the old evolutionary age of ciliates, signaling mechanisms are quite similar to those in animals and—with exceptions—in plants. Similarities encompass the role of monomeric GTP-binding proteins (G-proteins acting as GTPases) (Bright et al. 2010), H^+-ATPase, SNARE proteins and their chaperone, NSF, as well as the regulation of membrane fusion by a local Ca^{2+} signal (Plattner 2010). The importance of luminal acidification of trafficking vesicles is derived from the observation that a transmembrane signal generated by the conformational change of H^+-ATPase intramembranous V0 part causes binding of GTPase modulators (Hurtado-Lorenzo et al. 2006), thus facilitating docking and membrane fusion. Specificity of vesicle interaction is finally mediated by SNAREs (Plattner 2010) and GTPases (Bright et al. 2010). Sequences encoding GTPases and GTPase modulators, such as GAP

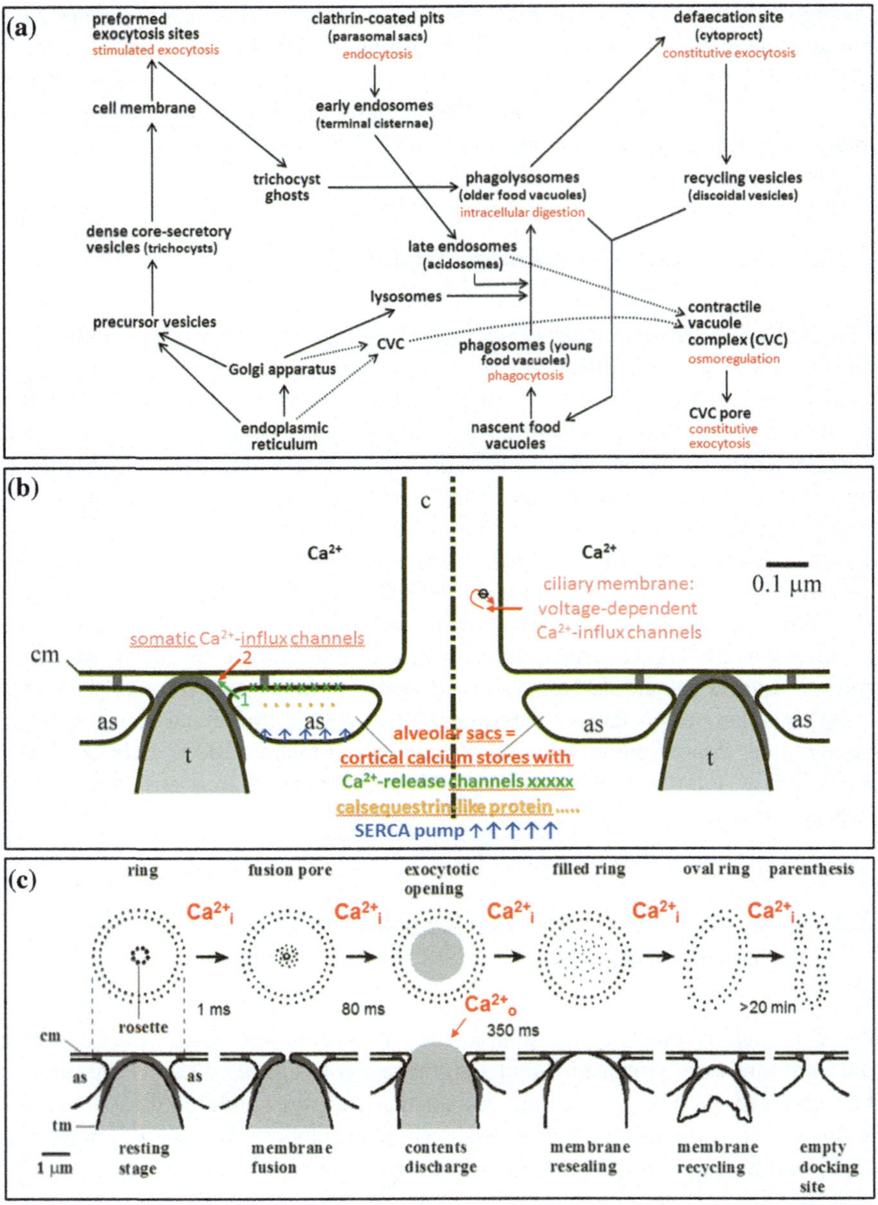
(a)
preformed exocytosis sites
stimulated exocytosis
clathrin-coated pits (parasomal sacs)
endocytosis
defaecation site (cytoproct)
constitutive exocytosis
cell membrane
early endosomes (terminal cisternae)
trichocyst ghosts
phagolysosomes (older food vacuoles)
intracellular digestion
recycling vesicles (discoidal vesicles)
dense core-secretory vesicles (trichocysts)
late endosomes (acidosomes)
lysosomes
contractile vacuole complex (CVC)
osmoregulation
precursor vesicles
CVC
Golgi apparatus
phagosomes (young food vacuoles)
phagocytosis
CVC pore
constitutive exocytosis
endoplasmic reticulum
nascent food vacuoles
(b)
c
Ca2+
Ca2+
0.1 µm
ciliary membrane: voltage-dependent Ca2+-influx channels
somatic Ca2+-influx channels
cm
2
1
as
as
as
as
t
t
alveolar sacs = cortical calcium stores with
Ca2+-release channels xxxxx
calsequestrin-like protein .....
SERCA pump ↑↑↑↑↑
(c)
ring
fusion pore
exocytotic opening
filled ring
oval ring
parenthesis
Ca2+i
Ca2+o
rosette
1 ms
80 ms
350 ms
>20 min
cm
as
as
tm
1 µm
resting stage
membrane fusion
contents discharge
membrane resealing
membrane recycling
empty docking site

◀ **Fig. 2.1** Signaling pathways in the *Paramecium* cell. **a** Vesicle trafficking pathways encompass different main streams, such as the exocytotic, the endocytotic, the phagocytotic pathway and less overt trafficking in the contractile vacuole complex. *Dotted arrows* are less well established, particularly membrane input into this organelle via acidosomes, as derived from various recent papers about other protists. Also for proteins passing or bypassing the Golgi apparatus has not yet been sufficiently specified in detail. **b** Cortial organelles, such as cilia and exocytosis sites are regulated separately. Depolatization induces ciliary beat reversal by Ca^{2+} influx via ciliary voltage-dependent Ca^{2+} channels, abolished via negative feedback (⊖) by intraciliary [Ca^{2+}] increase. CRCs in alveolar sacs, type RyR-LPs, are facing the plasmamembrane, opposite to the SERCA pump. Alveolar sacs contain a calsequestrin-like high capacity/low affinity CaBP. Trichocyst exocytosis is governed by a SOCE mechanism (store-operated Ca^{2+}-entry), i.e. Ca^{2+} release from alveolar sacs in a first step, followed by Ca^{2+} influx via somatic (non-ciliary) channels in a tightly coupled second step. **c** Summary of events during trichocyst exocytosis. *Top* Freeze-fracture images of fusion/resealing stages and their estimated duration, derived from synchronous stimulation/quenched-flow/rapid freezing analysis. Note decay of rosette particle aggregates and rapid formation of a fusion pore which expands and, thus, allows Ca^{2+} access to the secretory contents which triggers their explosive discharge by densondensation (stretching). *Below* Parallel situations seen on ultrathin sections. **a** Data pertinent to trichocyst processing are based on previous reviews (Plattner et al. 1993; Plattner 2014), those for endo-/phagocytotic trafficking are mainly derived from Allen and Fok (2000) and **c** trafficking in context of the contractile vacuole complex is based on recent reviews (Plattner 2015b, 2016a) **b** is modified from Plattner (2014), **c** is modified from Plattner et al. (1993, 1997)

(guanine nucleotide activation protein) and GEF (guanine nucleotide exchange factor), also occur in the *P. tetraurelia* database (Plattner and Kissmehl 2003b).

The Ca^{2+} signal is generated by intracellular CRCs of which different types are assigned to different organelles (Ladenburger and Plattner 2011; Plattner and Verkhratsky 2013). The Ca^{2+}-sensor causing fusion, as known from higher eukaryotes, is a low capacity/high affinity Ca^{2+}-binding protein (CaBP) which usually contains two high affinity Ca^{2+}-binding C2 domains (β-barrels with a Ca^{2+}-binding loop), such as synaptotagmin (Rizo et al. 2006; Südhof 2014). Although such CaBPs have not yet been specified in ciliates, equivalents of synaptotagmin occur in the *P. tetraurelia* database (Farrell et al. 2012). Extended synaptotagmins (e-syntag) with more than two C2 domains are known from some mammalian cells (Min et al. 2007), but they also occur in the *Paramecium* database (H. Plattner and R. Kissmehl, unpublished observations). Calmodulin (CaM) is another low capacity/high affinity CaBP, with four EF-hand type loops, each with high affinity Ca^{2+}-binding capacity. CaM operates at many sites also in ciliates. In the CaM molecule, the extensive conformational change upon hierarchical Ca^{2+} binding in the EF-hand loops I to IV represents the transduction of a chemical to a molecular-mechanical signal (Park et al. 2008). Thus, CaM can regulate a variety of surface influx channels (Saimi and Kung 2002), phagocytosis (Gonda et al. 2000) and probably endocytosis, also in ciliates.

Box 1 also shows that Ca^{2+} for activation may eventually also come from the outside medium for some specific effects, e.g. for activating some nucleotide cyclases, kinases and phosphatases, in the context of ciliary activity. This includes the signaling function of cyclic nucleotides, such as cyclic adenosine monophosphate (cAMP) and cyclic guanosine monophosphate (cGMP) and activation of the

Box 1. Kinds of cytoplasmic signals operating in ciliates – a survey

Electrical signals: depolarization, hyperpolarization → change of ciliary beat

Acidification of organelle lumen

H^+-ATPase/pump → binding of GTPases + their modulators → membrane docking and fusion by interference of SNARE proteins (SNARE = soluble N-ethylmaleimide sensitive factor [NSF] attachment protein receptors) and low capacity/high affinity Ca^{2+}-binding proteins (CaBPs) with C2-domains

Ca^{2+}, from outside or from intracellular stores (Ca^{2+}_o, Ca^{2+}_i)

→ activation of some nucleotide cyclases, some protein kinases and phosphatases, of CaBPs and of membrane fusion and fission (synaptotagmin with two C2 domains, extended synaptotagmin with more C2 domains): to be settled for ciliates

Nucleotides

cyclic adenosine monophosphate (cAMP) → protein kinase A (PKA)

cyclic guanosine monophosphate (cAMP) → protein kinase G (PKG)

→ protein phosphorylation, also in ciliates

cyclic adenosinediphosphoribose (cADPR) from NAD (nicotinamide adeninedinucleotide) → activation of some (ryanodine receptor-type) Ca^{2+}-release channels (CRCs) in mammalian cells: activators effective in ciliates, but not yet assigned to specific channels

nicotinic acid adenine dinucleotidephosphate (NAADP), formed from NADP, → activation of two-pore channels, in acidic compartments: also in ciliates?

Protein phosphorylation

PKA, PKG: as defined above

Ca^{2+}-dependent protein kinases (CDPKs) with integrated calmodulin- (CaM)-like motifs

Protein dephosphorylation by protein phosphatases (PPs) type PP1, PP2A, PP2B

PP2B (calcineurin)

with subunit (SU) A (catalytic SU, containing SU-B- and CaM-binding domain) and SU-B (regulatory SU, with Ca^{2+}-binding domain) also in ciliates

respective protein kinases, protein kinase A (PKA) and protein kinase G (PKG) (Bonini and Nelson 1990); for review see Plattner (2016a). Also some metabolic Ca^{2+} channel activators, such as *c*yclic *a*denosine*dip*hospho*r*ibose (cADPR) and *n*icotinic *a*cid *a*denine *d*inucleotide*p*hosphate (NAADP) are derived from nucleotides, i.e. nicotinamide adeninedinucleotide (NAD) and nicotinic acid adenine dinucleotidephosphate (NAADP), respectively, as known from vertebrates (Lee 2012). For cADPR and NAADP effects there is only circumstantial evidence in *Paramecium* (Plattner et al. 2012).

A total of 2600 kinases has been found in the *P. tetraurelia* genome (Bemm et al. 2009), thus contributing by 7 % to the macronuclear genome. In *T. thermophila* the proportion is 3.8 % (Tian et al. 2014). Both values stress their importance for signal transduction. The difference between the two genera may originate from whole genome duplication in *Paramecium.* A considerable difference between protein kinases in animal cells and in ciliates is the absence in the latter of a "CaM kinase", i.e. a kinase activated by a complex of calmodulin (CaM) and Ca^{2+}. Whereas such CaM-kinases in metazoans contribute to the regulation of neuronal activity, they are replaced in ciliates by "*C*a^{2+}-*d*ependent *p*rotein *k*inases" (CDPKs). These contain CaM-like sequences integrated in the kinase molecule (Kim et al. 1998).

Box 1 also indicates the occurrence in ciliates of protein phosphatases (PPs), e.g. PP1, PP2A and PP2B. PP2B, which is identical with calcineurin, encompasses two subunits, catalytical subunit A and regulatory subunit B, from ciliates (Fraga et al. 2010) to man where it regulates immune-response and long term potentiation, i.e. learning. In ciliates, multiple roles can be expected for calcineurin, including exo-/endocytosis regulation (Momayezi et al. 1987; Fraga et al. 2010).

2.3 Subcompartmentalization of Signaling Including Signaling in Cilia

Signals can be rather precisely restricted to subcompartments, e.g. cilia (Box 2), for which Box 3 shows details. Mechanical stimulation of a ciliated protozoan cell causes depolarization or hyperpolarization, depending on whether stimulation occurs at the anterior or posterior part of the cell (Eckert and Brehm 1979; Machemer 1988a, b). This is enabled by a graded differential distribution of specific ion channels over the somatic (non-ciliary) cell membrane. The respective receptor potential formed by different ion conductances activates different mechanisms in cilia. For instance, depolarization activates voltage-dependent Ca^{2+}-channels selectively occurring in cilia (Machemer and Ogura 1979) and, thus, a Ca^{2+}-carried action potential. (This signaling occurs no more in metazoans beyond Ctenophores.) Increased intraciliary Ca^{2+} shuts off this Ca^{2+} influx (Brehm and Eckert 1978). Hyperpolarization accelerates forward swimming (Preston et al. 1992).

During de- and hyperpolarization, different cyclic nucleotides are formed, activating PKG and PKA, respectively (Box 3). Ciliary activation mechanisms are independent of Ca^{2+}-activated processes during exocytosis, except when massive exocytosis stimulation entails an exuberant Ca^{2+} signal (Husser et al. 2004). In

Box 2. Examples of subcompartmentalization of signals in ciliates

External mechanical stimulation

→ Ca^{2+} signaling in cortical domains, e.g. for receptor potential formation

anterior stimulation → depolarization/action potential → ciliary reversal

posterior stimulation→ hyperpolarization → accelerated forward swimming

for details of ciliary activity, see Box 3

Activation of non-ciliary (somatic) membrane phenomena for exocytosis: see Box 5

Constitutive local intracellular Ca^{2+} signaling via organelle-specific Ca^{2+}-release channels (CRCs) in all trafficking vesicles

Organelle specific protein phosphorylation processes for other activation mechanisms: see Box 8

Box 3. Ciliary beat activity in ciliates

(A) Anterior mechanical stimulus (hitting an obstacle)

somatic cell membrane → receptor potential: depolarization by activation of anterior mechanosenstive Ca^{2+}-channels, repolarization by K^+-efflux

Effect of depolarization on cilia: action potential by

activation of voltage-dependent (ΔV) Ca^{2+} influx channels in ciliary membrane

→ guanylate cyclase activation → cGMP formation → PKG activation

→ phosphorylation of target proteins in ciliary axoneme;

in parallel: activation of axonemal calmodulin (CaM) → different effects

No Ca^{2+} spillover into cell soma

Inactivation of ciliary reversal by closing ciliary ΔV-channels by a Ca^{2+}/CaM complex and binding of excessive Ca^{2+} to immobile buffer (CaM, centrin)

(B) Posterior mechanical stimulus

Hyperpolarization by K^+ efflux (somatic cell membrane)

→ hyperpolarization-activated Ca^{2+}-channels (somatic cell membrane),

adenylate cyclase activation → cAMP formation → PKA activation → phosphorylation of target proteins in ciliary axoneme

(C) Gravikinesis/gravitaxis

Positive gravitaxis: "statocyst"-mediated intracellular signal perception (*Loxodes*)

Negative gravitaxis: very much predominant form of gravitaxis (*Paramecium*, *Tetrahymena*)

activation by hyperpolarization via posteriorly enriched K^+-channels; postulated link to cortical F-actin → upward movement summary, a mechanical signal is transformed into a long-range electrical signal generated at the somatic cell membrane that is transduced into cilia where it causes short-range Ca^{2+} signaling and a mechanical response in ciliary activity.

Mechanisms described for basic ciliary activity (Fig. 2.1b) also apply to chemotaxis and to gravitaxis/gravikinesis (Box 3). Chemotaxis requires the activation of distinct ion conductances to achieve specific behavioral responses (Saimi and Kung 2002; Bell et al. 2007; Yano et al. 2015). Positive gravitaxis is rare in ciliates where negative gravitaxis, causing upward swimming in the gravity field, by far prevails. For this, *Paramecium* is the best analyzed example (Machemer et al. 1998; Hemmersbach and Braun 2006; Machemer 2014). Accordingly cAMP and PKA are assumed to be involved in negative gravitaxis (Hemmersbach et al. 2002). Investigators assume that, for sensing gravity, channels have to be linked to filamentous actin (F-actin) (Machemer 2014). In fact, actin has been localized to the cell cortex (Sehring et al. 2007) and, even more precisely, to the narrow space between cell membrane and alveolar sacs (Kissmehl et al. 2004).

2.4 Organelle Trafficking Signals

2.4.1 Molecular Background

Long-range signals, such as firmly installed microtubules, can guide vesicles to their target sites (Box 4). This is true of trichocysts (Aufderheide 1978; Plattner et al. 1982) and organelles of the phagocytotic cycle (Allen and Fok 2000). Short-range signals involved are GTPases, SNAREs, H^+-ATPase, as outlined in Sect. 2.1, together with actin. For GTPases (Bright et al. 2010) and the other key players, organelle specific isoforms are available (Plattner 2010). The multimeric H^+-ATPase molecule is composed of an intramembranous V0 basepiece and a catalytic head, V1, which may dis- and re-assemble by interaction with an elongate, variable a-SU (Sun-Wada and Wada 2015). Considering the key role of H^+-ATPase (Sect. 2.2), the unsurpassed number of 17 a-subunits in *Paramecium* may mediate adjustment to local requirements (Wassmer et al. 2005, 2006, 2009). Among SNAREs, longin-type sequences in *Paramecium's* "synaptobrevins" may contribute to organelle specificity, in addition to the usual domain sequences (Schilde et al. 2006, 2010). In *P. tetraurelia*, plasmalemmal Syntaxin 1 (*Pt*Syx1) is engaged in trichocyst exocytosis (Kissmehl et al. 2007). For more details, see Plattner (2010, 2016a).

Vesicles undergoing trafficking are endowed with CRCs identical with, or related to $InsP_3Rs$ and RyRs (Ladenburger and Plattner 2011; Plattner and Verkhratsky 2013); see Box 5. An exception are trichocysts which seem to be devoid of luminal Ca^{2+}, in contrast to what is known from some other dense core-secretory vesicles, endosomes and phagocytotic organelles of higher eukaryotes (Hay 2007). The presence of the key players mentioned above, including

Box 4. Organellar trafficking signals

Long-range signals

microtubular "rails" as firmly established guidelines (emanating from oral cavity [for phagocytosis] and from ciliary basal bodies [for trichocyst docking], respectively)

Short-range signals/molecular recognition sites

cooperative role of H^+-ATPase molecules (acidification of organelle lumen) → binding of organelle-specific small GTPases (+ modifying proteins, e.g. GAP = guanosine nucleotide activation protein, GEF = guanosine nucleotide exchange factor, as known from higher eukaryotes)

vesicle docking: SNAREs + GTPases

→ docking at target organelle/membrane: organelle-specific GTPase-binding proteins yet to be identified

local Ca^{2+} signal and Ca^{2+} sensor activation → membrane fusion

Vesicle budding: coatamer proteins (COPs), clathrin, adaptor proteins

Golgi apparatus: ill-defined molecular cues and signals in ciliates awaiting srutiny

CRCs, in the endo-/phagocytotic cycle of *Paramecium* may reflect the intensity and multitude of vesicle trafficking known from ultrastructural studies (Allen and Fok 2000) [In *Paramecium*, not all of these vesicles are acidic (Wassmer et al. 2009), and not all lysosomal enzymes have an acidic pH-optimum (Fok and Paeste 1982; Fok 1983)]. Appropriate CRCs may drive membrane interactions in concert with, or independently from other key players. The importance of local availability and regulation of Ca^{2+} during membrane docking and fusion is discussed in the accompanying paper (Plattner 2016b). The numerous members of the six CRC subfamilies found in *Paramecium* may fine tune Ca^{2+} signals and membrane interactions depending on local requirements.

2.4.2 Dense Core-Secretory Vesicle Exocytosis

Ca^{2+} regulation of trichocyst exocytosis involves three steps (Box 5, Fig. 2.1b, c): (i) Ca^{2+} release from alveolar sacs via RyR-like proteins and (ii) immediately superimposed Ca^{2+} influx from the outside medium (Klauke and Plattner 1997; Ladenburger and Plattner 2011; Plattner 2014). Both mechanisms acting in concert are called store-operated Ca^{2+} entry, SOCE—a mechanism maintained up to mammals. A large excess of Ca^{2+}, much more than seen by fluorochromes, has to flood trichocyst exocytosis sites to become activated, just as in some

Box 5. Why a multiplicity of Ca^{2+}-release channels?

Paramecium contains 34 genes for Ca^{2+}-release channels (CRCs, 6 subfamilies), to be assigned to the superfamily of $InsP_3R$/Ryanodine receptor (RyR) type CRCs, distributed over different trafficking organelles for local signaling

Example A - RyR-like channels: alveolar sacs (cortical Ca^{2+} stores)

activation by RyR activators caffeine, 4-chloro-m-cresol or by polyamines (AED)

→ Ca^{2+} release from alveolar sacs by RyR-LP of CRC-IV subfamily, superimposed by Ca^{2+}-entry from the outside medium (store-operated Ca^{2+}-entry, SOCE) → trichocyst exocytosis based on

(i) membrane fusion (depending on $[Ca^{2+}]_i$ increase by SOCE) and →

(ii) Ca^{2+}_o access to trichocyst contents for inducing decondensation (vigorous ejection)

membrane resealing and ghost detachment (exocytosis-coupled endocytosis), also driven by $[Ca^{2+}]_i$ increase

Example B - $InsP_3Rs$: occurring in the contractile vacuole complex; serving

(i) for fine-tuning of $[Ca^{2+}]$ in the cytosol and

(ii) for internal membrane restructuring (hypothetic) during contraction cycles

$InsP_3R$/RyR- type channels also include mixed types, in compartments undergoing trafficking

Additional Ca^{2+}-release channels in ciliates

two-pore channels NAADP-activated, in acidic stores? Occurrence likely

TRP-type and mechanosensitive channels: also not yet specified in ciliates

neuroendocrine cells (for details, see Plattner 2016a). (iii) Discharge of contents follows formation of an exocytotic opening and requires the entry of Ca^{2+} from the outside and binding to some secretory components, thus causing decondensation by conformational change (Plattner et al. 1997; Klauke et al. 1998; Plattner 2014). This in turn depends on proper processing of secretory protein precursors (Pouphile et al. 1986; Bowman et al. 2005).

2.4.3 *The Phagocytotic Cycle*

This aspect is reviewed here in more detail, as it demonstrates the complex sequence of interacting signaling molecules although these are only partially known.

The phagocytotic cycle in *Paramecium* requires multiple signaling (Allen and Fok 2000), including firmly established microtubules as long-range signals and variable stage-dependent short-range signals. In detail the sequence is as follows. (i) At the cytopharynx, at the bottom of the cytostome, vesicles recycling from advanced stages of food vacuoles, together with vesicles from the cytoproct, deliver membrane material for a bulging nascent food vacuole. Thus, a phagosome is formed at converging microtubular rails, the "postoral fibers". (ii) After detachment, acidosomes (late endosomes) fuse with the phagosome, thus endowing it with H^+-ATPase molecules for luminal acidification. (iii) This is followed by fusion with lysosomes, thus forming phagolysosomes. (iv) Lysosomal enzymes are retrieved later on during cyclosis, (v) as are parts of the membrane for delivery to the cytopharynx. (vi) The contents of spent food vacuoles are released by exocytosis at the cytoproct and membranes are recycled as indicated for step (i) (Allen and Fok 2000).

In *Paramecium tetraurelia*, key players for signaling in the different stages (Box 6) encompass exchanging sets of SNAREs (Schilde et al. 2006, 2010; Kissmehl et al. 2007), subunits (SU) of H^+-ATPase (Wassmer et al. 2005, 2006), and actin, as outlined in a separate chapter (Plattner 2016b). In *Tetrahymena*, different types of GTPases are exchanged during cyclosis (Bright et al. 2010). In *Paramecium*, the exchange of numerous actin isoforms, types 1, 3, 6, 8, 11–14 and 16 as well as their patchy or unilateral arrangement in some stages is a most striking phenomenon (Sehring et al. 2007). This may serve propulsion of the organelle and/or regulation of accessibility to fusion and/or budding of vesicles. All this documents a series of interacting long- and short-range signaling during cyclosis.

Box 6. The phagocytotic cycle in ciliates

At cytopharynx

cell membrane enlargement by fusion of recycling vesicles → association with actin → nascent food vacuole pinches off to form a phagosome

During cyclosis

→ fusion with acidosomes (late endosomes) providing H^+-ATPase → luminal acidification → fusion with lysosomes to form a phagolysosome (mature food vacuole)

endowment with varying SNAREs, small GTPases, and actin coats (for details see text), whereas Ca^{2+}-release channels are throughout of type CRC-III ($InsP_3R$-type)

selective membrane input and retrieval, contents digestion

pH gradually increasing to ~7

At cytoproct

contents discharge by exocytosis and formation of recycling ("discoidal") vesicles

2.4.4 *The Contractile Vacuole Complex*

Surprisingly, the contractile vacuole complex contains all components relevant for vesicle trafficking, except actin, in even higher variability and with strict localization to specific substructures, such as the vacuole, the pore and the meshwork of the smooth spongiome (Box 7). The organelle has a very complex design (Allen and Naitoh 2002). It not only can expel fluid with an excess of Ca^{2+} and other ions (Stock et al. 2002), but it also shows some reflux of Ca^{2+} into the cytosol via constitutively active $InsP_3Rs$ (Ladenburger et al. 2006). This may serve not only for fine tuning of cytosolic Ca^{2+} but also to drive the extensive membrane fusion and fission processes within the organelle during systole/diastole cycles (Plattner 2015b).

Box 7. Signaling in the contractile vacuole complex

Signals assumed for self-assembly of new anlagen: centrin, CaM, γ-tubulin, NIMA kinase, as discussed elsewhere (Plattner 2015b)

Signals assumed for organelle growth: delivery of vesicle with specific v-/R- and t-/Q SNAREs

Local tubularization (spongiome) and reversible planar-tubular transitions: F-BAR proteins (hypothetic), as discussed elsewhere (Plattner 2015b)

Acidification by V-type H^+-ATPase

- ΔH^+ exploited for Ca^{2+} sequestration (hypothetic $X^+/H^+/Ca^{2+}$ exchanger, in addition to Ca^{2+}-ATPase (see Plattner 2016a) and expulsion of H_2O, Ca^{2+} and other ions by periodic exocytotic release at the pore
- CRCs of type $InsP_3R$ for constitutive partial Ca^{2+} reflux into cytosol: for $[Ca^{2+}]_i$ fine tuning and probably for spongiome restructuring

Pore for periodic contents release by exocytosis: with specific SNAREs and CRCs at the pore

- periodic signal for vacuole contents release: mechanosensitive channels (suggested by occurrence of stomatin [Reuter et al. 2013] and in agreement with other systems [Plattner 2015b]), in conjunction with pore-specific SNAREs and CRCs

2.4.5 *Additional Signals*

Little is known about other types of Ca^{2+} release channels in ciliates, such as two pore-channels (TPC) and transient receptor potential-channels (TRPC) and their activators (Box 5). Particularly metabolic CRC activators (Lee 2012), such as cADPR, NAADP, remain to be assigned to different channels and organelles in ciliates. Such channels have to be expected also in ciliates, based on microinjection studies (Plattner et al. 2012).

Vesicle budding at the Golgi apparatus and other organelles as well as at the plasmamembrane requires a set of additional proteins, such as coatamer proteins (COPs) and clathrin, together with their adaptor proteins known from higher eukaryotes up to mammals (Rothman 2014). In ciliate cells, coatamer coats are suggested to occur by electron microscopy in the cis- and trans-side of the Golgi apparatus (Allen and Fok 1993; Garreau De Loubresse 1993) and clathrin coats in addition by molecular biology according to Elde et al. (2005) who also reported the expression of adaptor proteins, AP-1, AP-2, AP-3 and AP-4 in *T. thermophila*. While none of them appear important for lysosome biogenesis (Briguglio et al. 2013), AP-2 is important for endocytosis via coated pits (Elde et al. 2005). Sequences encoding all these adaptor proteins have also been found in the *P. tetraurelia* database, in addition to the ARF/SAR-type G-protein known as a target of the drug, brefeldin A (Plattner and Kissmehl 2003b). The same is true of clathrin heavy chains and of COPs.

In summary, for vesicle trafficking ciliates have at their disposal most of the signaling components known from multicellular organisms. Note, however, that $InsP_3R$/RyR-like molecules are absent from higher plants (Plattner and Verkhratsky 2015), whereas they occur in some green algae (Wheeler and Brownlee 2008). Globally a ciliate's signaling machinery closely resembles that of metazoans.

2.5 Protein Phosphorylation for Activation and Deactivation of Signaling Processes

2.5.1 Phosphorylation Processes

As mentioned in Sect. 2.3, signaling in cilia includes PKA and PKG activity for enhanced forward and backward swimming, respectively (Kim et al. 1998; Kutomi et al. 2012). Activating cyclic nucleotides are generated within one ciliary stroke (Yang et al. 1997). Together with CDPKs they belong to the superfamily of Seryl/Threonyl kinases (Box 8). Phosphoproteins are substrates of the different phosphatases. Among them, PP1 dephosphorylates a ciliary phosphoprotein formed during ciliary reversal in *Paramecium* (Klumpp et al. 1990). PP2B/calcineurin probably has a broad spectrum of activity, depending on its A-subunit, whereas the two genes for the B-SU in *Paramecium* result in an identical translation product, with a well conserved binding domain in the A-SU (Fraga et al. 2010).

As indicated in Box 8 and discussed in more detail somewhere else (Plattner 2016a), the occurrence of Tyrosyl phosphorylation may be largely restricted in ciliates to cell cycle and mitosis regulation. Work with mammalian cells exposed to

Box 8. Protein phosphorylation for signaling and activation processes in ciliates

(A) Seryl/Threonyl phosphorylation

Protein kinases (PK)

protein kinase A, PKA (cAMP-activated)

protein kinase G, PKG (cGMP-activated)

CDPK (Ca^{2+}-dependent protein kinase, with a CaM-like domain) substituting for CaM-kinase (activated by a separate Ca^{2+}/CaM-complex) in animal cells

Protein phosphatases (PP): PP1, PP2A, PP2B (Ca^{2+}/CaM-dependent PP = calcineurin)

(B) Dedicated Tyrosyl phosphorylation

Some predicted for ciliates from proteomic analysis (still to be confirmed)

mainly concerning cell cycle and mitosis regulation (MAPKs = mitogen-activated protein kinases)

new aspects emerging from phosphoproteomic analysis

Euplotes gamones indicates signaling via a mitogen-activated protein kinase (MAPK) cascade with Tyrosine phosphorylation (Vallesi et al. 2010; Cervia et al. 2013). See also chapter by Luporini.

2.5.2 *Signal Downregulation*

Also ciliates possess different ways to downregulate signals (Box 9). Cyclic nucleotides are deactivated by diesterases and phosphoproteins are dephosphorylated by protein phosphatases. For instance, the association of calcineurin with parasomal sacs (Momayezi et al. 2000), the clathrin-coated pits in ciliates, is compatible with dynamin dephosphorylation known from mammalian coated pits.

Ca^{2+} signals are downregulated by different mechanisms with different kinetics (Box 9). The most rapid is binding to centrin (Sehring et al. 2009) a CaBP with high capacity/low affinity (in addition to low capacity/high affinity) binding sites localized in the cell cortex of *Paramecium* (see Plattner 2016a). This is orders of magnitude more rapid than downregulation by Ca^{2+}-ATPases/pumps (Plattner 2016a) of which type Sarcoplasmic/Endoplasmic Reticulum Ca^{2+}-ATPase (SERCA) (Hauser et al. 1998) or plasmamembrane Ca^{2+}-ATPase (PMCA) (Elwess and Van Houten 1997) have been analyzed in *Paramecium.* PMCA also occurs in cilia of *Tetrahymena* (Dentler 1988) and *Paramecium* (Yano et al. 2013). These two are P-type ATPases because they autocatalytically form a phospho-intermediate which then dephosphorylates itself. Ca^{2+} exchangers, though not yet identified, show up in ciliate databases; they are driven by a H$^+$-gradient formed by a H$^+$-ATPase (V-type, in vesicles) operating without a phospho-intermediate formation. Although such

Box 9. Shut-down of signalling in ciliates

Inactivation of Ca^{2+}

binding to high capacity/low affinity Ca^{2+}-binding proteins, e.g. centrin

reduction by pumps and transporters

Ca^{2+} extrusion and sequestration by Ca^{2+}-ATPases/pumps

PMCA (plasmamembrane Ca^{2+}-ATPase)

SERCA (sarcoplasmic/endoplasmic reticulum Ca^{2+}-ATPase)

hypothetical: X^+/Ca^{2+} exchangers, e.g. H^+/Ca^{2+} antiporter (postulated specifically for contractile vacuole complex)

Inactivation of cyclic nucleotides by diesterases

Reversion of phosphorylation state

protein phosphatase PP1, possibly also PP2C, for deactivation of ciliary reversal

PP2B/calcineurin: pleiotropic effects to be expected, e.g. dynamin activation for vesicle budding and regulation of Ca^{2+} stores, e.g. by effects on CRCs

exchangers urgently call for scrutiny in ciliates it appears that they are much more efficient in signal downregulation than the pumps (Ladenburger et al. 2006; Plattner 2016a).

2.6 Signaling by Surface Receptors

These aspects are summarized in Box 10. The occurrence of trimeric GTP-binding proteins (G-proteins) is likely (De Ondarza et al. 2003; Lampert et al. 2011), but not firmly established in protozoa in general (Krishnan et al. 2015) and in ciliates in particular since important details have not been examined yet, as discussed in more detail elsewhere (Plattner 2016a). The same is true of G-protein-coupled receptors (GPCRs). All this also applies to the secretagogue, aminoethyldextran, which, in *Paramecium,* is most efficient in activating highly synchronous exocytosis (Plattner et al. 1985; Plattner and Kissmehl 2003; Knoll et al. 1991) by a SOCE mechanism for trichocyst exocytosis (Hardt and Plattner 2000; Plattner 2014). For hints to MAPK activity and Tyrosyl phosphorylation, see Sect. 2.5.

Purinergic receptors can be assumed to occur in *Paramecium* as these cells, upon exposure to ≥10 μM GTP, perform periodic back- and forward swimming accompanied by depolarization (Clark et al. 1993) and Ca^{2+} waves oscillating with the same periodicity (Sehring and Plattner 2004). This is unusual insofar as purinergic receptors normally respond to ATP or, less common, to UTP. We assume a function in keeping cells from dispersal to low density which is known to inhibit cell division and maintenance of the population.

Box 10. Surface receptor signalling in ciliates

Trimeric GTP-binding proteins (G-proteins) and G-protein-coupled receptors (GPCRs)

existence in ciliates under considerable debate, fragmentary information

Mitogen-activated protein kinase (MAP kinase; MAPK): related activities are currently

assumed also for ciliates

Effects of exogenous GTP (≥10 μM): Ca^{2+} oscillations in parallel to de-/repolarizations

the first $[Ca^{2+}]_i$ peak (larger than subsequent periodic peaks) requires Ca^{2+}_o

→ further (smaller) cyclic activity peaks in ~8 s oscillations supported by Ca^{2+} from internal stores (type of store for internal Ca^{2+} mobilization during GTP activation: unknown) → ongoing periodic activation → desensitization and signal downregulation (mechanism unknown) → decaying Ca^{2+} signal

Unknown: purinergic receptors, Ca^{2+}/polyvalent cation-sensing receptor (a GPCR?)

Chemotaxis chemoreceptors

Operating via specific ion conductances (see text)

2.7 Conclusions

Intracellular signaling by pheromones (gamones) in ciliates (Luporini et al. 2014) is summarized separately in this volume. Epigenetic signaling is also covered separately in this volume by Nowacki; for surveys, also see Chalker et al. (2013) and Simon and Plattner (2014). Most of the other signaling mechanisms described here seem to be evolutionarily old and maintained from protozoa on, particularly ciliates, up to top-ranking metazoans. The impressive complexity of ciliate cells and their elaborate trafficking system may have required a complex signaling system—an old heritage from early eukaryotic ancestors (Dacks and Field 2007; Plattner and Verkhratsky 2015).

Acknowledgements Experimental work by the author cited herein has been supported by the German Research Council.

References

Allen RD, Fok AK (1993) Nonclathrin vesicle coats and filament networks in the transition zone and trans-Golgi region of the Golgi complex of *Paramecium*. J Struct Biol 110:215–226

Allen RD, Fok AK (2000) Membrane trafficking and processing in *Paramecium*. Int Rev Cytol 198:277–318

Allen RD, Naitoh Y (2002) Osmoregulation and contractile vacuoles of protozoa. Int Rev Cytol 215:351–394

Allen PD, Tominaga T, Naitoh Y (2009) The contractile vacuole complex and cell volume control in protozoa. In: Evans DH (ed) Osmotic and ionic regulation: cells and animals. CRC Press, Taylor and Francis Group, Boca Raton, FL, pp 69–106
Aufderheide KJ (1978) Motility events of trichocyst insertion in *Paramecium tetraurelia*. J Protozool 25:362–365
Beisson J (2008) Preformed cell structure and cell heredity. Prion 2:1–8
Bell WE, Preston RR, Yano J, Van Houten JL (2007) Genetic dissection of attractant-induced conductances in *Paramecium*. J Exp Biol 210:357–365
Bemm F, Schwarz R, Forster F, Schultz J (2009) A kinome of 2600 in the ciliate *Paramecium tetraurelia*. FEBS Lett 583:3589–3592
Bonini NM, Nelson DL (1990) Phosphoproteins associated with cyclic nucleotide stimulation of ciliary motility in *Paramecium*. J Cell Sci 95:219–230
Bowman GR, Elde NC, Morgan G, Winey M, Turkewitz AP (2005) Core formation and the acquisition of fusion competence are linked during secretory granule maturation in *Tetrahymena*. Traffic 6:303–323
Brehm P, Eckert R (1978) Calcium entry leads to inactivation of calcium channel in *Paramecium*. Science 202:1203–1206
Bright LJ, Kambesis N, Nelson SB, Jeong B, Turkewitz AP (2010) Comprehensive analysis reveals dynamic and evolutionary plasticity of Rab GTPases and membrane traffic in *Tetrahymena thermophila*. PLoS Genet 6:e1001155
Briguglio JS, Kumar S, Turkewitz AP (2013) Lysosomal sorting receptors are essential for secretory granule biogenesis in *Tetrahymena*. J Cell Biol 203:537–550
Cervia D, Catalani E, Belardinelli MC, Perrotta C, Picchietti S, Alimenti C, Casini G, Fausto AM, Vallesi A (2013) The protein pheromone Er-1 of the ciliate *Euplotes raikovi* stimulates human T-cell activity: involvement of interleukin-2 system. Exp Cell Res 319:56–67
Chalker DL, Meyer E, Mochizuki K (2013) Epigenetics of ciliates. Cold Spring Harb Perspect Biol 5:a017764
Clark KD, Hennessey TM, Nelson DL (1993) External GTP alters the motility and elicits an oscillating membrane depolarization in *Paramecium tetraurelia*. Proc Natl Acad Sci USA 90:3782–3786
Dacks JB, Field MC (2007) Evolution of the eukaryotic membrane-trafficking system: origin, tempo and mode. J Cell Sci 120:2977–2985
Dentler WL (1988) Fractionation of *Tetrahymena* ciliary membranes with triton X-114 and the identification of a ciliary membrane ATPase. J Cell Biol 107:2679–2688
De Ondarza J, Symington SB, Van Houten JL, ClarK JM (2003) G-protein modulators alter the swimming behavior and calcium influx of *Paramecium tetraurelia*. J Eukaryot Microbiol 50:349–355
Eckert R, Brehm P (1979) Ionic mechanisms of excitation in *Paramecium*. Ann Rev Biophys Bioeng 8:353–383
Elde NC, Morgan G, Winey M, Sperling L, Turkewitz AP (2005) Elucidation of clathrin-mediated endocytosis in *Tetrahymena* reveals an evolutionarily convergent recruitment of dynamin. PLoS Genet 1:e52
Elwess NL, Van Houten JL (1997) Cloning and molecular analysis of the plasma membrane Ca^{2+}-ATPase gene in *Paramecium tetraurelia*. J Eukaryot Microbiol 44:250–257
Farrell A, Thirugnanam S, Lorestani A, Dvorin JD, Eidell KP, Ferguson DJP, Anderson-White BR, Duraisingh MT, Marth GT, Gubbels MJ (2012) A DOC2 protein identified by mutational profiling is essential for apicomplexan parasite exocytosis. Science 335:218–221
Flötenmeyer M, Momayezi M, Plattner H (1999) Immunolabeling analysis of biosynthetic and degradative pathways of cell surface components (glycocalyx) in *Paramecium* cells. Eur J Cell Biol 78:67–77
Fok AK (1983) An inhibition and kinetic study of acid phosphatase in *Paramecium caudatum* and *Paramecium tetraurelia*. J Protozool 30:14–20
Fok AK, Paeste RM (1982) Lysosomal enzymes of *Paramecium caudatum* and *Paramecium tetraurelia*. Exp Cell Res 139:159–169

Fraga D, Sehring IM, Kissmehl R, Reiss M, Gaines R, Hinrichsen R, Plattner H (2010) Protein phosphatase 2B (PP2B, calcineurin) in *Paramecium*: partial characterization reveals that two members of the unusually large catalytic subunit family have distinct roles in calcium-dependent processes. Eukaryot Cell 9:1049–1063

Frankel J (2000) Cell biology of *Tetrahymena thermophila*. Meth Cell Biol 62:27–125

Garreau de Loubresse N (1993) Early steps of the secretory pathway in *Paramecium*: Ultrastructural, immunocytochemical, and genetic analysis of trichocyst biogenesis. In: Plattner H (ed) Membrane traffic in protozoa. JAI Press, Greenwich, CT, USA, pp 27–59

Gonda K, Komatsu M, Numata O (2000) Calmodulin and Ca^{2+}/calmodulin-binding proteins are involved in *Tetrahymena thermophila* phagocytosis. Cell Struct Funct 25:243–251

Hardt M, Plattner H (2000) Sub-second quenched-flow/X-ray microanalysis shows rapid Ca^{2+} mobilization from cortical stores paralleled by Ca^{2+} influx during synchronous exocytosis in *Paramecium* cells. Eur J Cell Biol 79:642–652

Hauser K, Pavlovic N, Kissmehl R, Plattner H (1998) Molecular characterization of a sarco (endo)plasmic reticulum Ca^{2+}-ATPase gene from *Paramecium tetraurelia* and localization of its gene product to sub-plasmalemmal calcium stores. Biochem J 334:31–38

Hay JC (2007) Calcium: a fundamental regulator of intracellular membrane fusion? EMBO Rep 8:236–240

Hemmersbach R, Braun M (2006) Gravity-sensing and gravity-related signaling pathways in unicellular model systems of protists and plants. Sign Transduct 6:432–442

Hemmersbach R, Wilczek M, Stieber C, Bräucker R, Ivanova K (2002) Variable acceleration influences cyclic AMP levels in *Paramecium biaurelia*. J Gravit Physiol 9:P267–P268

Hurtado-Lorenzo A, Skinner M, El Annan J, Futai M, Sun-Wada GH, Bourgoin S, Casanova J, Wildeman A, Bechoua S, Ausiello DA, Brown D, Marshansky V (2006) V-ATPase interacts with ARNO and Arf6 in early endosomes and regulates the protein degradative pathway. Nat Cell Biol 8:124–136

Husser MR, Hardt M, Blanchard MP, Hentschel J, Klauke N, Plattner H (2004) One-way calcium spill-over during signal transduction in *Paramecium* cells: from the cell cortex into cilia, but not in the reverse direction. Cell Calcium 36:349–358

Kim K, Messinger LA, Nelson DL (1998) Ca^{2+}-dependent protein kinases of *Paramecium*—cloning provides evidence of a multigene family. Eur J Biochem 251:605–612

Kissmehl R, Sehring IM, Wagner E, Plattner H (2004) Immunolocalization of actin in *Paramecium* cells. J Histochem Cytochem 52:1543–1559

Kissmehl R, Schilde C, Wassmer T, Danzer C, Nuehse K, Lutter K, Plattner H (2007) Molecular identification of 26 syntaxin genes and their assignment to the different trafficking pathways in *Paramecium*. Traffic 8:523–542

Klauke N, Plattner H (1997) Imaging of Ca^{2+} transients induced in *Paramecium* cells by a polyamine secretagogue. J Cell Sci 110:975–983

Klauke N, Kissmehl R, Plattner H, Haga N, Watanabe T (1998) An exocytotic mutant of *Paramecium caudatum*: membrane fusion without secretory contents release. Cell Calcium 23:349–360

Klumpp S, Cohen P, Schultz JE (1990) Okadaic acid, an inhibitor of protein phosphatase 1 in *Paramecium*, causes sustained Ca^{2+}-dependent backward swimming in response to depolarizing stimuli. EMBO J 9:685–689

Knoll G, Braun C, Plattner H (1991) Quenched flow analysis of exocytosis in *Paramecium* cells: time course, changes in membrane structure, and calcium requirements revealed after rapid mixing and rapid freezing of intact cells. J Cell Biol 113:1295–1304

Krishnan A, Mustafa A, Sällman Almén M, Fredriksson R, Williams MJ, Schiöth HB (2015) Evolutionary hierarchy of vertebrate-like heterotrimeric G protein families. Mol Phylogenet Evol 91:27–40

Kutomi O, Hori M, Ishida M, Tominaga T, Kamachi H, Koll F, Cohen J, Yamada N, Noguchi M (2012) Outer dynein arm light chain 1 is essential for controlling the ciliary response to cyclic AMP in *Paramecium tetraurelia*. Eukaryot Cell 11:645–653

Ladenburger EM, Plattner H (2011) Calcium-release channels in *Paramecium*. Genomic expansion, differential positioning and partial transcriptional elimination. PLoS ONE 6:e27111

Ladenburger EM, Korn I, Kasielke N, Wassmer T, Plattner H (2006) An Ins(1,4,5)P_3 receptor in *Paramecium* is associated with the osmoregulatory system. J Cell Sci 119:3705–3717

Ladenburger EM, Sehring IM, Korn I, Plattner H (2009) Novel types of Ca^{2+} release channels participate in the secretory cycle of *Paramecium* cells. Mol Cell Biol 29:3605–3622

Lampert TJ, Coleman KD, Hennessey TM (2011) A knockout mutation of a constitutive GPCR in *Tetrahymena* decreases both G-protein activity and chemoattraction. PLoS ONE 6:e28022

Lee HC (2012) Cyclic ADP-ribose and nicotinic acid adenine dinucleotide phosphate (NAADP) as messengers for calcium mobilization. J Biol Chem 287:31633–31640

Luporini P, Alimenti C, Vallesi A (2014) Ciliate mating types and pheromones. In: Hausmann K, Radek R (eds) Cilia and flagella, ciliates and flagellates. Schweizerbart Science Publishers, Stuttgart, pp 95–118

Machemer H (1988a) Electrophysiology. In: Görtz HD (ed) Paramecium. Springer, Berlin, Heidelberg, pp 185–215

Machemer H (1988b) Motor control of cilia. In: Görtz HD (ed) Paramecium. Springer, Berlin, Heidelberg, pp 216–235

Machemer H (2014) How do protists keep up? In: Hausmann K, Radek R (eds) Cilia and flagella, ciliates and flagellates. Schweizerbart Science Publishers, Stuttgart, pp 133–146

Machemer H, Ogura A (1979) Ionic conductances of membranes in ciliated and deciliated *Paramecium*. J Physiol 296:49–60

Machemer H, Bräucker R, Machemer-Röhnisch S, Nagel U, Neugebauer DC, Weskamp M (1998) The linking of extrinsic stimuli to behaviour: roles of cilia in ciliates. Eur J Protistol 34:254–261

Min SW, Chang WP, Südhof TC (2007) E-Syts, a family of membranous Ca^{2+}-sensor proteins with multiple C2 domains. Proc Natl Acad Sci USA 104:3823–3828

Momayezi M, Lumpert CJ, Kersken H, Gras U, Plattner H, Krinks MH, Klee CB (1987) Exocytosis induction in *Paramecium tetraurelia* cells by exogenous phosphoprotein phosphatase in vivo and in vitro: possible involvement of calcineurin in exocytotic membrane fusion. J Cell Biol 105:181–189

Momayezi M, Kissmehl R, Plattner H (2000) Quantitative immunogold localization of protein phosphatase 2B (calcineurin) in *Paramecium* cells. J Histochem Cytochem 48:1269–1281

Neher E (1998) Vesicle pools and Ca^{2+} microdomains: new tools for understanding their roles in neurotransmitter release. Neuron 20:389–399

Park HY, Kim SA, Korlach J, Rhoades E, Kwok LW, Zipfle WR, Waxham MN, Webb WW, Pollack L (2008) Conformational changes of calmodulin upon Ca^{2+} binding studied with a microfluidic mixer. Proc Natl Acad Sci USA 105:542–547

Plattner H (2010) Membrane trafficking in protozoa SNARE proteins, H^+-ATPase, actin, and other key players in ciliates. Int Rev Cell Mol Biol 280:79–184

Plattner H (2014) Calcium regulation in the protozoan model, *Paramecium tetraurelia*. J Eukaryot Microbiol 61:95–114

Plattner H (2015a) Molecular aspects of calcium signalling at the crossroads of unikont and bikont eukaryote evolution—the ciliated protozoan *Paramecium* in focus. Cell Calcium 57:174–185

Plattner H (2015b) The contractile vacuole complex of protists—new cues to function and biogenesis. Crit Rev Microbiol 41:218–227

Plattner H (2016a) Signalling in ciliates: Long- and short-range signals and molecular determinants for cellular dynamics. Biol Rev (in press). doi: 10.1111/brv.12218

Plattner H (2016b) Signals regulating vesicle trafficking in *Paramecium* cells. This volume

Plattner H, Klauke N (2001) Calcium in ciliated protozoa: sources, regulation, and calcium-regulated cell functions. Int Rev Cytol 201:115–208

Plattner H, Kissmehl R (2003a) Dense-core secretory vesicle docking and exocytotic membrane fusion in *Paramecium* cells. Biochim Biophys Acta (Mol Cell Res) 1641:183–193

Plattner H, Kissmehl R (2003b) Molecular aspects of membrane trafficking in *Paramecium*. Int Rev Cytol 232:185–216

Plattner H, Verkhratsky A (2013) Ca^{2+} signalling early in evolution—all but primitive. J Cell Sci 126:2141–2150
Plattner H, Verkhratsky A (2015) The ancient roots of calcium signalling evolutionary tree. Cell Calcium 57:123–132
Plattner H, Westphal C, Tiggemann R (1982) Cytoskeleton-secretory vesicle interactions during the docking of secretory vesicles at the cell membrane in *Paramecium tetraurelia* cells. J Cell Biol 92:368–377
Plattner H, Stürzl R, Matt H (1985) Synchronous exocytosis in *Paramecium* cells. IV. Polyamino-compounds as potent trigger agents for repeatable trigger-redocking cycles. Eur J Cell Biol 36:32–37
Plattner H, Knoll G, Pape R (1993) Synchronization of different steps of the secretory cycle in *Paramecium tetraurelia*: trichocyst exocytosis, exocytosis-coupled endocytosis, and intracellular transport. In: Plattner H (ed) Membrane traffic in protozoa. JAI Press, Greenwich (CT) and London, pp 123–148
Plattner H, Braun C, Hentschel J (1997) Facilitation of membrane fusion during exocytosis and exocytosis-coupled endocytosis and acceleration of "ghost" detachment in *Paramecium* by extracellular calcium. A quenched-flow/freeze-fracture analysis. J Membr Biol 158:197–208
Plattner H, Sehring IM, Mohamed IK, Miranda K, De Souza W, Billington R, Genazzani A, Ladenburger EM (2012) Calcium signaling in closely related protozoan groups (Alveolata): non-parasitic ciliates (*Paramecium, Tetrahymena*) vs. parasitic Apicomplexa (*Plasmodium, Toxoplasma*). Cell Calcium 51:351–382
Pouphile M, Lefort-Tran M, Plattner H, Rossignol M, Beisson J (1986) Genetic dissection of the morphogenesis of exocytosis sites in *Paramecium*. Biol Cell 56:151–162
Preston RR, Saimi Y, Kung C (1992) Calcium current activated upon hyperpolarization of *Paramecium tetraurelia*. J Gen Physiol 100:233–251
Reuter AT, Stuermer CAO, Plattner H (2013) Identification, localization, and functional impliclations of the microdomain-forming stomatin family in the ciliated protozoan Paramecium tetraurelia. Eukaryot Cell1 2:529–544
Rizo J, Chen X, Arac D (2006) Unraveling the mechanisms of synaptotagmin and SNARE function in neurotransmitter release. Trends Cell Biol 16:339–350
Rothman JE (2014) The principle of membrane fusion in the cell (Nobel lecture). Angew Chemie Int Ed 53:12676–12694
Saimi Y, Kung C (2002) Calmodulin as an ion channel subunit. Annu Rev Physiol 64:289–311
Schilde C, Wassmer T, Mansfeld J, Plattner H, Kissmehl R (2006) A multigene family encoding R-SNAREs in the ciliate *Paramecium tetraurelia*. Traffic 7:440–455
Schilde C, Schönemann B, Sehring IM, Plattner H (2010) Distinct subcellular localization of a group of synaptobrevin-like SNAREs in *Paramecium tetraurelia* and effects of silencing SNARE-specific chaperone NSF. Eukaryot Cell 9:288–305
Sehring IM, Plattner H (2004) Ca^{2+} oscillations mediated by exogenous GTP in *Paramecium* cells: assessment of possible Ca^{2+} sources. Cell Calcium 36:409–420
Sehring IM, Reiner C, Mansfeld J, Plattner H, Kissmehl R (2007) A broad spectrum of actin paralogs in *Paramecium tetraurelia* cells display differential localization and function. J Cell Sci 120:177–190
Sehring IM, Klotz C, Beisson J, Plattner H (2009) Rapid downregulation of the Ca^{2+}-signal after exocytosis stimulation in *Paramecium* cells: essential role of a centrin-rich filamentous cortical network, the infraciliary lattice. Cell Calcium 45:89–97
Simon M, Plattner H (2014) Unicellular eukaryotes as models in cell and molecular biology: critical appraisal of their past and future value. Int Rev Cell Mol Biol 309:141–198
Stock C, Grønlien HK, Allen RD (2002) The ionic composition of the contractile vacuole fluid of *Paramecium* mirrors ion transport across the plasma membrane. Eur J Cell Biol 81:505–515
Südhof TC (2014) The molecular machinery of neurotransmitter release (Nobel lecture). Angew Chemie Int Ed 53:12696–12717
Sun-Wada GH, Wada Y (2015) Role of vacuolar-type proton ATPase in signal transduction. Biochim Biophys Acta 1847:1166–1172

Tian M, Chen X, Xiong Q, Xiong J, Xiao C, Ge F, Yang F, Miao W (2014) Phosphoproteomic analysis of protein phosphorylation networks in *Tetrahymena thermophila*, a model single-celled organism. Mol Cell Proteom 13:503–519

Vallesi A, Di Pretoro B, Ballarini P, Apone F, Luporini P (2010) A novel protein kinase from the ciliate *Euplotes raikovi* with close structural identity to the mammalian intestinal and male-germ cell kinases: characterization and functional implications in the autocrine pheromone signaling loop. Protist 161:250–263

Wassmer T, Froissard M, Plattner H, Kissmehl R, Cohen J (2005) The vacuolar proton-ATPase plays a major role in several membrane-bounded organelles in *Paramecium*. J Cell Sci 118:2813–2825

Wassmer T, Kissmehl R, Cohen J, Plattner H (2006) Seventeen a-subunit isoforms of *Paramecium* V-ATPase provide high specialization in localization and function. Mol Biol Cell 17:917–930

Wassmer T, Sehring IM, Kissmehl R, Plattner H (2009) The V-ATPase in *Paramecium*: functional specialization by multiple gene isoforms. Eur J Physiol 457:599–607

Wheeler GL, Brownlee C (2008) Ca^{2+} signalling in plants and green algae—changing channels. Trends Plant Sci 13:506–514

Yang WQ, Braun C, Plattner H, Purvee J, Van Houten JL (1997) Cyclic nucleotides in glutamate chemosensory signal transduction of *Paramecium*. J Cell Sci 110:2567–2572

Yano J, Rajendran A, Valentine MS, Saha M, Ballif BA, Van Houten JL (2013) Proteomic analysis of the cilia membrane of *Paramecium tetraurelia*. J Proteom 78:113–122

Yano JY, Valentine MS, Van Houten JL (2015) Novel insights into the development and function of cilia using the advantages of the *Paramecium* cell and its many cilia. Cells 4:297–314

Zhen Y, Stenmark H (2015) Cellular functions of Rab GTPases at a glance. J Cell Sci 128:3171–3176

Chapter 3
RNA-Guided Genome Editing

Sarah Allen and Mariusz Nowacki

Abstract Ciliates have evolved highly complex and intricately controlled pathways to ensure the precise and complete removal of all genomic sequences not required for vegetative growth. At the same time, they retain a reference copy of all their genetic information for future generations. This chapter describes how different ciliates use RNA-mediated DNA comparison processes to form new somatic nuclei from germline nuclei. While these processes vary in their precise mechanisms, they all use RNA to target genomic DNA sequences—either for retention or elimination. They also all consist of more than one individual pathway acting cooperatively—the two subsets of small RNAs in *Paramecium* and the guide RNAs and Piwi-interacting RNAs in *Oxytricha*—to ensure a strong belt-and-braces approach to consistent and precise somatic nucleus development. Nonetheless, this genome comparison approach to somatic nucleus development provides an elegant method for trans-generational environmental adaptation. Conceptually, it is easy to imagine how somatic changes that occur during vegetative growth could be transferred to meiotic offspring, while an unaltered germline genome is retained. Further research in this area will have far-reaching implications for the trans-generational adaptation of more distantly related eukaryotes, such as humans.

3.1 Introduction

3.1.1 Ciliates as a Model for Trans-Generational Epigenetic Inheritance

All sexually-reproducing organisms face the problem of how to separate their germline DNA from their somatic DNA. The germline is precious, as it carries all of the information required to make a new copy of the organism. It must therefore be protected from mutation, as germline mutations will affect the entire lineage of the

S. Allen · M. Nowacki (✉)
Institute of Cell Biology, University of Bern, Bern, Switzerland
e-mail: mariusz.nowacki@izb.unibe.ch

G. Witzany and M. Nowacki (eds.), *Biocommunication of Ciliates*,
DOI 10.1007/978-3-319-32211-7_3

individual. It must also retain the ability to form an entirely new organism, and thus must remain largely unaltered from one generation to the next. In multicellular organisms such as humans, the germ and somatic lines are separated into different cells. Germline cells remain pluripotent, as each gamete must have the ability to develop into every type of cell in the organism. They are also largely dormant, avoiding any possibly mutagenic activity that may compromise the fitness of future generations. Somatic cells perform specialised functions depending on their tissue, and their gene expression profiles are thus widely different from each other. Since they do not contribute to the next generation, they can pick up mutations in day-to-day functioning without compromising the evolutionary fitness of the organism. In the differentiation process, genes are switched on and off more or less permanently, using chromatin modifications and DNA methylation. Crucially though, with few exceptions, all of the DNA sequence present in the pluripotent germ cells is also present in each somatic cell. While it was long thought that inheritance worked solely through a strict Darwinian model whereby fitness was dependent purely on the sequence of the genes, it is now becoming clear that DNA marks laid down during the lifetime of a parent in response to environmental or other stimuli can be inherited through several generations. This is known as epigenetic inheritance and is changing the way we look at information transfer through generations (Figs. 3.1, 3.2 and 3.3).

Ciliates present an interesting case for the study of the relationship between germline and somatic genomes, and the transfer through generations of information about the environment. They are single-celled organisms that separate their somatic DNA from their germline DNA by containing two separate nuclei in each cell. Each time a ciliate undergoes sexual reproduction, the old somatic nucleus is degraded and a new one is formed from one of the meiotic products of the germline nucleus. The somatic nucleus is large and transcriptionally active, and is called the macronucleus (MAC). The germline nucleus is by comparison small, compact and transcriptionally silent, and is called the micronucleus (MIC). In the process of developing from a germline to a somatic nucleus, large quantities of non-functional DNA are cut out and removed in a huge genome reorganisation event. This eliminated DNA consists of transposons, repetitive elements, non-coding sequences and short intragenic sequences known as internal eliminated sequences, or IESs. The genes are then assembled on short chromosomes and amplified manifold. The macronucleus thus has a lower total information content than the micronucleus, but is much larger due to the amplification of the gene-coding sequences.

3.1.2 Ciliate Life Cycle and Nuclear Development

Ciliates reproduce sexually using a mating process known as conjugation, in which two cells of opposite mating types come together and partially fuse such that cytoplasmic exchange can take place. They then both undergo meiosis and exchange a haploid nucleus. Following fusion with a maternal haploid nucleus, the

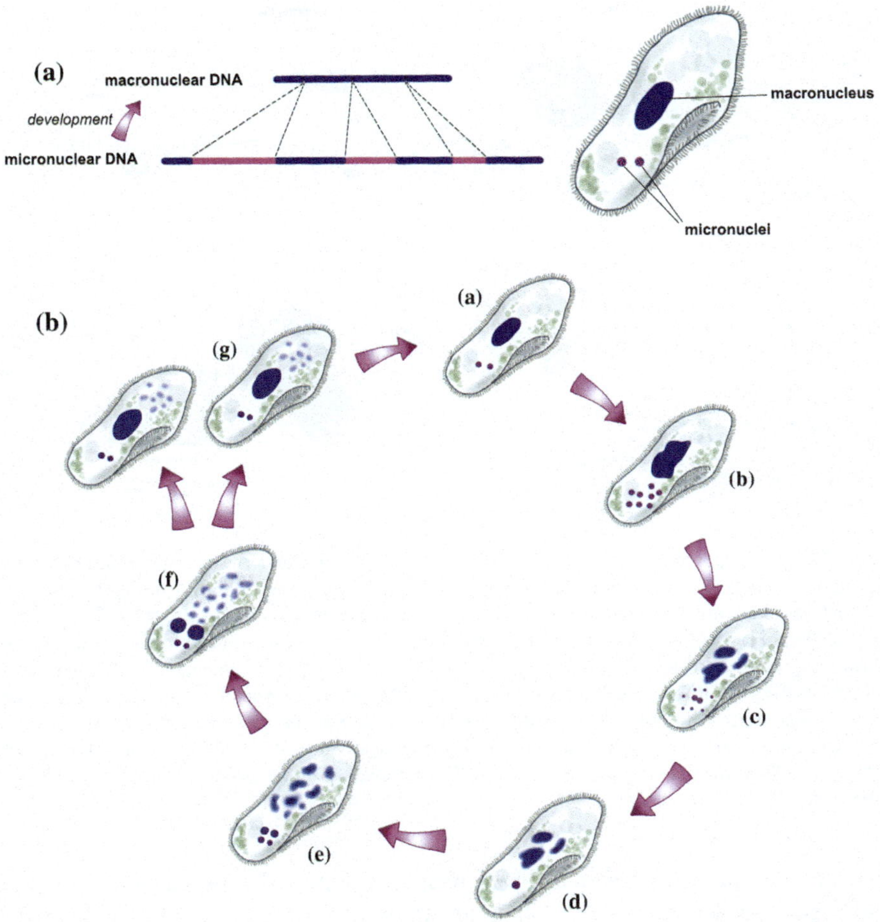

Fig. 3.1 The ciliate sexual reproduction cycle. **a** IESs (*magenta*) are excised from macronuclear destined sequences (*dark blue*) as a new macronucleus develops from a parental micronucleus. **b** (*a*). A vegetatively dividing ciliate has a large somatic macronucleus and one or two small germline micronuclei. Following conjugation or the initiation of autogamy (*Paramecium*), the micronuclei undergo meiosis to form four haploid products each (*b*). Seven out of eight haploid micronuclei degrade and the remaining one undergoes one mitotic division. In case of autogamy the two nuclei fuse or one of them is exchanged between two pairing cells in case of conjugation (*c*, *d*). The diploid zygotic nucleus undergoes mitosis twice, forming four diploid products (*e*), two of which will remain silent as the daughter micronuclei and two of which will develop into the daughter macronuclei. The DNA in the developing macronucleus is pruned to remove IESs and other non-coding sequences, and amplified to a high copy number (*f*). The cell is then ready to divide and begin further vegetative cycles (Color figure online)

resulting diploid nucleus undergoes mitosis and can start to form the new macronucleus of the daughter cell. In *Paramecium* species, cells can also self-fertilise is a process called autogamy. Autogamy occurs in individual cells in

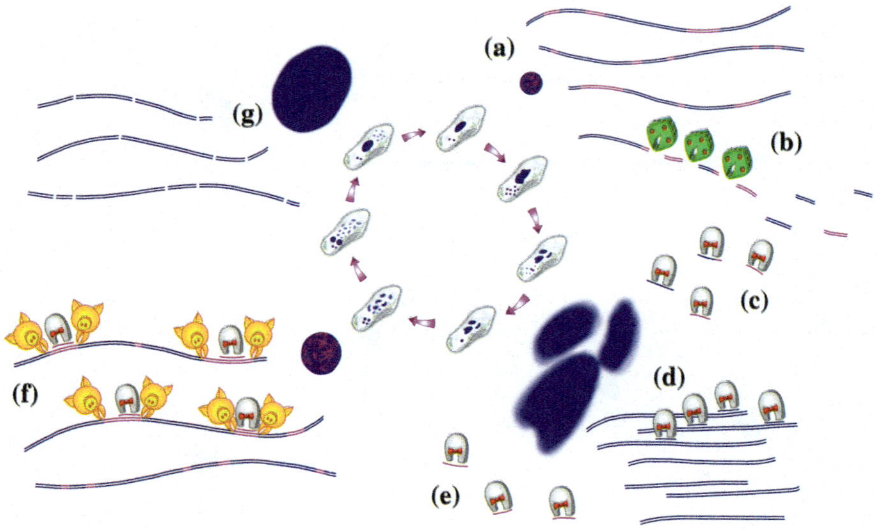

Fig. 3.2 The scanning model for *Paramecium* and *Tetrahymena* macronuclear development. Double-stranded RNA is transcribed from the parental micronucleus during the early stages of sexual development (*a*), then cleaved by Dicer-like proteins (*b*) into ~25 nt scnRNAs. The scnRNAs are loaded onto Piwi proteins (*c*) and transferred to the parental macronucleus (*d*), where they are compared to the macronuclear DNA and scnRNAs that match are degraded. Non-matching scnRNAs correspond to IESs and other germline-specific sequences. These are transported along with their Piwi proteins (*e*) back to the nascent macronucleus, which is in the process of developing from one of the micronuclear meiotic products. Here they guide Piggymac/Tbp2-mediated DNA cleavage and IES removal (*f*). The DNA is then amplified (*g*) to form the new macronucleus

response to starvation, and involves meiosis followed by production of a new macronucleus from two of the meiotic micronuclear products. An interesting aspect of the ciliate life cycle is that the newly formed cells resulting from a sexual reproduction are immature and cannot themselves undergo meiosis. They require a maturation period during which they divide vegetatively for a number of cycles—from around 20 divisions in *Paramecium* to 80 in *Tetrahymena*—before they can conjugate again. During these vegetative divisions the micronuclei divide through mitosis, but the macronuclei lack centromeres and so a mitotic spindle cannot form. Instead they divide in a process termed amitosis, in which the chromosomes are apparently randomly segregated. Due to the high copy number of each short macronuclear chromosome it seems that this process reliably produces functional macronuclei, even over many generations (Morgens and Cavalcanti 2015). It is also possible that a mechanism exists for assessing the copy number of the macronuclear chromosomes and correcting any discrepancies following vegetative division. Nonetheless, ciliates cannot divide vegetatively forever, they need to renew their macronucleus periodically.

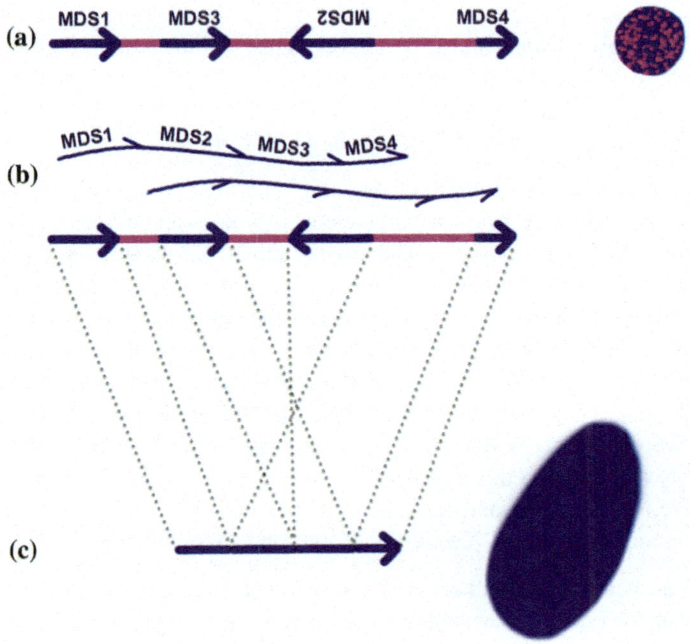

Fig. 3.3 *Oxytricha* macronuclear development. In *Oxytricha*, the macronucleus-destined sequences (MDSs, interspersed with IESs) can be scrambled and even reversed (*a*), making their rearrangement into coding nanochromosomes a formidable task. Long RNAs are transcribed bidirectionally from the parental macronucleus, and are then used as a template for the rearrangement of the DNA in the developing daughter macronucleus (*b*). Following IES removal and, when necessary, sequence rearrangement, the new macronuclear DNA is amplified (*c*)

3.1.3 Deleted DNA: Transposons and IESs

As part of the streamlining of the macronuclear genome during development, the number of chromosomes is increased through fragmenting of the micronuclear chromosomes, with de novo telomere addition to the ends. The degree of fragmentation varies between ciliates (see boxes 1–3), with the five *Tetrahymena* micronuclear chromosomes fragmenting into approximately 200 chromosomes in the macronucleus (Eisen et al. 2006), while *Oxytricha* macronuclei contain a whopping 16,000 chromosomes (Swart et al. 2013).

> *Paramecium* is historically the best-studied of the ciliates, and is useful experimentally in part due to the cells' ability to self-fertilise and undergo meiosis without the need for a conjugation partner. It contains between 30 and 60 micronuclear chromosomes, which fragment into up to 188 chromosomes in the macronucleus. During this process, chromosomes are amplified uniformly to approximately 800 n, and around 45,000 IESs are

excised. Many *Paramecium* IESs are contained within protein-coding genes, and as a result the excision must be highly precise so as not to compromise evolutionary fitness.

Reference: (Aury et al. 2006; Arnaiz et al. 2012)

Tetrahymena is, like *Paramecium*, an oligohymenophore. It has five micronuclear chromosomes, which have centromeres and divide by mitosis at each vegetative cell division. The micronuclear chromosomes fragment at specific chromosome breakage sequences to around 250–300 macronuclear chromosomes, each one maintained at a copy number of ~45. During this process, around 30 % of the micronuclear genome is deleted, the deleted regions corresponding to non-coding DNA, transposons and IESs. *Tetrahymena* IESs are longer than those of *Paramecium*, and are almost exclusively present in non-coding regions or introns. Their ends are less precisely defined than those of *Paramecium*, with microheterogeneity in both telomere addition sites and IES excision boundaries.

References: (Yao and Yao 1994; Hamilton et al. 2005; Eisen et al. 2006)

Oxytricha belongs to the class of *Spirotrichea* and was established as a model organism for genome rearrangements much more recently than *Paramecium* or *Tetrahymena*. Its micronuclear genome is possibly the most complex eukaryotic genome known, with 225,000 individual segments, many of which are scrambled and interwoven, coming together to form 16,000 chromosomes in the new macronucleus. The short macronuclear chromosomes, known as nanochromosomes, typically contain a single gene flanked by telomeres, and are amplified to copy numbers in a manner that appears to correlate with their expression. During macronuclear development around 90 % of the MIC genome is deleted and many of the coding segments are rearranged and interchanged. IES boundaries are highly precise.

References: (Swart et al. 2013; Chen et al. 2014)

In *Paramecium* and *Tetrahymena*, the DNA that is deleted during the development of the macronucleus can be divided broadly into two types: long, repetitive regions containing transposons and other repeats such as minisatellites, and internal eliminated sequences (IESs). The former are deleted concomitantly with the chromosomal fragmentation, which occurs reproducibly at specific sites, albeit with microheterogeneity in the specific breakpoints (Hamilton et al. 2005; Le Mouël et al. 2003). The latter are single-copy, short sequences that may exist within protein-coding genes as well as in intergenic regions. *Paramecium* in particular has 45,000 unique IESs, 80 % of which are within protein-coding genes. These are mainly short—90 % are under 150 bp—display a striking periodicity in length, and are all flanked by TA repeats that are required for their excision (Arnaiz et al. 2012). *Tetrahymena* has fewer IESs, in the order of 8000, and very few are contained within protein-coding genes (Fass et al. 2011). Many *Tetrahymena* IESs have microheterogeneous boundaries, however those within genes have highly specific boundaries, as is expected. IESs are thought to be derived from transposons, and to be under selective pressure for their ability to be excised using the machinery involved in genome rearrangement.

In *Oxytricha*, over 90 % of the micronuclear genome is deleted during MAC development, so development can be seen as a process of sorting a small number of MAC-destined sequences (MDSs) from a large amount of unwanted DNA. MDSs are often very short—the shortest are 0 bp linkers between other MDSs—and frequently scrambled and/or nested within other groups of MDSs. Both transposon-derived sequences and IESs are known to interrupt protein-coding genes, and the fragmentation and IES removal is highly precise (Chen et al. 2014).

3.2 RNAs in *Tetrahymena* and *Paramecium* Development

3.2.1 The Scanning Model

The most widely accepted model for the method by which *Tetrahymena* and *Paramecium* are able to identify which sequences are genes necessary for vegetative growth and which are 'junk' to be eliminated is the scanRNA (scnRNA) model. In this model, small RNAs—25–32 nt depending on species—corresponding to the entire micronuclear genome are produced from the micronucleus during the initial stages of conjugation or autogamy. These RNAs are then transported to the old macronucleus, where they are compared to the sequences present there, and those scnRNAs that match macronuclear sequences are degraded. The remaining scnRNAs, now corresponding to the undesired, micronucleus-specific IESs, are transported to the post-meiotic nucleus that is destined to develop into the new macronucleus. Here they identify the sequences to be deleted, following which the remaining DNA is amplified. By this process, a copy of the parental macronucleus is formed.

Elegant though this model is, it does not provide the whole story as to how IESs are eliminated from the developing MAC. In *Paramecium*, approximately two thirds of IESs are eliminated spontaneously, even if they are present in the maternal MAC. This was demonstrated by experiments involving microinjection of DNA corresponding to IESs into the maternal MAC, and assessment following autogamy as to the presence of these IESs in the new MAC (Duharcourt et al. 1998). This leads to one third of IESs being retained in the new MAC following autogamy, with a detrimental effect on the survival of the daughter cells. These are referred to as maternally-controlled IESs. The remaining two thirds, however, are excised reliably even when they are present in the maternal MAC. How these non-maternally controlled IESs are recognised is not clear. For the maternally-controlled IESs, the degree of retention varies from IES to IES. It is thought that the sequence of the IES plays a role in determining whether or not its excision requires the scnRNA pathway.

3.2.2 scnRNA Production and Selection

The scnRNA model was first proposed in 2002 in a paper identifying a new class of short RNAs and a *Tetrahymena*-specific Piwi homologue, *TWI1* (Mochizuki et al. 2002). ScnRNAs are produced from the micronucleus during the early stages of conjugation, through transcription of the entire micronuclear sequence by RNA polymerase II (Mochizuki and Gorovsky 2004a) followed by cleavage of the resulting dsRNA transcripts by dicer-like proteins. The dicer-like proteins responsible for scnRNA production—Dcl1 in *Tetrahymena* and Dcl2/3 in *Paramecium* (Malone et al. 2005; Sandoval et al. 2014)—have RNase III domains like Dicer proteins, but do not have helicase or DEAD-box domains and little is known about their mode of action. *Tetrahymena* scnRNAs are 26–32 nt long and display a strong 5′ U-bias along with a weaker A-bias 3 bases from the 3′ end. These base preferences are most likely a result of dicer cleavage, which cuts dsRNA at 5′ U, with a 2 nt 3′ overhang (Mochizuki and Kurth 2013). In *Paramecium*, scnRNAs are much more uniform: they are precisely 25 nt long and have a strong 5′ UNG bias. The two Dicer-like proteins Dcl2 and Dcl3 are unusually short and have apparently complementary functions in producing scnRNAs. Knockdown of Dcl2 results in a loss of length specificity, with scnRNAs between 24 and 30 nt being produced, conversely knockdown of Dcl3 results in 25 nt long scnRNAs but without the 5′ UNG bias (Sandoval et al. 2014).

Initially it was thought that scnRNAs were produced from the uniform transcription of the entire micronucleus, with the ‘scanning’ selection process being the only way to enrich for IES-matching scnRNAs. However, recently in *Tetrahymena* it has been shown that the transcription of scnRNA precursors in the MIC is biased towards non-MAC destined sequences (Schoeberl et al. 2012). It is not entirely clear how this transcriptional regulation is mediated, although it has been suggested that it may be through grandparental scnRNAs that localise to the new MIC after selection.

In conjunction with their production, the scnRNAs are complexed with Piwi proteins—Twi1p in *Tetrahymena* and Ptiwi01/09 in *Paramecium* (Mochizuki and Gorovsky 2004b; Bouhouche et al. 2011)—and transferred to the parental macronucleus. scnRNAs matching macronuclear sequences are eliminated during the course of development, and it is thought that some form of sequence comparison and targeting for degradation must take place while the scnRNAs are in the parental MAC (Mochizuki and Gorovsky 2004b; Sandoval et al. 2014). How this comparison, or ‘scanning’, takes place is not clear, but it has been proposed that the scnRNA-associated Piwi protein binds to a non-coding RNA produced from the old MAC, and that this binding provides the information required to trigger MAC-matching scnRNA degradation. This is based on the observation that Ema1p, a putative RNA helicase, is necessary both for the binding of the scnRNA-Twi1p to parental MAC chromatin in developing *Tetrahymena*, and for the removal of MAC-matching scnRNAs (Aronica et al. 2008). In the same paper, ncRNAs corresponding to parental MAC sequences were detected by RT-PCR, and shown to bind scnRNA-Twi1p in an Ema1p-dependent manner.

3.2.3 scnRNAs in MAC Development

Following, or perhaps in conjunction with, the scanning process, scnRNAs in association with their piwi proteins are transferred to the nascent macronucleus. The transport of the Twi1p-scnRNA complex in Tettrahymena is dependent both on the Slicer activity of the Twi1p, possibly in association with its role in passenger strand removal, and on a Twi1p-binding protein named Giw1p (Noto et al. 2010). This protein is required both for the transfer of new scnRNAs to the old MAC, and for the accumulation in the new MAC following selection.

Once in the developing MAC, the scnRNAs identify the sequences that must be eliminated for a functional new MAC to form. The mechanism of this identification process may involve the transcription of the new MAC, with scnRNAs binding to nascent transcripts as is thought to occur during scnRNA selection in the parental MAC. Sequences containing IESs in the zygotic MAC have been shown to be transcribed prior to IES excision in *Paramecium* (Lepère et al. 2008), and a transcription factor that localises to the new MAC in early development has been shown to be necessary for the excision of a large number of maternally-controlled IESs (Maliszewska-Olejniczak et al. 2015).

In *Tetrahymena*, the sequences identified for elimination gain histone methylation on histone H3 lysine 9 and lysine 27 (H3K9me and H3K27me). The mechanism for this process is not clear, however it has been shown to be mediated by Polycomb group protein Ezl1, and it is dependent on the scnRNA-associated proteins Twi1 and Dcl1 (Taverna et al. 2002; Liu et al. 2007). In non-ciliate eukaryotes, H3K9me and H3k27me are associated with heterochromatin, the tightly packaged and condensed DNA of silenced genes and regions. In most organisms, heterochromatin remains in the cell, however in ciliates it marks sequences for elimination. Following H3 methylation a global genome reorganisation event takes place, during which the 6000 or so IESs from all over the genome come together into distinct foci. These foci merge during later development into large elimination bodies, and are the location for the subsequent DNA elimination. Recent work indicates that the elimination bodies may be formed subsequent to the DNA cleavage required for IES elimination, and that their role may be in aiding DNA repair (Cheng et al. 2010; Shieh and Chalker 2013), but the sequence of events that takes place is not known. The chromatin reorganisation requires a number of different proteins, including Pdd1, a chromodomain protein that appears to bind directly to the H3K9 and H3K27 (Madireddi et al. 1996). Pdd1 is the necessary signal for IES excision, as shown by the fact that artificially tethering Pdd1 to DNA leads to DNA elimination even in the absence of histone methylation.

The role of heterochromatin in *Paramecium* IES excision is less well studied, but recently a chromatin assembly factor, PtCAF1, was identified which is necessary for H3K9me and H3K27me in the maternal and zygotic macronuclei (Ignarski et al. 2014). Interestingly, the effect that knockdown of PtCAF1 has on scnRNAs is to prevent the elimination of MAC genome-matching scnRNAs, resulting in the failure to correctly identify IESs. This suggests that chromatin modifications are somehow involved in the scanning process itself, as well as the identification for elimination of IES sequences in the new MAC.

The DNA cleavage in IES elimination is carried out by a domesticated PiggyBac transposase, named PiggyMac (Pgm) in *Paramecium* and Tbp2 in *Tetrahymena*. This transposase is necessary both for the excision of IESs and the fragmentation of chromosomes, indicating that these processes are connected (Baudry et al. 2009; Cheng et al. 2010; Vogt and Mochizuki 2013a, b). In *Tetrahymena*, Tbp2 has been shown to directly interact with the N-terminal tail of histone H3, indicating that it may bind directly to heterochromatic regions. The IESs are removed in one piece and are at least in some cases circularised (Saveliev and Cox 1994, 2001; Yao and Yao 1994). The ends of the MAC-destined DNA are then annealed using the non-homologous end joining DNA repair (NHEJ) pathway, with Ku70/Ku80 heterodimers binding the broken DNA ends and Ligase IV with its partner Xrcc4p mediating the reannealing of the ends (Kapusta et al. 2011; Marmignon et al. 2014; Lin et al. 2012). Cleavage is coordinated, with both ends of an IES cleaved in an interdependent manner that requires the NHEJ machinery to be in place as well as the transposase. The nature of this cleavage-reannealing process is intriguing: the DNA must be cleaved with base-pair precision so as not to interrupt the reading frame of the MAC-destined sequence, yet the heterochromatic marks that have shown to directly mediate DNA excision are by nature accurate only down to the nearest nucleosome. It is thought that the explanation may lie in the sequence consensus of the IES ends. In *Paramecium*, all IESs are flanked by a TA sequence; one of these flanking TAs is retained in the macronuclear sequence after IES excision. PiggyMac cleavage leads to double-strand breaks with four-base 5′ overhangs, centred on the TA (Gratias and Bétermier 2003), similar to the cleavage preference for PiggyBac class II transposons (TTAA). The *Tetrahymena* transposase Tbp2 also has sequence preferences, as demonstrated by experiments shifting IES boundaries, but the sequence varies from IES to IES and a common consensus has not been found (Chalker et al. 1999; Vogt and Mochizuki 2013b). The current model for the precise recognition and cleavage of IES ends is a combinatorial one: scnRNAs guide histone methylation which recruits Pdd1p and PiggyMac/Tbp2, the latter of which then cleaves at the closest consensus site—the assumption being that together this provides sufficient information for the required precision. However, it is clear that this cannot be the whole story. Firstly, in the AT-rich genomes of ciliates, the loose consensus sequences so far identified do not provide enough information for reproducible identification of IES ends. Indeed, in longer IESs TA or TTAA repeats will occur more than once within the IES sequence. Why, if PiggyMac/Tbp2 binds directly to H3K9/K27me, does it only cleave at IES ends? It has been speculated that this may have to do with Pdd1p blocking PiggyMac/Tbp2 binding, or the density of the heterochromatin preventing cleavage, but so far it has not been explained (Vogt and Mochizuki 2013b). Additionally, in *Paramecium* it has recently been shown that different IESs of different lengths and with different small RNA requirements (see section on iesRNAs) have differing consensus sequences at their ends (Swart et al. 2014). This shows that the role of the small RNAs cannot be restricted to guiding chromatin modifications, as this would not explain these differences.

3.2.4 *iesRNAs*

In addition to scnRNAs, a second class of small RNAs that target IESs was recently discovered in *Paramecium*. These are 27–28 nt long, slightly longer than *Paramecium* scnRNAs, and appear later in development in the newly developing MAC (Sandoval et al. 2014). These are termed iesRNAs and they are produced in the new MAC by a further dicer-like protein, Dcl5. iesRNAs map exclusively to IESs, with much higher density at IES ends, and are not involved in any scanning process. Their production is mysterious: since they map exclusively within IESs it has been suggested that they are produced from excised IESs (Sandoval et al. 2014), but there are problems with this hypothesis. Many iesRNA-requiring IESs are very short—28 nt—and iesRNAs map precisely to these IESs, including the ends. Production of sRNAs from an excised IES would require first loading of RNA polymerase onto such a short molecule, then Dcl5 processing of the resulting transcript, with no loss of nucleotides at either end. This is not possible given what is known about RNA polymerase and dicer-like protein requirements.

The fact that iesRNAs occur later in development than scnRNAs, and in many cases correspond to IESs that require scnRNAs for exision, initially led to the hypothesis that iesRNAs function as a kind of 'proof-reading' mechanism for scnRNA excision (Sandoval et al. 2014). However, recent sequence data indicates that this cannot be the whole story. While many of the same IESs require Dcl5 for their excision as require Dcl2/3, there are subsets of IESs that require Dcl2/3 but not Dcl5 (i.e. scnRNAs but not iesRNAs), and conversely Dcl5 but not Dcl2/3 (Swart, unpublished). The differences in small RNA requirements for excision is related to the sequences of the IES ends, with different consensuses for different subsets of IESs. This suggests that the two pathways have evolved separately, and that either, both, or neither pathway can be involved in the removal of a given IES.

3.2.5 *Early and Late scnRNAs in* Tetrahymena

Recent research indicates that the *Tetrahymena* scnRNA pathway also involves two subsets of IES-targeting RNAs. These have been termed Early- and Late-scnRNAs (Noto et al. 2015), and similarly to the *Paramecium* iesRNA pathway, the Late-scnRNAs bind to a distinct and separate Piwi protein, Twi11p. However, they are thought to be produced by the same Dicer-like protein as the Early-scnRNAs, Dcl1p. The two classes of scnRNA appear to be derived from different genomic regions and types of IES, with Early-scnRNAs mapping mainly to transposons and repetitive DNA, and Late-scnRNAs to more gene-rich regions. Transposons generally represent more recently acquired IESs, indicating that the pathways may have evolved to control newly-acquired transposons and more ancient IESs respectively. Interestingly, the production of Late-scnRNAs is dependent on the Early-scnRNA pathway; knocking down the latter inhibits the former (Noto et al. 2015). This is reminiscent of the iesRNA pathway in *Paramecium* in which iesRNAs are generated from excised scnRNA-dependent IESs, but in *Tetrahymena* the Late-scnRNAs are formed in cis from the Early-scnRNAs, prior to IES excision and concomitant to Early-scnRNA-mediated heterochromatin formation.

3.3 RNAs in *Oxytricha* Development

3.3.1 Pointers and Guide RNAs

The scrambled nature of the *Oxytricha* genome means that assembling protein-coding genes onto nanochromosomes is not just a matter of removing non-coding DNA, it also requires the ordering and in some cases reversing of MAC-destined sequences (MDSs). The scnRNA model of *Tetrahymena* and *Paramecium* is not sufficient to explain this phenomenon. Similarly to the TA repeats flanking *Paramecium* IESs, *Oxytricha* MDSs have short repeat sequences termed 'pointers' which link MDSs, with a repeat at the 3′ end of one MDS matching that at the 5′ end of the next MDS. Just like the TA repeats in *Paramecium*, one copy of the pointer sequence is retained in the coding gene sequence while the other is eliminated with the IES, and the pointer is required for the correct excision of the IES (Landweber et al. 2000). The pointers are between 2 and 20 nt long and are not always unique (Prescott and DuBois 1996; Landweber et al. 2000), meaning that they are not sufficient in themselves to direct the rearrangement of the genome.

It was demonstrated in 2007 that the rearrangement of the *Oxytricha* genome is guided by long template RNAs that are transcribed from the parental MAC during early conjugation (Nowacki et al. 2008). These RNAs, known as 'guide' RNAs, appear to be transcribed bidirectionally from the entire maternal nanochromosome. They are sufficient to guide the rearrangement of the chromosome, as demonstrated by experiments in which a long RNA consisting of an alternatively rearranged gene was injected and led to the rearrangement of the chromosome in daughter cells following conjugation (Nowacki et al. 2008). The machinery involved in the transcription, transport and function of these RNAs is not known, however it seems likely that they must be transported to the developing macronucleus in order to act as a template for the chromosomal rearrangements.

While guide RNAs direct the rearrangements of the gene segments, the pointer sequences are also important. Scrambled segments of genes contain on average longer pointer sequences than segments of non-scrambled genes (Chen et al. 2014). All known gene segments contain a pointer sequence, though some are as short as two bp, and in the experiments causing gene rearrangements via an artificial guide RNA, alternative pointers were engineered into the guide RNAs (Nowacki et al. 2008). This suggests that the repeats might contribute in some way to the cleavage or reannealing of the ends.

The *Oxytricha* genome rearrangements, like those of *Tetrahymena* and *Paramecium*, are carried out using domesticated transposase proteins. However unlike the oligohymenophores, *Oxytricha* uses at least three different transposases, and these are transcribed from the micronuclear genome at the start of conjugation (Nowacki et al. 2009). The micronuclear genomes of *Tetrahymena* and *Paramecium* are thought to be transcriptionally silent.

3.3.2 *Piwi-Interacting Small RNAs in* Oxytricha *Development*

More recently, a second class of non-coding RNAs involved in *Oxytricha* development were discovered (Zahler et al. 2012; Fang et al. 2012). These are 27 nt long and match exclusively to the MAC genome. They bind to an Argonaute protein with homology to *Tetrahymena* Twi1p, termed Otiwi1, and they lead to the retention of their matching sequences in the new MAC (Fang et al. 2012). These piRNAs thus have the opposite effect to the scnRNAs found in *Tetrahymena* and *Paramecium*—rather than targeting sequences for elimination in the new MAC, they identify sequences to be retained. How the *Oxytricha* piRNAs are produced and/or selected and amplified is unknown, but Otiwi1 localisation studies suggest that they are transcribed in the maternal MAC, and move to the new MAC during development.

References

Arnaiz O, Mathy N, Baudry C et al (2012) The paramecium germline genome provides a niche for intragenic parasitic DNA: evolutionary dynamics of internal eliminated sequences. PLoS Genet 8:e1002984. doi:10.1371/journal.pgen.1002984

Aury JM, Jaillon O, Duret L et al (2006) Global trends of whole-genome duplications revealed by the ciliate *Paramecium tetraurelia*. Nature 444(7116):171–178

Aronica L, Bednenko J, Noto T et al (2008) Study of an RNA helicase implicates small RNA–noncoding RNA interactions in programmed DNA elimination in *Tetrahymena*. Genes Dev 22:2228–2241. doi:10.1101/gad.481908

Baudry C, Malinsky S, Restituito M et al (2009) PiggyMac, a domesticated PiggyBac transposase involved in programmed genome rearrangements in the ciliate *Paramecium tetraurelia*. Genes Dev 23:2478–2483. doi:10.1101/gad.547309

Bouhouche K, Gout J-F, Kapusta A et al (2011) Functional specialization of Piwi proteins in *Paramecium tetraurelia* from post-transcriptional gene silencing to genome remodelling. Nucleic Acids Res. doi:10.1093/nar/gkq1283

Chalker DL, La Terza A, Wilson A et al (1999) Flanking regulatory sequences of the *Tetrahymena* R deletion element determine the boundaries of DNA rearrangement. Mol Cell Biol 19:5631–5641

Cheng C-Y, Vogt A, Mochizuki K, Yao M-C (2010) A domesticated PiggyBac transposase plays key roles in heterochromatin dynamics and DNA cleavage during programmed DNA deletion in *Tetrahymena thermophila*. Mol Biol Cell 21:1753–1762. doi:10.1091/mbc.E09-12-1079

Chen X, Bracht JR, Goldman AD et al (2014) The Architecture of a scrambled genome reveals massive levels of genomic rearrangement during development. Cell 158:1187–1198. doi:10.1016/j.cell.2014.07.034

Duharcourt S, Keller A-M, Meyer E (1998) Homology-dependent maternal inhibition of developmental excision of internal eliminated sequences in *Paramecium tetraurelia*. Mol Cell Biol 18:7075–7085

Eisen JA, Coyne RS, Wu M et al (2006) Macronuclear genome sequence of the ciliate *Tetrahymena thermophila*, a model eukaryote. PLoS Biol 4:e286. doi:10.1371/journal.pbio.0040286

Fang W, Wang X, Bracht JR et al (2012) Piwi-interacting RNAs protect DNA against loss during *Oxytricha* genome rearrangement. Cell 151:1243–1255. doi:10.1016/j.cell.2012.10.045

Fass JN, Joshi NA, Couvillion MT, et al (2011) Genome-scale analysis of programmed DNA elimination sites in *Tetrahymena thermophila*. G3 Genes Genomes Genet 1:515–522. doi:10.1534/g3.111.000927

Gratias A, Bétermier M (2003) Processing of double-strand breaks is involved in the precise excision of *Paramecium* internal eliminated sequences. Mol Cell Biol 23:7152–7162. doi:10.1128/MCB.23.20.7152-7162.2003

Hamilton E, Bruns P, Lin C et al (2005) Genome-Wide Characterization of *Tetrahymena thermophila* chromosome breakage sites. I. Cloning and identification of functional sites. Genetics 170:1611–1621. doi:10.1534/genetics.104.031401

Ignarski M, Singh A, Swart EC et al (2014) *Paramecium tetraurelia* chromatin assembly factor-1-like protein PtCAF-1 is involved in RNA-mediated control of DNA elimination. Nucleic Acids Res 42:11952–11964. doi:10.1093/nar/gku874

Maliszewska-Olejniczak Kamila, Gruchota Julita, Gromadka Robert et al (2015) TFIIS-dependent non-coding transcription regulates developmental genome rearrangements. PLoS Genet. doi:10.1371/journal.pgen.1005383

Kapusta A, Matsuda A, Marmignon A et al (2011) Highly precise and developmentally programmed genome assembly in *Paramecium* requires ligase IV–dependent end joining. PLoS Genet 7:e1002049. doi:10.1371/journal.pgen.1002049

Landweber LF, Kuo T-C, Curtis EA (2000) Evolution and assembly of an extremely scrambled gene. Proc Natl Acad Sci 97:3298–3303. doi:10.1073/pnas.97.7.3298

Le Mouël A, Butler A, Caron F, Meyer E (2003) Developmentally regulated chromosome fragmentation linked to imprecise elimination of repeated sequences in paramecia. Eukaryot Cell 2:1076–1090. doi:10.1128/EC.2.5.1076-1090.2003

Lepère G, Bétermier M, Meyer E, Duharcourt S (2008) Maternal noncoding transcripts antagonize the targeting of DNA elimination by scanRNAs in *Paramecium tetraurelia*. Genes Dev 22:1501–1512. doi:10.1101/gad.473008

Lin I-T, Chao J-L, Yao M-C (2012) An essential role for the DNA breakage-repair protein Ku80 in programmed DNA rearrangements in *Tetrahymena thermophila*. Mol Biol Cell 23:2213–2225. doi:10.1091/mbc.E11-11-0952

Liu Y, Taverna SD, Muratore TL et al (2007) RNAi-dependent H3K27 methylation is required for heterochromatin formation and DNA elimination in *Tetrahymena*. Genes Dev 21:1530–1545. doi:10.1101/gad.1544207

Madireddi MT, Coyne RS, Smothers JF et al (1996) Pdd1p, a novel chromodomain-containing protein, links heterochromatin assembly and DNA elimination in *Tetrahymena*. Cell 87:75–84. doi:10.1016/S0092-8674(00)81324-0

Malone CD, Anderson AM, Motl JA et al (2005) Germ line transcripts are processed by a dicer-like protein that is essential for developmentally programmed genome rearrangements of *Tetrahymena thermophila*. Mol Cell Biol 25:9151–9164. doi:10.1128/MCB.25.20.9151-9164.2005

Marmignon A, Bischerour J, Silve A et al (2014) Ku-mediated coupling of DNA cleavage and repair during programmed genome rearrangements in the ciliate *Paramecium tetraurelia*. PLoS Genet 10:e1004552. doi:10.1371/journal.pgen.1004552

Mochizuki K, Fine NA, Fujisawa T, Gorovsky MA (2002) Analysis of a piwi-related gene implicates small RNAs in genome rearrangement in *Tetrahymena*. Cell 110:689–699. doi:10.1016/S0092-8674(02)00909-1

Mochizuki K, Gorovsky MA (2004a) RNA polymerase II localizes in *Tetrahymena thermophila* meiotic micronuclei when micronuclear transcription associated with genome rearrangement occurs. Eukaryot Cell 3:1233–1240. doi:10.1128/EC.3.5.1233-1240.2004

Mochizuki K, Gorovsky MA (2004b) Conjugation-specific small RNAs in *Tetrahymena* have predicted properties of scan (scn) RNAs involved in genome rearrangement. Genes Dev 18:2068–2073. doi:10.1101/gad.1219904

Mochizuki K, Kurth HM (2013) Loading and pre-loading processes generate a distinct siRNA population in *Tetrahymena*. Biochem Biophys Res Commun 436:497–502. doi:10.1016/j.bbrc.2013.05.133

Morgens DW, Cavalcanti ARO (2015) Amitotic chromosome loss predicts distinct patterns of senescence and non-senescence in ciliates. Protist 166:224–233. doi:10.1016/j.protis.2015.03.002

Noto T, Kataoka K, Suhren JH et al (2015) Small-RNA-mediated genome-wide trans-recognition network in *Tetrahymena* DNA elimination. Mol Cell 59:229–242. doi:10.1016/j.molcel.2015.05.024

Noto T, Kurth HM, Kataoka K et al (2010) The *Tetrahymena* argonaute-binding protein Giw1p directs a mature argonaute-siRNA complex to the nucleus. Cell 140:692–703. doi:10.1016/j.cell.2010.02.010

Nowacki M, Higgins BP, Maquilan GM et al (2009) A Functional role for transposases in a large eukaryotic genome. Science 324:935–938. doi:10.1126/science.1170023

Nowacki M, Vijayan V, Zhou Y et al (2008) RNA-mediated epigenetic programming of a genome-rearrangement pathway. Nature 451:153–158. doi:10.1038/nature06452

Prescott DM, DuBOIS ML (1996) Internal eliminated segments (IESs) of Oxytrichidae1. J Eukaryot Microbiol 43:432–441. doi:10.1111/j.1550-7408.1996.tb04502.x

Sandoval PY, Swart EC, Arambasic M, Nowacki M (2014) Functional diversification of dicer-like proteins and small RNAs required for genome sculpting. Dev Cell 28:174–188. doi:10.1016/j.devcel.2013.12.010

Saveliev SV, Cox MM (2001) Product analysis illuminates the final steps of IES deletion in *Tetrahymena thermophila*. EMBO J 20:3251–3261. doi:10.1093/emboj/20.12.3251

Saveliev SV, Cox MM (1994) The fate of deleted DNA produced during programmed genomic deletion events in *Tetrahymena thermophila*. Nucleic Acids Res 22:5695–5701

Schoeberl UE, Kurth HM, Noto T, Mochizuki K (2012) Biased transcription and selective degradation of small RNAs shape the pattern of DNA elimination in *Tetrahymena*. Genes Dev 26:1729–1742. doi:10.1101/gad.196493.112

Shieh AWY, Chalker DL (2013) LIA5 Is required for nuclear reorganization and programmed DNA rearrangements occurring during *Tetrahymena* macronuclear differentiation. PLoS ONE 8:e75337. doi:10.1371/journal.pone.0075337

Swart EC, Bracht JR, Magrini V et al (2013) The *Oxytricha* trifallax macronuclear genome: a complex eukaryotic genome with 16,000 tiny chromosomes. PLoS Biol 11:e1001473. doi:10.1371/journal.pbio.1001473

Swart EC, Wilkes CD, Sandoval PY et al (2014) Genome-wide analysis of genetic and epigenetic control of programmed DNA deletion. Nucleic Acids Res 42:8970–8983. doi:10.1093/nar/gku619

Taverna SD, Coyne RS, Allis CD (2002) Methylation of histone H3 at lysine 9 targets programmed DNA elimination in *Tetrahymena*. Cell 110:701–711. doi:10.1016/S0092-8674(02)00941-8

Vogt A, Mochizuki K (2013a) A domesticated PiggyBac transposase interacts with heterochromatin and catalyzes reproducible DNA elimination in *Tetrahymena*. PLoS Genet 9:e1004032. doi:10.1371/journal.pgen.1004032

Vogt A, Mochizuki K (2013b) A domesticated PiggyBac transposase interacts with heterochromatin and catalyzes reproducible DNA elimination in *Tetrahymena*. PLoS Genet 9:e1004032. doi:10.1371/journal.pgen.1004032

Yao MC, Yao CH (1994) Detection of circular excised DNA deletion elements in *Tetrahymena thermophila* during development. Nucleic Acids Res 22:5702–5708

Zahler AM, Neeb ZT, Lin A, Katzman S (2012) Mating of the stichotrichous ciliate *Oxytricha* trifallax induces production of a class of 27 nt small RNAs derived from the parental macronucleus. PLoS ONE 7:e42371. doi:10.1371/journal.pone.0042371

Chapter 4
Intracytoplasmic Signaling from Cilia in Ciliates

Peter Satir and Birgit H. Satir

Abstract The ciliates are model organisms for pioneering studies of ciliary signaling as it controls cell behavior through second messengers such as Ca^{2+} and cAMP. Signaling is initiated via special receptors uniquely or significantly localized to the ciliary membrane. The receptors initiate a cascade of molecular changes in the ciliary matrix and in certain cases molecules move from the ciliary matrix into the cytoplasm, and sometimes enter the cell nucleus to alter gene expression. Like the cell nucleus, the cilium is specialized compartment of the cytoplasm. The entrance to the ciliary membrane or matrix is defined by the ciliary necklace barriers—ciliary pores—whose composition and function seem related to the nuclear pores. A well-defined signaling cascade can be traced from the cilia-localized receptor tyrosine kinase (TtPTK1) (NEK1) in *Tetrahymena.* In this cascade, Rad 51 which localizes to both the cilia and the nucleus, presumably may shuttle from the cilium to the nucleus to affect DNA replication and repair. The signaling scaffold protein parafusin (PFUS) may also represent this class of molecules. PFUS is localized to *Paramecium* dense core secretory vesicles (DSCVs) but also to the cilium and the nucleus. Knockdown of PFUS shuts off overall DSCV production, suggesting an effect on gene expression. Localization of signaling molecules such as Rad 51, Rsk and PFUS to both cilium and nucleus is found in both ciliates and mammalian cells, but the dynamics of movement between compartments is generally unknown and needs further elucidation.

4.1 Introduction

The ciliate protists, particularly *Tetrahymena* and *Paramecium*, but also *Euplotes, Oxytricha* and many other genera have contributed extensively to our understanding of molecular and developmental cell biology. Often principles or observations first

P. Satir (✉) · B.H. Satir
Department of Anatomy and Structural Biology, Albert Einstein College of Medicine, Bronx, NY 10461, USA
e-mail: peter.satir@einstein.yu.edu

G. Witzany and M. Nowacki (eds.), *Biocommunication of Ciliates*,
DOI 10.1007/978-3-319-32211-7_4

discovered in these organisms have later been shown to apply generally throughout the metazoan phylogenetic tree and specifically to mammalian and human cells. In addition, the ciliates have illuminated unsolved problems in nuclear and cytoplasmic organization. In this chapter, we discuss how early and sometimes incomplete information on ciliary signaling processes in ciliates has led to unanticipated new ideas and questions concerning intracytoplasmic signaling and information flow in cells.

4.2 Cilia as Signal Generators

It is unsurprising that cilia would be useful as sensory, as well as motor, organelles. In highly ciliated cells such as the ciliates, the ciliary membrane represents the first contact with the environment and is best positioned to sample environmental changes that could be favorable or detrimental to cell survival. Response to environmental conditions relies on signals generated by the ciliary membrane that invade the axoneme to control instantaneous behavior or that move onward into the cell to influence cytoplasmic organization and/or gene expression. The initial understanding of the importance of specific ciliary membrane receptors and channels was made in ciliates, originally in terms of cell swimming. The broad outline of response is rather similar to bacterial swimming and depends on cilia operating in a low Reynolds number hydrodynamic regime so that stoppage or directional changes have virtually no inertial component. Receptors for both attractants and repellent chemicals are present in the membrane but sometimes the chemical can directly diffuse into the axoneme to affect behavior (Van Houten 1998).

4.2.1 Axonemal Second Messengers

In *Paramecium*, a voltage dependent Ca^{2+} channel, specific to the ciliary membrane (Ogura and Takahashi 1976; Dunlap 1977), opens under certain mechanical or chemical stimuli, to depolarize the cell and raise axonemal Ca^{2+} momentarily to produce backward swimming (Naitoh and Eckert 1969; Preston et al. 1991; Kleene and Van Houten 2014). Further, a transmembrane adenylyl cyclase (a K+ channel) specific to the ciliary membrane (Schultz et al. 1992), when stimulated, for example by chemicals such as glutamate (Van Houten 2000) produces an increase in axonemal cAMP that leads to faster forward swimming (Bonini and Nelson 1988; Hamasaki et al. 1991; Satir et al. 1993).

4.2.2 Co-signaling via Ciliary Attachment

Although swimming behavioral responses to ciliary receptors in ciliates, or at least in *Paramecium*, have a long history of study, much less attention has been devoted to responses initiated in ciliate cilia that affect cell organization and gene expression. This is of major interest for mammalian cilia function. In ciliates, study of the coupling of ciliary and cellular response is most obvious and has been best examined for mating behavior, i.e. conjugation. Attachment between cilia of opposite mating types in ciliates including *Tetrahymena, Paramecium* and *Euplotes* initiates conjugation. Attachment can be reproduced with isolated cilia or ciliary membrane vesicles (Luporini and Dallai 1980; Love and Rotheim 1984; Watanabe 1990; Wolfe et al. 1993). The mating inducing factors (gamones) have been well studied, particularly by Luporini's group (Vallesi et al. 2005). The gamone (pheromone) receptors are glycoproteins of the ciliary membrane whose interaction with their respective gamones can be blocked by concanavalin A or competing gamones. The limited information available suggests that after binding and clustering, the receptor-gamone complex is internalized by endocytosis. Signaling specifics are not well studied but mating events that follow ciliary pairing require remodeling of the cell surfaces, including ciliary resorption, cell membranes fusion, micronuclear migration, subsequent meiotic and somatic genome reorganizations and cell division (Martindale et al. 1982). Signaling via ciliary contact of the sort exemplified by ciliate conjugation is found in mammalian primary cilia (Ott et al. 2012).

4.3 The Ciliary Compartment

The receptors for intracytoplasmic signal generation in swimming and mating in ciliates are strictly functional only when placed in the ciliary membrane. Presumably after synthesis in ER membranes they move through the Golgi and are present in vesicles directed to the ciliary membrane by ciliary localization signals, the simplest of which are RVxP and VxPx (Geng et al. 2006) or AX[S/A]XQ (Berbari et al. 2008). The cilium, like the nucleus, is a privileged cellular compartment whose membrane and matrix components are a highly selective set of molecules. Entrance to the cilium is controlled by a barrier, the ciliary necklace (Gilula and Satir 1972), whose organization, protein structure and function are reminiscent of the nuclear pore (Kee and Verhey 2013), possibly because the cilium and nucleus evolved around the same time in early eukaryotic cell evolution (Satir et al. 2015). Molecules such as certain ciliate guanine nucleotide exchange factors (GEFs) (Bell et al. 2009) and signaling scaffold proteins such as parafusin (PFUS) are localized both to the cilium and the nucleus and molecules (e.g. Rad 51, Gli proteins) move from cilium to nucleus to change gene expression. Homologous scaffold proteins used for membrane trafficking processes were utilized in the development of the barrier organization and function in the two organelles (Jekely and Arendt 2006).

The ciliary necklace of ciliates is well described in freeze fracture (Satir et al. 1976; Bardele 1981) and some of the important barrier proteins involved in human ciliopathies, such as meckelin (TMEM 67) (Leightner et al. 2013), are present (Picariello et al. 2014) in ciliates.

4.3.1 Ciliate Ciliogenesis

Once past the barrier, the cilium grows by intraflagellar transport (IFT) and membrane receptors move either by coupling to the IFT motors or by diffusion. IFT, originally described in *Chlamydomonas* is present in ciliates and many key IFT components, such as IFT 57 (HIPPI) of *Paramecium* (Shi 2013) or the homodimeric IFT kinesin (Kin5) of *Tetrahymena* (Awan et al. 2004, 2009), are known to be involved in ciliate ciliogenesis. However, information as to the interaction of these proteins with ciliary membrane receptors in signaling pathways well known from vertebrate studies is very fragmentary. Nevertheless, since the genomes of several major ciliate species are complete, comprehensive studies on ciliary membrane proteins can be undertaken. One example is from *Paramecium* tetraurelia (Yano et al. 2013).

4.4 Status of Major Ciliary Signaling Pathways in Ciliates

Do the major signaling pathways that affect cytoplasmic organization and gene expression known for mammalian cilia such as polycystin signaling, hedgehog (Hh) signaling, G-protein coupled receptor (GPCR) signaling or receptor tyrosine kinase (RTK) signaling run through ciliate cilia?

4.4.1 Polycystin Signaling

Polycystin 2 (PC2), the Ca^{2+} channel is present along the length of *Paramecium* cilia. Import into the ciliary membrane requires BBS proteins (Valentine et al. 2012). However, it is unclear if the PC2 signal in *Paramecium* cilia has a mechanoreceptor component or how the signal influences the cell.

4.4.2 Hedgehog Signaling

Hedgehog signaling proteins such as Smo, Ptc and Gli are apparently absent from the *Paramecium* genome, but the BBS protein B9 is present and IFT122 is present in *Tetrahymena* (Tsao and Gorovsky 2008a, b). These proteins are involved in controlling protein movement into and out of the cilium and may play this role in ciliates before hedgehog signaling pathways evolved. IFT57, IFT 122 and the BBS proteins have been implicated in hedgehog signaling in various vertebrates (Houde et al. 2006; Qin et al. 2011; Zhang et al. 2012), largely through control of movement of the Hh signaling proteins.

4.4.3 GPCR Signaling

Gene sequences and partial cloning suggests that GPCRs are most likely present in several ciliates. In *Tetrahymena,* a macronuclear knockout of a putative constitutively active GPCR (GPCR6) has been carefully studied (Lampert et al. 2011). The knockout has loss of chemoattraction. Microsomal [35S]GTPγS binding assays show a decrease in wild type basal G-protein activity in the GPCR6 knockout to the same level as with pertussis toxin (a G-protein inhibitor). A similar gene is present in *Paramecium.* A β adrenergic receptor has also been described in *Paramecium* (Wiejak et al. 2004). Rhodopsin immunoanalogs are present in many ciliates including *Blespharisma* and *Stentor* (Fabczak et al. 2008). These proteins have been localized to the cell membrane, but none specifically to cilia.

4.4.4 RTK Signaling

Various peptide growth factors including insulin signal via receptor tyrosine kinases (RTKs) (Fig. 4.1). In many instances the receptors for these substances are largely or exclusively localized to cilia (Christensen et al. 2012). One well studied example is PDGFRα, a specific isoform of the platelet-derived growth factor receptor found in mammalian fibroblasts which dimerizes and signals exclusively from the primary cilium (Schneider et al. 2005). RTKs were not known for ciliates, but several studies concluded that growth factors such as insulin were signaling molecules that had profound effects on aspects of cell survival (Christensen et al. 1996; Rasmussen and Wheatley 2007).

4.4.4.1 A Receptor Tyrosine Kinase (RTK) in Ciliate Cilia

Christensen et al. (2003) showed that antibodies raised against the tyrosine kinase domain of the human insulin β chain localize to cilia of *Tetrahymena thermophila.* Further, treatment with human recombinant insulin increased the localization, specifically to a 66 kDa protein, which they cloned and named TtPTK1. This protein is 80 % identical to NEK1, one of many NIMA like proteins found in the cilia of unicellular organisms (Quarmby and Mahjoub 2005), most of which are thought to be serine threonine kinases, rather than tyrosine kinases. *Tetrahymena* may have as many as 39 NEKs, some of which regulate ciliary morphogenesis (Wloga et al. 2006). TtPTK1 (or *Tetrahymena thermophila* NEK1) has a kinase domain that contains specific motifs conserved for tyrosine kinases and a ca. 30 % identity and 65 % homology to metazoan insulin β chain kinase domains. The physiological ligand for the complete TtPTK1 receptor is not known and is probably not insulin itself. Rasmussen and Wheatley (Rasmussen and Wheatley 2007) have purified a small peptide (TCSF1) that prolongs cell survival in minimal media that could be an insulin substitute but Fig. 4.2 shows that isolated *Tetrahymena* cilia bind fluorescently labeled mammalian insulin. Additionally, an IRα reactive protein and an IRS1 ortholog localize to isolated cilia (Guerra 2007), supporting the suggestion that TtPTK1 is the β chain of a complete functional insulin-like receptor in the ciliary membrane.

A pathway for signaling from an insulin-like growth factor to control gene expression for the regulation of cell growth has been proposed for mammalian cells (Trojanek et al. 2003). Figure 4.1 adapts this scheme to a hypothetical signaling pathway from the TtPTK1 receptor in *Tetrahymena* in order to demonstrate which molecules could carry the signal from cilium to nucleus. Using the TIGR database

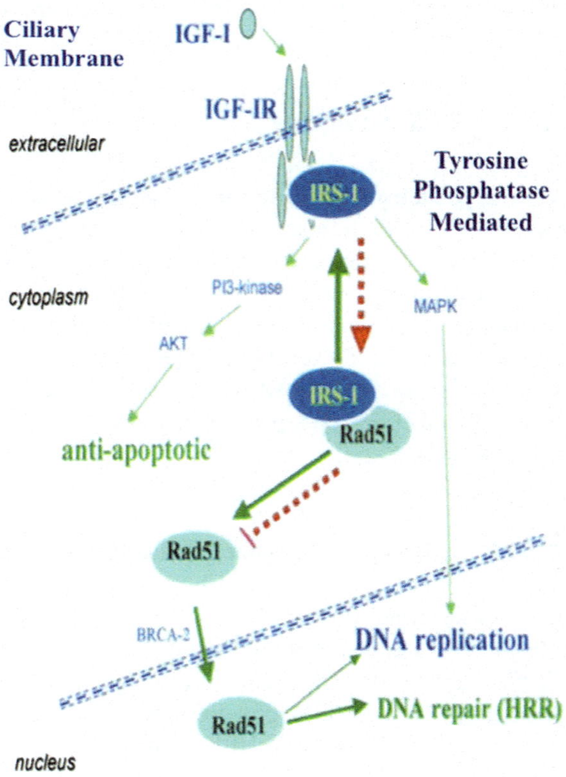

Fig. 4.1 Proposed RTK Signal Transduction Pathway —modified after Trojanek et al. (2003). Cascade of key molecules moving from the Tetrahymena Insulin-binding receptor (TtPTK1 = IGF − 1R) in the ciliary membrane into and through the cytoplasm into the nucleus to affect DNA repair or replication (modified from Guerra 2007)

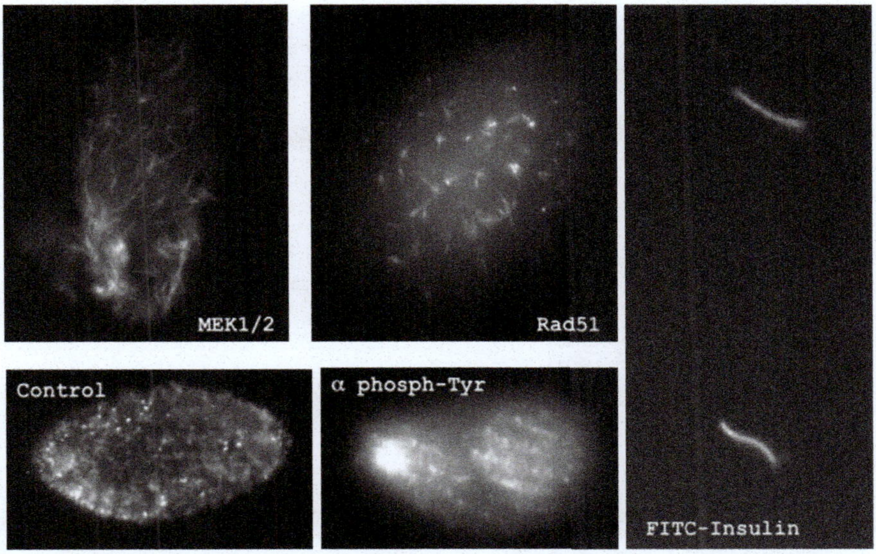

Fig. 4.2 Ciliary localization of key molecules on the TtPTK1 transduction pathway (modified from Guerra 2007)

for *Tetrahymena,* Guerra (2007) searched for putative orthologues of such signaling molecules. With RT-PCR, we could show that genes for important signaling proteins in the insulin response pathway are expressed in the *Tetrahymena* genome and are related to their mammalian counterparts. Then deciliation and regrowth were used to show that production of RNA for these pathway proteins is upregulated during ciliogenesis, while RNA for heat shock protein (HSP) is not (Fig. 4.3). Conversely, heat upregulates HSP but not ciliary pathway proteins.

4.4.4.2 Pathway Proteins Localize to Cilia

Some of the key pathway proteins, namely Mek1/2 and Rad 51, can be directly localized to *Tetrahymena* cilia (Fig. 4.2). Rad 51 also localizes to the macronucleus (Campbell and Romero 1998) where it functions in DNA replication and repair during the cell cycle (Smith et al. 2004).

From this information, we conclude that an RTK signaling pathway from cilium to macronucleus is present in *Tetrahymena*. Molecules such as Rad 51 sequestered in the ciliary tip presumably respond to a ligand that binds to the TtPTK1 membrane receptor by moving out of the cilium, through the cytoplasm and into the nucleus to affect the cell cycle.

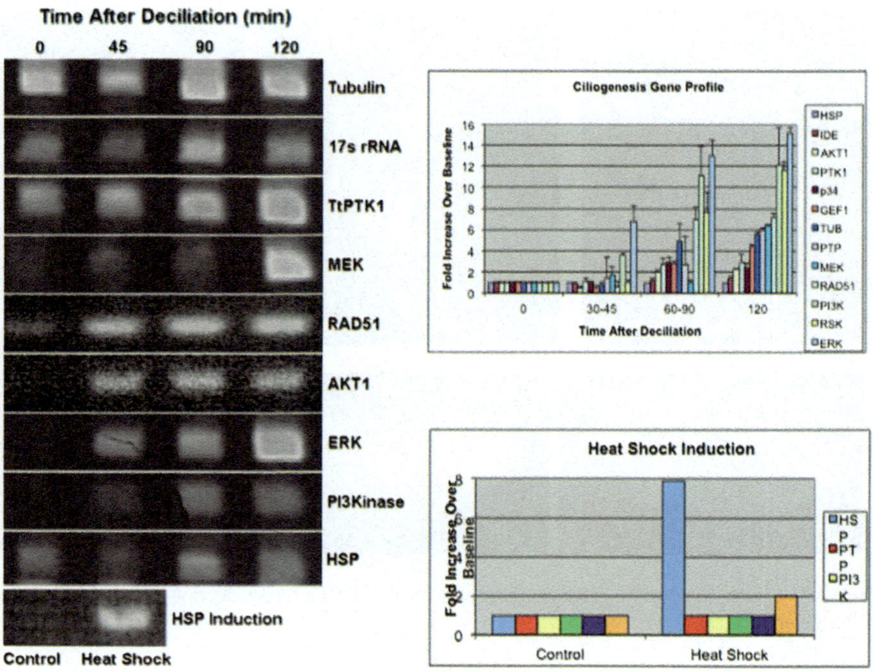

Fig. 4.3 Upregulation of RNA of key signaling molecules on the TtPTK1 transduction pathway after deciliation. Note that Heat shock protein RNA is not upregulated by deciliation and that heat shock does not upregulate the TtPTK1 signaling pathway RNAs (modified from Guerra 2007) (Color figure online)

4.5 Parafusin: A Scaffold Molecule in Intracytoplasmic Signaling?

Parafusin (PFUS), a 63 kDa member of the phosphoglucomutase (PGM) superfamily without significant PGM activity, is a well described, cloned and sequenced signaling scaffold protein of *Paramecium* (Subramanian et al. 1994), also found in *Tetrahymena,* other apicomplexa such as *Toxoplasma* (Matthiesen et al. 2001), as well as in mammalian tissues (Wyroba et al. 1995) and cells in culture. A specific antibody to PFUS that does not recognize PGM has been generated for localization studies. In *Paramecium*, PFUS exists with several different post-translational modifications, particularly serine phosphorylation and phosphoglucosylation (Subramanian and Satir 1992). Phosphoglucosylated PFUS is particularly localized to the cytoplasmic side of the membrane of trichocysts, the cell's dense core secretory vesicles (DCSV's) (Zhao and Satir 1998). When the cell is subject to mechanical or chemical stimulation, such that Ca^{2+} channels in the membrane open and Ca^{2+} enters the cytoplasm, the DCSV's exocytose explosively as their content expands upon

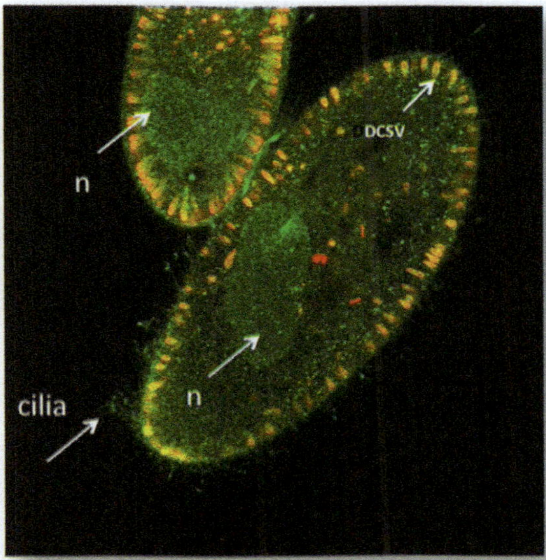

Fig. 4.4 Localization of PFUS (*green*) in *Paramecium*. PFUS is a cytoplasmic molecule. Phosphoglucosylated PFUS forms part of the membrane scaffold around DSCV content (*red*). PFUS is also found throughout the macronucleus (n) and along cilia, concentrated at cilia tips (*arrow*) (modified from Liu et al. 2011) (Color figure online)

release and their membrane is incorporated into the cell membrane or recycled. Exocytosis requires Ca^{2+}dependent deglucosylation of PFUS, which presumably affects the integrity of the DSCV membrane scaffold, and exocytosis fails in mutants where deglucosylation is blocked (Satir and Zhao 1991). The deglucosylated molecule becomes cytoplasmic and presumably can then become phosphorylated for additional functions. PFUS is also significantly localized to the ciliary base and axoneme, sometimes at the ciliary tip, and to the nucleus (Fig. 4.4) (Satir et al. 2015).

In *Paramecium,* knockdown of PFUS using siRNA blocks overall DSCV production (Liu et al. 2011), suggesting that nuclear localization of PFUS affects gene expression of a subset of membrane and perhaps DSCV content genes. The possibility that PFUS is signaling within the nucleus is further supported by the finding that in mouse embryo fibroblasts (MEFs) PFUS localizes specifically to the nucleolus (Fig. 4.5a, b). In MEFs PFUS also localizes to the ciliary base (Fig. 4.5c, d).

We do not know whether the localizations of PFUS to cilia and nucleus are dependent on specific posttranslational states of PFUS. It is clear that both in *Paramecium* and MEFs, PFUS as synthesized is a cytoplasmic, peripheral membrane signaling molecule. As with the other signaling molecules we have been discussing that are localized to both cilia and the nucleus, it would be interesting to know whether subsequent movement of PFUS into the nucleus is dependent on its previous location in the cilium or at the ciliary base or ciliary pocket.

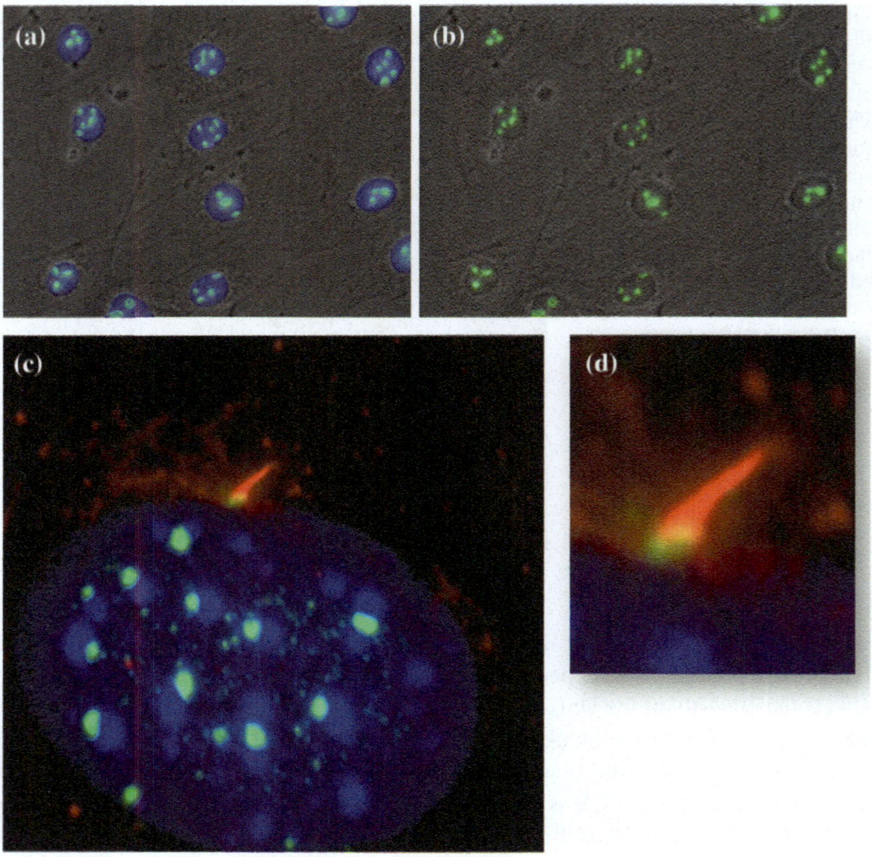

Fig. 4.5 **a** Nucleoli of MEF nuclei concentrate DAPI (*blue*). **b** Localization of PFUS (*green*) to the nucleoli in these MEFs. **c** Localization of PFUS (*green*) to the ciliary pocket, ciliary base and to nucleoli. MEF nucleus stained by DAPI (*blue*). Acetylated α tubulin (*red*) localized along primary cilium. **d** Detail of (**c**). *Yellow color* at ciliary base indicates co-localization of PFUS and acetylated α tubulin (modified from Satir et al. 2015) (Color figure online)

4.6 Conclusions

Intracytoplasmic signaling from cilia probably first arose as an evolutionary adaptation to escape predation. Building on this, ciliates used intracytoplasmic signaling from ciliary membrane receptors and channels to second messenger systems largely within the axoneme to change ciliary beat direction and swimming speed. Further, the cilium and the nucleus evolved together in the early eukaryotic ancestor with common molecular pathways of entry from the cytoplasm through barrier pores. Signals from ciliary membrane receptors passed through the barriers to reach the nucleoplasm and effect gene changes leading to cytoplasmic reorganization and mating. From these behavioral responses, more complicated signaling pathways such as the polycystin, GPCR and RTK pathways evolved in ciliates. It seems likely that many important signaling molecules including Rad 51, Rsk, and probably metabolic proteins such as PFUS, move from cilium to nucleus to alter gene expression in ciliates. Similar signaling from cilium to nucleus is preserved in mammalian cells.

References

Awan A, Bell AJ, Satir P (2009) Kin5 knockdown in Tetrahymena thermophila using RNAi blocks cargo transport of Gef1. PLoS ONE 4:e4873

Awan A, Bernstein M, Hamasaki T, Satir P (2004) Cloning and characterization of Kin5, a novel Tetrahymena ciliary kinesin II. Cell Motil Cytoskeleton 58:1–9

Bardele CF (1981) Functional and phylogenetic aspects of the ciliary membrane: a comparative freeze-fracture study. Biosystems 14:403–421

Bell AJ, Guerra C, Phung V, Nair S, Seetharam R, Satir P (2009) GEF1 is a ciliary Sec7 GEF of Tetrahymena thermophila. Cell Motil Cytoskeleton 66:483–499

Berbari NF, Johnson AD, Lewis JS, Askwith CC, Mykytyn K (2008) Identification of ciliary localization sequences within the third intracellular loop of G protein-coupled receptors. Mol Biol Cell 19:1540–1547

Bonini NM, Nelson DL (1988) Differential regulation of Paramecium ciliary motility by cAMP and cGMP. J Cell Biol 106:1615–1623

Campbell C, Romero DP (1998) Identification and characterization of the RAD51 gene from the ciliate Tetrahymena thermophila. Nucleic Acids Res 26:3165–3172

Christensen ST, Quie H, Kemp K, Rasmussen L (1996) Insulin produces a biphasic response in Tetrahymena thermophila by stimulating cell survival and activating proliferation in two separate concentration intervals. Cell Biol Int 20:437–444

Christensen ST, Clement CA, Satir P, Pedersen LB (2012) Primary cilia and coordination of receptor tyrosine kinase (RTK) signalling. J Pathol 226:172–184

Christensen ST, Guerra CF, Awan A, Wheatley DN, Satir P (2003) Insulin receptor-like proteins in Tetrahymena thermophila ciliary membranes. Curr Biol 13:R50–R52

Dunlap K (1977) Localization of calcium channels in Paramecium caudatum. J Physiol 271:119–133

Fabczak H, Sobierajska K, Fabczak S (2008) A rhodopsin immunoanalog in the related photosensitive protozoans Blepharisma japonicum and Stentor coeruleus. Photochem Photobiol Sci 7:1041–1045

Geng L, Okuhara D, Yu Z, Tian X, Cai Y, Shibazaki S, Somlo S (2006) Polycystin-2 traffics to cilia independently of polycystin-1 by using an N-terminal RVxP motif. J Cell Sci 119:1383–1395

Gilula NB, Satir P (1972) The ciliary necklace: a ciliary membrane specialization. J Cell Biol 53:494–509

Guerra C (2007) Ciliary signaling systems for motility and growth control in Tetrahymena. Ph.D. thesis, Albert Einstein College of Medicine. New York

Hamasaki T, Barkalow K, Richmond J, Satir P (1991) cAMP-stimulated phosphorylation of an axonemal polypeptide that copurifies with the 22S dynein arm regulates microtubule translocation velocity and swimming speed in Paramecium. Proc Natl Acad Sci USA 88:7918–7922

Houde C, Dickinson RJ, Houtzager VM, Cullum R, Montpetit R, Metzler M, Simpson EM, Roy S, Hayden MR, Hoodless PA, Nicholson DW (2006) Hippi is essential for node cilia assembly and Sonic hedgehog signaling. Dev Biol 300:523–533

Jekely G, Arendt D (2006) Evolution of intraflagellar transport from coated vesicles and autogenous origin of the eukaryotic cilium. BioEssays 28:191–198

Kee HL, Verhey KJ (2013) Molecular connections between nuclear and ciliary import processes. Cilia 2:11

Kleene SJ, Van Houten JL (2014) Electrical signaling in motile and primary cilia. Bioscience 64:1092–1102

Lampert TJ, Coleman KD, Hennessey TM (2011) A knockout mutation of a constitutive GPCR in Tetrahymena decreases both G-protein activity and chemoattraction. PLoS ONE 6:e28022

Leightner AC, Hommerding CJ, Peng Y, Salisbury JL, Gainullin VG, Czarnecki PG, Sussman CR, Harris PC (2013) The Meckel syndrome protein meckelin (TMEM67) is a key regulator of cilia function but is not required for tissue planar polarity. Hum Mol Genet 22:2024–2040

Liu L, Wyroba E, Satir BH (2011) RNAi knockdown of parafusin inhibits the secretory pathway. Eur J Cell Biol 90:844–853

Love B, Rotheim MB (1984) Cell surface interactions in conjugation: Tetrahymena ciliary membrane vesicles. Mol Cell Biol 4:681–687

Luporini P, Dallai R (1980) Sexual interaction of Euplotes crassus: differentiation of cellular surfaces in cell-to-cell union. Dev Biol 77:167–177

Martindale DW, Allis CD, Bruns PJ (1982) Conjugation in Tetrahymena thermophila: a temporal analysis of cytological stages. Exp Cell Res 140:227–236

Matthiesen SH, Shenoy SM, Kim K, Singer RH, Satir BH (2001) A parafusin-related Toxoplasma protein in Ca2+-regulated secretory organelles. Eur J Cell Biol 80:775–783

Naitoh Y, Eckert R (1969) Ionic mechanisms controlling behavioral responses of paramecium to mechanical stimulation. Science 164:963–965

Ogura A, Takahashi K (1976) Artificial deciliation causes loss of calcium-dependent responses in Paramecium. Nature 264:170–172

Ott C, Elia N, Jeong SY, Insinna C, Sengupta P, Lippincott-Schwartz J (2012) Primary cilia utilize glycoprotein-dependent adhesion mechanisms to stabilize long-lasting cilia-cilia contacts. Cilia 1:3

Picariello T, Valentine MS, Yano J, Van Houten J (2014) Reduction of meckelin leads to general loss of cilia, ciliary microtubule misalignment and distorted cell surface organization. Cilia 3:2

Preston RR, Kink JA, Hinrichsen RD, Saimi Y, Kung C (1991) Calmodulin mutants and Ca2(+)-dependent channels in Paramecium. Annu Rev Physiol 53:309–319

Qin J, Lin Y, Norman RX, Ko HW, Eggenschwiler JT (2011) Intraflagellar transport protein 122 antagonizes Sonic Hedgehog signaling and controls ciliary localization of pathway components. Proc Natl Acad Sci USA 108:1456–1461

Quarmby LM, Mahjoub MR (2005) Caught Nek-ing: cilia and centrioles. J Cell Sci 118:5161–5169

Rasmussen MI, Wheatley DN (2007) Purification and characterisation of cell survival factor 1 (TCSF1) from Tetrahymena thermophila. J Cell Commun Signal 1:185–193

Satir B, Sale WS, Satir P (1976) Membrane renewal after dibucaine deciliation of Tetrahymena: freeze-fracture technique, cilia, membrane structure. Exp Cell Res 97:83–91

Satir BH, Zhao H (1991) Parafusin and calcium-induced signal transduction in exocytosis. Protoplasma 206:228–233

Satir BH, Wyroba E, Liu L, Lethan M, Satir P, Christensen ST (2015) Evolutionary implications of localization of the signaling scaffold protein parafusin to both cilia and the nucleus. Cell Biol Int 39:136–145

Satir P, Barkalow K, Hamasaki T (1993) The control of ciliary beat frequency. Trends Cell Biol 3:409–412

Schneider L, Clement CA, Teilmann SC, Pazour GJ, Hoffmann EK, Satir P, Christensen ST (2005) PDGFRalphaalpha signaling is regulated through the primary cilium in fibroblasts. Curr Biol 15:1861–1866

Schultz JE, Klumpp S, Benz R, Schurhoff-Goeters WJ, Schmid A (1992) Regulation of adenylyl cyclase from Paramecium by an intrinsic potassium conductance. Science 255:600–603

Shi L (2013). Dual Role of Ift57 in Cilia and Nucleus in Paramecium. Ph.D. thesis, Universite Paris-Sud, Paris

Smith JJ, Cole ES, Romero DP (2004) Transcriptional control of RAD51 expression in the ciliate Tetrahymena thermophila. Nucleic Acids Res 32:4313–4321

Subramanian SV, Satir BH (1992) Carbohydrate cycling in signal transduction: parafusin, a phosphoglycoprotein and possible Ca(2+)-dependent transducer molecule in exocytosis in Paramecium. Proc Natl Acad Sci USA 89:11297–11301

Subramanian SV, Wyroba E, Andersen AP, Satir BH (1994) Cloning and sequencing of parafusin, a calcium-dependent exocytosis-related phosphoglycoprotein. Proc Natl Acad Sci USA 91:9832–9836

Trojanek J, Ho T, Del Valle L, Nowicki M, Wang JY, Lassak A, Peruzzi F, Khalili K, Skorski T, Reiss K (2003) Role of the insulin-like growth factor I/insulin receptor substrate 1 axis in Rad51 trafficking and DNA repair by homologous recombination. Mol Cell Biol 23:7510–7524

Tsao CC and Gorovsky MA (2008a) Different effects of Tetrahymena IFT172 domains on anterograde and retrograde intraflagellar transport. Mol Biol Cell 19:1450–1461

Tsao CC and Gorovsky MA (2008b) Tetrahymena IFT122A is not essential for cilia assembly but plays a role in returning IFT proteins from the ciliary tip to the cell body. J Cell Sci 121:428–436

Valentine MS, Rajendran A, Yano J, Weeraratne SD, Beisson J, Cohen J, Koll F, Van Houten J (2012) Paramecium BBS genes are key to presence of channels in Cilia. Cilia 1:16

Vallesi A, Ballarini P, Di Pretoro B, Alimenti C, Miceli C, Luporini P (2005) Autocrine, mitogenic pheromone receptor loop of the ciliate Euplotes raikovi: pheromone-induced receptor internalization. Eukaryot Cell 4:1221–1227

Van Houten JL (1998) Chemosensory transduction in Paramecium. Eur J Protistol 34:301–307

Van Houten JL (2000) Chemoreception in microorganisms. In: Finger TE, Silver WL, Restrepo D (eds) Neurobiology of taste and smell. Wiley, New York, pp 11–40

Watanabe T (1990) The role of ciliary surfaces in mating in Paramecium. In: Bloodgood RA (ed) Ciliary and flagellar membranes. Plenum Press, New York, pp 149–171

Wiejak J, Surmacz L, Wyroba E (2004) Dynamin-association with agonist-mediated sequestration of beta-adrenergic receptor in single-cell eukaryote Paramecium. J Exp Biol 207:1625–1632

Wloga D, Camba A, Rogowski K, Manning G, Jerka-Dziadosz M, Gaertig J (2006) Members of the NIMA-related kinase family promote disassembly of cilia by multiple mechanisms. Mol Biol Cell 17:2799–2810

Wolfe J, Mpoke S, Tirone SF (1993) Cilia, ciliary concanavalin A-binding proteins, and mating recognition in Tetrahymena thermophila. Exp Cell Res 209:342–349

Wyroba E, Widding Hoyer A, Storgaard P, Satir BH (1995) Mammalian homologue of the calcium-sensitive phosphoglycoprotein, parafusin. Eur J Cell Biol 68:419–426

Yano J, Rajendran A, Valentine MS, Saha M, Ballif BA, Van Houten JL (2013) Proteomic analysis of the cilia membrane of Paramecium tetraurelia. J Proteomics 78:113–122

Zhang Q, Seo S, Bugge K, Stone EM, Sheffield VC (2012) BBS proteins interact genetically with the IFT pathway to influence SHH-related phenotypes. Hum Mol Genet 21:1945–1953

Zhao H, Satir BH (1998) Parafusin is a membrane and vesicle associated protein that cycles at exocytosis. Eur J Cell Biol 75:46–53

Chapter 5
Chemotaxis as an Expression of Communication of *Tetrahymena*

Laszlo Köhidai

Abstract Migratory responses of cells belong to the most basic cell physiological activities of evolution of intercellular communication. In this process we distinguish two main phases: (i) in the first period a large set of "pure" physical effects were only present therefore several very essential locomotor responses developed to these physical factors; (ii) much later the complexity of chemical and biological signals appeared and these responses (e.g. chemotaxis) were/are far enough to perform the project. In the present chapter our objective was to discuss the following problems with a special respect to the chemotactic activity of the eukaryotic ciliate *Tetrahymena* pyriformis, a well known model-cell of chemotaxis: (i) characterization of chemotactic ligands (inorganic, amino acids, oligo- and polypeptides, lipids and carbohydrates) of *Tetrahymena*; (ii) signalling mechanisms of *Tetrahymena* used in chemotaxis (ligand-'chemotaxis receptor' relations, second messengers, kinase-cascades); (iii) significance of paracrine and autocrine activity in chemotaxis; (iv) special phenomena based on chemotaxis (chemotactic range fitting; chemotactic selection; chemotactic drug targeting). Practical approaches (test systems, bioreactors) based on protozoan motility (e.g. bioindicator of the freshwater quality and production organic compounds on industrial level) are also discussed.

5.1 Phylogeny of Motility

Migratory responses of cells are belonging to the most essential cell physiological activities. The ability to detect and respond to the continuously changing characteristics of the environment embodies one of the most important capability of living organisms. This way of communication has a significant role to determine the biological entity of living organisms on unicellular and multicellular levels as well.

L. Köhidai (✉)
Department of Genetics, Cell- and Immunobiology, Semmelweis University,
Nagyvárad tér 4, Budapest 1089, Hungary
e-mail: kohlasz2@gmail.com

G. Witzany and M. Nowacki (eds.), *Biocommunication of Ciliates*,
DOI 10.1007/978-3-319-32211-7_5

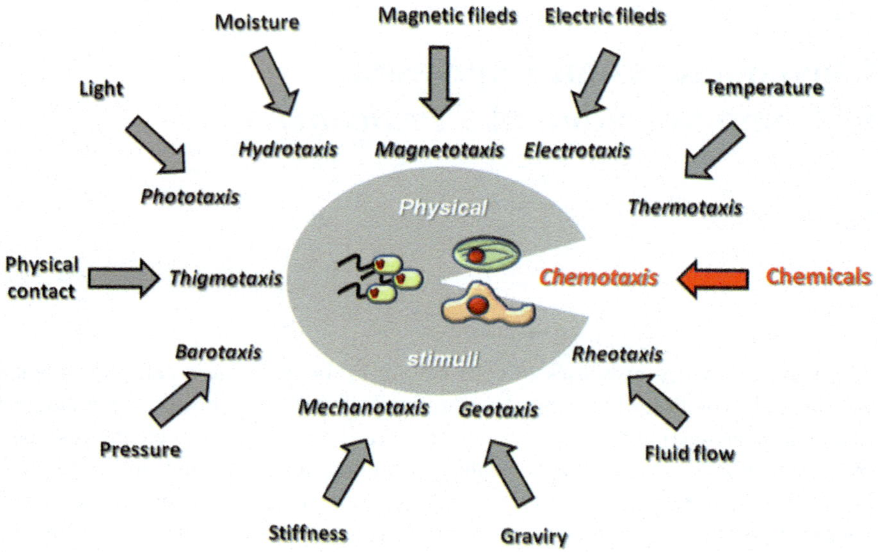

Fig. 5.1 Environmental effects with migratory response inducer potential—Decisive role of physical stimuli in the early levels of evolution with overlaying significance of chemical and biological signals

The migratory behaviour of cells is influenced by several exogenous and endogenous factors, among others physical environmental effects like pressure, light, temperature, electric field etc. (Charras and Sahai 2014) and the diverse set of bioactive regulatory molecules like hormones or enzymes released by the cells or itself the phase of the cell cycle. In respect of diversity of the long list of different environmental factors described as responsible for induction a type of migration we can see that majority of these effects belongs to the physical ones, while chemically or biologically induced responses are much rarer (Fig. 5.1). This fact points to that the communication between the cell and its environment is a good mirror of the long process of evolution. According to our understanding the evolution of communication resulting motility had two significantly diverse phases: (i) in the first period a large set of "pure" physical effects were only present therefore several very essential migratory responses developed to these physical factors (Charras and Sahai 2014); (ii) much later the complex chemical and biological signals appeared—these significantly less but more complex responses (e.g. chemotaxis or quorum sensing systems) were/are far enough to perform the project. For sure in the second phase more or less physical effect guided mechanisms are also present, these background activities are increasing the complexity of the communication. Regularly a network of influences is present with a combination of more or less dominant factors—they give the characteristic feature of the motile behaviour of the cell. A tight connection of prokaryotic-eukaryotic levels is presented in predator or chemotaxis inducer systems where diversities in the bacterial quorum sensing results well tuned cellular responses e.g. Pseudomonas—*Tetrahymena* (Friman et al. 2013) or Pseudomonas—neutrophil granulocyte (Kahle et al. 2013) relations. As comprehensive studies documented (Kim et al. 2009, 2010) it means that complexity of cell motility is based on more than one signal-response relation; however, it is regularly hard to detect and evaluate the real significance of the non-dominant, accessory—regularly physical—factors.

The relationship between the motile cell and its environment is rather complex, in some cases very simple biological characteristics (e.g. size) of the cell can significantly influence all the three basic ways (intra-; inter- and trans-organismic) of communication. Differences of migratory behaviour are well expressed in different levels of phylogeny when two essentially diverse migratory mechanism developed in free swimming prokaryotes (e.g. bacteria) and in eukaryotic cells ($\sim$10–100 μm) using surface bound (e.g. Dictyostelium) or ciliary movement (e.g. *Tetrahymena*). While the relatively big eukaryotic cells—due to their size—are able to detect axial concentration gradients between the two poles of the cell, the significantly smaller ($\sim$1–5 μm) prokaryotes are too small to detect such gradients. This relatively simple diversity in communication between the cell and its environment was enough to result completely different migratory strategies: while eukaryotes are waiting in standing pose to detect the triggering effect, prokaryotes have to scan the given space by relatively random runs until they get positive or negative signal which will guide them in the next period. Nevertheless, in the backgrounds of the two mechanisms described above more complex, even mathematically described processes were found which are even formulated as (1) temporal (e.g. bacteria) and (2) spatial (e.g. yeast) forms of sensing (DeLisi et al. 1982).

$$\frac{C(t2) - C(t1)}{t2 - t1} = \frac{dC}{dt} \tag{5.1}$$

if $\frac{\mathrm{dC}}{\mathrm{dt}} > 0$, straith run, else tumble

$$\frac{C(x2) - C(x1)}{x2 - x1} = \frac{dC}{dx} \tag{5.2}$$

if $\frac{\mathrm{dC(x2)}}{\mathrm{dx}} > 0$, project in direction.

Finally, we have to mention that in some relatively big prokaryotes (e.g. in bipolar flagellated vibrioid bacteria (size: 2 × 6 μm) mixed forms of the temporal and spatial sensing were described, which data support further that the cell size has a critical significance in respect of chemotactic sensing in unicellular level (Thar and Kuh 2003).

In respect to the directness of the motile cells we distinguish two main types of responses: (i) kinesis—with rather random and (ii) taxis—with vectorial migratory responses. (The third main type—tropism—is not discussed here as it results characteristically changes in position or shape but there is no locomotor activity.) Among others the most frequently observed forms of kinesis are the otrokinesis and klin-iokinesis when the speed or the frequency of movement is changing in a good relation to the inducer stimulus. In taxis the concentration dependent chemotaxis and haptotaxis are described as most important inducers of the movement, the chief difference of the two forms is that while chemotaxis is induced by a gradient developed in fluid phase, the haptotaxis requires more fixed, surface associated gradients. In the last two decades it was proved that several migrations described originally as "chemotaxis" are in fact haptotactic events (e.g. vascular transmigration of cells) and even some forms of swimming behaviour of protozoa (e.g. creeping) was documented as a surface and concentration range influenced process. Significance of these processes in research is portrayed by the significantly increased number of publications dealing with chemotaxis/haptotaxis in total as well as in the last 10 years when the number of these publications was increased in 40 % (National Center for Biotechnology Information, PubMed.gov. http://www.ncbi.nlm.nih.gov/pubmed/?term=chemotaxis). Nevertheless, not only chemotaxis and haptotaxis are known as chemically or biologically induced forms of motility. Aerotaxis induced by oxygen or necrotaxis elicited by substances released from dead cells are still significant forms of migration in unicellular or higher ranked levels of phylogeny.

5.2 *Tetrahymena* as a Proper Model-Cell of Chemotaxis Studies

Chemotaxis research has several model-cells like (*Escherichia coli, Bacillus substilis, Saccharomices cerevisiae, Dictyostelium discoideum*, polymorphonuclear cells or sperms). Each cell type offers special advantages compared to the others (sensibility, type of migration, significance in physiological/pathological processes, technical approaches), however, *Tetrahymena pyriformis* represents a very special model with preferred moieties (Kőhidai 1999). In theoretical aspects *Tetrahymena* embodies a favourable eukaryotic chemotaxis-model as its motility and chemotaxis are cell physiologically essential, basic properties of this cell (Fig. 5.2).

The wide range sensibility to different groups of ligands (biogenic amines, peptides, carbohydrates, lipid derivatives or synthetic substances—please see their detailed discussion below) is well documented as well as their receptors and signalling mechanisms offer a wide range basis of comparative studies. *Tetrahymena* served as a model for description and characterisation some chemotaxis related mechanism and techniques like chemotactic range-fitting (Köhidai et al. 2003d) or chemotactic selection (Köhidai and Csaba 1998). In practical approaches we have to mention the short cell cycle time (~150 min) which makes possible to follow even hundreds of offspring generations; the easy handling and the relatively cost-saving way of culturing these cells. Development of reliable qualitative and quantitative chemotaxis assays (e.g. multi-channel assays, impedimetry based counting) of *Tetrahymena* chemotaxis research also support to use *Tetrahymena* chemotaxis as one of the most objective assays of cell physiology (Kőhidai et al. 1995b).

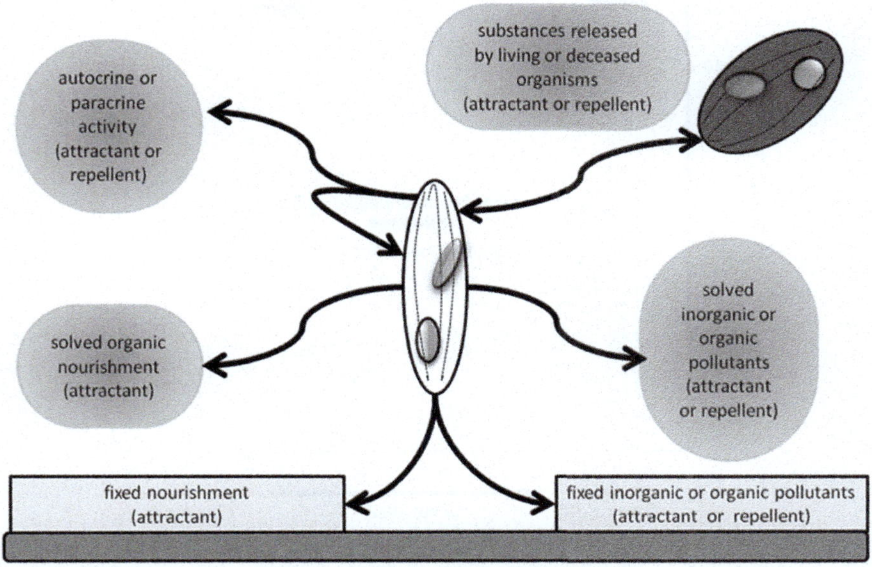

Fig. 5.2 Main components of chemotactic communication between *Tetrahymena* and its environment. The characteristic type of chemotactic responses are shown in parentheses

In the present chapter our objective was to discuss the following problems with a special focus on *Tetrahymena* chemotaxis:

- Characterization of chemotactic ligands of *Tetrahymena*
- Signalling mechanisms of *Tetrahymena* used in chemotaxis
- Significance of paracrine and autocrine activity in chemotaxis
- Special phenomena based on chemotaxis
- Practical approaches (test system, bioreactors) based on protozoan motility.

5.3 Chemotactic Ligands Acting in *Tetrahymena*

Chemotactic responsiveness belongs to one of the most basic cell physiological activities. The specific moiety of this cellular response compared to the other activities like proliferation, metabolic processes or phagocytosis is embodied in the locomotion of the cell which provides an active role to select the optimal environment. By recognition of chemoattractant or chemorepellent substances motile cells obtained the special ability to tune their position with high selectivity which was essential even at early phases of phylogeny. References proved that changes even in the ionic milieu (NaCl, KCl, $CaCl_2$) can change dramatically the swimming behaviour of these cells (Tanabe et al. 1980). A close correlation between molecular evolution and cellular responses is well represented by the relationship of

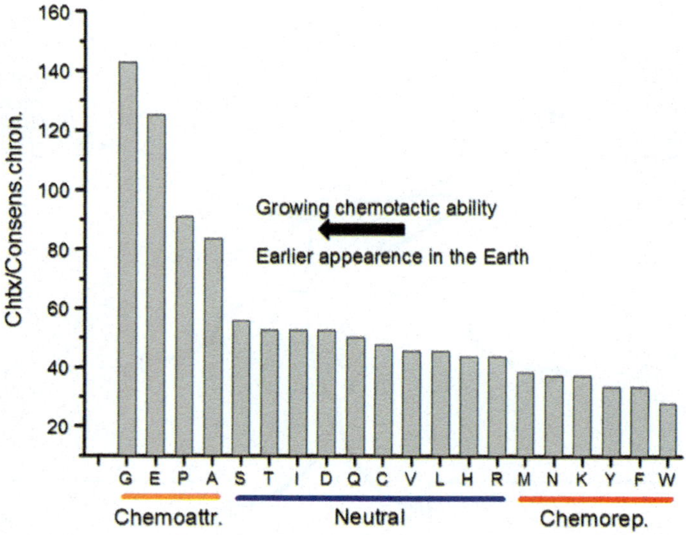

Fig. 5.3 Correlation of the chemotactic ability of amino acids in *Tetrahymena* and consensus sequences of the amino acids in the primordial soup

phylogeny of amino acids and their chemotactic behaviour. Consensus sequence of amino acids describes the sequence of appearance of these relatively simple organic substances in the primordial soup. According to comparative studies of this ranking of amino acids and their chemotactic ability expressed in the ciliate protozoan *Tetrahymena* pyriformis a strong chemoattractant property (Gly, Glu, Pro, Ala) proved to be advantaged—these molecules appeared at first in the primordial soup, while the chemorepellent characteristic is associated to the late onset members (Trp, Phe, Tyr, Lys, Asn, Met) of the genetically determined list of amino acids (Kőhidai et al. 2003d) (Fig. 5.3).

High level selectivity of chemotactic signalling in *Tetrahymena* was also proved by comparative works on L- and D-enantiomers of amino acids. Investigations on the concentration dependence of chemotactic effects elicited by 10 amino acids (Ala, Asp, Glu, Leu, Met, Phe, Pro, Ser, Thr and Trp) proved that 50 % of the tested substances resulted diverse (chemoattreactant vs. chemorepellent) responses, while Met and Ser (chemoattractant) and Leu, Phe, Thr (chemorepellent) worked still in partially identical (some other concentrations their effects were diverse) (authors not published data).

However, not only amino acids are effective ligands with chemotactic moiety in *Tetrahymena*. There is a wide range of ligands which molecules are thought to have biological or clinical significance only on higher levels of phylogeny, but they have characteristic chemotactic effects in *Tetrahymena*. A representative selection of these substances is presented in Tables 5.1 and 5.2.

Tables 5.1 and 5.2 show that ligands acting on chemotaxis of vertebrates (e.g. fMLF or IL-8) are chemotactic in *Tetrahymena* as well as chief regulators of

Table 5.1 Main groups of molecules possessing chemotactic ability in *Tetrahymena* pyriformis

Type of chemotactic ligand	Reference	Type of chemotactic ligand	Reference
Inorganic substances	Tanabe et al. (1980)	*Peptide carriers of targeted drug delivery*	
Amino acids and derivatives	Kőhidai et al. (2003d)	Polylysine based ligands	Szabó et al. (2003)
Peptide or protein type ligands		Oligotuftsin based ligands	Mező et al. (2006)
Di- and oligopeptides	Kőhidai (1999)	GnRH based ligands	Lajkó et al. (2013)
Formyl-peptides	Kőhidai et al. (1994, 2003a, b, c, d, e, f)	Cyclodextrin carriers	Kőhidai et al. (2005)
Peptides with 'EWS' motive	Illyés et al. (2002), Kőhidai et al. (2003b), Láng et al. (2012)	*Non peptidic ligands*	
ECM peptides	Kőhidai et al. (2004)	Steroids	Kőhidai et al. (2003c)
Peptide hormones	Kőhidai and Csaba (1996)	Volatile oils	Kőhidai et al. (1995b)
Vasoactive peptides	Kőhidai and Csaba (1995), Kőhidai et al. (1995a, 2002a, 2015)	Odorants and tastants	Láng et al. (2011)
Neuropeptides	O'Neill et al. (1988)	Melatonin	Kőhidai et al. (2002c)
Cytokines	Kőhidai and Csaba (1998)	Mono-, di- and oligosaccharides	Szemes et al. (2015)
Lectins	Kőhidai and Csaba (1996), Kőhidai et al. (2003a)	Pharmaceutical substances	Láng and Kőhidai (2012)
Pheromones	Kőhidai et al. (2007)		

Table 5.2 Representative list of chemotactic ligands in *Tetrahymena* pyriformis

Ligand	Chemotactic effect	Optimal concentration [M]	Reference
fMLF	Chemoattractant	10^{-9}–10^{-8}	Kőhidai et al. (2003e)
Insulin (porcine, amorphus)	Chemoattractant	10^{-12}–10^{-6}	Csaba et al. (1994)
Bradykinin (1–9)	Chemoattractant	10^{-11}	Kőhidai et al. (2002a)
	Chemorepellent	10^{-12}, 10^{-10}, 10^{-9}, 10^{-6}	
Oxytocin	Chemorepellent	10^{-11}–10^{-6}	Csaba et al. (2000)
Endothelin-1	Chemoattractant	10^{-14}–10^{-12}, 10^{-10}	Kőhidai and Csaba (1995), Kőhidai et al. (2001)
Atrial natruretic peptide	Chemoattractant	10^{-15}–10^{-11}, 10^{-8}–10^{-7}	Kőhidai et al. (1995a)
Adrenomedullin	Chemoattractant	10^{-8}	Kőhidai et al. (2015)
ACTH	Chemoattractant	10^{-9}, 10^{-6}	Kőhidai and Csaba (1996)
	Chemorepellent	10^{-12}, 10^{-8}	
Interleukin-8	Chemoattractant	1 ng/ml	Kőhidai and Csaba (1998)
	Chemorepellent	20; 50 ng/ml	

metabolic processes (e.g. insulin, ACTH) or ligands acting as controller and modulator elements of the vascular system (bradykinin, endothelin, atrial natriuretic peptide or adrenomedullin) on higher levels of phylogeny.

It is also to be mentioned that chemotactic communication between *Tetrahymena* and its environment embodies relatively distant relations but not only in the ciliate—vertebrate axis. As it was mentioned above there is a well tuned chemotaxis induced by bacterial di-, tri- and tetrapeptides containing the formyl-residue (chemoattractants: fMV, fMP, fNLF, fMLF; chemorepellents: fMS, fMMM, fMLFF) (Kőhidai et al. 1994) which has overlapping effects to the PMN cell responses described by Schowell (Showell et al. 1976) Several plant origin lectins (Con-A, Lens, Glycine max, PHA etc.) were reported as wide concentration range (10^{-12}–10^{-6} M) effective chemoattractants in *Tetrahymena* (Kőhidai et al. 2003a) which data contribute to the wide substantiation of the theory that lectins are chief forerunner molecules of the immune system with accomplishing some functions of the antibody network (Quesenberry et al. 2003). Finally we have to mention pheromones where not only high level of selectiveness but clear interspecies activity was also detected when cell physiological effects (chemotaxis, swimming behaviour, proliferation, phagocytosis) of Er1 and Er2 pheromones of *Euplotes raikovi* were investigated (Kőhidai et al. 2007). However, pheromones are thought to have species specific activity which is supported by the diverse ways of signalling (Er1—autocrine, non-clustering receptor dimers; Er2—paracrine, clustering receptor dimers) the reported investigations proved that the two Euplotes origin pheromones have characteristic effects also in *Tetrahymena* (Er1—chemorepellent, Er2—chemoattractant) which observation raises the possibility that in the background of chemotactic signalling there is a—still not described—complex and rather dynamic network with molecular patterns.

The above described variety of ligands is further widened by the non-peptide type substances like odorants and tastants (e.g. isoamyl acetate, propyl isobutyrate, methyl butyrate and furfuryl derivatives) (Láng et al. 2011), simple sugars (e.g. glucose, mannose, galactose) (Szemes et al. 2015) and several lipids like steroids (e.g. testosterone, progesterone, hydrocortisone) (Kőhidai et al. 2003c) and volatile oils (e.g. linalool, citronellal, citronellol) (Kőhidai et al. 1995b). The colourful group and the ligand specific chemotactic activities of the above listed substances gives a good support that the above suggested complex network of chemicals acting on motility of the cells is a complete one, all the possible chemical forms of the ligands have a good chance to be represented with characteristic ability to modulate chemotactic behaviour of free swimming unicellular organisms.

In respect ligand-receptor interactions of chemotactic signalling we have to mention those molecular level, physico-chemical characteristics of the ligand that have been confirmed as determining or influencing factors of signalling. In these studies some model ligands and libraries of ligands were very suitable, like amino acids (Kőhidai et al. 2003d), mirror variants of Pro containing dipeptides (Kőhidai et al. 1997), formyl peptides (Kőhidai et al. 2003e) and the WSXWS and SXWS libraries (Illyés et al. 2002; Láng et al. 2012) which sequences are present as characteristic domains of cytokine I-type receptors but they are detectable in free forms as components of the soluble forms of the receptor. Comparative analysis of amino acids showed a decreased pK (–COOH), an increased pK (–NH2), and a decrease in solvent exposed areas (SEA) and hydropathy indexes in chemoattractant amino acids compared to chemorepellent ones. The accessibility of the receptor is also influenced by some physicochemical moieties of the ligand. Among others, size of the residual volume and lipofilicity of the N-terminal aminoacids in Pro-X dipeptides determines the chemotactic effect with an advantage of the low values. The (W)-SXWS peptide family provides perhaps actually one of the best model molecule where we can read a large set of physicochemical characteristics determining chemotactic moieties. These are (i) expression of amide or carboxyl group on the C-terminus; (ii) presence of aromatic W on the N-terminus; (iii) intramolecular distance between X_{4C}-S_{4C}; (iv) mass/SEA ratio of X_4 residue (Illyés et al. 2002; Kőhidai et al. 2003b; Láng et al. 2012). Nevertheless, up to now ratios considering solubility or mass values/solvent exposed area (SEA) of the ligand (e.g. amino acids, formyl peptides, SXWS peptides) proved to be the most widely applicable physicochemical characteristics, where high values predict chemorepellent while low values predict chemoattractant activity of the peptide type ligand. Finally, some recent papers referred that in some cases not only the receptor-ligand interaction is responsible for the chemotactic responses but the active penetration of the surface membrane. This process is mainly influenced by low values of the topological polar surface area (TPSA) and high values of the partition coefficient (XLogP) and determines internalization of ligands (e.g. amino acids, simple sugars) and in this way effects on intracellular targets like cytoskeleton or metabolic processes (Szemes et al. 2015; Diaz et al. 2015).

5.4 Signalling Mechanisms Used to Accomplish Chemotaxis in *Tetrahymena*

In *Tetrahymena*—as it was described above—majority of the chemotactic ligands accomplish signalling via G-protein coupled receptors (GPCR), nevertheless, data are also available about reciprocal activation of these receptors, too (Lampert et al. 2011). Some of these GPCR receptors are still characterized like insulin (Leick et al. 2001; Christopher and Sundermann 1995), bradykinin (Kőhidai et al. 2002a) and endothelin-1 (Kőhidai et al. 2001), however, in several others they are only supported by circumstantial evidences like effect of inhibitors. Studies on chemotaxis elicited by beta-cyclodextrin carrier bound steroids has proved that *Tetrahymena* membrane has some components ready to interact in a rapid signalling pathway with steroids which ligands original targets are in the cytoplasm and in the nucleus (Kőhidai et al. 2003c). In protozoa a large group of ligands (e.g. amino acids, oligopeptides) is thought to act via "chemotaxis receptors" of the surface membrane, where ligand dependent modulator activities will determine the "chemoattractant" or "chemorepellent" outcome of the induction (Kim et al. 1998).

The intracellular signalling pathways triggered by chemotactic ligands are rather complex even in *Tetrahymena.* One of the most essential mechanisms is the molecular cascade which links the afferent and efferent intracellular pathways between the receptor and the cilia. A long list of cytoplasmic transmitters are present in the cascade, the most significant members are: (i) cyclic nucleotides (cAMP, cGMP), (ii) protein kinases (PKA, PKG) and (iii) cytoskeleton with specific end-point activator enzymes. Over the above described "classical" way of signalling the ciliary membrane itself and the submembranous molecular complexes including basal bodies represent a motion specialized functional unit. The first signal transmitters of transmembrane receptors of the cilia are cAMP, cGMP and Ca^{2+}; however, receptors can activate tyrosine phosphorylation cascade in a direct way, too. The next level signalling targets are protein kinases, they have significant role also in determining the direction of motion (PKA—forward swimming, PKG backward swimming). The dominant members of the implementer proteins are calmodulin, centrin, TCBP 23/25, kinesins and dyneins. The described network of signalling molecules provides the possibility of an orchestrated interaction of cytoskeletal and signal transmitter proteins which serves as a basis of the well tuned chemotaxis (Awan et al. 2004; Guerra et al. 2003; Seetharam and Satir 2008; Nam et al. 2009).

Over the above mentioned molecular interactions inositol-1,4,5-trisphosphate 3-kinase (IP3K) was also reported as a chief regulator of intracellular chemotactic signalling in eukaryotic model cells including *Tetrahymena.* The central role of this enzyme is due to that it serves as intracellular linker between several receptors of chemotactic ligands (PDGF, SDF-1, eotaxin, MCP, IGF, EGF etc.) and the target enzyme cascades like Akt-MAPK, ERK3-MEF2, PKB, RAC1-PAK1 or PTK-Ras-Raf1-MEK1-ERK1-2. In *Tetrahymena* the two most frequently used specific inhibitors of PI3K are wortmannin and LY294002, they are used to test involvement of the enzyme cascades in chemotaxis. Investigations on a selected group of ligands (insulin, bradykinin, endothelin-1, atrial natriuretic peptide, concanavalin-A) eliciting chemoattractant responses on diverse pathways in *Tetrahymena* have proved that both wortmannin and LY294002 (50 and 100 μg/ml) could dramatically diminish chemotactic responses (Kőhidai et al. in preparation-a), while other types of ligands (odorants and tastants) the signalling remained intact which implies that some other pathways may be involved (Láng et al. 2011).

5.5 Paracrine and Autocrine Activity as a Regulatory Element of Chemotaxis

As it was already mentioned *Tetrahymena* has a rather active paracrine and autocrine activity which synthetic capabilities make these cell able to regulate their own biological processes as well as to set the chemical environment optimal and communicate with the close cell population(s) with most frequently mating or alarming signals.

Table 5.3 Synthesis of vertebrate hormone like endogenous substances and their chemotactic effects in *Tetrahymena* pyriformis

Endogenous substance	Chemotactic effect	Optimal concentration [M]	Reference
ACTH	Chemoattractant	10^{-9}, 10^{-6}	LeRoith et al. (1982)
	Chemorepellent	10^{-12}, 10^{-8}	
Endothelin-1	Chemoattractant	10^{-14}–10^{-12}, 10^{-10}	Kőhidai and Csaba (1995)
Histamine	Chemoattractant	10^{-10}, 10^{-8}	Lajkó et al. (2012)
Insulin (crystalline)	Chemoattractant	10^{-10}	Csaba et al. (1999)
	Chemorepellent	10^{-9}–10^{-6}	
IL-6	Chemorepellent	10^{-6}–10^{-1}	Kőhidai et al. (2000)
Melatonin	In light chemoattractant	10^{-12}	Kőhidai et al. (2002b, 2003f)
	In darkness chemorepellent	10^{-12}–10^{-7}	
NT-proANP	Chemoattractant	10^{-15}–10^{-11}, 10^{-8}, 10^{-7}	Kőhidai et al. in preparation-b
Serotonin	Chemorepellent	10^{-9}	Lajkó et al. (2011)
Di- and Triiodothyronine (T2 and T3)	Chemoattractant	10^{-8}	Schwabe et al. (1983)
	Chemorepellent	10^{-12}–10^{-10}	

The assortment of molecules released is rather wide actually it is hard to assess both quantity and quality of these substances. However, a set of vertebrate like molecules were already identified as endogenous and released products of *Tetrahymena* (Table 5.3).

As Table 5.2 represents the substances synthesized by *Tetrahymena* are not simply many but their cell physiological effects have good correlations with their higher level (vertebrate) activities—see well tuned effects of metabolism modulator ACTH and insulin, chemorepellence of IL-6 or the illumination dependent synthesis and action of melatonin. We have to consider that even the number of the identified substances is much more (e.g. beta-endorphin, relaxin) (LeRoith et al. 1982; Schwabe et al. 1983) with a special dominance of biogenic amines (e.g. histamine, serotonin, epinephrine, dopamine, melatonin) (Csaba 2015), at the same time we are indebted to the answer why are these molecules present in such a wide variety but relatively far from the well understood biological functions. Actually there are three theories with interpretations involving motility, too. (i) According to the first explanation molecular level phylogeny of all these hormones or other bioactive molecules has started much earlier as we had previously thought it—therefore all of these substances are present with full of their biological functions—including chemotaxis—, in *Tetrahymena*, too. (ii) The other interpretation is that the wide range of molecules represents only an attempt of the synthetic capacity on the eukaryotic unicellular level. This very rich set of substances has not obligatory biological functions in each member but provided a wide assortment to be used. (iii) Finally it is also conceivable that the enormous synthetic capacity of eukaryotic ciliates produces a wide range of organic molecules with substances in surplus. However, these junk products are synthesized but not used by the cell. According to the detailed investigations on the biological significance of the endogenous molecules of *Tetrahymena* theory (i) and (ii) are more supported than theory (iii)—almost all of the detected vertebrate type substances were proved as cell physiologically (chemotaxis, proliferation, phagocytosis etc.) active ligands.

5.6 Special Phenomena Based on Chemotaxis

In addition to the above-mentioned characteristics of *Tetrahymena* chemotaxis there are some phenomena where the way of chemotactic communication provides a unique choice to investigate specific signalling properties in respect of the chemotactic ligand and its receptor. In the present chapter three of these phenomena are reviewed in brief.

(i) Chemotactic range fitting (Kőhidai et al. 2003d)—Chemotactic responses elicited by the ligand-receptor interactions are distinguished generally upon the optimal effective concentration(s) of the ligand. Nevertheless, correlation of the amplitude elicited and ratio of the responder cells compared to the total number are also characteristic features of the chemotactic signalling. Investigations of ligand families (e.g. amino acids or oligopeptides) proved that there is a fitting of ranges (amplitudes; number of responder cells) and chemotactic activities: chemoattractant moiety is accompanied with wide, while chemorepellent character narrow ranges.

(ii) Chemotactic selection (Kőhidai and Csaba 1998)—Chemotaxis receptors are expressed in the surface membrane with diverse dynamics, some of them have long-term characteristics as they are determined genetically, others have short-term moiety as their assembly is induced ad hoc in the presence of the ligand. The diverse feature of the chemotaxis receptors and ligands provides the possibility to select chemotactic responder cells with a simple chemotaxis assay. By chemotactic selection we can determine whether a still not characterized molecule acts via the long- or the short-term receptor pathway. Recent results proved that chemokines (e.g. IL-8, RANTES) are working on long-term chemotaxis receptors, while vasoactive peptides (e.g. endothelin-1) act more on the short-term ones. Term chemotactic selection is also used to design a technique which separates eukaryotic or prokaryotic cells upon their chemotactic responsiveness to selector ligands.

(iii) Chemotactic drug targeting (CDT) (Mező et al. 2006; Láng et al. 2003)—This is a special, reversal form of drug delivery where modulation of active migratory behaviour of the target cells is used to achieve targeted effects. The model experiments of CDT were carried out on *Tetrahymena* cells.

The general components of the conjugates are designed as follows: (a) carrier —regularly possessing promoter effect also on internalization into the cell; (b) chemotactically active ligands (e.g. fMLF or SXWS derivatives) acting on the target cells; (c) drug (e.g. methotrexate) to be delivered in a selective way and (iv) a ligand specific spacer sequence which joins drug molecule to the carrier and due to its enzyme labile moiety makes possible the intracellular compartment specific release of the drug. Careful selection of chemotactic component of the ligand not only the chemoattractant character could be expressed, however, chemorepellent ligands are also valuable as they are useful to keep away cell populations degrading the conjugate containing the drug. The novel way of drug delivery is under intensive investigations, on the basis of model experiments its significance is potentially high in the treatment of some tumors.

5.7 *Tetrahymena* Chemotaxis as a Practical Tool

The chemotaxis based communication of *Tetrahymena* has not only theoretical values but it offers several applications where biological advantages of both *Tetrahymena* and chemotaxis are exploited. Very first of these application is itself the model value of *Tetrahymena* chemotaxis, which is frequently used to characterize chemotactic moieties of ligands. This practical application of chemotaxis requires reliable test systems, in the case of ciliates regularly capillary techniques (e.g. application of high precision multichannel micropipette) are advantageous as reversible systems fit very well to the free-swimming behaviour of *Tetrahymena* (Kőhidai et al. 1995b; Leick and Helle 1983). Nevertheless, application of computer based tracking analysis provides significantly more information about characteristics of the motion as it helps to distinguish swimming patterns (linear swimming, turns, creeping), not mentioning about the very recent impedimetry based analysis which measures creeping frequency and detects even changes in position of cells on the surface of the electrode (Kőhidai 2012).

Reliable techniques to measure chemotaxis made possible to use *Tetrahymena* chemotactic responsiveness as an environmental sensor. As this ciliate is a bioindicator of the freshwater quality, its cell physiological responses—most frequently chemotaxis and proliferation—are used test systems. The rapid reactions as well as the cost-effectiveness made *Tetrahymena*-chemotaxis a reference tool of whole cell freshwater pollution screening. Due to the high sensitivity of *Tetrahymena*—it detects pollutants even in 10^{-15}–10^{-12} M range—, the most frequently detected substances are aquatic contaminant human pharmaceuticals (NSAIDs, antibiotics, beta-blockers and a frequently used X-ray contrast media) (Láng and Kőhidai 2012). To evaluate levels of decontamination in water by different technologies (e.g. vacuum ultraviolet photolysis) *Tetrahymena* proved also a good tool, therefore it is used to evaluate also industrial waste waters (Arany et al. 2014).

Chemotactic sensibility to native and synthetic compounds make *Tetrahymena* chemotaxis available as a first-line screening cell of drugs. Evaluation of drug-peptide conjugates (e.g. oligotuftsin-Mtx, GnRH-Dox) in CDT studies proved this model as a tool to analyse complex responses including chemotaxis, phagocytosis, internalization and proliferation (Mező et al. 2006; Lajkó et al. 2013).

Ability of *Tetrahymena* to synthesize a wide variety of vertebrate/human type substances (hormones or enzymes) raised the possibility that these cell cultures offer an inexpensive and well regulated way of production organic compounds on industrial level. *Tetrahymena* cultures were proved to be proper systems to produce several enzymes (e.g. phospholipase A1 precursor and mature human DNaseI) (Weide et al. 2006) and were successfully applied to reduce cholesterol level of milk in bioreactors (Noseda et al. 2007). The migratory ability of *Tetrahymena* cultured in bioreactors or in water cleaning systems represents an additional value as application of proper chemical of physical signals makes considerably easier separation of work phases or recovery of working mass of cell cultures.

5.8 Concluding Remarks

The chapter above was dedicated to give a succinct coverage of a special way of communication—chemotactic responsiveness—in biology and to understand phylogenetical approaches on molecular and cellular levels in the case of *Tetrahymena* pyriformis as a target and responder in one. Our observations show that chemotaxis may be considered as a biological consequence of several physically determined forms of motility. The high level diversity of ligands as well as intracellular signalling pathways show the plastic and dynamic feature of chemotactic communication which is well tuned by the physicochemical characteristics of the ligands and further specialized by the active auto- and paracrine activity of *Tetrahymena* cells. This complex network allows us to use *Tetrahymena* chemotaxis as a well defined model to investigate novel and old problems of the molecular level expressed cell biology.

Acknowledgments Author would like to express his deep gratitude to Professor György Csaba, Professor Péter Kovács, Dr. Orsolya Láng and Dr. Eszter Lajkó for their valuable help, useful critiques and contribution to his research work in the last decade(s).

Author would also like to extend his grateful thanks to all members of Chemotaxis Research Group, Department of Genetics, Cell- & Immunobiology, Semmelweis University for their tireless work focused on *Tetrahymena* chemotaxis.

Section 5.6—Special phenomena based on chemotaxis—, is a modified and re-edited material of the corresponding, free Wikipedia pages ('Chemotactic range-fitting', 'Chemotactic selection', 'Chemotactic drug-targeting'), generated and edited by the Author.

References

Arany E, Láng J, Somogyvári D, Láng O, Alapi T, Ilisz I, Gajda-Schrantz K, Dombi A, Kőhidai L, Hernádi K (2014) Vacuum ultraviolet photolysis of diclofenac and the effects of its treated aqueous solutions on the proliferation and migratory responses of Tetrahymena pyriformis. Sci Total Environ 468–469:996–1006

Awan A, Bernstein M, Hamasaki T, Satir P (2004) Cloning and characterization of Kin5, a novel Tetrahymena ciliary kinesin II. Cell Motil Cytoskeleton 58:1–9

Charras G, Sahai E (2014) Physical influences of the extracellular environment on cell migration. Nat Rev Mol Cell Biol 15:813–824

Christopher GK, Sundermann CA (1995) Isolation and partial characterization of the insulin binding sites of Tetrahymena pyriformis. Biochem Biophys Res Commun 212:515–523

Csaba G (2015) Biogenic amines at a low level of evolution: production, functions and regulation in the unicellular Tetrahymena. Acta Microbiol Immunol Hung 62:93–108

Csaba G, Gaál A, Kovács P, Simon G, Kőhidai L (1999) Prolonged elevation of insulin content in the unicellular Tetrahymena after insulin treatment: induction of insulin production or storage? Cell Biochem Funct 17:165–173

Csaba G, Kovács P, Kőhidai L (1994) Tetrahymena cells distinguish insulins according to their amorphous and crystalline condition or their bovine and porcine origin. Study of imprinting in aspects of hormone binding and chemotaxis. Microbios 80:215–221

Csaba G, Kovács K, Kőhidai L (2000) Effect of oxytocin and its analogues on the chemotaxis of Tetrahymena: evolutionary conclusions. Acta Protozool 39:345–347

DeLisi C, Marchetti F, Del Grosso G (1982) A theory of measurement error and its implications for spatial and temporal gradient sensing during chemotaxis. Cell Biophys 4:211–229

Diaz E, Zacarias AK, Pérez S, Vanegas O, Kőhidai L, Padrón-Nieves M, Ponte-Sucre A (2015) Effect of aliphatic, monocarboxylic, dicarboxylic, heterocyclic and sulphur-containing amino acids on Leishmania spp. chemotaxis. Parasitology 23:1–10

Friman VP, Diggle SP, Buckling A (2013) Protist predation can favour cooperation within bacterial species. Biol Lett 9:20130548

Guerra C, Wada Y, Leick V, Bell A, Satir P (2003) Cloning, localization, and axonemal function of Tetrahymena centrin. Mol Biol Cell 14:251–261

Illyés E, Sz Bősze, Láng O, Kőhidai L, Sebestyén F, Hudecz F (2002) A new class of chemotactic peptides containing EWS motif: a minireview. Chim Oggi 20:55–61

Kahle NA, Brenner-Weiss G, Overhage J, Obst U, Hänsch GM (2013) Bacterial quorum sensing molecule induces chemotaxis of human neutrophils via induction of p38 and leukocyte specific protein 1 (LSP1). Immunobiology 218:145–151

Kim DH, Casale D, Kőhidai L, Kim MJ (2009) Galvanotactic and phototactic control of Tetrahymena pyriformis as a microfluidic workhorse. Appl Phys Lett 94:163901

Kim DH, Cheang UK, Kőhidai L, Byun D, Kim MJ (2010) Artificial magnetotactic motion control of Tetrahymena pyriformis using ferromagnetic nanoparticles: a tool for fabrication of microbiorobots. Appl Physics Lett 97:173702

Kim JY, Borleis JA, Devreotes PN (1998) Switching of chemoattractant receptors programs development and morphogenesis in Dictyostelium: receptor subtypes activate common responses at different agonist concentrations. Dev Bio. 197:117–128

Kőhidai L (1999) Chemotaxis: the proper physiological response to evaluate phylogeny of signal molecules. Acta Biol Hung 50:375–394

Kőhidai L (2012) Tetrahymena pyriformis in motion. In: Kim MJ, Steager E, Agung AJ (eds) Microbiorobotics—biologically inspired microscale robotic systems. Elsevier, Oxford, pp 55–84

Kőhidai L, Csaba G (1995) Effects of the mammalian vasoconstrictor peptide, endothelin-1, on Tetrahymena pyriformis GL, and the immunocytological detection of endogenous endothelin-like activity. Comp Biochem Physiol 111C:311–316

Kőhidai L, Csaba G (1996) Different and selective chemotactic responses of Tetrahymena pyriformis to two families of signal molecules: lectins and peptide hormones. Acta Microbiol Immunol Hung 43:83–91

Kőhidai L, Csaba G (1998) Chemotaxis and chemotactic selection induced with cytokines (IL-8, RANTES and TNF-alpha) in the unicellular Tetrahymena pyriformis. Cytokine 10:481–486

Kőhidai L, Kovács P, Csaba G (1994) Chemotactic response of unicellular Tetrahymena to leukocyte attractant peptide and its repellent derivate: evolutionary conclusions. Cell Biol Int 18:119–121

Kőhidai L, Csaba G, Karsa J (1995a) Effect of atrial natriuretic peptide (ANP) on the unicellular model Tetrahymena pyriformis. Microbios 82:27–40

Kőhidai L, Lemberkovits É, Csaba G (1995b) Molecule dependent chemotactic responses of Tetrahymena pyriformis elicited by volatile oils. Acta Protozool 34:181–185

Kőhidai L, Soós P, Csaba G (1997) Effects of dipeptides containing the amino acid, proline on the chemotaxis of Tetrahymena pyriformis. Evolutionary conclusions on the formation of hormone receptors and hormones. Cell-Biol Internat Rep 21:341–345

Kőhidai L, Kovács P, Lázár-Molnár E, Csaba G (2000) Presence, uptake and localization of the cytokine, interleukine 6/IL-6/ in the unicellular Tetrahymena pyriformis. Cell Biol Int 24:749–755

Kőhidai L, Tóth K, Ruskoaho H, Csaba G (2001) Effect of vasoactive peptides on Tetrahymena. chemotactic properties of endothelins (ET-1, ET-2, ET-3, fragment 11-21 of ET-1 and big endothelin-1): a short-term inducible signalling mechanism of chemotaxis. Cell Biol Int 25:1173–1177

Kőhidai L, Kovács K, Csaba G (2002a) Direct chemotactic effect of bradykinin and related peptides—significance of amino- and carboxyterminal character of oligopeptides in chemotaxis. Cell Biol Int 26:55–62

Kőhidai L, Vakkuri O, Keresztesi M, Pállinger É, Leppaluoto J, Csaba G (2002b) Melatonin in the unicellular Tetrahymena: effects of different lighting conditions regarding possible functions of melatonin at this low level of evolution. Cell Biochem Funct 20:269–272

Kőhidai L, Vakkuri O, Keresztesi M, Pállinger É, Leppäluoto J, Csaba G (2002c) Impact of melatonin on the cell division, phagocytosis and chemotaxis of Tetrahymena pyriformis. Acta Protozool 41:85–89

Kőhidai L, Bánky Cs, Csaba G (2003a) Comparison of lectin induced chemotactic selection and chemical imprinting in Tetrahymena pyriformis. Acta Protozool 42:91–97

Kőhidai L, Bősze Sz, Soós P, Illyés E, Láng O, Mák M, Sebestyén F, Hudecz F (2003b) Chemotactic activity of oligopeptides containing EWS motif on Tetrahymena pyriformis. The effect of amidation of the C-terminal residue. Cell Biochem Funct 21:113–120

Kőhidai L, Katona J, Csaba G (2003c) Effects of steroid hormones on five functional parameters of Tetrahymena: evolutionary conclusions. Cell Biochem Funct 21:19–26

Kőhidai L, Láng O, Csaba G (2003d) Chemotactic-range-fitting of amino acids and its correlations to physicochemical parameters in Tetrahymena pyriformis–evolutionary consequences. Cell Mol Biol (Noisy-le-grand) 49(Online Pub):OL487–495

Kőhidai L, Török K, Illyés E, Tamási J, Sebestyén F, Láng O, Csaba Gy, Hudecz F (2003e) Characterization of chemotactic ability to peptides containing N-formyl-methionyl residue in Tetrahymena model—for-Met-Leu-Phe (f-MLP) as a targeting ligand. Cell Biol Int 27:695–700

Kőhidai L, Vakkuri O, Keresztesi M, Leppäluoto J, Csaba G (2003f) Induction of melatonin synthesis in Tetrahymena pyriformis by hormonal imprinting—a unicellular "factory" of the indoleamine. Cell Mol Biol 49:521–524

Kőhidai L, Kun L, Pállinger É, Csaba G, Mihala N, Majer Z, Süli-Vargha H (2004) Cell-physiological effects of elastin derived (VGVAPG)n oligomers in a unicellular model system. J Pept Sci 10:427–438

Kőhidai L, Szabó L, Láng O, Katona J, Csaba G (2005) Chemotactic drug targeting—significance of cyclodextrin carriers for targeting SXWS peptides and steroids. Glycoproteomics: Protein modificationsfor versatile functions. Dubrovnik, Croatia P32

Kőhidai L, Gál G, Banchetti R (2007) Interspecific effect of Er-1 and Er-2 Euplotes pheromones in Tetrahymena. Acta Zool Sin 52:1125–1132

Kőhidai L, Tóth K, Samotik P, Ranganathan K, Láng O, Tóth M, Ruskoaho H (2015) Effect of vasoactive peptides in Tetrahymena: chemotactic activities of adrenomedullin, proadrenomedullin N-terminal 20 peptide (PAMP) and calcitonin gene-related peptide (CGRP). Mol Cell Biochem Epub ahead of print. doi:10.1007/s11010-015-2589-6

Kőhidai L, Kun L, Amouzegar AS, Ranganathan K, Samotik P, Tóth M, Csaba G Phosphatidyl-inositol-3-kinase dependence of chemotactic responsiveness in the unicellular Tetrahymena. Effect of vasoactive hormones and growth factor type ligands (in preparation-a)

Kőhidai L, Vakkuri O, Leppäluoto J, Tóth M, Ruskoaho H Endogenous NT-proANP content of Tetrahymena pyriformis—time and environment dependent synthesis and release (in preparation-b)

Lajkó E, Pállinger É, Csaba G (2011) Investigations on the triiodothyronine (T3)-specificity of thyrotropic (TSH) and gonadotropic (HCG) hormone in the unicellular Tetrahymena. Acta Microbiol Immunol Hung 58:85–91

Lajkó E, Pállinger E, Csaba G (2012) Durable effect of heat-stress on the hormone production of Tetrahymena. Effect of insulin on the consequences of stress. Acta Microbiol Immunol Hung 59:249–256

Lajkó E, Szabó I, Andódy K, Pungor A, Mező G, Kőhidai L (2013) Investigation on chemotactic drug targeting (chemotaxis and adhesion) inducer effect of GnRH-III derivatives in Tetrahymena and human leukemia cell line. J Pept Sci 19:46–58

Lampert TJ, Coleman KD, Hennessey TM (2011) A knockout mutation of a constitutive GPCR in Tetrahymena decreases both G-protein activity and chemoattraction. PLoS ONE 6:e28022

Láng J, Kőhidai L (2012) Effects of the aquatic contaminant human pharmaceuticals and their mixtures on the proliferation and migratory responses of the bioindicator freshwater ciliate Tetrahymena. Chemosphere 89:592–601

Láng J, Rákász V, Magyar A, Pállinger É, Kőhidai L (2011) Chemotactic effect of odorants and tastants on the ciliate Tetrahymena pyriformis. J Rec Sign Transduct Res 31:423–433

Láng O, Illyés E, Menyhárd DK, Láng J, Sebestyén F, Hudecz F, Kőhidai L (2012) Chemotaxis induced by SXWS tetrapeptides in Tetrahymena—Overlapping chemotactic effects of SXWS sequences and their identical amino acids. J Mol Recogn 25:24–31

Láng O, Török K, Mező G, Hudecz F, Kőhidai L (2003) Chemotactic conjugates New Aspects in Drug-Targeting. Cellular transport Strategies for Targeting of Epitopes, drugs and Reporter Molecules. Budapest, Hungary, p. 118

Leick V, Bøg-Hansen TC, Juhl HA (2001) Insulin/FGF-binding ciliary membrane glycoprotein from Tetrahymena. J Membr Biol 181:47–53

Leick V, Helle J (1983) A quantitative assay for ciliate chemotaxis. Anal Biochem 135:466–469

LeRoith D, Liotta AS, Roth J, Shiloach J, Lewis ME, Pert CB, Krieger DT (1982) Corticotropin and beta-endorphin-like materials are native to unicellular organisms. Proc Natl Acad Sci USA 79:2086–2090

Mező G, Láng O, Jakab A, Bai KB, Szabó I, Schlosser G, Láng J, Kőhidai L, Hudecz F (2006) Synthesis of oligotuftsin-based branched oligopeptide conjugates for chemotactic drug targeting. J Pept Sci 12:328–336

Nam SW, Kim ST, Lee KM, Kim SH, Kou S, Lim J, Hwang H, Joo MK, Jeong B, Yoo SH, Park S (2009) N-Methyl-D-aspartate receptor-mediated chemotaxis and Ca(2+)signaling in Tetrahymena pyriformis. Protist 160:331–342

Noseda DG, Gentili HG, Nani ML, Nusblat A, Tiedtke A, Florin-Christensen J, Nudel CB (2007) A bioreactor model system specifically designed for Tetrahymena growth and cholesterol removal from milk. Appl Microbiol Biotechnol 75:515–520

O'Neill JB, Pert CB, Ruff MR, Smith CC, Higgins WJ, Zipser B (1988) Identification and characterization of the opiate receptor in the ciliated protozoan Tetrahymena. Brain Res 450:303–315

Quesenberry MS, Ahmed H, Elola MT, O'Leary N, Vasta GR (2003) Diverse lectin repertoires in tunicates mediate broad recognition and effector innate immune responses. Integr Comp Biol 43:323–330

Schwabe C, LeRoith D, Thompson RP, Shiloach J, Roth J (1983) Relaxin extracted from protozoa (Tetrahymena pyriformis). Molecular and immunologic properties. J Biol Chem 258:2778–2781

Seetharam RN, Satir P (2008) Coordination of outer arm dynein activity along axonemal doublet microtubules. Cell Motil Cytoskeleton 65:572–580

Showell HJ, Freer RJ, Zigmond SH, Schiffmann E, Aswanikumar S, Corcoran B, Becker EL (1976) The structure-activity relations of synthetic peptides as chemotactic factors and inducers of lysosomal secretion for neutrophils. J Exp Med 143:1154–1169

Szabó R, Mező G, Hudecz F, Kőhidai L (2003) Effect of the polylysine based polymeric on the growth and chemotaxis of Tetrahymena pyriformis. J Bioact Compat Polymers 17:399–415

Szemes Á, Lajkó E, Láng O, Kőhidai L (2015) Chemotactic effect of mono- and disaccharides on the unicellular Tetrahymena pyriformis. Carbohydr Res 407:158–165

Tanabe H, Kurihara K, Kobatake Y (1980) Changes in membrane potential and membrane fluidity in Tetrahymena pyriformis in association with chemoreception of hydrophobic stimuli: fluorescence studies. Biochemistry 19:5339–5344

Thar R, Kuhl M (2003) Bacteria are not too small for spatial sensing of chemical gradients: an experimental evidence. Proc Natl Acad Sci USA 100:5748–5753

Weide T, Herrmann L, Bockau U, Niebur N, Aldag I, Laroy W, Contreras R, Tiedtke A, Hartmann MW (2006) Secretion of functional human enzymes by Tetrahymena thermophila. BMC Biotechnol 6:19.

Chapter 6
Signals Regulating Vesicle Trafficking in *Paramecium* Cells

Helmut Plattner

Abstract Most data available from *Paramecium*, fewer from *Tetrahymena*, disclose essentially the same principles of signaling as in metazoans up to man. Microtubules serve for long-range signaling, whereas SNARE proteins, H^+-ATPase, GTPases and actin provide short-range molecular signals, with Ca^{2+} as a most efficient, locally and spatially restricted signal particularly for membrane fusion. This is enabled by the strategic positioning of Ca^{2+}-release channels, type $InsP_3$ receptors and ryanodine-receptor-like proteins, also in ciliates. In *Paramecium*, the most evident trafficking routes encompass exo-/endocytosis, endo-/phagocytosis and the contractile vacuole complex. The high specificity, precision and efficiency of vesicle trafficking regulation in ciliates is facilitated by their highly regular, epigenetically controlled cell structure, with firmly installed microtubular rails.

6.1 Basic Aspects of Vesicle Trafficking and Signaling Pathways in Ciliates

Ciliates are well designed cells, with a cell cortex composed of unit fields called kinetids, about 1 × 2 μm in size. Each kinetid possesses a cilium (or two) in its center and intermittently positioned dense core-secretory organelles. These are trichocysts in *Paramecium* and mucocysts in *Tetrahymena*—the most intensely studies genera. This epigenetically controled surface pattern (Beisson and Sonneborn 1965; Wloga and Frankel 2012) entails important consequences for signaling, e.g. by the universal second messenger, Ca^{2+}, as it allows for alternative activation of cilia and of vesicle exocytosis, respectively, due to strict local and temporal signal confinement (Klauke and Plattner 1997; Husser et al. 2004). Exocytosis is an example of short-range signaling enabled by a pre-assembled signaling machinery and a local Ca^{2+} impulse.

H. Plattner (✉)
Department of Biology, University of Konstanz, P.O. Box M625, 78457 Konstanz, Germany
e-mail: helmut.plattner@uni-konstanz.de

G. Witzany and M. Nowacki (eds.), *Biocommunication of Ciliates*,
DOI 10.1007/978-3-319-32211-7_6

Microtubules emanating from ciliary basal bodies act as long-range signals for the transport of trichocysts to docking sites (Aufderheide 1977; Glas-Albrecht et al. 1991). Close to basal bodies, parasomal sacs serve for formation of clathrin-coated endosomes (Elde et al. 2005), alternating with constitutive exocytosis of surface coat materials (Flötenmeyer et al. 1999). Other landmarks in the cell cortex are the cytostome from where phagosomes pinch off, fed by acidosomes (late endosomes) and recycling vesicles originating from the cytoproct—the site of spent food vacuole release (Allen and Fok 2000). Again, microtubules form a long-range signaling system in this "ventral" part of the cell by guiding vesicles to the nascent food vacuole. The outlets of the two contractile vacuole complexes are positioned "dorsally", also under epigenetic control.

Three vesicle trafficking pathways are recognizable: (i) from endoplasmic reticulum via Golgi apparatus to the cell surface for the exocytotic route, paralleled by a reverse endocytotic pathway via endosomes for membrane recycling and/or degradation in lysosomes; (ii) the phagocytotic pathway, coupled to endocytosis and recycling, for digestion of food particles; finally, (iii) the contractile vacuole complex for periodic expulsion of an excess of water and ions (Allen and Naitoh 2002). This counteracts the rapid permeation of water and Ca^{2+} and—beyond pulsation activity—also involves less overt membrane dynamics and signaling (Plattner 2015c).

Before genomic databases became available, Allen and Fok (2000) have pioneered the field of vesicle trafficking in *Paramecium* by using organelle-specific monoclonal antibodies for immunolocalization studies (http://www5.pbrc.hawaii.edu/allen/). Subsequently molecular biology allowed for more stringent molecular identification, localization and gene silencing of key players in signal-based vesicle trafficking. Short-range signals include SNARE proteins, the SNARE-specific chaperone NSF, H^+-ATPase/pump, actin, Ca^{2+}-release channels (CRC), as analyzed in *Paramecium* (reviewed by Plattner 2010a), and small GTPases analyzed in *Tetrahymena* (Bright et al. 2010). SNAREs stand for "*s*oluble *N*-ethylmaleimide sensitive factor (NSF) *a*ttachment protein *re*ceptors", NSF being a SNARE-specific chaperone (Rothman 2014; Südhof 2014). SNAREs, together with Rab-type monomeric GTP-binding proteins/GTPases and their organelle-specific binding proteins (Grosshans et al. 2006) and high sensitivity Ca^{2+}-binding proteins are important for short-range signaling enabling membrane interaction and fusion (Rothman 2014; Südhof 2014).

Note that genes and proteins of *Paramecium tetraurelia*, are designated with the prefix "Pt", e.g. PtSyb for synaptobrevin. As there are subfamilies of variable size, a designation may be PtSyb5-1. The prefix Pt is important since synaptobrevins are difficult to correlate with their equivalents up to mammals with regard to localization and function. Although this is more easy, e.g. with syntaxins, we maintained this nomenclatorial principle throughout. For CRCs we used a comparable designation, for instance, PtCRC-IV-1 for a subfamily member. Considering the number of paralogs (also called ohnologs when arisen by whole genome duplications) in a protein/gene family of *P. tetraurelia* it was not possible to analyse them all and their localization and function may be similar or even identical. Therefore, in this review, the types of subfamily members are usually not specified. More details can be retrieved from the original publications and reviews cited, e.g. Plattner (2010a, b) for SNAREs and Plattner and Verkhratsky (2013) for CRCs.

Figure 6.1a illustrates the general principle of signalling for vesicle trafficking, Fig. 6.1b the requirement of a strictly confined Ca^{2+} signal for membrane fusion, Fig. 6.1c the molecular interactions during intracellular vesicle trafficking, as exemplified by food vacuole (phagolysosome) formation, and Fig. 6.1d the particular situation of an exocytosis site in ciliates.

6.2 Long-Range Trafficking Signals

Since biogenesis of the surface membrane complex requires proper positioning of components, such as cilia, parasomal sacs and trichocysts, one may ask for cause and consequence. On the one hand one may ask for the causation of this regular arrangement and, on the other hand, for its consequences for signalling processes. Reversing this arrangement by spontaneous misplacement in *Euplotes* (Frankel 1973) or experimentally by microsurgery in *Paramecium* (Beisson and Sonneborn 1965) maintains and transfers such local disorientation, thus indicating epigenetic regulation of surface pattern formation and resulting signalling (Plattner 2015a).

Analysis of surface pattern formation has revealed several crucial components of ciliary basal bodies (Beisson and Jerka-Dziadosz 1999). Saltatory docking of trichocysts (Aufderheide 1977) along microtubules originating at ciliary basal bodies (Plattner et al. 1982; Glas-Albrecht et al. 1991), quite unusually, occurs in plus-to-minus direction of microtubules—a phenomenon maintained in evolution, e.g. in mammalian epithelia and T-lymphocytes (Griffiths et al. 2010). Some possible mechanisms relevant for the distinct positioning of the contractile vacuole complexes, also microtubule-based, have been discussed recently (Plattner 2013, 2015c).

Important parts of the phagocytotic pathways and delivery of recycling vesicles in ciliates are also guided by microtubules. Crucial work from Allen and Fok (2000) revealed that recycling vesicles travel from the cytoproct to the cytostome for reuse, as shown with *Paramecium multimicronucleatum* (Fok and Allen 1993) and *Tetrahymena thermophila* (summarized by Wloga and Frankel 2012).

In conclusion, pre-established microtubular arrays, together with appropriate motor proteins, allow for long-range signaling and, thus, render vesicle trafficking very efficient. Since many types of vesicles bounce around in the cytoplasm one has to ask what molecular signal makes a vesicle jump onto the bandwagon suitable for delivery to its final address. In the end this is determined by vesicle-resident proteins, e.g. an "early endosome antigen" (EEA)—a short-range signal, as described below.

Prerequisite to the strict signalling phenomena in the cell cortex is the targeted delivery of vesicles and their membrane components to proper sites. The selective expression pattern during reciliation following deciliation, e.g. in *T. thermophila*

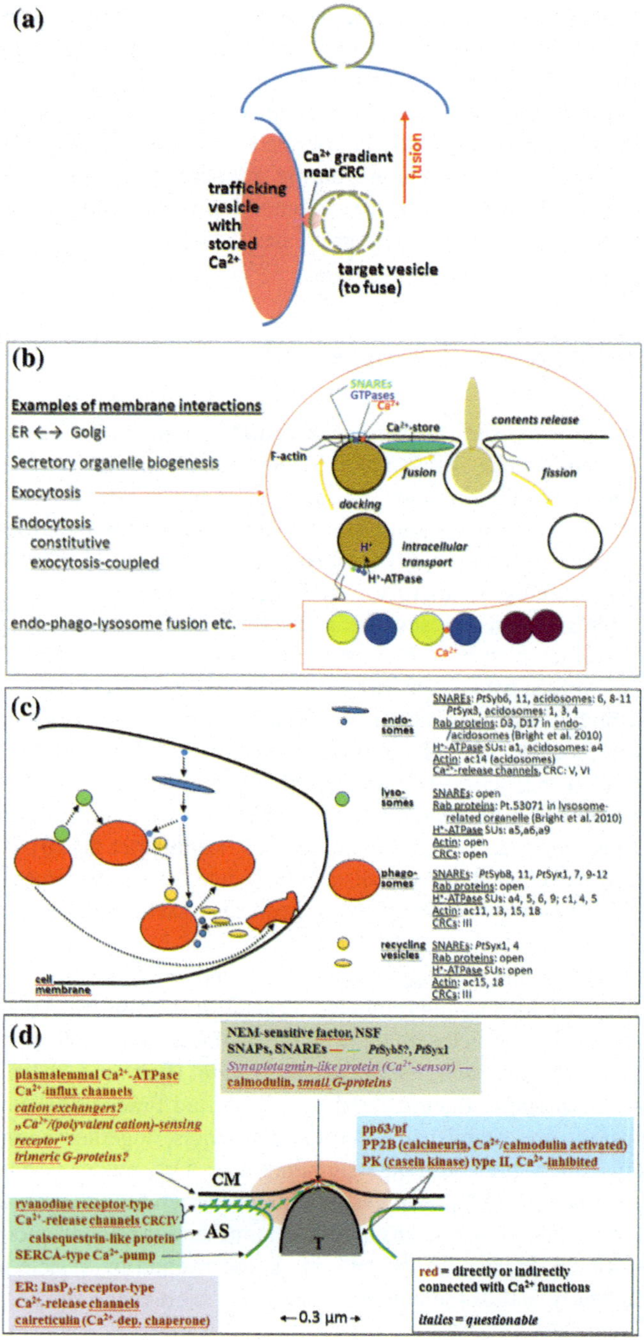
(a)
fusion
Ca²⁺ gradient near CRC
trafficking vesicle with stored Ca²⁺
target vesicle (to fuse)
(b)
Examples of membrane interactions
ER ←→ Golgi
Secretory organelle biogenesis
Exocytosis
Endocytosis
constitutive
exocytosis-coupled
endo-phago-lysosome fusion etc.
SNAREs
GTPases
Ca²⁺
Ca²⁺-store
contents release
F-actin
fusion
fission
docking
H⁺
intracellular transport
H⁺-ATPase
Ca²⁺
(c)
endo-somes
SNAREs: PtSyb6, 11, acidosomes: 6, 8-11
PtSyx8, acidosomes: 1, 3, 4
Rab proteins: D3, D17 in endo-/acidosomes (Bright et al. 2010)
H⁺-ATPase SUs: a1, acidosomes: a4
Actin: ac14 (acidosomes)
Ca²⁺-release channels, CRC: V, VI
lyso-somes
SNAREs: open
Rab proteins: Pt.53071 in lysosome-related organelle (Bright et al. 2010)
H⁺-ATPase SUs: a5,a6,a9
Actin: open
CRCs: open
phago-somes
SNAREs: PtSyb8, 11, PtSyx1, 7, 9-12
Rab proteins: open
H⁺-ATPase SUs: a4, 5, 6, 9; c1, 4, 5
Actin: ac11, 13, 15, 18
CRCs: III
recycling vesicles
SNAREs: PtSyx1, 4
Rab proteins: open
H⁺-ATPase SUs: open
Actin: ac15, 18
CRCs: III
cell membrane
(d)
NEM-sensitive factor, NSF
SNAPs, SNAREs — — PtSyb5?, PtSyx1
Synaptotagmin-like protein (Ca²⁺-sensor) —
calmodulin, small G-proteins
plasmalemmal Ca²⁺-ATPase
Ca²⁺-influx channels
cation exchangers?
„Ca²⁺/(polyvalent cation)-sensing receptor"?
trimeric G-proteins?
pp63/pf
PP2B (calcineurin, Ca²⁺/calmodulin activated)
PK (casein kinase) type II, Ca²⁺-inhibited
CM
ryanodine receptor-type
Ca²⁺-release channels CRCIV
AS
calsequestrin-like protein
SERCA-type Ca²⁺-pump
T
ER: InsP₃-receptor-type
Ca²⁺-release channels
calreticulin (Ca²⁺-dep. chaperone)
←0.3 µm→
red = directly or indirectly connected with Ca²⁺ functions
italics = questionable

◀ **Fig. 6.1** **a** Principles of vesicle/membrane interaction, with the involvement of SNAREs, GTPases, H^+-ATPase, actin and Ca^{2+}. Acidification of the vesicle lumen by the organellar H^+-ATPase signals binding of additional components, leading to vesicle docking at a target membrane and fusion upon a Ca^{2+} signal. This principle applies not only to exocytosis (*top*) but also to intracellular membrane interactions (*bottom*). **b** Illustrates the local restriction of Ca^{2+} signals for membrane fusion, e.g. by release via local CRCs. **c** Vesicle trafficking and molecules involved in the endo-/phagocytotic process. At the right, key players relevant for the respective steps are listed. Note, for instance, the exchange of actin, CRC and SNARE isoforms. **d** Components assembled at and around a trichocyst docking site include not only proteins relevant for docking and fusion, but also for Ca^{2+} storage in, and release from alveolar sacs (AS), as well as Ca^{2+}-pumps, Ca^{2+}-binding proteins and proteins relevant for energetic aspects, such as pp63/pf, PP2B and a casein kinase, shown at the *right/center*. *Green arrows* indicate Ca^{2+} flux upon stimulation. Some components, listed with a question mark, require identification in ciliates. AS = alveolar sac, CM = cell membrane, T = trichocyst. For further details, see text and Plattner (2016) (Color figure online)

(Miao et al. 2009), can provide insight into the relevance of individual components. Frequently this is pinpointed by following up in ciliates the relevance of mutations underlying mammalian ciliary pathology, as summarized previously (Simon and Plattner 2014). It should be noted that vesicles are not delivered into cilia in any system analysed. Unexpectedly we observed a ring of PtSyb10 around the ciliary basis (Schilde et al. 2010). Since synaptobrevins are normally harbored in vesicles, rather than in target membranes (see below) this might indicate retention due to excessive vesicle delivery, whereas this normally seems to be counterbalanced by membrane retrieval.

Another aspect is that Ca^{2+} can separately regulate widely different processes, such as trichocyst exocytosis and ciliary activity. Why is a cell then not confused when stimulated? Normally there is no spill-over of Ca^{2+} from cilia to exocytosis sites which, therefore, remain inactive (Husser et al. 2004). However, the opposite process does take place when a predator attacks a *Paramecium* cell. In consequence, explosive local trichocyst release keeps the predator at a distance and ciliary reversal allows the potential prey to rapidly escape (Knoll et al. 1991a). Both processes depend on Ca^{2+} and spillover of Ca^{2+} into cilia (Husser et al. 2004) allows, in this case, to bypass the physiological activation of ciliary reversal. Normally this is induced by cell membrane depolarization and Ca^{2+} influx selectively into cilia via voltage dependent channels [For review see Plattner and Klauke (2001) and Plattner (2014)]. Beyond this exceptional situation of a long-range Ca^{2+} signal there are several mechanisms to confine a Ca^{2+} signal very strictly (Sect. 6.3).

6.3 Short-Range Trafficking Signals

6.3.1 General Aspects

Evidently integral membrane proteins can serve as local landmarks for specific interaction with a partner membrane (Fig. 6.1a). This applies to SNAREs, anchored by their carboxy-terminal region, as well as to C2-type high sensitivity Ca^{2+}-sensor proteins, type synaptotagmin, inserted by their amino-terminal region. Both are crucial for membrane fusion (In *Paramecium* we have found only synaptotagmin-like proteins, still to be characterized; see Sect. 6.3.3). For a summary of key players in membrane recognition and fusion, see Plattner (2010a). Also some soluble proteins, type Rab, have to be locally available. As mentioned, they bind to organelle-specific integral membrane proteins (essentially unknown in ciliates). All these molecules have to match with their counterparts in the target membrane.

Ca^{2+} for signaling originates either from release from a nearby store, or interacting vesicles contain Ca^{2+} themselves and are able to release Ca^{2+} via CRCs integrated in their membrane (Plattner 2015b); see Fig. 6.1b. Cytosolic Ca^{2+} concentration has to rise very locally and for a very restricted time, for different reasons (Plattner and Verkhratsky 2015). First, Ca^{2+} at too high concentrations is toxic; second, Ca^{2+} diffuses rapidly; third its concentration is rapidly downregulated by binding to Ca^{2+}-binding proteins, and over longer times by sequestration and extrusion from the cell; fourth, its effect depends of concentration in a supralinear function (Neher 1998a). All this calls for microdomain regulation in ciliates (Klauke and Plattner 1997; Plattner and Klauke 2001), just as in neurons (Neher 1998b). Local Ca^{2+} signals for membrane fusion are so short-lived that their actual size and duration can only be extrapolated from intracellularly applied Ca^{2+} chelators with different binding and time constant (Neher 1998a, b). In *Paramecium*, Ca^{2+} for trichocyst exocytosis comes from alveolar sacs (Sect. 6.3.3), the well established cortical stores (Stelly et al. 1991; Hardt et al. 2000; Plattner and Klauke 2001). Local values of $\sim 5\ \mu M$ are required for exocytosis (Klauke and Plattner 1997). Considering that, in mammalian cells, most trafficking organelles contain Ca^{2+} themselves (Hay 2007) and the occurrence of CRCs in such organelles (except trichocysts) in *Paramecium* (Ladenburger and Plattner 2011), one can reasonably assume that Ca^{2+} signaling in microdomains occurs throughout vesicle trafficking pathways also in ciliates. For any further discussion, see Plattner (2014).

Some essential aspects of signaling are outlined in Fig. 6.1a. A vesicle contains SNARE proteins (called v- or R-SNAREs due to an R [Arg] residue in the center of the SNARE domain). For docking, a target membrane has to contain a matching set of t-/Q-SNARES (Q = Glu). SNAP-25 is an additional SNARE that, because of its wide distribution throughout the *Paramecium* cell is hardly appropriate to further specify short-range signaling. As mentioned, additional specificity is provided by the reversible binding of monomeric GTP-binding proteins (GTPases type Rab for vesicle trafficking), as analysed in *Tetrahymena* (Bright et al. 2010). These in turn bind to organelle-specific residents proteins (hardly identified in ciliates). A total of 44 SNAREs, v-/R- (Schilde et al. 2006, 2010) and t-/Q-SNARES (Kissmehl et al. 2007) have been assigned in *Paramecium* to specific organelles, some containing more than one type, including some rather similar ohnologs (Plattner 2010b) due to recent whole genome duplications. 57 Rabs are reported from *Tetrahymena* (Bright et al. 2010).

The activity of Rabs is governed by activating and inhibitory proteins (hardly known from ciliates), respectively. In higher eukaryotes their attachment to the respective target membrane depends on the luminal pH value (Hurtado et al. 2006). Concomitantly, vesicles undergoing trafficking contain a H^+-ATPase (Mellman 1992). The conformational change of the H^+-ATPase/pump resulting from the actual luminal acidification enables the binding of Rab regulators and Rab proteins, probably also in ciliates. Here, we also find this multimeric H^+-ATPase in the organelles undergoing trafficking (Wassmer 2005, 2006, 2009). Particularly remarkable is the unprecedented number of a-subunit isoforms which allows for the attachment of the catalytic V1 part to the membrane-integrated H^+-conducting base piece, V0; both, V0 and V1 are a complex of variable monomers. The high number of theoretical combinations with the a-subunit may allow for a range of subunit combinations for local requirements. Similarly organelle-specific CRCs may account for specific local requirements. Work with higher eukaryotes suggests that targeting of some of the components under consideration for short-range signalling is mutually interdependent, as reviewed previously (Plattner 2010a).

6.3.2 Signaling During the Endo-Phagolysosomal Cycle

Figure 6.1c outlines vesicle trafficking and signalling molecules involved in food vacuole (phagolysosome) formation. Already before the molecular era, Allen and Fok (2000) have elucidated the principal pathways of phagocytosis in *Paramecium multimicronucleatum*.

Formation of food vacuoles starts at the cytostome. Here, the cell membrane seems to bulge in; in reality this part of the cell membrane—in the absence of alveolar sacs, ciliary basal bodies and of trichocysts—is extended by recycling vesicles. These are delivered from spent food vacuoles after contents discharge at the cytoproct and by other recycling vesicles originating from progressed stages of food vacuoles. The nascent vacuole becomes a phagosome which is a non-acid organelle devoid of H^+-ATPase. Only after pinching off, late endosomes, called acidosomes because of their endowment with H^+-ATPase, fuse with the phagosome (Allen et al. 1993). On their way through the cell (cyclosis) food vacuoles fuse with lysosomes and, thus, become phagolysosomes which digest food particles. At a later stage, lysosomal components are retrieved and the spent food vacuole releases indigestible materials at the cytoproct (Allen and Fok 2000). When stained with permeable acidity markers, which are retained in protonated form and change color depending on luminal pH, a color change indicates the sequence: neutral → acid → neutral (Wassmer et al. 2009).

The sequence of events described is supported by long-range and short-range, molecular signaling. Here, like in other acidic compartments, such as endosomes, occurrence of an H^+-ATPase was demonstrated (Fok et al. 2002; Wassmer et al. 2009). As mentioned, changing acidification also implies change of Rab protein binding (Hurtado-Lorenzo et al. 2006) at the organelle surface. Changing binding partners, from endosome to phagolysosome, are indicated in Fig. 6.1c. Whereas this background information mainly comes from *Paramecium* (*P. multimicronucleatum* for classical cell biology, *P. tetraurelia* for molecular biology), data concerning changing association of Rab proteins throughout the digesting cycle come from *T. thermophila*, with a side glance on *P. tetraurelia* (Bright et al. 2010).

Figure 6.1c summarizes what is known about the changing isoforms of SNAREs, H^+-ATPase subunits, actin and CRCs along the endo-phagolysosomal pathway, as determined in our laboratory, supplemented by data on changing Rab proteins collected by the Turkewitz laboratory (Bright et al. 2010). This comprehensive work reveals extensive specification of Rabs in compartments designated as follows: endocytotic vesicles, posterior potential recycling vesicles, lyso/phagosomes, all phagosomes, selected phagosomes and cytoproct-associated phagosomes with 4, 4, 1, 1, 6, and 3 Rab types, respectively, complemented by five Rabs associated with the oral apparatus (including the cytosome) and three associated with the cytoproct. Considering that, in *P. tetraurelia*, usually several similar ohnologs occur, the number of Rabs may be several times higher, although probably with redundant localizations and functions.

6.3.3 Dense Core-Secretory Vesicle Biogenesis, Docking and Signaling for Exocytosis

In ciliates, the endoplasmic reticulum is very elaborate, whereas the Golgi apparatus is inconspicuous; in *Paramecium* the Golgi apparatus is split in several hundred Golgi fields (dictyosomes) with very few membrane stacks (Allen 1988). Biogenesis of both, mucocysts and trichocyst, follows this classical assembly line. Precursor vesicles fuse and secretory proteins are trimmed (Briguglio et al. 2013; Gautier et al. 1994). When not posttranslationally cleaved, assembly to crystalline contents and transfer to the cell membrane are inhibited with both, mucocysts (Briguglio et al. 2013) and trichocysts (Pouphile et al. 1986; Gautier et al. 1994). Therefore, there must be a luminal signal, linked to a surface signal, for delivery along microtubules to the cell membrane.

Upon arrival at the cell membrane trichocysts induce restructuring of the docking site by formation of a "fusion rosette" (Beisson et al. 1976). The identity of the proteins forming a rosette is not clear (Plattner 2010a) although we know that its assembly requires the activity of the SNARE chaperone, NSF (Froissard et al.

2002). One of the problems is that one rosette particle, to account for its diameter, would have to contain ~70 syntaxins (Plattner 2010b). Nevertheless, the t-/Q-PtSyx1 can be safely assumed to occur at exocytosis sites although, by immunolocalization, it is scattered over the entire cell membrane (Kissmehl et al. 2007). No V0 parts of the H^+-ATPase are seen at exocytosis sites (Wassmer et al. 2005), thus excluding a membrane fusion model involving these molecules (Plattner and Kissmehl 2003). The only v-/R-SNARE detected on the trichocyst membrane is PtSyb5 (Schilde et al. 2010). In *T. thermophila*, the Rab protein TtRabD41 has been detected at exocytosis sites (Bright et al. 2010).

A Ca^{2+}-sensor, type synaptotagmin, still requires identification and localization in ciliates. Such molecule is mandatory for rapid exocytosis—considering that all trichocysts can be rapidly released within 80 ms upon stimulation (Knoll et al. 1991b, Plattner et al. 1993)—the highest rate ever found with dense core-secretory organelles (Plattner and Kissmehl 2003). Candidate Ca^{2+}-sensors found in the *Paramecium* genomic database have eight C2 domains (Kissmehl and Plattner unpubl. observ.), rather than the usual two, for high sensitivity Ca^{2+} binding (Südhof 2014). Such extended synaptotagmins (e-syts) also occur in other systems, up to mammalians, and they allow membrane-to-membrane links over a broader interspace (Pérez-Lara and Jahn 2015). This would fit the trichocyst-cell membrane interspace at docking sites. Another cursory note based on database mining indicates in *P. tetraurelia* occurrence of a homolog of DOC2.1—a two C2 domain-bearing Ca^{2+}-binding protein in apicomplexan parasites (Farrell et al. 2012).

Figure 6.1d illustrates which additional proteins are assembled at, and around a trichocyst docking site. Calmodulin is required for the assembly of SNAREs (Kerboeuf et al. 1993). Specific isoforms of actin (Sehring et al. 2007a, b), together with a 63 kDa phosphoprotein (pp63/parafusin, pf), phosphatase 2B (calcineurin), and a Ca^{2+}-inhibited protein kinase are distributed around trichocyst tips and the nearby subplasmalemmal space. They are considered important for energetic aspects of signaling during stimulated trichocyst exocytosis (Plattner and Kissmehl 2003). A plasmalemmal Ca^{2+}-ATPase/pump serves for keeping subplasmalemmal Ca^{2+} concentration low, whereas a SERCA-type Ca^{2+}-ATPase/pump refills the stores (see Plattner 2014). Upon stimulation, release of Ca^{2+} from alveolar sacs, the cortical Ca^{2+}-stores (Stelly et al. 1991; Hardt and Plattner 2000), is mediated by ryanodine receptor-like protein (RyR-LP) channels (Ladenburger et al. 2009; Plattner 2015b). This microanatomical arrangement forces Ca^{2+} flux over trichocyst tips, thus mediating rapid response, as required for predator defence (Harumoto and Miyake 1991).

The CRCs, as well as the luminal high capacity/low affinity Ca^{2+}-binding proteins are different in endoplasmic reticulum and in alveolar sacs, respectively. Other molecules of potential interest, such as a surface Ca^{2+}/polyvalent cation-sensing receptor and trimeric G-proteins are under debate. Such details concerning signaling in ciliates are discussed in a more extensive review (Plattner 2016).

Once an exocytotic pore is formed, Ca^{2+} can rapidly flow into the trichocyst lumen where it triggers the explosive decondensation (stretching) and release of the crystalline contents (Bilinski et al. 1981) by binding to specific secretory proteins (Klauke et al. 1998).

The Ca^{2+} signal flushing over exocytosis sites suffices to drive also exocytosis-coupled endocytosis of trichocyst "ghosts" and their internalization (Plattner et al. 1997), thus making these sites again available for docking of new trichocysts. After stimulation, in the cytosol, Ca^{2+} is most rapidly downregulated by binding to the Ca^{2+}-binding protein, centrin (Sehring et al. 2009), whereas the pumps are primarily devoted to overall household regulation (Plattner 2016).

6.3.4 Contractile Vacuole Complex

This complex organelle is most dynamic, as it contracts and expands in ≤10 s intervals. From the contractile vacuole, radial canals emanate which are associated with a flexible three-dimensional network, the spongiome (Allen and Naitoh 2002). Its part proximal to the canals is smooth and periodically collapses after contraction (systole) and, thus, may represent a membrane reservoir for the expansion of the canals and of the vacuole during diastole. Canals are disconnected from the vacuole before every systole (Tominaga et al. 1998). In the electron microscope the peripheral part of the spongiome appears decorated by the head parts (V1) of the H^+-ATPase (Fok et al. 2002; Wassmer et al. 2005) and, therefore, is called the decorated spongiome.

SNAREs (PtSyx2, 14, 15, and PtSyb2, 6, 9) and Ca^{2+}-release channels (type PtCRC-II, i.e. inositol 1,4,5-trisphosphate receptors [$InsP_3Rs$]) occur in different parts of the organelle in *P. tetraurelia*, together of PtSyx2 and 15 as well as PtSyb2 and 9 at the vacuole outlet (the pore). Mechanosensitive Ca^{2+} channels occur in the database and are also localized to the contractile vacuole complex in other species. All this suggests the following signaling scenario (Plattner 2013, 2015c). (i) A proton gradient enables osmotic filling with water and ions from the cytosol (Allen and Naitoh 2002), including an excess of Ca^{2+} (Stock et al. 2002). (ii) Increasing turgor causes reversible rearrangement of membranes in the smooth spongiome with the help of SNAREs (Plattner 2013). (iii) Partial constitutive reflux of Ca^{2+} from the organelle through the CRC-II/$InsP_3R$-type channels into the nearby cytosol (Ladenburger et al. 2006) can drive membrane rearrangements in the smooth spongiome and vesicle fusion for ongoing protein replacement. (iv) Mechanosensitive channels may also mediate fusion of the vacuole with the cell membrane, thus opening the pore. (v) In *T. thermophila*, three Rab proteins have been detected, TtRabD2, 10 and 14 (Bright et al. 2010).

Before cytokinesis, de novo formation of additional contractile vacuole complexes is observed. Localization studies with ciliates and with other protozoa suggest signaling by proteins many of which are known to govern cytokinesis (Plattner 2015c).

6.4 Conclusions and Open Questions

Despite considerable progress, there are important gaps in our knowledge about vesicle trafficking and underlying signaling mechanisms in ciliates although they now belong to the best analysed protozoa. In *Paramecium*, the abundance of ohnologs serves for a kind of gene amplification, but some may undergo neo-functionalization. It would be particularly rewarding to analyze the functional relevance of the unprecedented number of H^+-ATPase a-subunits. May they serve for fine tuning of local signalling? Also important is the establishment of the identity of Rab target proteins. We also still have to learn a lot about the Golgi apparatus in ciliates.

Beyond these and some other gaps, many details of general validity about vesicle trafficking and underlying signaling mechanisms have been established specifically in ciliates, mainly with *Paramecium* and *Tetrahymena*. With P. *tetraurelia* our primary goal was to sound out, on a broad scale, the inventory available, while punctual scrutiny of some more important aspects was a secondary goal. Work may now be extended to detailed analysis of ciliate-specific aspects, as has been done for some important aspects, mainly with *T. thermophila*. Both systems together have taught us that signaling mechanisms underlying vesicle trafficking and beyond have been invented already at this level of evolution. The redundancy of structural elements, particularly in the cell cortex, and amplification effects of synchronized processes, such as synchronous exocytosis induction and synchronous reciliation after deciliation, can greatly facilitate the analysis of signalling processes.

References

Allen RD (1988) Cytology. In: Görtz HD (ed) Paramecium. Springer, Berlin, pp 4–40

Allen RD, Fok AK (2000) Membrane trafficking and processing in *Paramecium*. Int Rev Cytol 198:277–318

Allen RD, Naitoh Y (2002) Osmoregulation and contractile vacuoles of protozoa. Int Rev Cytol 215:351–394

Allen RD, Ma L, Fok AK (1993) Acidosomes: recipients of multiple sources of membrane and cargo during development and maturation. J Cell Sci 106:411–422

Aufderheide KJ (1977) Saltatory motility of uninserted trichocyst and mitochondria in *Paramecium tetraurelia*. Science 198:299–300

Beisson J, Jerka-Dziadosz M (1999) Polarities of the centriolar structure: morphogenetic consequences. Biol Cell 91:367–378

Beisson J, Sonneborn TM (1965) Cytoplasmic inheritance of the organization of the cell cortex in *Paramecium aurelia*. Proc Natl Acad Sci USA 53:275–282

Beisson J, Lefort-Tran M, Pouphile M, Rossignol M, Satir B (1976) Genetic analysis of membrane differentiation in *Paramecium*. Freeze-fracture study of the trichocyst cycle in wild-type and mutant strains. J Cell Biol 69:126–143

Bilinski M, Plattner H, Matt H (1981) Secretory protein decondensation as a distinct, Ca^{2+}-mediated event during the final steps of exocytosis in *Paramecium* cells. J Cell Biol 88:179–188

Bright LJ, Kambesis N, Nelson SB, Jeong B, Turkewitz AP (2010) Comprehensive analysis reveals dynamic and evolutionary plasticity of Rab GTPases and membrane traffic in *Tetrahymena* thermophila. PLoS Genet 6:e1001155

Briguglio JS, Kumar S, Turkewitz AP (2013) Lysosomal sorting receptors are essential for secretory granule biogenesis in *Tetrahymena*. J Cell Biol 203:537–550

Elde NC, Morgan G, Winey M, Sperling L, Turkewitz AP (2005) Elucidation of clathrin-mediated endocytosis in *Tetrahymena* reveals an evolutionarily convergent recruitment of dynamin. PLoS Genet 1:e52

Farrell A, Thirugnanam S, Lorestani A, Dvorin JD, Eidell KP, Ferguson DJP, Anderson-White BR, Duraisingh MT, Marth GT, Gubbels MJ (2012) A DOC2 protein identified by mutational profiling is essential for apicomplexan parasite exocytosis. Science 335:218–221

Flötenmeyer M, Momayezi M, Plattner H (1999) Immuno-labeling analysis of biosynthetic and degradative pathways of cell surface components (glycocalyx) in *Paramecium* cells. Eur J Cell Biol 78:67–77

Fok AK, Allen PD (1993) Membrane flow in the digestive cycle of *Paramecium*. In: Plattner H (ed) Membrane Traffic in Protozoa. JAI Press, Greenwich (CT), London, pp 311–337

Fok AK, Yamauchi K, Ishihara A, Aihara MS, Ishida M, Allen RD (2002) The vacuolar-ATPase of *Paramecium multimicronucleatum*: gene structure of the B subunit and the dynamics of the V-ATPase-rich osmoregulatory membranes. J Eukaryot Microbiol 49:185–196

Frankel J (1973) Dimensions of control of cortical patterns in *Euplotes* - Role of preexisting structure, clonal life-cycle, and genotype. J Exp Zool 183:71–94

Froissard M, Kissmehl R, Dedieu JC, Gulk-Krzywicki T, Plattner H, Cohen J (2002) N-ethylmaleimide-sensitive factor is required to organize functional exocytotic microdomains in *Paramecium*. Genetics 161:643–650

Gautier MC, Garreau De Loubresse N, Madeddu L, Sperling L (1994) Evidence for defects in membrane traffic in *Paramecium* secretory mutant unable to produce functional storage granules. J Cell Biol 124:893–902

Glas-Albrecht R, Kaesberg B, Knoll G, Allmann K, Pape R, Plattner H (1991) Synchronised secretory organelle docking in *Paramecium*. Saltatory movement along microtubules transiently formed from ciliary basal bodies and selective exclusion of microinjected heterologous organelles. J Cell Sci 100:45–54

Griffiths GM, Tsun A, Stinchcombe JC (2010) The immunological synapse: a focal point for endocytosis and exocytosis. J Cell Biol 189:399–406

Grosshans BL, Ortiz D, Novick P (2006) Rabs and their effectors: achieving specificity in membrane traffic. Proc Natl Acad Sci USA 103:11821–11827

Hardt M, Plattner H (2000) Sub-second quenched-flow/X-ray microanalysis shows rapid Ca^{2+} mobilization from cortical stores paralleled by Ca^{2+} influx during synchronous exocytosis in *Paramecium* cells. Eur J Cell Biol 79:642–652

Harumoto T, Miyake A (1991) Defensive function of trichocysts in *Paramecium*. J Exp Zool 260:84–92

Hay JC (2007) Calcium: a fundamental regulator of intracellular membrane fusion? EMBO Rep 8:236–240

Hurtado-Lorenzo A, Skinner M, El Annan J, Futai M, Sun-Wada GH, Bourgoin S, Casanova J, Wildeman A, Bechoua S, Ausiello DA, Brown D, Marshansky V (2006) V-ATPase interacts with ARNO and Arf6 in early endosomes and regulates the protein degradative pathway. Nature Cell Biol 8:124–136

Husser MR, Hardt M, Blanchard MP, Hentschel J, Klauke N, Plattner H (2004) One-way calcium spill-over during signal transduction in *Paramecium* cells: from the cell cortex into cilia, but not in the reverse direction. Cell Calcium 36:349–358

Kerboeuf D, Le Berre A, Dedieu JC, Cohen J (1993) Calmodulin is essential for assembling links necessary for exocytotic membrane fusion in *Paramecium*. EMBO J 12:3385–3390

Kissmehl R, Schilde C, Wassmer T, Danzer C, Nühse K, Lutter K, Plattner H (2007) Molecular identification of 26 syntaxin genes and their assignment to the different trafficking pathways in *Paramecium*. Traffic 8:523–542

Klauke N, Plattner H (1997) Imaging of Ca^{2+} transients induced in *Paramecium* cells by a polyamine secretagogue. J Cell Sci 110:975–983

Klauke N, Kissmehl R, Plattner H, Haga N, Watanabe T (1998) An exocytotic mutant of *Paramecium caudatum*: membrane fusion without secretory contents release. Cell Calcium 23:349–360

Knoll G, Haacke-Bell B, Plattner H (1991a) Local trichocyst exocytosis provides an efficient escape mechanism for *Paramecium* cells. Eur J Protistol 27:381–385

Knoll G, Braun C, Plattner H (1991b) Quenched flow analysis of exocytosis in *Paramecium* cells: time course, changes in membrane structure and calcium requirements revealed after rapid mixing and rapid freezing of intact cells. J Cell Biol 113:1295–1304

Ladenburger EM, Plattner H (2011) Calcium-release channels in *Paramecium*. Genomic expansion, differential positioning and partial transcriptional elimination. PLoS ONE 6(11): e27111

Ladenburger EM, Korn I, Kasielke N, Wassmer T, Plattner H (2006) An Ins(1,4,5)P_3 receptor in *Paramecium* is associated with the osmoregulatory system. J Cell Sci 119:3705–3717

Ladenburger EM, Sehring IM, Korn I, Plattner H (2009) Novel types of Ca^{2+}-release channels participate in the secretory cycle of *Paramecium* cells. Mol Cell Biol 29:3605–3622

Mellman I (1992) The importance of being acid: the role of acidification in intracellular membrane traffic. J Exp Biol 172:39–45

Miao W, Xiong J, Bowen J, Wang W, Liu Y, Braguinets O, Grigull J, Pearlman RE, Orias E, Gorovsky MA (2009) Microarray analyses of gene expression during the *Tetrahymena thermophila* life cycle. PLoS ONE 4:e4429

Neher E (1998a) Usefulness and limitations of linear approximations to the understanding of Ca^{++} signals. Cell Calcium 24:345–357

Neher E (1998b) Vesicle pools and Ca^{2+} microdomains: new tools for understanding their roles in neurotransmitter release. Neuron 20:389–399

Pérez-Lara A, Jahn R (2015) Extended synaptotagmins (e-syts): architecture and dynamics of membrane contact sites revealed. Proc Natl Acad Sci USA 112:4837–4838

Plattner H (2010a) Membrane trafficking in protozoa: SNARE proteins, H^+-ATPase, actin, and other key-players in ciliates. Int Rev Cell Mol Biol 280:79–184

Plattner H (2010b) How to design a highly organized cell: an unexpectedly high number of widely diversified SNARE proteins positioned at strategic sites in the ciliate, *Paramecium tetraurelia*. Protist 161:497–516

Plattner H (2013) The contractile vacuole complex—its expanding protein inventory. Int Rev Cell Mol Biol 306:371–416

Plattner H (2014) Calcium regulation in the protozoan model cell, *Paramecium tetraurelia*. J Eukaryot Microbiol 61:95–114

Plattner H (2015a) Calcium signalling in the ciliated protozoan model, *Paramecium*: strict signal localisation by epigenetically controlled positioning of different Ca^{2+}-channels. Cell Calcium 57:203–213

Plattner H (2015b) Molecular aspects of calcium signalling at the crossroads of unikont and bikont eukaryote evolution—the ciliated protozoan *Paramecium* in focus. Cell Calcium 57:174–185

Plattner H (2015c) The contractile vacuole complex—new cues to function and biogenesis. Crit Rev Microbiol 41:218–227

Plattner H (2016) Signaling in ciliates: long- and short-range signals and molecular determinants for cellular dynamics. Biol Rev (in press). doi:10.1111/brv.12218

Plattner H, Kissmehl R (2003) Dense-core secretory vesicle docking and exocytotic membrane fusion in *Paramecium* cells. Biochim Biophys Acta (Mol Cell Res) 1641:183–193

Plattner H, Klauke N (2001) Calcium in ciliated protozoa: sources, regulation, and calcium-regulated cell functions. Int Rev Cytol 201:115–208

Plattner H, Verkhratsky A (2013) Ca^{2+} signalling early in evolution—all but primitive. J Cell Sci 126:2141–2150

Plattner H, Verkhratsky A (2015) The ancient roots of calcium signalling evolutionary tree. Cell Calcium 57:123–132

Plattner H, Knoll G, Pape R (1993) Synchronization of different steps of the secretory cycle in *Paramecium tetraurelia*: trichocyst exocytosis, exocytosis-coupled endocytosis and intracellular transport. In: Plattner H (ed) Membrane traffic in protozoa. JAI Press, Greenwich (CT), London, pp 123–148

Plattner H, Braun C, Hentschel J (1997) Facilitation of membrane fusion during exocytosis and exocytosis-coupled endocytosis and acceleration of "ghost" detachment in *Paramecium* by extracellular calcium. A quenched-flow/freeze-fracture analysis. J Membr Biol 158:197–208

Plattner H, Westphal C, Tiggemann R (1982) Cytoskeleton-secretory vesicle interactions during the docking of secretory vesicles at the cell membrane in *Paramecium tetraurelia* cells. J Cell Biol 92:368–377

Pouphile M, Lefort-Tran M, Plattner H, Rossignol M, Beisson J (1986) Genetic dissection of the morphogenesis of exocytosis sites in *Paramecium*. Biol Cell 56:151–162

Rothman JE (2014) The principle of membrane fusion in the cell (Nobel lecture). Angew Chem Int Ed 53:12676–12694

Schilde C, Wassmer T, Mansfeld J, Plattner H, Kissmehl R (2006) A multigene family encoding R-SNAREs in the ciliate *Paramecium tetraurelia*. Traffic 7:440–455

Schilde C, Schönemann B, Sehring IM, Plattner H (2010) Distinct subcellular localization of a group of synaptobrevin-like SNAREs in *Paramecium tetraurelia* and effects of silencing SNARE-specific chaperone NSF. Eukaryot Cell 9:288–305

Sehring IM, Reiner C, Mansfeld J, Plattner H, Kissmehl R (2007a) A broad spectrum of actin paralogs in *Paramecium tetraurelia* cells display differential localization and function. J Cell Sci 120:177–190

Sehring IM, Mansfeld J, Reiner C, Wagner E, Plattner H, Kissmehl R (2007b) The actin multigene family of *Paramecium tetraurelia*. BMC Genomics 8:82

Sehring IM, Klotz C, Beisson J, Plattner H (2009) Rapid downregulation of the Ca^{2+}-signal after exocytosis stimulation in *Paramecium* cells: essential role of a centrin-rich cortical filamentous network, the infraciliary lattice. Cell Calcium 45:89–97

Simon M, Plattner H (2014) Unicellular eukaryotes as models in cell and molecular biology: critical appraisal of their past and future value. Int Rev Cell Mol Biol 309:141–198

Stelly N, Mauger JP, Claret M, Adoutte A (1991) Cortical alveoli of *Paramecium*: a vast submembranous calcium storage compartment. J Cell Biol 113:103–112

Stock C, Grønlien HK, Allen RD (2002) The ionic composition of the contractile vacuole fluid of *Paramecium* mirrors ion transport across the plasma membrane. Eur J Cell Biol 81:505–515

Südhof TC (2014) The molecular machinery of neurotransmitter release (Nobel lecture). Angew Chem Int Ed 53:12696–12717

Tominaga T, Allen RD, Naitoh Y (1998) Electrophysiology of the in situ contractile vacuole complex of *Paramecium* reveals its membrane dynamics and electrogenic site during osmoregulatory activity. J Exp Biol 201:451–460

Wassmer T, Froissard M, Plattner H, Kissmehl R, Cohen J (2005) The vacuolar proton-ATPase plays a major role in several membrane-bounded organelles in *Paramecium*. J Cell Sci 118:2813–2825

Wassmer T, Kissmehl R, Cohen J, Plattner H (2006) Seventeen a-subunit isoforms of *Paramecium* V-ATPase provide high specialization in localization and function. Mol Biol Cell 17:917–930

Wassmer T, Sehring IM, Kissmehl R, Plattner H (2009) The V-ATPase in *Paramecium*: Functional specialization by multiple gene isoforms. Pflugers Arch Eur J Physiol (invited review) 457:599–607

Wloga D, Frankel J (2012) From molecules to morphology: cellular organization of *Tetrahymena thermophila*. Meth Cell Biol 109:83–140

Chapter 7
How Do Cysts Know When to Hatch? The Role of Ecological Communication in Awakening Latent Life

John R. Bracht, Emily M. Ferraro and Kathryn A. Bracht

Abstract When environmental conditions deteriorate, many ciliates enter a state of suspended animation known as a resting cyst. The cyst is a survival strategy: protected within a thick cell wall, the organism can survive harsh conditions and emerge once conditions improve. A significant literature describes the formation of cysts in the laboratory but surprisingly little is known about the signals that reawaken the dormant organisms. In this review, we discuss signals that awaken cysts in the context of their environment and ecology, and integrate these findings into a conceptual framework. We also discuss the role of cysts (and their awakening signals) in the pathogenicity of human parasitic protozoan diseases: *Giardia, Plasmodium, Toxoplasma, Cryptosporidium, Balantidium,* and *Entamoeba.*

7.1 Introduction

As microbial eukaryotes, ciliates face an uncertain future. Single-celled aquatic organisms, they are extremely dependent on capricious environmental conditions. In response to this vulnerability, many ciliates have developed a remarkable survival strategy: secreting a thick cell wall, they enter a dormant, non-motile resting stage known as a cyst (or resting cyst) and wait for conditions to improve (Gutierrez et al. 2001; Verni and Rosati 2011). The wait can be long—the literature contains examples of viable ciliates recovered from permafrost over 25,000 years old (Stoupin et al. 2012; Shatilovich et al. 2015). At a metabolic near-standstill but not dead, the resting cyst is an exemplar of a new category of living matter: 'cryptobiosis' or hidden life (Keilin 1959; Gutierrez et al. 2001). Occupying a foundational position in the food chain, ciliates and their resting cysts are both abundant and

J.R. Bracht (✉) · E.M. Ferraro
Department of Biology, American University, Washington DC 20016, USA
e-mail: jbracht@american.edu

K.A. Bracht
Department of Sociology, American University, Washington DC 20016, USA

G. Witzany and M. Nowacki (eds.), *Biocommunication of Ciliates*,
DOI 10.1007/978-3-319-32211-7_7

dynamic (Vaque et al. 1997; Muller and Wunsch 1999; Ekelund et al. 2002; Muller et al. 2002; Chao et al. 2013; Grattepanche et al. 2014).

Resting cysts are stress-resistant, able to survive starvation, desiccation, heat, and repeated cycles of freeze-thaw (Corliss and Esser 1974; Shatilovich et al. 2015). Stress is the general trigger for cyst formation (encystment): most commonly, starvation (Corliss and Esser 1974; Gutierrez et al. 2001; Verni and Rosati 2011), though other stresses have been reported to induce encystment, including low temperature (Weisse 2004), desiccation (detected as changes in salt concentration) (Yamaoka et al. 2004), or overcrowding (Barker and Taylor 1931; Beers 1947; Jeffries 1956; Corliss and Esser 1974; Ekelund et al. 2002).

7.2 Cyst Types

The ability to form cysts is a source of adaptational plasticity that has been exploited by microbial eukaryotes as they adapt to diverse niches, and there can be more than one cyst type in a lifecycle. Not only can cysts take on a remarkable variety of forms (Corliss 2001), but several parasitic species have developed at least two different cyst forms that play different functional roles. For example, *Toxoplasma*, an animal parasite, uses one type of cyst to embed within tissues, such as muscle or brain (thus, named a 'tissue cyst' or bradyzoite) and a different, thicker-walled oocyst for environmental release and invasion of new hosts (Hill and Dubey 2002). In the intestinal parasite *Cryptosporidium* there are also two cyst forms; a thick-walled infective cyst for finding a new host, and a thinner-walled cyst used for re-infection of the same host (Bouzid et al. 2013). Two forms of cyst have also been reported for some non-pathogenic ciliates—*Woodruffia metabolica* and *Bursaria truncatella,* of class Colpodea—one 'stable' and thick-walled, and one 'unstable' and thin-walled, poised to excyst easily, being triggered even by distilled water (Johnson and Evans 1940; Beers 1948; Stout 1955). The ecological significance of the two cyst forms in this case remains unclear and unfortunately the role of alternative cyst-stages has been neglected in more recent work. Adding to the complexity are the 'reproductive' cysts, which as their name implies, provide temporary life-stage in which cell division occurs and are commonly found among the Colpodean ciliates and in some parasites (Stuart et al. 1939; Corliss 2001; Matuschewski 2006; Lynn 2008; Long and Zufall 2010; Coyne et al. 2011).

Complicating matters further, there are reports of additional 'digestion' cysts in the *Woodruffia* (a Colpodean ciliate) which, as their name implies, facilitate the processing of ingested food, and are involved in neither reproduction nor responses to stressful environmental conditions (Johnson and Evans 1940; Johnson and Evans 1941). While all these types of cyst are intriguing and merit further investigation, in this manuscript we will discuss the resting cysts, which protect the ciliate from unsuitable environmental conditions. These cysts are responsive to signals from the environment for their dissolution, making them interesting cases of bio-communication. Engaged in continuous decision-making between rest or hatch fates, resting cysts are in constant bio-communication with their environment. Consistent with this constant decision-making process, in *Pseudourostyla cristata* components of known cell signaling pathways are among the most strongly cyst-expressed proteins (Gao et al. 2015).

7.3 Excystment of Resting Cysts: Balancing the Tension Between Protection and Sensing

There are many excellent reviews of the general phenomenon of resting cysts in ciliates (Corliss and Esser 1974; Corliss 2001; Gutierrez et al. 2001; Verni and Rosati 2011) and for more information the reader is referred to them, and to references therein. Our focus is on the signals responsible for excystment—or, how does the cyst know when to hatch? There is a fundamental tension between competing goals of a cyst: to protect the inhabitant from external conditions that may be lethal versus the need to detect improved conditions under which the cyst itself should be abolished and its inhabitant set free. This 'sensing' versus 'protecting' tension has very real consequences for the ciliate, since hatching under toxic conditions would lead to immediate death and removal of the organism from the gene pool. However, failure to excyst under conditions favorable to growth is just as effective as death in weeding out an otherwise viable organism. Remarkably little is known about how excystment occurs and is regulated, as noted in several reviews of the literature (Corliss and Esser 1974; Corliss 2001; Verni and Rosati 2011).

7.4 The Semi-permeable Barrier: The Cyst Wall Structure and Composition

In terms of biocommunication and environmental signaling, the cyst wall must meet competing requirements to keep out harmful agents (biological and chemical) while serving as a first line of cellular reception for signal molecules that induce hatching. This structure therefore has a level of sophistication which is not well understood at the molecular level, partly due to the difficulty inherent in studying insoluble protein aggregates. Molecular studies have been carried out only in limited cases, but so far glycoproteins have been identified in cyst walls of Colpodea (Izquierdo et al. 1999, 2000), the Oligohymenophorean *Opisthonecta henneguyi* (Calvo et al. 2003), and in some parasitic protozoan cysts (Samuelson et al. 2013). Analysis of amino acids from two *Colpoda* species showed an enrichment for glycine and hydrophobic amino acids (Izquierdo et al. 2000), consistent with the composition of keratin, which was recently identified in a mass-spectroscopic analysis of resting cysts of *Euplotes encysticus* (Chen et al. 2014). The polysaccharide chitin has been identified as a structural component in multiple cyst walls (Mulisch and Hausmann 1989; Gutierrez et al. 2001; Calvo et al. 2003; Foissner et al. 2005; Chatterjee et al. 2009; Samuelson et al. 2013).

7.4.1 Cyst Wall Layers

While cysts are found in a dazzling diversity of forms (Corliss 2001), four distinct layers can be resolved in most cyst walls (Grimes 1973; Walker et al. 1975; Gutierrez et al. 1983; Calvo et al. 1986; Calvo et al. 2003; Verni and Rosati 2011). These are, from the inside: the granular layer, endocyst, mesocyst, and ectocyst. Occasionally the ectocyst is decorated with structures like spines (Gutierrez et al. 1983; Ricci et al. 1985), grooves (Grimes 1973), hollow spheres (lepidosomes) (Foissner et al. 2005; Foissner and Pichler 2006), or even glass granules (biomineralized silica) (Foissner 2009; Foissner et al. 2009). Aside from making cysts aesthetically appealing, the role of these structures remains obscure: to enhance adhesion or allow bacteria to grow or perhaps to resist predation. The role of cysts in protecting against predation has in general been neglected in the literature, perhaps in part due to the challenge of re-creating the relevant portions of the food chain in the laboratory to monitor inter-species interactions. However, ingestion of cysts is of high medical relevance, as the human protozoan parasites *Balantidium coli, Entamoeba*, *Cryptosporidium, Toxoplasma,* and *Giardia* are infective during the cyst stage, only emerging once ingested by a suitable host (Marshall et al. 1997). We describe each of these cases in more detail in the Sect. 7.9 at the end of this chapter, summarizing what we know about the excystment signals in each case. Non-pathogenic ciliates, however, are better off *not* being ingested by other organisms; in these cases natural selection has likely honed the (sometimes elaborate) ectocyst structures (Corliss 2001) toward predation deterrence.

7.5 Cysts and Genome Rearrangement

Ciliates are famous for astounding genome rearrangements, labeled by the late David Prescott, a pioneer of the molecular study of ciliates, as "genome gymnastics" (Prescott 2000). These molecular contortions, occurring during the sexual cycle of certain species as they construct a new somatic genome, are remarkable and have been described in detail in many excellent reviews (including this volume; see chapter by Nowacki and coworkers, and Prescott 1994, 2000; Nowacki et al. 2011; Bracht et al. 2013; Betermier and Duharcourt 2014) and will not be discussed further here. The formation and dissolution of resting cysts in response to environmental conditions has not been linked to genome rearrangement phenomena with two tantalizing exceptions: (1) in *Oxytricha*, an early report described nuclear division and resorption events during excystment (Grimes 1973) but it was not reproducibly observed in later work (Adl and Berger 1997), and (2) in *Colpoda inflata* the excystment process is accompanied by extrusion of genomic DNA, perhaps material that was unable to be re-activated in the transition to vegetative growth (Chessa et al. 2001). The exact nature and sequence of the genomic material extruded, and the functional role of this phenomenon, remain unknown.

7.6 Cysts, Time, and Forecasting

Overcrowding as an encystment signal highlights an important aspect of cyst biology: forecasting. If the ciliates are crowded then clearly their *current* environment is favorable to their survival and reproduction. Yet overcrowding is a condition that suggests an impending crash in the food population, along with the buildup of waste products and toxic metabolites. Notice that here encystment is induced in response to *future*, not current, conditions. Ciliates have mastered the art of forecasting their environment's future conditions and take preemptive action if there is trouble on the horizon. These organisms utilize sophisticated models of the environment that extrapolate from current conditions and recent history into a prediction about what the future will be.

This in turn requires a mechanism for sensing and tracking time, and these built-in clocks have enabled some remarkable examples of environmental adaptation. In a famous example, the ciliate *Strombidium oculatum* occupies tidal pools that are regularly flooded at high tide. It therefore has adapted to a tidal rhythm, only swimming freely for 6 h of low tide (when pools are isolated from the ocean), and forming anchored cysts minutes before the pool floods with seawater, thereby avoiding being washed out to sea (Jonsson 1994; Montagnes et al. 2002). Excystment is coordinated with low tide, however the ciliate waits a full cycle of low-high-low tide before re-emerging. The ciliate therefore swims freely for ~6 h, is encysted for ~19 h, and the rhythm continues in laboratory culture for several days (Jonsson 1994; Montagnes et al. 2002).

In several examples, ciliates track the seasons: resting cysts of *Bursaria truncatella* that formed in the autumn were unresponsive for 11 months, but then exhibited morphological changes and thereafter excysted efficiently in the laboratory; spring-formed cysts were able to excyst within a week (Beers 1948). Late winter and early spring-formed cysts of *Strombidium concium* excysted with a delay of 1–2 months relative to cysts formed at other times, which could hatch within days (Kim and Taniguchi 1995).

Excystment of *Histriculus cavicola* required several days' maturation (Nakamura and Matsusaka 1991) and a week-long development of cyst wall structure was documented in *Colpoda cucullus* (Chessa et al. 2002). The mechanism by which an encysted protozoan keeps track of time, and the mechanism by which time impinges on cyst properties (like hatching delays) are intriguing avenues for future work.

7.7 The Excystment Signal: Bacteria or Metabolites?

Of the eleven ciliate classes, eight have been reported to form resting cysts (Fig. 7.1) (Lynn 2008). Bacteria or algae, the food organism of the ciliates, would, in theory, serve as excellent signals for excystment. However, a review of the spotty and scattered literature shows that in most cases, excystment is actually induced by the presence of metabolites, rather than actual food organisms (Figs. 7.1 and 7.2a–c). The reasons for this remain obscure, and though some cases of bacterially-induced excystment are known (and discussed below), in general they are the minority; we found no literature reports of direct algae-induced excystment, even though many ciliates happily feed on algal cells. This surprising state of affairs is quite likely significant and might be rooted in practicalities. The cyst wall is thick, comprised of insoluble networks of glycoprotein fibers that may repel water (Gutierrez et al. 2001), and certainly form an effective barrier to large particles like bacteria and algae. However, metabolites might more easily diffuse through the cyst wall and contact the cell membrane, thereby interacting with hypothetical receptor molecules.

In this essay we propose a 'future prediction' model for excystment: *excystment occurs as a prediction of future food.* This implies that ciliates are optimized for environmental forecasting and acting accordingly, similar to the processes of encystment discussed above. Ciliates read the signs of their environments, perhaps more than the immediate conditions. The predictive cues are quite interesting cases of biocommunication, in which the environment signals its future condition by chemical signals interpreted by the encysted ciliate. For example, the presence of complex metabolites is an excellent sign of bacterial presence in the future, since most habitats of ciliates are not sterile, so excystment makes sense for bacterivorous ciliates (Fig. 7.2a). In the next sections we will discuss some examples from the literature, including some perplexing examples that truly appear maladaptive.

7.7.1 Sources of Excystment Signals

A common method for culturing ciliates is the use of hay infusion, which is produced by boiling hay or straw in water, then filtering and sterilizing the resulting extract. Once exposed to the air or deliberately inoculated with a single bacterial strain, this media is generally very good at growing dense bacterial cultures that support robust ciliate growth. In general, encystment occurs once the bacteria population is exhausted by the ciliates. In several cases, it has been reported that cysts hatch upon addition of fresh, *sterile* hay infusion. Other (sterile) plant extracts such as Cerophyl (wheat grass infusion), lawn grass extracts, or waterweed (*Elodea*) extract, have been used to induce cyst hatching and probably function similarly to hay infusion. For examples, see studies on *Colpoda cucullus* (Barker and Taylor 1933; Haagen-Smit and Thimann 1938), *Colpoda duodenaria* (Prater

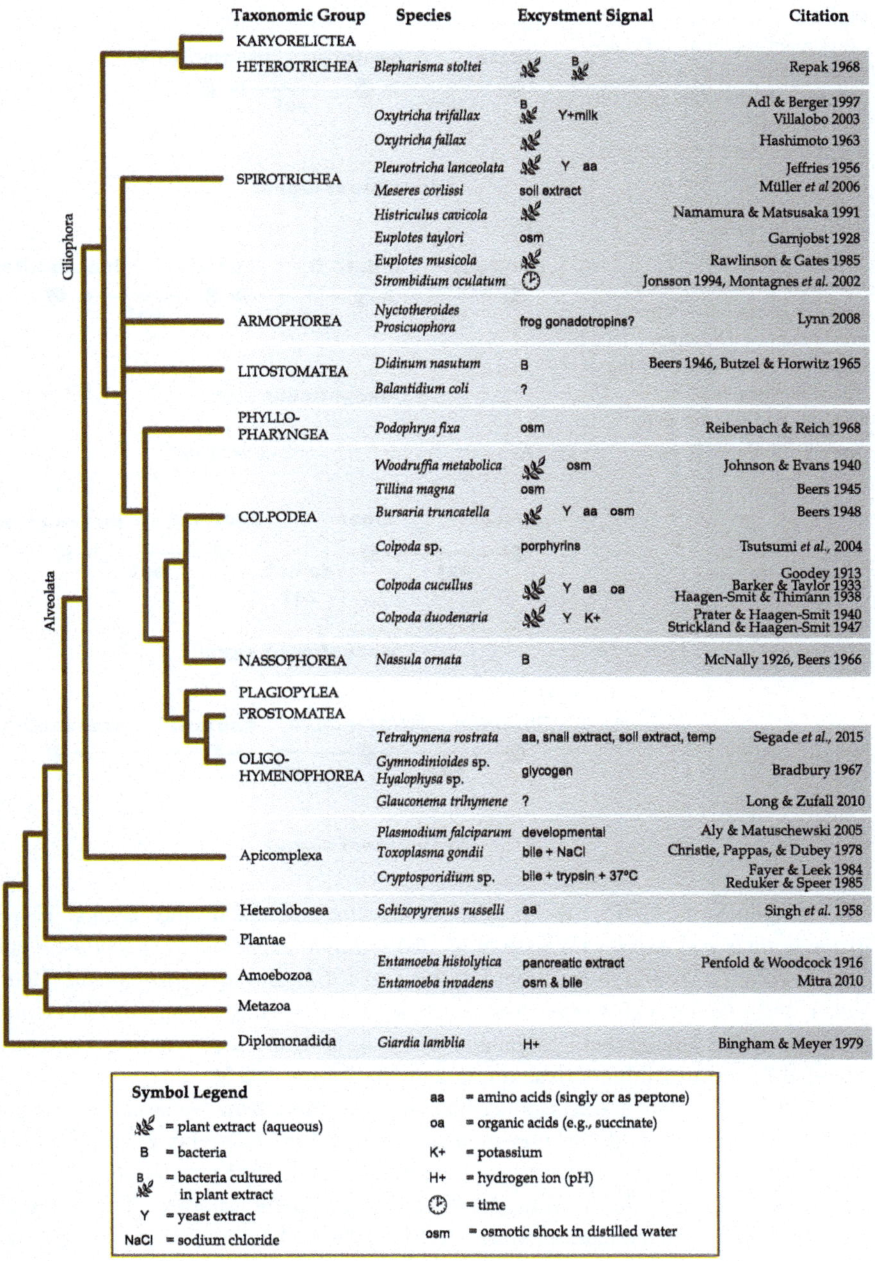

Fig. 7.1 Cladogram table of the species described in the text, with information on the excystment signal(s), when known, for each case. Tree branching after Baldauf et al. (2000), with ciliates after Lynn (2008). The superphylum Alveolata and phylum Ciliophora are indicated in vertical text at their divergence points. The eleven ciliate classes are denoted by names in all capital letters, with genus and species listed, in italics, within the grey boxed regions

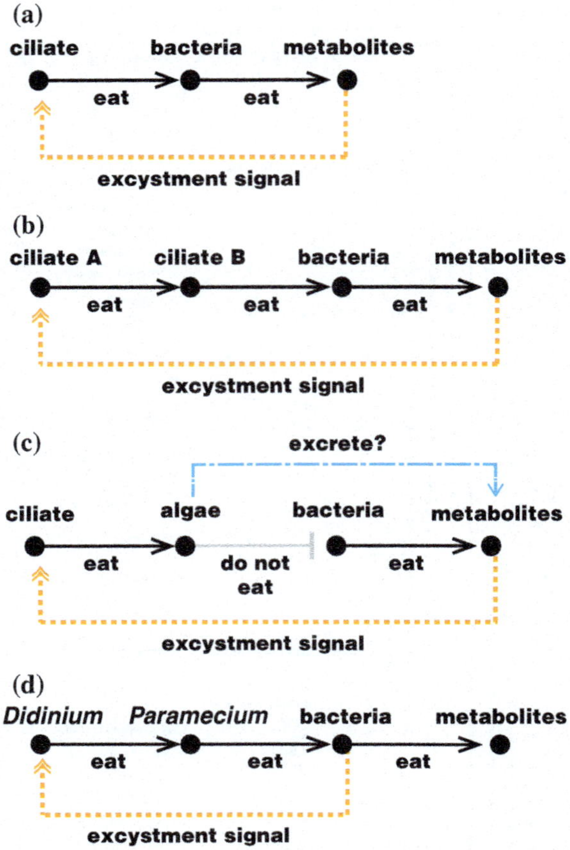

Fig. 7.2 Models for excystment signaling (*orange dotted arrow*) in several ciliates. **a** Metabolite-induced cyst hatching. **b** Triply-removed excystment signal. **c** Triply-removed broken-chain excystment signal. *Blue broken line* indicates controversial excretion of metabolites by algae. **d** Bacterially-induced excystment of *Didinium nasutum*, which feeds on the ciliate *Paramecium*, not bacteria. See text for details (Colour figure online)

and Haagen-Smit 1940), *Woodruffia metabolica* (Johnson and Evans 1940), *Bromeliothrix metopoides* (Weisse et al. 2013) *Euplotes musicola* (Rawlinson and Gates 1985), *Oxytricha fallax* Stein (Hashimoto 1963), *Bursaria truncatella* (Beers 1948), and *Blepharisma stoltei* (Repak 1968) (Fig. 7.1). Excystment of *Histriculus cavicola* was induced with vegetable baby food from the Wakodo company, intended for human infants (Nakamura and Matsusaka 1991).

In some cases, plant extracts induce hatching *even though the bacteria they support are not the food of the ciliates they induce to excyst* (Fig. 7.2b). For example, sterile lettuce leaf extract was an effective excystment inducer for the Spirotrich *Pleurotricha lanceolata,* (Fig. 7.1) but its direct food was *Tetrahymena* (another ciliate), not the bacteria that grow in the lettuce extract (Jeffries 1956). The presence of plant-derived metabolites is a triply-removed sign for food: lettuce extract is the food of bacteria, which are the food of *Tetrahymena,* which are the food of *Pleurotricha* (Fig. 7.2b). *Woodruffia metabolica* also uses a triply-removed food-signal (Fig. 7.2b), as it was cultured on the ciliate *Paramecium* but its cysts hatch when exposed to plant extracts, including hay infusion, lawn grass extracts, or water-weed extracts (Johnson and Evans 1940). An even more puzzling case was reported for *Meseres corlissi*, which lives in the soil and feeds on algae there (Muller et al. 2006). The cysts of this species hatch upon the addition of sterile 'soil extract', a supernatant obtained by saturation of commercial garden soil with distilled water, autoclaving, and filtering to ensure both solid material and bacteria are removed. The presence or absence of algal food was irrelevant to both encystment and excystment, which were regulated instead by the presence (or absence) of soil extracts (Muller et al. 2006). This is a particularly puzzling case since soil extracts are not necessary for the growth of the direct algal food nor are the bacteria that the extract might nurture used as food by the ciliate. We might describe this situation as a 'broken-chain' triply-removed excystment signal (Fig. 7.2c). The authors did not discuss whether more traditional plant extracts such as hay infusions or lettuce extracts might also contain the excystment-inducing mystery factor, but this is plausible since garden soil contains large amounts of decomposing plant detritus. There is a body of literature describing the excretion of organic compounds by algae in marine and freshwater environments (Hellebust 1965; Watt 1966), and some of the compounds released, such as amino acids, match those found to induce excystment of some species of ciliates (see 'Isolation of Single Active Compounds' below), suggesting that this may 'close the loop' of this particular ecological mystery (broken blue arrow, Fig. 7.2c). However, later authors suggested that organic excretion studies were largely artifactual and the actual release of compounds by healthy algal cells is in fact minimal (Sharp 1977). This is consistent with the fact that the algal food was unable to induce excystment of *Meseres corlissi* and thus apparently do not excrete sufficient metabolites in this case (Muller et al. 2006). In the view of Sharp (1977) and Duursma (1963), the release of organic compounds into aquatic environments is largely through dead and decaying or lysed cells, consistent with the role of 'soil extract' (containing organics derived from dead plant matter) in the excystment of *Meseres* (Muller et al. 2006). This raises the question: why would ciliates want to use *dead* plant matter as a barometer of environmental suitability, when live algae are the actual food? Might the death and lysis of algal cells point to a toxic or harmful environment, rather than a healthful one? For now, the broken blue arrow shown in Fig. 7.2c must be considered hypothetical until this important question is addressed by future work. One implication of these observations is that at times, the ciliates might excyst into non-ideal environmental conditions. We will discuss examples of excystment into harmful conditions in a later section.

7.7.2 Excystment in Animal or Yeast Extracts

Plant extracts are not the only effective inducers of excystment: peptone (an enzymatic digest of meat), or yeast extracts are also effective in many cases. Both peptone and yeast extract were demonstrated to cause excystment in *Pleurotricha lanceolata* (Jeffries 1956), *Bursaria truncatella* (Beers 1948), and *Colpoda cucullus* (Barker and Taylor 1933). Cysts of the parasitic *Tetrahymena rostrata* were found to excyst in standard growth media containing both peptone and tryptone, an enzymatic digest of casein (milk) proteins (Segade et al. 2015) (see Sect. 7.7.5.3, below). Yeast extract was successfully used for *Colpoda duodenaria* (Strickland and Haagen-Smit 1947); peptone was not tested. Villalobo et al. (2003) reported that a milk and yeast extract mixture was a good excystment inducer in *Oxytricha trifallax*, but a more thorough analysis of excystment induction in this organism has not been reported in the literature, though two reports used bacterized plant extracts to induce excystment (lettuce medium, Grimes (1973) or wheat-grass infusion, Adl and Berger (1997)).

7.7.3 Isolation of Single Active Compounds

Hay infusions and other plant extracts, yeast extracts, and peptone are all complex chemical mixtures that can induce excystment, but what is the active component? Perhaps not surprisingly, multiple different answers can be found in the literature, and in general different studies with different species of ciliate have yielded quite diverse answers. Several studies point to amino acids as excystment inducers. In *Pleurotricha lanceolata*, the single amino acid glycine was an effective inducer of excystment (Jeffries 1956). The excystment of bacterivorous non-ciliate amoeba *Schizopyrenus russelli* also is triggered by isolated amino acids (Fig. 7.1), and glycine was among the active compounds (Singh et al. 1958). Glycine accelerated the excystment in *Tillina magna,* however organic compounds are not strictly required, as distilled water is an effective inducer of excystment for this organism (Beers 1945). The convergence of several studies around the amino acid glycine is striking, yet glycine is not a universal excystment inducer: in *Colpoda cucullus* five individual amino acids were tested, including glycine, and found ineffective (Thimann and Barker 1934). Instead, Haagen-Smit and Thimann (1938) isolated the active compounds from hay infusion and concluded that several neutralized simple organic acids were the excystment-inducing substances. These included malate, citrate, fumarate, succinate, and lactic acids, among others, and their activity was strongly enhanced by sugar mixture 'co-factors' that contained no activity by themselves but could greatly increase the activity of the organic acids (Haagen-Smit and Thimann 1938) (Fig. 7.1). In the related *Colpoda duodenaria* potassium ions were the inducing signal isolated from yeast extract (Strickland and Haagen-Smit 1947) (Fig. 7.1). Another study of a non-specified *Colpoda* species reported that porphyrins (heterocyclic ring structures related to both heme and chlorophyll) were the active excystment inducers from fractionated cereal infusion (Tsutsumi 2004) (Fig. 7.1). The diversity is startling: amino acids (glycine), simple organic acids, porphyrins, and potassium ions have very little in common and suggest that ciliates have adapted to a remarkable diversity of signals, sometimes differing even between closely related species (Fig. 7.1).

7.7.4 *Amino Acids, Taste, and Excystment Signals*

The five basic human tastes are sweet, sour, bitter, salty, and umami (savory) (Yarmolinsky et al. 2009). Surprisingly, umami's savor comes predominantly from the amino acid glutamate, though other amino acids also stimulate the umami response, including glycine (Nelson et al. 2002). Given that glycine can also induce excystment of some ciliates (see 'Isolation of Single Active Compounds', above), it is tempting to speculate that these cysts 'taste' their environments, searching for glycine (or other amino acids) as an emergence signal. However conceptually appealing, this analogy cannot extend to the molecular level since the particular receptors for amino acids in the mammalian nervous system, the T1R family of G-protein coupled receptors (GPCR), have not been found outside of vertebrates (Yarmolinsky et al. 2009). However, GPCRs are an ancient class of proteins dating back to the origin of eukaryotes (de Mendoza et al. 2014; Krishnan and Schioth 2015) so it remains quite possible that some non-T1R member of the GPCR family has become adapted in ciliates to sense excystment signals.

In the umami response there is a strong potentiating effect driven by purine nucleotides including Adenosine, Guanosine, or Inosine Monophosphate (AMP/GMP/IMP). These nucleotides cannot induce umami signaling on their own but can increase the cellular response to glutamate significantly—up to 30-fold (Li et al. 2002; Nelson et al. 2002). There are echoes here of the role of sugars in potentiating—but not themselves being—the excystment signal of *Colpoda cucullus* (Haagen-Smit and Thimann 1938) (see Sects. 7.7.3 and 7.7.6). In the case of umami taste, the mechanism involves co-binding of the GPCR by both nucleotide and glutamate at different sites, with the nucleotide helping stabilize the receptor in an active conformation (Zhang et al. 2008). Interestingly, sweet taste sensing relies also on the T1R family of receptors; nucleotide potentiation has been reported there also (Zhang et al. 2008; Li 2009). The T1R family of GPCRs establishes a mechanistic paradigm for cell receptor-mediated signal transduction, potentiated by additional compounds, which may undergird at least some excystment signals in ciliates—though this model awaits experimental testing.

7.7.5 *Unusual Cases*

7.7.5.1 Crab Symbionts

Apostome ciliates are symbionts of crustaceans, associating as resting cysts with their hosts' exoskeletons, and excysting upon molting or death of their hosts (Chatton and Lwoff 1935). The hermit crab symbionts *Gymnodinioides* and *Hyalophysa* feed upon the exuvial fluids left in the crab's discarded exoskeleton post-molt (Bradbury and Trager 1967). The excystment trigger is glycogen, which builds up in the crab's blood just prior to the molt—an excellent example of future forecasting by the encysted organism (Bradbury and Trager 1967) (Fig. 7.1).

7.7.5.2 Frog Parasites

Prosicuophora and *Nyctotheroides* are the symbionts of frogs' intestines, and their transmission appears to be via ingestion of cysts by the tadpole (Lynn 2008). Intriguingly, the hatching of the cysts within the intestine might be regulated by the developmental hormones of the amphibian host, producing active ciliates in tadpoles but not adults, or vice versa (Lynn 2008) (Fig. 7.1).

7.7.5.3 Snail Parasites

The oligohymenophorean ciliates *Tetrahymena thermophila,* along with *Paramecium tetaurelia* are the two most well-studied groups of ciliates, but neither can form cysts (Lynn 2008). However, of 41 described *Tetrahymena* species, six are listed as producing resting cysts (Lynn and Doerder 2012); these six examples are generally under-studied; for example in one case cysts were reported but the workers were unable to induce hatching (Nanney and McCoy 1976). However, a recent study of *Tetrahymena rostrata*, a facultative parasite of snails, demonstrated that excystment occurs in the presence of snail (host) kidney extract or mucus extract, soil extract, *T. rostrata* (self) extract, the growth medium (ATCC 357, rich in amino acids from peptone and tryptone), and (with lower efficiency) mineral water (Segade et al. 2015). Interestingly, merely shifting temperature from 30 °C to 18 °C was also an effective excystment inducer. The ability of very rich media to induce excystment supports the notion that some organic compound is the signal, but no single compound was identified.

7.7.6 *Inactive or Negative Effects*

Bacterial growth was generally neutral in excystment studies, as described in previous sections. In striking contrast, Jeffries (1956) demonstrated a *negative* effect of bacteria on excystment of *Pleurotricha*, though the nature of the inhibitory effect was not explored further.

Sugars have never induced excystment in ciliates. Species tested include *Pleurotricha* (Jeffries 1956), *Colpoda cucullus* (Barker and Taylor 1933) *Colpoda duodenaria* (Strickland and Haagen-Smit 1947), and *Didinium nasutum* (Beers 1946). However, Haagen-Smit and Thimann (1938) working with cysts of *Colpoda cucullus*, reported that sugars (combined sucrose, dextrose, lactose, and maltose) could *potentiate* the excystment activity of other active compounds—neutralized organic acids—even though the sugars have no excystment-inducing activity themselves.

The concentration of hydrogen ions in the media (pH) is of special concern in ciliate studies, especially when bacteria are co-culturants. However, the pH of the sterile media, over a wide range of values, was found to have no appreciable detrimental effect on excystment in *Pleurotricha, Colpoda,* and *Woodruffia* (Goodey 1913; Barker and Taylor 1933; Johnson and Evans 1940; Jeffries 1956). *Didinium nasutum,* which is one of two cases that require bacteria for excystment (see Sect. 7.7.7), could not be induced to excyst by altering the pH of spring water, consistent with the general inactivity of sterile media towards this organism (Beers 1946). Consistent with the above, the rate of cyst formation was unaffected by pH in *Colpoda* (Barker and Taylor 1931).

7.7.7 A Case of Bona Fide Bacterial Induction of Excystment: An Exception to the Rule?

Contrasting with many of the studies described above, *Didinium nasutum* and *Nassula ornata* excyst only in the presence of live bacteria and not in sterile media (Beers 1946, 1966; Butzel and Horwitz 1965). In the case of *Didinium,* this is perplexing because bacteria are not its food; it feeds voraciously on other ciliates and is a notorious *Paramecia*vore. Oddly enough *Paramecium* washed of bacteria did not induce excystment in *Didinium*, so the direct food organism is still not the 'excyst' signal (Beers 1946) (Fig. 7.2d). *Nassula ornata* excystment requires bacterial growth, but the inducing species were motile rods, not the filamentous cyanobacteria *Oscillatoria* that is consumed by the ciliate (Beers 1966). This suggests a curious laboratory-versus-nature mismatch, but an early study (McNally 1926) showed that fresh water plus *Oscillatoria* was sufficient to induce excystment of *Nassula ornata*, a result not reproducible in Beers (1966). If McNally's results are correct (and the *Oscillatoria* culture was not contaminated with other bacterial species), this represents the only case, to our knowledge, in which the ciliate's direct food organism was also the excystment signal.

7.8 Excystment Under Harmful Conditions

Based on the discussion above, we conclude that in every case in which the excystment signal is known, ciliates are sensitive to an *earlier, non-food* component of the food chain for excystment. For example, if the food are *Paramecia*, the excyst signal is bacteria (the *Paramecia* food) (Fig. 7.2d) while if the ciliate's food is bacteria, the excyst signal consists of inorganic compounds that can nourish bacterial growth (Fig. 7.2a). It seems a risky evolutionary strategy to excyst in the presence of the food of one's food, since there are more chances for noise to interfere with the signal. Indeed, this appears to be the case. In a recent examination of the bromeliad-dwelling *Bromeliothrix metopoides*, cysts were induced to hatch under conditions of poor food (in the presence of the wrong algae) or even no food (media only) (Weisse et al. 2013). Even more extreme examples can be found: *Didinium* can excyst under conditions that immediately cause death from bacterial overload (Beers 1946) and *Tillina magna* and *Bursaria truncatella* can emerge into unbuffered distilled water, where they die of osmotic shock (Beers 1945, 1948; Bridgman 1948). We identified five cases of 'osmotic-shock' excystment, induced by exposure of cysts to distilled water, from classes Spirotrichida, Colpodea, and Phyllopharyngia, in the species *Euplotes taylori, Podophrya fixa, Woodruffia metabolica, Tillina magna,* and *Bursaria truncatella* (Garnjobst 1928; Johnson and Evans 1940; Beers 1945, 1948; Bridgman 1948; Reibenbach and Reich 1968) (Fig. 7.1).

Given the relatively tenuous connection between the excystment signal and direct ciliate food (Fig. 7.2a–d), these sorts of misfire events in excystment are perhaps inevitable. However, they raise questions about why ciliates are not 'tuned' to a better signal. Why has natural selection not weeded out these cases of apparent maladaptation? This 'bio-miscommunication of ciliates' suggests that we are missing fundamental knowledge of the microenvironment in which ciliates enter and exit cysts. As an example of this, consider the excystment of ciliates in distilled water. Dehydration is an important step in cyst formation, producing a state of suspended animation (Corliss 2001; Gutierrez et al. 2001; Verni and Rosati 2011). The (highly artificial) osmotic shock of distilled water should simply create an inrush of water into the cyst, thereby directly reversing the cryptobiotic state and bypassing the normal signaling mechanisms. Species of cysts not induced to hatch by distilled water may simply have more impregnable cell walls or membranes; at any rate the usefulness of these data are somewhat unclear at the moment. Much more study is needed to clarify these important questions, though we emphasize that integrating the study of ciliates with plausible 'natural' environments (even better: integrated, co-cultured food chains) is likely to be more fruitful than standard laboratory approaches using nutrient-rich media for cultivating bacteria, or experiments using distilled water to induce excystment.

7.9 Medical Relevance

Only one ciliate species is known to infect humans: the intestinal parasite *Balantidium coli*, whose presence can be asymptomatic in some but cause severe intestinal distress in others, including abdominal pain, bloating and cramping, diarrhea, and bloody stool (Schuster and Ramirez-Avila 2008). Only rarely is the disease fatal, unless combined with other infectious agents or immunocompromise (Schuster and Ramirez-Avila 2008). Within its subclass Trichostomatia (class: Litostomatea), this is the only cyst-forming ciliate (Lynn 2008), though many of *Balantidium*'s Trichostome relatives are also endosymbionts of vertebrates, feeding on plant and animal material in the hosts' digestive tracts (Lynn 2008). The ability of *Balantidium* to encyst is not incidental, since the cyst is the infective stage of the life cycle (Schuster and Ramirez-Avila 2008). Infection occurs by ingestion of feces-contaminated drinking water in areas with poor sanitation, and seems to transmit from pigs to humans (Schuster and Ramirez-Avila 2008). It is possible to isolate the parasite in the laboratory from fecal material (Schuster and Ramirez-Avila 2008; Barbosa Ada et al. 2015), but it remains an under-studied organism, and no definitive report of excystment signals has been published.

7.9.1 *Giardia Lamblia*

The protozoan parasite *Giardia lamblia* is a member of one of the earliest diverging lineages of eukaryotes, and appears devoid of many eukaryotic accoutrements, having simplified protein machinery and many bacteria- or archaea-like enzymes consistent with its basal position on the eukaryotic tree (Sogin et al. 1989). Giardiasis, or infection with *Giardia,* is one of the most prevalent parasitic infections in the United States, with an estimated 400,000 to 2 million people infected (Hlavsa et al. 2005). *Giardia* infects the lumen of the small intestine and, while sometimes asymptomatic, it can cause severe gastrointestinal symptoms such as diarrhea, bloating, and weight loss (Marshall et al. 1997). The life cycle of *Giardia* consists of two stages: an actively dividing and motile trophozoite and a protective cyst (Marshall et al. 1997). Ciliate and *Giardia* cysts are structurally similar: both have polysaccharide and protein-rich cyst walls (Manning et al. 1992; Izquierdo et al. 2000; Gutierrez et al. 2001; Gerwig et al. 2002; Sun et al. 2003), though *Giardia* has two cyst wall layers in contrast to most ciliates' four (Erlandsen et al. 1996; Verni and Rosati 2011). The disease cycle begins by an infected animal contaminating the food and water of other animals (or humans) with its fecal matter loaded with *Giardia* cysts. The durable cyst serves as the primary mode of transmission of the disease, allowing it to survive environmental conditions during transmission, similar to the function of the cyst in cryptobiotic ciliates. *Giardia* is an anaerobic organism (bearing vistigial mitochondria (Regoes et al. 2005)) and the cyst wall protects the organism both from oxygen and (in the new host) the extremely acidic conditions of the stomach. Excystment of *Giardia* must be tightly controlled due to its very specific environmental needs. Bingham and Meyer (1979) reported that *Giardia* excystation was induced by acid in vitro. Interestingly, a series of pH changes were most effective, consistent with passage through the stomach (pH of 2) followed by a more alkaline environment in the small intestine (pH 8) (Rice and Schaefer 1981). The mechanism by which the cysts records its 'acid history'—the stomach-like drop in pH and the intestine-like rise in pH—is unknown.

7.9.2 *Entamoeba*

This intestinal parasite causes amoebic dysentery and is transmitted as a cyst, ingested by the host from fecal-contaminated water (Marshall et al. 1997). In the reptile parasite *Entamoeba invadens*, excystment was efficiently induced by first soaking in distilled water for 6 h, then incubation in a combination of sodium bicarbonate plus bile (Mitra et al. 2010). In early work on the human parasite *E. histolytica,* pancreatic extract could induce excystment (Penfold and Woodcock 1916). The cyst wall of *E. invadens* has only one structural layer detected by electron microscopy (Chavez-Munguia et al. 2007), but chitin is a major component (Arroyobegovich et al. 1980) consistent with many ciliate cyst walls as described previously.

7.9.3 Plasmodium

Malaria is an infectious disease that kills up to one million people each year (Cai et al. 2012). Malaria is caused by a protozoan parasite belonging to the genus *Plasmodium* and clustering with ciliates in the superphylum Alveolata (Fig. 7.1): species infecting humans include *P. falciparum, P. vivax, P. malariae, P. ovale* and *P. knowlesi* (Calderaro et al. 2013). Malaria is transmitted to humans by *Anopheles* mosquitoes rather than moving directly from human to human (Sinden 2002). *Plasmodium* parasites undergo a complex life cycle in both mosquito and human hosts. Most relevant to our discussion, the mosquito portion of the *Plasmodium* life cycle includes a cyst stage which is crucial for producing infective parasites (sporozoites) for transmission to a human host during the mosquito's next blood-meal (Matuschewski 2006). The cyst (or oocyst) comprises the longest single stage of the *Plasmodium* life cycle, lasting for 14 days, and is a key link between the mosquito gut-stage and mosquito salivary-gland stage (Matuschewski 2006). The oocyst provides a safe harbor for development of hundreds of sporozoites within a single oocyst capsule: a bilayered barrier comprised of the basal lamina of the mosquito gut plus a layer of circumsporozoite protein (CSP) (Posthuma et al. 1988). Release of the newly developed sporozoites appears to occur by developmentally-regulated expression of of cysteine proteases by the sporozoites, leading to degradation of the CSP capsule and parasite escape into the mosquito hemolymph (Aly and Matuschewski 2005; Matuschewski 2006). Therefore, cyst hatching in this case is under developmental control, and is not linked to sensing environmental conditions as in other cyst examples described in this chapter. The oocyst remains, however, an under-studied potential target for antimalarial therapies (Sinden 2002; Matuschewski 2006).

7.9.4 Toxoplasma

An apicomplexan relative of *Plasmodium*, *Toxoplasma gondii* is a widespread parasite of humans and other animals, and may infect up to 40 % of people in the US, though rates of infection can be much higher in Central and South America or Europe (Hill and Dubey 2002). While symptoms appear to be negligible in healthy individuals, severe disease sometimes occurs in immunocompromised patients, especially those with HIV/AIDS, or children who acquire the infection from their mothers during pregnancy (Hill and Dubey 2002). *T. gondii* can infect any organ, is frequently noted in muscle or brain tissue, and can generate a wide variety of symptoms including mental retardation, vison problems, encephalitis, hydrocephalus (Hill and Dubey 2002), or even in rare cases, schizophrenia (Torrey and Yolken 2003).

Cats comprise the definitive host for *T. gondii*; an infected cat can release tens of millions of infectious cysts (oocysts) per day in its feces (Dubey 2001), placing a significant parasite burden on the envornment (Lilly and Wortham 2013). The two primary pathways of human infection are the ingestion of these oocysts or alternatively, consumption of tainted, undercooked meat containing 'tissue cysts' (bradyzoites) which are a second cyst type, illuminating the complexity of the *T. gondii* lifecycle (Hill and Dubey 2002). The infectious oocyst is highly resistant to evironmental insult, including bleach, 2 % sulfuric acid, or UV light, and is able to survive months or years in the environment (Hill and Dubey 2002; Belli et al. 2006). This protection is presumably due to the structure of its cyst wall, which has two layers (Belli et al. 2006; Dumetre et al. 2013) and is lipid, polysaccharide, and protein-rich (Samuelson et al. 2013). One study found that bile salts were effective at inducing excystment (Christie et al. 1978) (Fig. 7.1), similarly to *Cryptosporidium* (Sect. 7.9.5) and *Entamoeba invadens* (Sect. 7.9.2); however the mechanisms and signals surrounding excystment remain under-studied and specific chemical inducer(s) have not been identified.

7.9.5 *Cryptosporidium*

Cryptosporidium is a family of a cyst-forming protozan parasites that cause gastrointestinal symptoms in a range of vertebrates hosts including humans, and is evolutionarily related closely to *Plasmodium* and more distantly to ciliates within the superphylum Alveolata (Fig. 7.1). However, *Cryptosporidium* does not require an insect vector for transmission like *Plasmodium*. Instead, *Cryptosporidium* is transmitted from human to human by the fecal-oral route through contaminated water, and the infectious agents are cysts (oocysts) (Bouzid et al. 2013). After ingestion, oocysts release vegetative sporozites in the small intestine, causing gastrointestinal symptoms: most commonly, watery diarrhea and abdominal pain, though symptoms can be more severe especially in immunocompromised individuals (Bouzid et al. 2013). Several excystment signals have been identified and they are consistent with exposure to a mammalian gastrointestinal tract: elevated temperature of 37 °C (Fayer and Leek 1984) and trypsin plus bile (Fayer and Leek 1984; Reduker and Speer 1985) (Fig. 7.1). As for *Toxoplasma* and *Entamoeba*, specific chemical inducer(s) of excystment in *Cryptosporidium* have not been identified.

7.10 Conclusions

Perhaps surprisingly, a key insight into the vagaries of ciliate life comes from the ancient philosopher Aristotle. A keen observer of nature, Aristotle's discussion of taste and touch in his *De Anima* offers a helpful way to conceptualize ciliate excystment, forecasting, and cryptobiosis. In book III, chapter 13, he highlights the immediacy, and thus centrality, of touch for animals: while other senses operate through a medium and therefore at a distance, touch involves direct contact between the organism and its environment (Aristotle et al. 1993). In this passage Aristotle also notes that other senses may be destroyed through an excess of a certain stimulus (i.e. an individual may be deafened by too loud a noise, or blinded by too bright a light), but an extreme in the sense of touch may destroy not only the sense itself, but the entire organism (Aristotle et al. 1993). Uniquely vulnerable to their environments, and unable to avoid touching their surroundings, ciliates have developed the capacity to enter a dormant cyst phase in which a thick cyst wall limits their sense of touch, preserving the life of the organism as conditions worsen. Balancing protection with sensing, the encysted organism leaves itself the possibility of reawakening through the action of specific stimuli which indicate a more favorable environment.

The presence of the cyst wall as a medium through which only a few elements are allowed to pass places sensing in an encysted ciliate closer to Aristotle's notion of taste. Classed as a kind of touch, taste allows the organism to detect whether what it has interiorized is nourishing or harmful, whether it will contribute to its flourishing or be detrimental [*De Anima* III.12, (Aristotle et al. 1993)]. This mediated exposure isn't fool-proof, but it does allow the ciliate to make an educated guess—which is perhaps the best we all can do.

References

Adl SM, Berger JD (1997) Timing of life cycle morphogenesis in synchronous samples of Sterkiella histriomuscorum.1. The vegetative cell cycle. Eur J Protistology 33:99–109

Aly ASI, Matuschewski K (2005) A malarial cysteine protease is necessary for Plasmodium sporozoite egress from oocysts. J Exp Med 202:225–230

Aristotle, Hamlyn DW and Shields CJ (1993) De anima: books II and III with passages from book 1. Oxford University Press, Oxford

Arroyobegovich A, Carabeztrejo A, Ruizherrera J (1980) Identification of the structural component in the cyst wall of entamoeba-invadens. J Parasitol 66:735–741

Baldauf SL, Roger AJ, Wenk-Siefert I, Doolittle WF (2000) A kingdom-level phylogeny of eukaryotes based on combined protein data. Science 290:972–977

Barbosa Ada S, Bastos OM, Uchoa CM, Pissinatti A, Ferreira Filho PR, Dib LV, Azevedo EP, de Siqueira MP, Cardozo ML, Amendoeira MR (2015) Isolation and maintenance of Balantidium coli (Malmsteim, 1857) cultured from fecal samples of pigs and non-human primates. Vet Parasitol 210:240–245

Barker HA, Taylor CV (1931) A study of the conditions of encystment of Colpoda cucullus. Physiol Zool 4:620–634

Barker HA, Taylor CV (1933) Studies on the excystment of Colpoda cucullus. Physiol Zool 6:127–136

Beers CD (1945) Some factors affecting excystment in the Ciliate Tillina magna. Physiol Zool 18:80–99

Beers CD (1946) Excystment in Didinium nasutum, with special reference to the role of bacteria. J Exp Zool 103:201–231

Beers CD (1947) The relation of density of population to encystment in Didinium nasutum. J Elisha Mitchell Sci Soc Chapel Hill N C 63:141–154

Beers CD (1948) Excystment in the Ciliate Bursaria truncatella. Biol Bull 94:86–98

Beers CD (1966) The excystment process in the Ciliate Nassula Ornata Ehrbg. J Protozool 13:79–83

Belli SI, Smith NC, Ferguson DJP (2006) The coccidian oocyst: a tough nut to crack! Trends Parasitol 22:416–423

Betermier M, Duharcourt S (2014) Programmed rearrangement in ciliates: paramecium. Microbiol Spectr 2(6)

Bingham AK, Meyer EA (1979) Giardia excystation can be induced in vitro in acidic solutions. Nature 277:301–302

Bouzid M, Hunter PR, Chalmers RM, Tyler KM (2013) Cryptosporidium pathogenicity and virulence. Clin Microbiol Rev 26:115–134

Bracht JR, Fang W, Goldman AD, Dolzhenko E, Stein EM, Landweber LF (2013) Genomes on the edge: programmed genome instability in ciliates. Cell 152:406–416

Bradbury PC, Trager W (1967) Excystation of apostome ciliates in relation to molting of their crustacean hosts. II Effect of glycogen. Biol Bull 133:310–316

Bridgman AJ (1948) Studies on some aspects of cystment in the ciliate Tillina magna. J Exp Zool 108:21–44

Butzel HM Jr, Horwitz H (1965) Excystment of Didinium nasutum. J Protozool 12:413–416

Cai H, Zhou Z, Gu J, Wang Y (2012) Comparative genomics and systems biology of Malaria Parasites. Curr Bioinform 7(4)

Calderaro A, Piccolo G, Gorrini C, Rossi S, Montecchini S, Dell'Anna ML, De Conto F, Medici MC, Chezzi C, Arcangeletti MC (2013) Accurate identification of the six human Plasmodium spp. causing imported malaria, including plasmodium ovale wallikeri and plasmodium knowlesi. Malaria J 12:321

Calvo P, Fernandez-Aliseda MC, Garrido J, Torres A (2003) Ultrastructure, encystment and cyst wall composition of the resting cyst of the peritrich ciliate opisthonecta henneguyi. J Eukaryot Microbiol 50:49–56

Calvo P, Torres A, Perezsilva J (1986) Ultrastructural and cytochemical study of the encystment in the hypotrichous ciliate histriculus-similis. Arch Protistenkd 132:201–211

Chao CF, Tsai AY, Ishikawa A, Chiang KP (2013) Seasonal dynamics of ciliate cysts and the impact of short-term change of salinity in a eutrophic coastal marine ecosystem. Terr Atmos Ocean Sci 24:1051–1061

Chatterjee A, Ghosh SK, Jang K, Bullitt E, Moore L, Robbins PW, Samuelson J (2009) Evidence for a "wattle and daub" model of the cyst wall of entamoeba. PLoS pathog 5(7):e1000498

Chatton EPLeo, Lwoff A (1935) Les ciliés apostomes; morphologie, cytologie, éthologie, évolution, systematique. Premiére partie. Aperçu historique et général, étude monographique des genres et des espèces. [Paris]: [H. Le Soudier]

Chavez-Munguia B, Omana-Molina M, Gonzalez-Lazaro M, Gonzalez-Robles A, Cedillo-Rivera R, Bonilla P, Martinez-Palomo A (2007) Ultrastructure of cyst differentiation in parasitic protozoa. Parasitol Res 100:1169–1175

Chen JW, Gao XX, Wang BZ, Chen FF, Wu N, Zhang YY (2014) Proteomic approach to reveal the proteins associated with encystment of the ciliate Euplotes encysticus. PloS one 9(5):e97362

Chessa MG, Gallus L, Tiano L, Trielli F, Corrado MUD (2001) Variations in macronuclear chromatin structure and chromatin extrusion in excystment from resting cysts of Colpoda inflata. Eur J Protistol 37:281–290

Chessa MG, Largana I, Trielli F, Rosati G, Politi H, Angelini C, Corrado MUD (2002) Changes in the ultrastructure and glycoproteins of the cyst wall of Colpoda cucullus during resting encystment. Eur J Protistol 38:373–381

Christie E, Pappas PW, Dubey JP (1978) Ultrastructure of excystment of Toxoplasma gondii oocysts. J Protozool 25:438–443

Corliss JO (2001) Protozoan cysts and spores. eLS, John Wiley & Sons, Ltd

Corliss JO, Esser SC (1974) Comments on the role of the cyst in the life cycle and survival of free-living protozoa. Trans Am Microsc Soc 93:578–593

Coyne RS, Hannick L, Shanmugam D, Hostetler JB, Brami D, Joardar VS, Johnson J, Radune D, Singh I, Badger JH, Kumar U, Saier M, Wang YF, Cai H, Gu JY, Mather MW, Vaidya AB, Wilkes DE, Rajagopalan V, Asai DJ, Pearson CG, Findly RC, Dickerson HW, Wu M, Martens C, Van de Peer Y, Roos DS, Cassidy-Hanley DM and Clark TG (2011) Comparative genomics of the pathogenic ciliate Ichthyophthirius multifiliis, its free-living relatives and a host species provide insights into adoption of a parasitic lifestyle and prospects for disease control. Genome Biol 12

de Mendoza A, Sebe-Pedros A, Ruiz-Trillo I (2014) The evolution of the GPCR signaling system in eukaryotes: modularity, conservation, and the transition to metazoan multicellularity. Genome Biol Evol 6:606–619

Dubey JP (2001) Oocyst shedding by cats fed isolated bradyzoites and comparison of infectivity of bradyzoites of the VEG strain Toxoplasma gondii to cats and mice. J Parasitol 87:215–219

Dumetre A, Dubey JP, Ferguson DJP, Bongrand P, Azas N, Puech PH (2013) Mechanics of the Toxoplasma gondii oocyst wall. P Natl Acad Sci USA 110:11535–11540

Duursma EK (1963) The production of dissolved organic matter in the sea, as related to the primary gross production of organic matter. Neth J Sea Res 2:85–94

Ekelund F, Frederiksen HB, Ronn R (2002) Population dynamics of active and total ciliate populations in arable soil amended with wheat. Appl Environ Microb 68:1096–1101

Erlandsen SL, Macechko PT, Keulen HV, Jarroll EL (1996) Formation of the giardia cyst wall: studies on extracellular assembly using immunogold labeling and high resolution field emission SEM. J Eukaryot Microbiol 43:416–430

Fayer R, Leek RG (1984) The effects of reducing conditions, medium, pH, temperature, and time on in vitro excystation of Cryptosporidium. J Protozool 31:567–569

Foissner W (2009) The stunning, glass-covered resting cyst of Maryna umbrellata (Ciliophora, Colpodea). Acta Protozool 48:223–243

Foissner W, Muller H, Weisse T (2005) The unusual, lepidosome-coated resting cyst of Meseres corlissi (Ciliophora: Oligotrichea): Light and scanning electron microscopy, cytochemistry. Acta Protozool 44:201–215

Foissner W, Pichler M (2006) The unusual, lepidosome-coated resting cyst of Meseres corlissi (Ciliophora: Oligotrichea): Genesis of four complex types of wall precursors and assemblage of the cyst wall. Acta Protozool 45:339–366

Foissner W, Weissenbacher B, Krautgartner WD, Lutz-Meindl U (2009) A cover of glass: first report of biomineralized silicon in a ciliate, Maryna umbrellata (Ciliophora: Colpodea). J Eukaryot Microbiol 56:519–530

Gao XX, Chen FF, Niu T, Qu RD and Chen JW (2015) Large-scale identification of encystment-related proteins and genes in Pseudourostyla cristata. Sci Rep 5:11360

Garnjobst L (1928) Induced Encystment and Excystment in Euplotes taylori, sp. nov. Physiol Zool 1:561–575

Gerwig GJ, van Kuik JA, Leeflang BR, Kamerling JP, Vliegenthart JF, Karr CD, Jarroll EL (2002) The Giardia intestinalis filamentous cyst wall contains a novel beta(1-3)-N-acetyl-D-galactosamine polymer: a structural and conformational study. Glycobiology 12:499–505

Goodey T (1913) The excystation of Colpoda cucullus from its resting cysts, and the nature and properties of the cyst membranes. Proc R Soc Lond B: Biol Sci 86:427–439

Grattepanche JD, Santoferrara LF, Andrade J, Oliverio AM, McManus GB and Katz LA (2014) Distribution and diversity of oligotrich and choreotrich ciliates assessed by morphology and DGGE in temperate coastal waters. Aquat Microb Ecol 71:211-U234

Grimes GW (1973) Differentiation during encystment and excystment in Oxytricha fallax. J Protozool 20:92–104

Gutierrez JC, Callejas S, Borniquel S, Benitez L, Martin-Gonzalez A (2001) Ciliate cryptobiosis: a microbial strategy against environmental starvation. Int Microbiology 4:151–157

Gutierrez JC, Torres A, Perezsilva J (1983) Fine-structure of the cyst wall of Laurentiella-Acuminata (Hypotrichida, Oxytrichidae). T Am Microsc Soc 102:55–59

Haagen-Smit AJ, Thimann KV (1938) The excystment of Colpoda cucullus. I. The chemical nature of the excysting factors in hay infusion. J Cell Comp Physiol 11:389–407

Hashimoto K (1963) Formation of ciliature in Excystment and Induced re-encystment of Oxytricha fallax stein. J Protozool 10:156–166

Hellebust JA (1965) Excretion of some organic compounds by marine phytoplankton. Limnol Oceanogr 10:192–206

Hill D, Dubey JP (2002) Toxoplasma gondii: transmission, diagnosis and prevention. Clin Microbiol Infec 8:634–640

Hlavsa MC, Watson JC, Beach MJ (2005) Giardiasis surveillance–United States, 1998–2002. MMWR Surveill Summ 54:9–16

Izquierdo A, Martin-Gonzalez A, Diaz S, Gutierrez JC (1999) Resting cyst wall proteins and glycoproteins of four Colpodid ciliates: a comparative analysis. Microbios 99:27–43

Izquierdo A, Martin-Gonzalez A, Gutierrez JC (2000) Resting cyst wall glycoproteins isolated from two Colpodid cilliates are glycine-rich proteins. Cell Biol Int 24:115–119

Jeffries WB (1956) Studies on excystment in the Hypotrichous ciliate Pleurotricha lanceolata. J Protozool 3:136–144

Johnson WH, Evans FR (1940) Environmental factors affecting cystment in Woodruffia metabolica. Physiol Zool 13:102–121

Johnson WH, Evans FR (1941) A further study of environmental factors affecting cystment in Woodruffia metabolica. Physiol Zool 14:227–237

Jonsson PR (1994) Tidal rhythm of cyst formation in the rock pool ciliate Strombidium-Oculatum Gruber (Ciliophora, Oligotrichida)—a description of the functional biology and an analysis of the tidal synchronization of encystment. J Exp Mar Biol Ecol 175:77–103

Keilin D (1959) The problem of anabiosis or latent life: history and current concept. Proc R Soc Lond. Series B, Biol Sci 150:149–191

Kim YO, Taniguchi A (1995) Excystment of the Oligotrich ciliate Strombidium-Conicum. Aquat Microb Ecol 9:149–156

Krishnan A, Schioth HB (2015) The role of G protein-coupled receptors in the early evolution of neurotransmission and the nervous system. J Exp Biol 218:562–571

Li XD (2009) T1R receptors mediate mammalian sweet and umami taste. Am J Clin Nutr 90:733S–737S

Li XD, Staszewski L, Xu H, Durick K, Zoller M, Adler E (2002) Human receptors for sweet and umami taste. P Natl Acad Sci USA 99:4692–4696

Lilly EL and Wortham CD (2013) High prevalence of Toxoplasma gondii oocyst shedding in stray and pet cats (Felis catus) in Virginia, United States. Parasit Vectors 6:266

Long H, Zufall RA (2010) Diverse modes of reproduction in the marine free-living ciliate Glauconema trihymene. BMC Microbiol 10:108

Lynn DH (2008) The ciliated protozoa: characterization, classification, and guide to the literature. Springer, New York

Lynn DH, Doerder FP (2012) The life and times of Tetrahymena. Methods Cell Biol 109:9–27

Manning P, Erlandsen SL, Jarroll EL (1992) Carbohydrate and amino acid analyses of Giardia muris cysts. J Protozool 39:290–296

Marshall MM, Naumovitz D, Ortega Y, Sterling CR (1997) Waterborne protozoan pathogens. Clin Microbiol Rev 10:67–85

Matuschewski K (2006) Getting infectious: formation and maturation of Plasmodium sporozoites in the Anopheles vector. Cell Microbiol 8:1547–1556

McNally E (1926) Life cycle of Nassula ornata and Nassula elegans: are these species valid? Biol Bull 51:237–244

Mitra BN, Pradel G, Frevert U, Eichinger D (2010) Compounds of the upper gastrointestinal tract induce rapid and efficient excystation of Entamoeba invadens. Int J Parasitol 40:751–760

Montagnes DJS, Wilson D, Brooks SJ, Lowe C, Campey M (2002) Cyclical behaviour of the tide-pool ciliate Strombidium oculatum. Aquat Microb Ecol 28:55–68

Mulisch M, Hausmann K (1989) Localization of chitin on ultrathin sections of cysts of 2 ciliated protozoa, Blepharisma-Undulans and Pseudomicrothorax-Dubius, using colloidal gold conjugated wheat-germ-agglutinin. Protoplasma 152:77–86

Muller H, Foissner W, Weisse T (2006) Role of soil in the life cycle of Meseres corlissi (Ciliophora: Oligotrichea): experiments with two clonal strains from the type locality, an astatic meadow pond. Aquat Microb Ecol 42:199–208

Muller H, Stadler P, Weisse T (2002) Seasonal dynamics of cyst formation of strombidiid ciliates in alpine Lake Mondsee, Austria. Aquat Microb Ecol 29:181–188

Muller H, Wunsch C (1999) Seasonal dynamics of cyst formation of pelagic strombidiid ciliates in a deep prealpine lake. Aquat Microb Ecol 17:37–47

Nakamura T, Matsusaka T (1991) Effects of Cyst Age on Excystment of the Ciliate Histriculus-Cavicola (Kahl, 1935) (Ciliophora—Stichotrichia). Eur J Protistology 27:375–380

Nanney DL, McCoy JW (1976) Characterization of the species of the Tetrahymena pyriformis complex. T Am Microsc Soc 95:664–682

Nelson G, Chandrashekar J, Hoon MA, Feng L, Zhao G, Ryba NJ, Zuker CS (2002) An amino-acid taste receptor. Nature 416:199–202

Nowacki M, Shetty K, Landweber LF (2011) RNA-mediated epigenetic programming of genome rearrangements. Annu Rev Genomics Hum Genet 12:367–389

Penfold WJ, Woodcock HM (1916) The excystation of Entamoeba Histolytica (Tetragena) as an indication of the vitality of the cysts. Br Med J 1:714–715

Posthuma G, Meis JF, Verhave JP, Hollingdale MR, Ponnudurai T, Meuwissen JH, Geuze HJ (1988) Immunogold localization of circumsporozoite protein of the malaria parasite Plasmodium falciparum during sporogony in Anopheles stephensi midguts. Eur J Cell Biol 46:18–24

Prater AN, Haagen-Smit AJ (1940) The excystment of protozoa. Isolation of crystalline excystment factors for colpoda duodenaria. J Cell Comp Physiol 15:95–101

Prescott DM (1994) The DNA of ciliated protozoa. Microbiol Rev 58:233–267

Prescott DM (2000) Genome gymnastics: unique modes of DNA evolution and processing in ciliates. Nat Rev Genet 1:191–198

Rawlinson NG, Gates MA (1985) The excystment process in the ciliate Euplotes Muscicola: an integrated light and scanning electron microscopic study 1. J Protozool 32:729–735

Reduker DW, Speer CA (1985) Factors influencing excystation in Cryptosporidium oocysts from cattle. J Parasitology 71:112–115

Regoes A, Zourmpanou D, Leon-Avila G, van der Giezen M, Tovar J, Hehl AB (2005) Protein import, replication, and inheritance of a vestigial mitochondrion. J Biol Chem 280:30557–30563

Reibenbach Z, Reich K (1968) Some observations on cysts and excystment of an Israel strain of Podophrya fixa (O. F. müller, 1786). J Protozool 15:259–262

Repak AJ (1968) Encystment and excystment of the heterotrichous ciliate Blepharisma Stoltei Isquith. J Protozool 15:407–412

Ricci N, Verni F, Rosati G (1985) The cyst of Oxytricha-Bifaria (Ciliata, Hypotrichida).1. morphology and significance. T Am Microsc Soc 104:70–78

Rice EW, Schaefer FW (1981) Improved invitro excystation procedure for Giardia-Lamblia cysts. J Clin Microbiol 14:709–710

Samuelson J, Bushkin GG, Chatterjee A, Robbins PW (2013) Strategies to discover the structural components of cyst and oocyst walls. Eukaryot Cell 12:1578–1587

Schuster FL, Ramirez-Avila L (2008) Current world status of Balantidium coli. Clin Microbiol Rev 21:626–638

Segade P, Garcia N, Garcia Estevez JM, Arias C and Iglesias R (2015) Encystment/excystment response and serotypic variation in the gastropod parasite Tetrahymena rostrata (Ciliophora, Tetrahymenidae). Parasitol Res

Sharp JH (1977) Excretion of organic matter by marine phytoplankton: do healthy cells do it? Limnol Oceanogr 22:381–399

Shatilovich A, Stoupin D, Rivkina E (2015) Ciliates from ancient permafrost: assessment of cold resistance of the resting cysts. Eur J Protistol 51:230–240

Sinden RE (2002) Molecular interactions between plasmodium and its insect vectors. Cell Microbiol 4:713–724

Singh BN, Mathew S, Anand N (1958) The role of Aerobacter sp., Escherichia coli and certain amino acids in the excystment of Schizopyrenus russelli. J Gen Microbiol 19:104–111

Sogin ML, Gunderson JH, Elwood HJ, Alonso RA, Peattie DA (1989) Phylogenetic meaning of the kingdom concept: an unusual ribosomal RNA from Giardia lamblia. Science 243:75–77

Stoupin D, Kiss AK, Arndt H, Shatilovich AV, Gilichinsky DA, Nitsche F (2012) Cryptic diversity within the choanoflagellate morphospecies complex Codosiga botrytis—Phylogeny and morphology of ancient and modern isolates. Eur J Protistology 48:263–273

Stout J (1955) Environmental factors affecting the life history of three soil species of Colpoda (Ciliata). Trans R Soc NZ 82:1165–1188

Strickland AGR, Haagen-Smit AJ (1947) Chemical substances inducing excystment of the resting cysts of Colpoda duodenaria. J Cell Comp Physiol 30:381–390

Stuart CA, Kidder GW, Griffin AM (1939) Growth studies on Ciliates. III. Experimental alteration of the method of reproduction in Colpoda. Physiol Zool 12:348–362

Sun CH, McCaffery JM, Reiner DS, Gillin FD (2003) Mining the Giardia lamblia genome for new cyst wall proteins. J Biol Chem 278:21701–21708

Thimann KV, Barker HA (1934) Studies on the excystment of Colpoda cucullus. II. The action of the excystment inducing substance. J Exp Zool 69:37–57

Torrey EF, Yolken RH (2003) Toxoplasma gondii and schizophrenia. Emerg Infect Dis 9:1375–1380

Tsutsumi T, Watoh T, Kumamoto K, Kotsuki H and Matsuoka T (2004) Effects of porphyrins on encystment and excystment in ciliated protozoan Colpoda sp. Jpn. J Protozool 37:119–127

Vaque D, Blough HA, Duarte CM (1997) Dynamics of ciliate abundance, biomass and community composition in an oligotrophic coastal environment (NW Mediterranean). Aquat Microb Ecol 12:71–83

Verni F, Rosati G (2011) Resting cysts: a survival strategy in Protozoa Ciliophora. Ital J Zool 78:134–145

Villalobo E, Moch C, Fryd-Versavel G, Fleury-Aubusson A, Morin L (2003) Cysteine proteases and cell differentiation: excystment of the ciliated protist Sterkiella histriomuscorum. Eukaryot Cell 2:1234–1245

Walker GK, Maugel TK, Goode D (1975) Some ultrastructural observations on encystment in Stylonychia mytilus (Ciliophora: Hypotrichida). T Am Microsc Soc 94:147–154

Watt WD (1966) Release of dissolved organic material from the cells of phytoplankton populations. Proc R Soc Lond B: Biol Sci 164:521–551

Weisse T (2004) Meseres corlissi: a rare oligotrich ciliate adapted to warm water and temporary habitats. Aquat Microb Ecol 37:75–83

Weisse T, Scheffel U, Stadler P, Foissner W (2013) Bromeliothrix metopoides, a boom and bust ciliate (Ciliophora, Colpodea) from tank bromeliads. Eur J Protistology 49:406–419

Yamaoka M, Watoh T, Matsuoka T (2004) Effects of salt concentration and bacteria on encystment induction in ciliated protozoan Colpoda sp. Acta Protozool 43:93–98

Yarmolinsky DA, Zuker CS, Ryba NJP (2009) Common sense about taste: from mammals to Insects. Cell 139:234–244

Zhang F, Klebansky B, Fine RM, Xu H, Pronin A, Liu HT, Tachdjian C, Li XD (2008) Molecular mechanism for the umami taste synergism. P Natl Acad Sci USA 105:20930–20934

Chapter 8
Hormonal Communication of *Tetrahymena*

György Csaba

Abstract The unicellular ciliate *Tetrahymena* has an almost complete hormonal system, similar to mammals. It produces, stores and secretes amino acid and polypeptide type hormones and does not synthesize steroid hormones. It has receptors for these hormones and also mammalian-like transducing systems. Using these factors it is able to send messages inside the population and deciphering them it can react to the hormones. The specificity of the response is high, however less than that is in mammals. The pros and cons of the abundant experimental data are discussed pointing to their evolutionary importance.

8.1 Introduction

The biological communication means the transmission of signals between members of a community (intracommunity communication) or between members of different communities (intercommunity communication). The community could be an organism with different organs and cells or a population of identical or similar cells e.g. bacterial or protozoan populations, or cells of an organ. The communication is the transfer of information, which is needed for recognizing each other, for organizing the common actions of the cell population and in an organism for the maintenance of homeostasis and coordination of life functions (Witzany 2008, 2010). For executing the communication tools are needed, which send the message, the system by which the message is coded, the channel in which the message is flowing—this could be liquid as well as gaseous—and the receiver, which is able to accept and decipher the message and transmit it to the responder by using signal pathways.

In the case of hormonal communication the sender is a cell, which synthesize and secretes the hormone into the liquid, the messenger, which contains the mes-

G. Csaba (✉)
Department of Genetics, Cell and Immunobiology, Semmelweis University, Budapest, Hungary
e-mail: csaba.gyorgy@med.semmelweis-univ.hu

G. Witzany and M. Nowacki (eds.), *Biocommunication of Ciliates*,
DOI 10.1007/978-3-319-32211-7_8

sage, the code of which is determined by genes. In higher ranked organisms the sender cells are usually accumulated in glands however single-celled forms are also present. The channels are formed mainly by the blood circulation (endocrine communication) however, paracrine and autocrine communication are also known. The receivers are the hormone receptors, which are localized in the cell membrane, cytoplasm or nucleus of the receiver cell. The receptor activates second messengers which transmit the deciphered information to the responder organelles.

In the communication-systems the noise also must be considered. This means that in the channel and in the neighborhood of the receiver cell a lot of disturbing molecules could be present, which can be similar or dissimilar to the messenger molecule (hormone in the present case). The selective ability of the receptor for specifically recognizing the message molecule is a very important factor in the exact transmission of the message.

As *Tetrahymena* is a sweet-water ciliate the water-soluble hormones (messengers) must be synthesized by the cells and secreted into the water (channel), in which these meet numerous molecules which represent the noise. The receivers (receptors) must be present on the surface of the cells (in the plasma membrane). As the aim of the present chapter the demonstration of the hormonal communication of *Tetrahymena*. The presence and function of hormones, receptors and their signal pathways and second messengers must be studied as well, as the mechanisms by which the content of the message is realized.

The *Tetrahymena* is not only a ciliated unicell, it is a single celled organism at the top of the unicellular evolution. So, the study of it can give answer not only to the possibility of communication by *Tetrahymena*, but enlightens evolutionary aspects.

8.1.1 Hormones of Higher Ranked Animals in Tetrahymena

8.1.1.1 Amino-Acid-Type Hormones

Bioamines. Histamine, serotonin, epinephrine, dopamine and melatonin is synthesized, stored in and secreted by *Tetrahymena* (Blum 1967; Csaba 1985, 1994, 2012a, b; Essman 1987; Csaba 2015). The histidine decarboxylase (HDC)-gene was demonstrated and it was found similar to the human one (Hegyesi et al. 1999) and different from the prokaryotic HDC-gene. The synthesis of serotonin and histamine are taking place also in nutrient-free physiological medium, even this condition enhances the production of amines (Csaba et al. 2010a, b, 2011). Long lasting starvation (in nutrient-free medium) elevates serotonin and histamine content of the cells with 50 % (Csaba et al. 2007a, b). However the cells are also able to take up the biogenic amines from the surrounding medium and localize them in the cytoplasm and intranuclearly (Csaba et al. 1983b), predominantly in a heterochromatic localization. There is also the possibility of combined synthesis, which means that e.g. tyrosin is transformed non-enzymatically in the medium to

L-DOPA, this is incorporated by the cells and enzymatically transformed to dopamine (Gundersen and Thompson 1985). Melatonin is also synthesized by the cells, which is influenced by the lighting conditions and previous encounter (hormonal imprinting) (Hardeland 1999; Kőhidai et al. 2002a, b, 2003).

Triiodothyronine. The hormone was demonstrated by using immunocytochemistry and flow cytometry (Csaba et al. 2010a, b; Lajkó et al. 2012). A maintenance in physiological salt solution, which mimic the natural conditions significantly increased the hormone synthesis of the cells. Heat stress increased the hormone level for two weeks and 24 h starvation, as stress caused a 50 % increase in the hormone content.

8.1.1.2 Peptide Hormones

Insulin

It is synthesized, stored in and secreted by *Tetrahymena* (Csaba 2013). In immunoassays or bioassays this insulin was identical to mammalian insulin (LeRoith et al. 1985), and its purified form's bioactivity was neutralized by anti-insulin antibody. In the logarithmic phase of growth the insulin content increased multifold. The *Tetrahymena* insulin stimulated the glucose uptake of rat diaphragm (Csaba et al. 1999), and increased lipogenesis (LeRoith et al. 1980). Not only rat, but guinea pig immunoreactivity of *Tetrahymena* insulin was studied and found real (de Pablo et al. 1986). The cells also can take up insulin from the surrounding medium. This is at first internalized and later partly released (Fülöp and Csaba 1991). It can enter to the nucleus, localizing heterochromatically. During long starvation in physiological saline the insulin level elevates with about 50 % (Csaba et al. 2007a, b). Single insulin treatment of *Tetrahymena* long term (at least for 200 generations) stimulate the insulin production. Though the insulin concentration of cells fluctuates from day to day the single-treated cells always content more insulin than the controls (Csaba and Kovács 1995).

Pituitary Hormones

Hormones coded by the POMC-gene, as adrenocorticotropin and endorphin were found as early, as 1982 (LeRoith et al. 1982, 1983). The *Tetrahymena* hormones were very similar to mammalian ones. Later the presence of follicle stimulating hormone, luteinizing hormone and thyrotropic hormone was also demonstrated (Csaba and Kovács 2000). The hormones were localized diffusely in the cell, and higher amount was found in the cortical region, in the cilia and in the oral field. An increase was caused by maintenance in salt media (Csaba et al. 2007a, 2010a). After treatment with any of the hormones each hormone content durably increased. Beta-endorphin is also present (Rodriguez et al. 2004; Reanaud et al. 1996) and localized in the oral field, cilia and nuclear envelope (Csaba and Kovács 1999).

Other Peptide Hormones

Epidermal growth factor also can be found, which is localized diffusely and mainly in the cytopharynx (Csaba et al. 2004). Salmon calcitonin-like compound, placental lactogen (hPL), relaxin and somatostatin immunoreactivity was demonstrated (Schwabe et al. 1983; Deftos et al. 1985), NT-proANP, the hormone of cardiomyocytes (Kőhidai 2015) and IL-6 (Kőhidai et al. 2000) was also found.

Steroid Hormones

Hydrocortisone, estradiol and testosterone are not present in *Tetrahymena* however, prolonged treatment of the cells whith these hormones provokes the production of them (Csaba et al. 1998). *Tetrahymena* has enzymes for synthesizing and transforming steroids (Lamontagne et al. 1976, 1977).

8.2 Reception of Hormones

As early as 1975 the effect of insulin on the glucose uptake of *Tetrahymena* was demonstrated (Csaba and Lantos 1975). This suggested the presence of insulin receptors, which was found at first in 1995, as a binding site, a ciliary membrane protein (62–67 kDa), which was immunologically similar to insulin (Christopher and Sundermann 1995a). Further studies suggested, that these binding sites behave similar to mammalian ones (Christopher and Sundermann 1992, 1995b, Christensen et al. 2003). The proteins of binding sites of ciliary membrane responded to an antibody against the beta-chain of human insulin receptor (Leick et al. 2001), and also to antibody for fibroblast growth factor receptors. Although receptors are present scattered on the body ciliature, they are extremely accumulated on the oral region (Fülöp and Csaba 1997). The receptors are present not only in the plasma membrane (Csaba et al. 1977), but in other places intracellularly and mainly on the nuclear envelope, to where insulin is bound with greater specificity, than to the plasma membrane (Hegyesi and Csaba 1992; Csaba and Hegyesi 1992, 1994). The receptor nature of the insulin binding sites are also justified by cytofluorimetric study of binding kinetics, which resembled to a true saturation curve (Kovács and Csaba 1990a).

Dopamine D1 receptor was demonstrated in *Tetrahymena thermophila*, using fluorescent ligands (Ud-Daula et al. 2012). The study showed that the receptors do not crossbind with D2 receptor selective antagonist. The D1 receptors are present not only in the cell membrane, but similar to the insulin receptors, intracellularly, mainly in the endoplasmic reticulum and endosomes.

Histamine binding sites were studied and found by using fluorescence technique (Kovács and Csaba 1980). These are localized on the cilia of the body and they are not present on the interciliary regions or the oral field.

Studying the regulation of phagocytosis, mu-like opioid (beta-endorphin and morphine) receptors were found (Chiesa et al. 1993).

For the related pituitary hormones thyrotropin (TSH) and gonadotropin (GTH) also have receptors on the *Tetrahymena*, stimulating triiodothyronine (T3) synthesis (Csaba and Pállinger 2011). The effect is not completely T3- specific, as the trop-hormones also reduced epinephrine production (Lajkó et al. 2011). The incompleteness of specificity is also demonstrated in the case of epinephrine, which elevates glucose uptake similar to insulin (Csaba and Lantos 1976a), in the case of insulin-epidermal growth factor (EGF) overlap (Csaba and Kovács 1991), and also in the case of vasopressin and oxytocin, manifested in the regulation of contractile vacuole (Csaba and Kovács 1992). The basis amino acids can simulate the receptorial effect of the hormone in many cases (Csaba and Darvas 1986).

Histamine receptors can differentiate between histamine antagonists. H1 and H2 antagonists can inhibit the binding of histamine only, when the structure of antagonist is similar to histamine. Antagonists, highly different from histamine are not able to inhibit histamine binding (Kovács et al. 1981).

8.3 The Noise

When other similar or dissimilar hormones are present in the milieu of *Tetrahymena*, the reception of a given signal is disturbed. This was demonstrated also in case of receptor memory (hormonal imprinting) (Csaba et al. 1983a). Glucose utilization caused by insulin treatment is enhanced by the presence of histamine (Kőhidai and Csaba 1985). Though insulin alone moderately increases the phagocytotic capacity of the cells, it antagonizes the effect of histamine in case of simultaneous exposure as well, as after two days (Csaba and Darvas 1992). A combined treatment with serotonin, histamine and insulin was different from the single treatment on the production of serotonin, T3 or ACTH (Csaba et al. 2010b).

8.4 Signal Transduction

All of the second messengers and signal pathways, characteristic to mammalian cells were searched and was found in *Tetrahymena*. Cyclic adenosine monophosphate and (cAMP) and cyclic guanosine monophosphate (cGMP) is present inside the cell, and is secreted to the milieu (Nandini-Kishore and Thompson 1980; Umeki and Nozawa 1996). The receptor is heterotrimeric and G-protein coupled (Shpakov et al. 2004). The adenylate cyclase enzyme is membrane-bound (Kassis and Kindler 1975) and influenced by K+ and Ca2+ concentrations (Nandini-Kishore and Thompson 1979; Schultz and Schönborn 1994; Derkach et al. 2010). Epinephrine, insulin, glucagon as well as serotonin and histamine increase cAMP level (Csaba and Nagy 1976; Csaba and Lantos 1976b, 1977; Csaba et al. 1976; Derkach et al.

2010). Calcium/calmodulin (Kovács et al. 1989) regulated guanylyl cyclase was also demonstrated (Linder and Schultz 2002; Derkach et al. 2010), localized inside the ciliary membrane (Schultz et al. 2005). Calmodulin is identical to brain calmodulin of mammals (Suzuki et al. 1981). However, the *Tetrahymena* adenylyl and guanylyl cyclase evolved from an ancestral cyclase, independently of the mammalian isoforms (Linder et al. 1999). Cyclic AMP level influences basic life functions, as phagocytosis (Csaba et al. 1978), movement (Kovács et al. 1994) etc. Externally given cAMP specifically bound to the cells increasing the activity of cyclases (Shpakov et al. 2012).

In addition to the adenylyl cyclase cAMP, cGMP and calcium-calmodulin systems, inositol phospholipids (Ryals 2009) and sphingolipid derived molecules as well as arachidonic acid (Munaron 2011) are participating in the transmission of signals (Kovács and Csaba 1990b, 1994, 1997; Kovács et al. 1996; Leonarditis and Galanopoulos 2011).

8.5 The Response

Histamine and serotonin increase the phagocytotic capacity (Fig. 8.1) of *Tetrahymena* pyriformis and the latter's relative, the plant hormone indoleacetic acid is discriminated by the receptors (Csaba and Lantos 1973). The effect of serotonin is concentration dependent as well, as the similar effect of catecholamines (Quinones-Maldonado and Renaud 1987). Low concentration of histamine

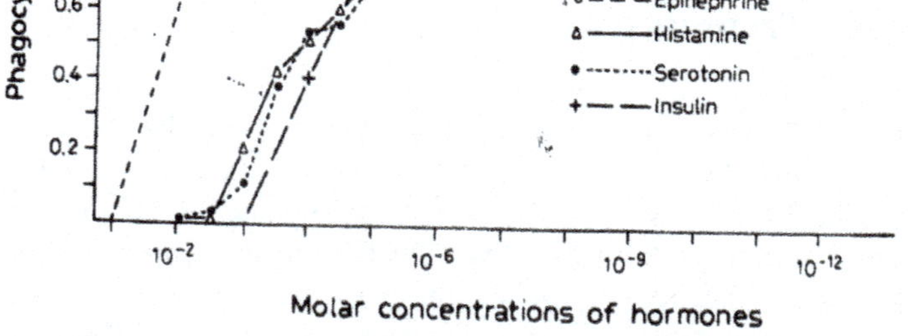

Fig. 8.1 Effects of epinephrine, histamine, serotonin and insulin on the phagocytic activity of *Tetrahymena*. Phagocyte coefficient: phagocytic capacity related to the control value of 0. Histamine has the strongest effect however, the other hormones are also effective

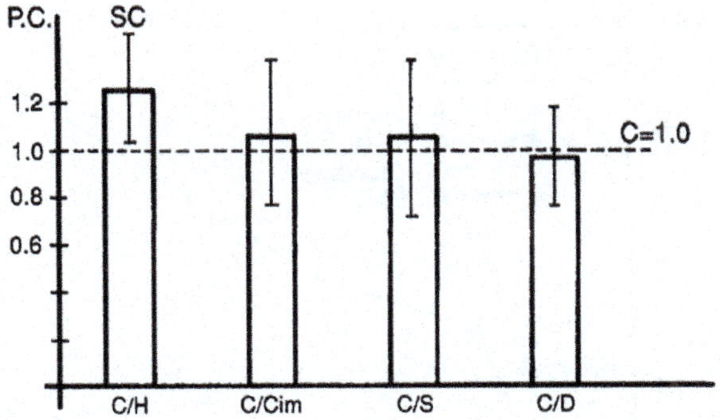

Fig. 8.2 Effect of low concentration of histamine and histamine antagonists on the phagocytosis of *Tetrahymena* related to the control as 1.0 (broken line). *P.C.* phagocyte coefficient, *SC* significance bar, *C* control, *H* histamine, *Cim* cimetidine, *S* chloropyramine, *D* tripellenamine. Only histamine elevated the phagocytic capacity

stimulates (Fig. 8.2), high concentration inhibits phagocytosis (Darvas et al. 1999). Melatonin increased the phagocytosis (Kőhidai et al. 2002b). Polypeptide hormones ACTH, a beta-endorphin-like substance, atrial natriuretic peptide (ANP) and insulin negatively influenced phagocytosis (Kőhidai et al. 1995a, b; Renaud et al. 1995). Lysosomal phosphatase activity was stimulated by histamine and depressed by serotonin (Csaba and Cserhalmi 1985).

Insulin and EGF had mitogenic effect in association with stimulation of DNA synthesis (Selivanova et al. 2002; Shemarova et al. 2002, 2007). Other polypeptide hormones, as TSH and glucagon enhanced RNA synthesis (Csaba and Ubornyák (1982) (Fig. 8.3), others reduced it. The effect of insulin was time and concentration dependent (Hegyesi and Csaba 1997). Other peptide hormones, as glucagon, TSH, PMSG, ACTH and the cytokine IL-6 (Csaba et al. 1995) also stimulated the cell growth and non-hormone polypeptides (BSA, protamine) simulated the effect Csaba et al. 1985a). T3, melatonin, serotonin and its relative, gramin enhanced the multiplication (Csaba et al. 1979; Leclerq et al. 2002).

Insulin and epinephrine enhanced glucose uptake (Csaba and Lantos 1975, 1976a), however there was some difference between the effects: if a surplus of sugar was given exogeneously, epinephrine always dropped the glucose level in all regions of the cell, while similar effect of insulin was less unequivocal (Csaba and Kovács 1979). Histamine treatment increased the glycogen content of the cells, but H2 receptor antagonist was more effective. H1 receptor antagonist was ineffective (Darvas and Csaba 1981).

Regeneration of cilia was increased by serotonin and insulin (Darvas et al. 1988). Serotonin effect is executed by a mechanism which involves cAMP and calcium (Rodriguez and Renaud 1980). Micromolar concentrations of catecholamines also enhanced the ciliary regeneration (Castrodad et al. 1988).

Oxytocin and vasopressin, two related pituitary hormones influenced the systolic contraction of the contractile vacuole (Csaba and Kovács 1992).

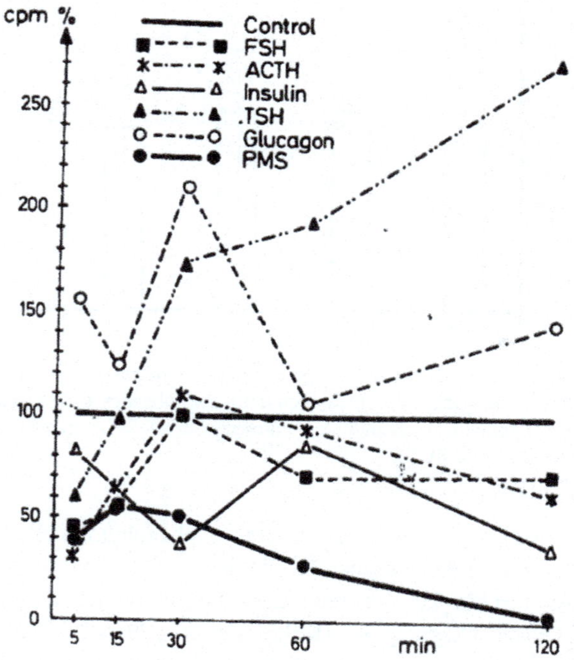

Fig. 8.3 Effect of polypeptide hormones on RNA synthesis, which is measured by 3H uridine uptake and expressed as a percentage of the control value defined as 100 cpm/105 cells PMS (pregnant mare serum). TSH and glucagon are stimulating, while the other hormones studied reduced RNA synthesis

The first encounter of *Tetrahymena* with an amino acid or peptide hormone provokes the hormonal imprinting, the effect of which is manifested in the progenies and can be observed at least up to the 1000th generation (Csaba 1985, 1994, 2012a, b; Kőhidai et al. 2012) (Fig. 8.4). This imprinting with changed (mostly

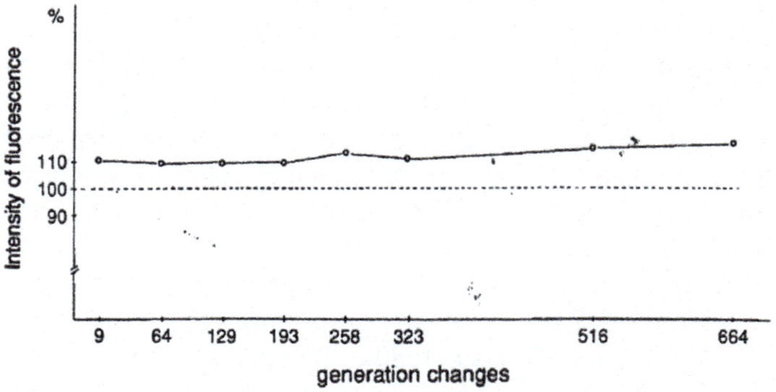

Fig. 8.4 FITC-insulin binding in the offspring generations of *Tetrahymena* treated with the hormone in a single occasion (control = 100 %). After 664 generations the effect of imprinting is expressed more strongly, than one day after treatment

increased) reception of the given hormone, is also a response to the signal. This is also provoked in case of steroid hormones, when imprinting induced the appearance of hormones and their receptors alike (Csaba et al. 1985b).

8.6 Conclusions

Considering the hormonal communication, the best known is the mammalian (first of all the human) communication. This is the reason—in addition to the anthropocentric aspect of the researcher—why the components and mechanisms of this higher ranked communication is searched also at a unicellular level. It can be established that almost all of the components of mammalian cell-to-cell communication, which were searched at all, can be found at this evolutionarily low level. *Tetrahymena* has all of the amino acid and peptide hormones which are present in man, have their receptors, signal pathways and second messengers and also respond —mostly similar to men—to the stimuli of hormones. However, though it is able to synthesize steroid hormones, it does not do it in normal conditions. It also does not have a feedback mechanism.

The absence of steroid hormones and their utilization is understandable. These hormones are lipid soluble however, *Tetrahymena* is living in a watery milieu. So, these hormones are not suitable for the intercellular communication in the water. In addition it seems to be likely that the intracellular receptors of steroid hormones presumably evolved by the engulfment of their membrane receptors however, if there is not stimulus for the fixation of membrane patterns suitable to bind the hormone, there is no receptor development. According to Lenhoff (1968) hormone receptors developed from nourishment (amino acid) receptors and we supposed earlier that hormonal imprinting is the provocator of the fixation of membrane patterns, which continuously and variably inquire the neighborhood (Csaba 1994, 2012b; Csaba and Kovács 1990; Csaba et al. 1982b). This is supported by the facts, that a hormone precursor of thyroxine is more effective in *Tetrahymena*, than the hormone acting in mammals (Csaba and Németh 1980) (Figs. 8.5 and 8.6), in contrast to serotonin, where the presursors are not hormones.

The feedback mechanism is needed in a mammalian organism to hinder the overproduction of a hormone in a closed system. *Tetrahymena* in natural conditions is living in an open system in which the dissolution of the hormone produced by it is extremely variable. In this case the hormone concentration is continuously fluctuate and between very broad limits the presence of the hormone has a meaning and not its amount. As the feedback regulates the amount of the hormone produced, this is meaningless in case of *Tetrahymena*.

Tetrahymena is at a very low level of phylogeny however, it is a eukaryotic animal cell. The most primitive cells are the prokaryotes. In these cells some elements of the communication also can be found (Stoka 1999). One of these is the quorum sensing, by which bacteria can get information on the density of population (Ng and Bassler 2009), which gives the possibility of a coordinated behavior

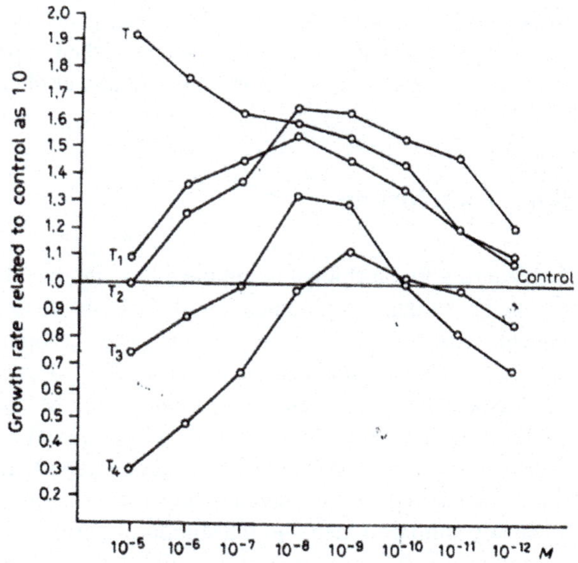

Fig. 8.5 Effect of thyroxine and its precursors on the growth of *Tetrahymena*. *T* tyrosine, T_1 monoiodotyrosine, T_2 diiodotyrosine, T_3 triiodothyronine, T_4 thyroxine. Tyrosine is a nourishment, which influences the cell growth concentration dependently, while the strongest effect of precursors is produced by the less developed but double iodinated T_2

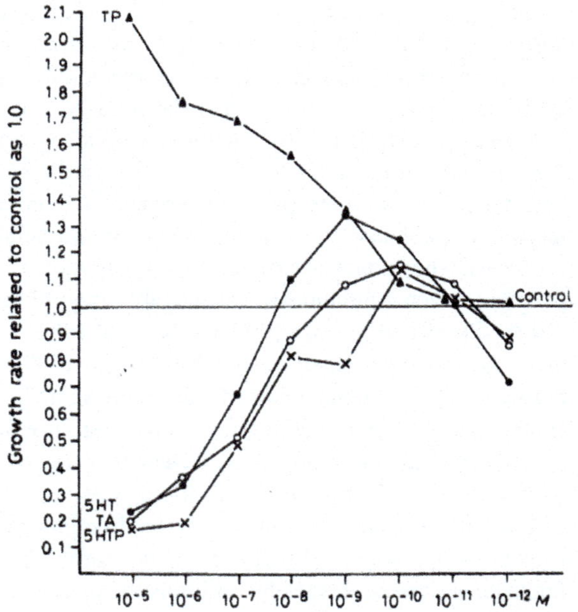

Fig. 8.6 Effect of serotonin and its precursor molecules on the growth rate of *Tetrahymena*. *TP* tryptophane, *TA* tryptamine, *5HTP* 5-hydroxytryptophane, *5HT* serotonin. In contrast to T_3 and T_4, serotonin, which is an universal hormone in the plant and animal world is the most effective

(Witzany 2008, 2010). However, many traces of the later endocrine system also can be found in bacteria (Pertseva 1991). A lot of bacteria contain hCG, Clostridium perfringens synthesize TSH and insulin is present in E.coli, Neurospora crassa, Bordatella pertussis, calcitonin was found in Candida albicans and E. coli etc. (LeRoith et al. 1986). Hormone binding proteins are also present (Lenard 1992) and by the help of these, bacterial behavior during diseases are influenced (Kaprelyants and Kell 1996; Lyte 2004). However, they have not a complete and organized hormonal system, as has the *Tetrahymena*.

In our modern, technically advanced age many sophisticated methods are used for demonstratring substances in cells, which earlier were not found. In mammalian organisms there is an endocrine system, the members (glands) of which chemically regulate the different functions under the direction of the nervous system. This is a classical definition. However the new methods demonstrated that a mass of cells which does not categorized as neuroendocrine, are synthesizing such hormones as insulin, epinephrine, TSH etc. (Csaba 2012a). Insulin was found in each immune cells and many other cells however, it is not cleared, whether these cells regulate specific life processes as it is done by the pancreatic Langerhans islets. This is an open problem similar to the insulin (or other hormone) production of *Tetrahymena*. Is this a result of the sophisticated methods, or these hormones are signals which regulates important functions? In *Tetrahymena*, are these hormones participating in the process of communication or they are side-products of protein synthesis, without any role in communication?

It is not dubious that these hormones can be signals and can regulate the life function of *Tetrahymena*, as cell multiplication, phagocytosis, sugar metabolism etc. However it is not known, that these hormones are really utilized, essentially needed or not. How can work these life functions in the absence of these hormones? In mammals the change of metabolism in the case of thyreoidectomy can be studied, however in *Tetrahymena* the effect of knocking out of the -unknown- TSH or thyroxine gene would be needed for demonstrating that TSH or thyroxin is the regulator of metabolism.

The presence of the complete hormonal system, signals, receptors, signal pathways, second messengers and the noise, strongly support the use of these elements in the real communication of *Tetrahymena*. In addition, the life saving function of a hormone (insulin) or other hormones shows that this type of communication is very important. The light-dependence of melatonin secretion also points to the operation of this system.

There is also a possibility, that the meaning of a hormone molecule for *Tetrahymena* is quite another, than it is in a mammalian organism. However, we are searching the multicellular (mammalian) meaning and we found this indeed, as the molecular structure is suitable to regulate the function, but this could be unimportant for *Tetrahymena* and the real sense is not known.

As the mentioned data shows, *Tetrahymena* can bind the exogenously given mammalian hormones (signals) and can decipher their information. The question is, whether the hormone produced by the cell is secreted into the milieu indeed and *Tetrahymena* uses this hormone for transmitting information? The answer to the first part of the question is yes. Insulin is secreted into the medium and this insulin can act on mammalian target cells (diaphragm) provoking typical reaction (Csaba et al. 1999). The answer to the second part is also yes, as the medium in which *Tetrahymena* was present, provoked hormonal imprinting in virgin *Tetrahymena* (Csaba et al. 1990). The information given by the hormone and used by

Tetrahymena can be so important as the protection of life. In the case of a low density of cells when they are dying in a synthetic nutrient medium, presence of insulin in 10–6 M concentration prevented cell death and promoted proliferation (Christensen 1993; Christensen et al. 1996).

The above reviewed data show that *Tetrahymena* uses the same code, as mammals for influencing cell functions. However, this is right inversely: the endocrine code is present already in *Tetrahymena* and this remain in use almost unchanged during the whole phylogeny. Even such high-level regulation can be present in *Tetrahymena*, as the trop-hormone—target-hormone relation, demonstrated by the effect of TSH on T3 level (Csaba and Pállinger 2011; Lajkó et al. 2011).

In natural conditions *Tetrahymena* is living in a watery milieu in which the cells are settled very dispersed and the secreted molecules are highly dissoluted. Probably this is the reason, why some hormones act at extremely low -10^{-21} M-concentration, when only few molecules are present around the cells. This shows the very high sensitivity of reception (Csaba et al. 2006, 2007b, 2008, 2011), which is helped by the hormonal imprinting, after which the reception capacity is multiplied (Csaba et al. 1982a; Csaba 1996, 2012b).

In *Tetrahymena* one cell produces each hormone, in contrast to multicellulars. It is not known, whether the hormone synthesis takes place in the same time or independent times, but the hormones can interact each other. This suggests that a hormonal network is present influencing the synthesis and reception of hormones alike (Csaba and Kovács 1994, 2005; Csaba and Pállinger 2008). However, the details of this process are not cleared.

Tetrahymena, as a ciliate is a top-product of the unicellular evolution (Zhang et al. 2010) similar to men, which is also a top-product of mammalian evolution. However, only very scarce data are at our disposal on the hormonal communication of non-ciliate unicells. So, we can not declare that the hormonal communication of *Tetrahymena* is a result of an evolutionary process inside the unicellular world. We also can not declare that an evolution of hormonal system was not needed, as at the very low level of evolution an almost complete hormonal communication system exists. Nevertheless, the similarities between the mammalian (human) and *Tetrahymena* hormonal communication are striking.

References

Blum JJ (1967) An adrenergic control system in Tetrahymena. Proc Natl Acad Sci USA 58:81–88

Castrodad FA, Renaud FL, Ortiz J, Phillips DM (1988) Biogenic amines stimulate regeneration of cilia in *Tetrahymena* thermophila. J Protozool 35:260–264

Chiesa R, Silva WI, Renaud FL (1993) Pharmacological characterization of an opioid receptor in the ciliate Tetrahymena. J Eukaryot Microbiol 40:800–804

Christensen ST (1993) Insulin rescues the unicellular eukaryote Tetrahymena from dying in a complete, synthetic nutrient medium. Cell Biol Int 17:833–837

Christensen ST, Kemp K, Quie H, Rasmussen L (1996) Cell death, survival and proliferation in Tetrahymena thermophila. Effects of insulin, sodium nitroprusside, 8-Bromo cyclic GMP, NG. methyl-arginine and methylene blue. Cell Biol Int 20(653):656

Christensen ST, Guerra CF, Awan A, Wheatley DN, Satir P (2003) Insulin receptor-like protein in Tetrahymena thermophila ciliary membranes. Curr Biol 13:R50–52
Christopher GK, Sundermann CA (1992) Conventional and confocal microscopic studies of insulin receptor induction in Tetrahymena pyriformis. Exp Cell Res 201:477–484
Christopher GK, Sundermann CA (1995a) Isolation and partial characterization of the insulin binding sites of Tetrahymena pyriformis. Biochem Biophys Res Commun 212:515–523
Christopher GK, Sundermann CA (1995b) Effect of long-term insulin exposure on insulin binding in Tetrahymena pyriformis. Tissue Cell 27:585–589
Csaba G (1985) The unicellular Tetrahymena as a model cell for receptor research. Int Rev Cytol 95:327–377
Csaba G (1994) Phylogeny and ontogeny of chemical signaling: origin and development of hormone receptors. Int Rev Cytol 155:1–48
Csaba G (1996) Evolutionary significance of the hormone receptor capacity in unicellular organisms. Development of hormone receptors. In: Csaba G, Müller WEG (eds) Signaling mechanisms of Protozoa and Invertebrates. Springer, Berlin, pp 1–28
Csaba G (2012a) The hormonal system of the unicellular Tetrahymena: a review with evolutionary aspects. Acta Microbiol Immunol Hung 59:131–156
Csaba G (2012b) Hormonal imprinting in the unicellular Tetrahymena: the proto-model of epigenetics. Acta Microbiol Immunol Hung 59:291–310
Csaba G (2013) Insulin at a unicellular eukaryote level. Cell Biol Int 37:267–275
Csaba G (2015) Biogenic amines at a low level of evolution. Production, functions and regulation in the unicellular Tetrahymena. Acta Microbiol Immunol Hung 62:93–108
Csaba G, Cserhalmi M (1985) Influence of biogenic amines histamine, serotonin on the function of lysosomal enzymes of the Tetrahymena. Acta Protozool 242:136–138
Csaba G, Darvas Z (1986) Receptor-level interelationships of amino acids and the adequate amino acid type hormones in Tetrahymena: a receptor evolution model. Biosystems 19:55–59
Csaba G, Darvas Z (1992) Insulin antagonizes the phagocytosis stimulating action of histamine in Tetrahymena. Biosci Rep 12:23–27
Csaba G, Hegyesi H (1992) Specificity of insulin binding by plasma and nuclear membrane receptors in Tetrahymena: similarities and dissimilarities at the two levels. Cytobios 70:154–158
Csaba G, Hegyesi H (1994) Immunocytochemical verification of the insulin receptor's specificity in the nuclear envelope of Tetrahymena. Comparison with receptors of the plasma membrane. Biosci Rep 14:25–31
Csaba G, Kovács P (1979) Complex cytophotometric analysis of hormone-induced alterations in the glucose utilization of Tetrahymena. Endokrinologie 73:116–119
Csaba G, Kovács P (1990) Impact of 5-azacytidine on insulin binding and insulin–induced receptor formation in Tetrahymena. Biochem Biophys Res Commun 168:709–713
Csaba G, Kovács P (1991) EGF receptor induction and insulin-EGF overlap in Tetrahymena. Experientia 47:718–721
Csaba G, Kovács P (1992) Oxytocin and vasopressin change the activity of the contractile vacuole in Tetrahymena: newer contributions to the phylogeny of hormones and hormone receptors. Comp Biochem Physiol Comp Physiol 102:353–355
Csaba G, Kovács P (1994) Effect of hormones and hormone-induced imprinting on the serotonin level in Tetrahymena: immunocytochemical studies. Microbios 80:155–163
Csaba G, Kovács P (1995) Insulin treatment (hormonal imprinting) increases the insulin production of the unicellular Tetrahymena long term. Is there a simultaneous formation of hormone receptor and hormone? Cell Biol Int 19:1011–1014
Csaba G, Kovács P (1999) Localization of beta-endorphin in Tetrahymena by confocal microscopy. Induction of the prolonged production of the hormone by hormonal imprinting. Cell Biol Int 23:695–702
Csaba G, Kovács P (2000) Human chorionic gonadotropin (HCG)-like hormones (FSH, LH, TSH) in Tetrahymena. A confocal microscopic analysis. Acta Protozool 39:191–198

Csaba G, Kovács P (2005) Hormonal interactions in Tetrahymena: effect of hormones on levels of epidermal growth factor (EGF). Cell Biol Int 29:301–305

Csaba G, Lantos T (1973) Effect of hormones on Protozoa. Studies on the phagocytotic effect of histamine, 5-hydroxytryptamine and indoleacetic acid in Tetrahymena pyriformis. Cytobiologie 7:361–365

Csaba G, Lantos T (1975) Effect of insulin on the glucose uptake of protozoa. Experientia 31:1097–1098

Csaba G, Lantos T (1976a) Effect of epinephrine on glucose metabolism in Tetrahymena. Endokrinologie 68:239–240

Csaba G, Lantos T (1976b) Effect of cyclic AMP and theophylline on phagocytotic activity of Tetrahymena pyriformis. Experientia 32:321–322

Csaba G, Lantos T (1977) An attempt to differentiate selection and amplification in hormone receptor development: the unicellular model. Differentiation 8:57–59

Csaba G, Nagy SU (1976) Effect of vertebrate hormones on the cyclic AMP level in Tetrahymena. Acta Biol Med Ger 35:1399–1401

Csaba G, Németh G (1980) Effect of hormones and their precursors on protozoa—the selective responsiveness of Tetrahymena. Comp Biochem Physiol B 65:387–390

Csaba G, Pállinger É (2008) Is there a hormonal network in Tetrahymena? A systematic investigation of hormonal effects on the hormone content. Cell Biochem Funct 26:303–308

Csaba G, Pállinger É (2011) Thyrotropin (TSH) regulates triiodothyronine (T3) production in the unicellular Tetrahymena. Acta Biol Hung 62:228–234

Csaba G, Ubornyák L (1982) Effect of polypeptide hormones (insulin, thyrotropin, gonadotropin, adrenocorticotropin) on RNA synthesis in Tetrahymena as assessed from incorporation of ^{3}H uridine. Acta Biol Acad Sci Hung 33:381–384

Csaba G, Nagy SU, Lantos T (1976) Are biogenic amines acting on Tetrahymena through a cyclic AMP mechanism? Acta Biol Med Germ 35:259–262

Csaba G, Sudár F, Nagy SU, Dobozy O (1977) Localization of hormone receptors in Tetrahymena. Protoplazma 91:179–189

Csaba G, Nagy SU, Lantos T (1978) Cyclic AMP and its functional relationship in Tetrahymena: a comparison between phagocytosis and glucose uptake. Acta Biol Med Ger 37:505–507

Csaba G, Németh G, Prohászka J (1979) Effect of hormones and related compounds on the multiplication of Tetrahymena. Exp Cell Biol 47:307–311

Csaba G, Németh G, Vargha P (1982a) Influence of hormone concentration and time factor on development of receptor memory in a unicellular (Tetrahymena) model system. Comp Biochem Physiol B 73:357–360

Csaba G, Németh G, Juvancz I, Vargha P (1982b) Involvement of selection and amplification mechanisms in hormone receptor development in a unicellular model system. Biosystems 15:59–63

Csaba G, Németh G, Vargha P (1983a) Attempt to disturb receptor memory in a unicelluler (Tetrahymena) model system. Acta Physiol Hung 61:121–1236

Csaba G, Sudár F, Ubornyák L (1983b) Comparative study of the internalization and nuclear localization of amino acid type hormones in Tetrahymena and rat lymphocytes. Exp Clin Endocrinol 82:61–67

Csaba G, Németh G, Kovács P, Vargha P, Vas Á (1985a) Chemical reception mechanisms at a low level of phylogeny. Influence of polypeptide hormones and non-hormone polypeptides on the growth of Tetrahymena. Biosystems 17:227–231

Csaba G, Inczefi-Gonda Á, Fehér T (1985b) Induction of steroid binding sites (receptors) and presence of steroid hormones in the unicellular Tetrahymena pyriformis. Comp Biochem Physiol A 82:567–570

Csaba G, Kovács P, László V (1990) Cell-to-cell transmission of hormonal imprinting persists long in Tetrahymena. Acta Microbiol Hung 37:277–280

Csaba G, Kovács P, Falus A (1995) Human cytokines interleukin (IL)-3 and IL-6 affect the growth and insulin binding of the unicellular organism Tetrahymena. Cytokine 7:771–774

Csaba G, Poteczin E, Fehér T, Kovács P (1998) Steroid hormone (hydrocortisone, oestradiol and testosterone) uptake, storage or induced synthesis in Tetrahymena. Cell Biol Int 22:875–878

Csaba G, Gaál A, Kovács P, Simon G, Kőhidai L (1999) Prolonged elevation of insulin content in the unicellular Tetrahymena after insulin treatment: induction of insulin production or storage? Cell Biochem Funct 17:165–173

Csaba G, Kovács P, Pállinger É (2004) Presence and localization of epidermal growth factor (EGF) and EGF-receptor-like immunoreactivity in Tetrahymena. Cell Biol Int 28:491–496

Csaba G, Kovács P, Tóthfalusi L, Pállinger É (2006) Effects of extremely low concentrations of hormones on the insulin binding of Tetrahymena. Cell Biol Int 30:957–962

Csaba G, Kovács P, Pállinger É (2007a) Increased hormone level in Tetrahymena after long-lasting starvation. Cell Biol Int 31:924–925

Csaba G, Kovács P, Pállinger É (2007b) How does the unicellular Tetrahymena utilise the hormones that it produces? Paying a visit to the realm of atto-and zeptomolar concentrations. Cell Tissue Res 327:199–203

Csaba G, Kovács P, Pállinger É (2008) Effect of femtomolar concentrations of hormones on insulin binding by Tetrahymena, as a function of time. Cell Biochem Funct 26:205–209

Csaba G, Lajkó E, Pállinger É (2010a) Comparison of the effect of hormones on the hormone synthesis of Tetrahymena in medium or salt solution. Cell Biol Int 34:1095–1098

Csaba G, Lajkó E, Pállinger É (2010b) Hormonal effects on Tetrahymena: change in case of combined treatment. Acta Microbiol Immunol Hung 57:393–399

Csaba G, Lajkó E, Pállinger É (2011) Effect of different concentrations of serotonin, histamine and insulin on the hormone (serotonin and ACTH) production of Tetrahymena in nutrient-free physiological milieu. Exp Parasitol 129:179–182

Darvas Z, Csaba G (1981) Effect of histamine and histamine antagonists on the glycogen content of Tetrahymena. Acta Physiol Acad Sci Hung 58:9–14

Darvas Z, Árva G, Csaba G, Vargha P (1988) Enhancement of cilia regeneration by hormone treatment of Tetrahymena. Acta Microbiol Hung 35:45–48

Darvas Z, Madarász B, László V (1999) Study of histamine effects on phagocytosis and enzyme secretion of Tetrahymena pyriformis. Acta Biol Hung 50:325–334

Deftos LJ, LeRoith D, Shiloach J, Roth J (1985) Salmon calcitonin-like immunoactivity in extracts of Tetrahymena pyriformis. Horm Metab Res 17:82–85

de Pablo F, Lesniak MA, Hernandez ER, LeRoith D, Shiloach J, Roth J (1986) Extracts of Protozoa contain materials that react specifically in the immunoassay for guinea pig insulin. Horm Metab Res 18:82–87

Derkach KV, Shpakov AO, Uspenskaja ZI, Iudin AL (2010) Functional characteristics of calcium-sensitive adenylyl cyclase of ciliate Tetrahymena pyriformis. Tsitologia 52:967–972

Essman EJ (1987) The serotonergic system in Tetrahymena pyriformis. Ric Clin Lab 17:77–82

Fülöp AK, Csaba G (1991) Turnover and intranuclear localization of 125I-insulin in Tetrahymena. An autoradiographic study. Acta Morphol Hung 39:71–77

Fülöp AK, Csaba G (1997) Accumulation of insulin-gold particles in the oral apparatus of Tetrahymena after insulin pretreatment (imprinting). Microbios 90:123–128

Gundersen RE, Thompson GA Jr (1985) Further studies of dopamine metabolism and function in Tetrahymena. J Protozool 32:25–31

Hardeland R (1999) Melatonin and 5-methoxytryptamine in non-metazoans. Reprod Nutr Dev 39:399–408

Hegyesi H, Csaba G (1992) Specific insulin binding by, and imprintability of, the nuclear membrane of Tetrahymena. Cytobios 72:153–157

Hegyesi H, Csaba G (1997) Time and concentration-dependence of the growth-promoting activity of insulin and histamine in Tetrahymena. Application of the MTT-method for the determination of cell proliferation in a protozoan model. Cell Biol Int 21:289–293

Hegyesi H, Szalai C, Falus A, Csaba G (1999) The histidine decarboxylase (HDC) gene of Tetrahymena pyriformis is similar to the mammalian one. A study of HDC expression. Biosci Rep 19:73–79

Kaprelyants AS, Kell DB (1996) Do bacteria need to communicate with each other for growth? Trends Microbiol 4:237–241

Kassis S, Kindler SH (1975) Dispersion of epinephrine sensitive and insensitive adenylate cyclase from the ciliate Tetrahymena pyriformis. Biochim Biophys Acta 391:513–516

Kovács P, Csaba G (1980) Detection of histamine binding sites (receptors) in Tetrahymena by fluorescence technique. Acta Biol Med Ger 39:237–241

Kovács P, Csaba G (1990a) Evidence of the receptor nature of the binding sites induced in Tetrahymena by insulin treatment. A quantitative cytofluorimetric technique for the study of binding kinetics. Cell Biochem Funct 8:49–56

Kovács P, Csaba G (1990b) Involvement of the phosphoinositol (PI) system in the mechanism of hormonal imprinting. Biochem Biophys Res Commun 170:119–126

Kovács P, Csaba G (1994) Effect of insulin on the incorporation of 3H-inositol into the inositol phospholipids (PI, PIP, PIP2) and glycosyl-phosphatidylinositols (GPIs) of Tetrahymena pyriformis. Biosci Rep 14:215–219

Kovács P, Csaba G (1997) PLA2 activity in Tetrahymena pyriformis. Effects of inhibitors and stimulators. 15:233–247

Kovács P, Darvas Z, Csaba G (1981) Investigation of histamine-antihistamine differentiation ability of Tetrahymena receptors, by means of lectins and antihistamine antibodies. Acta Biol Acad Sci Hung 32:111–117

Kovács P, Csaba G, Nagao S, Nozawa Y (1989) The regulatory role of calmodulin-dependent guanylate cyclase in association with hormonal imprinting in Tetrahymena. Microbios 59:123–128

Kovács P, Lovas G, Csaba G (1994) Influence of insulin on the movement of Tetrahymena pyriformis. Hormonal imprinting alters the velocity. Comp Biochem Physiol Comp Physiol 107:375–379

Kovács P, Csaba G, Ito Y, Nozawa Y (1996) Effect of insulin on the phospholipase-D activity of untreated and insulin-pretreated (hormonally imprinted) Tetrahymena. Biochem Biophys Res Commun 15:359–361

Kőhidai L (2015) NT-proANP in Tetrahymena. Unpublished

Kőhidai L, Csaba G (1985) Effects of insulin and histamine in themselves and in combination on the glucose metabolism of Tetrahymena. Acta Biol Hung 36:281–285

Kőhidai L, Csaba G, Karsa J (1995a) Effects of atrial natriuretic peptide on the unicellular Tetrahymena pyriformis model. Microbios 82:27–40

Kőhidai L, Lovas B, Csaba G (1995b) Effect of adrenocorticotropic hormone (ACTH) and insulin on the phagocytic capacity of Tetrahymena. Zoolog Sci 12:277–281

Kőhidai L, Kovács P, Lázár-Molnár E, Csaba G (2000) Presence, uptake and localization of an immunoreactively interleukin 6 (IL-6)-like molecule in Tetrahymena pyriformis. Cell Biol Int 24:749–755

Kőhidai L, Vakkuri M, Keresztesi M, Leppaluoto J, Csaba G (2002a) Melatonin in the unicellular Tetrahymena pyriformis: effects of different lighting conditions. Cell Biochem Funct 20:269–272

Kőhidai L, Vakkuri O, Keresztesi M, Pállinger É, Leppaluoto J, Csaba G (2002b) Impact of melatonin on the cell divisdion, phagocytosis and chemotaxis of Tetrahymena pyriformis. Acta Protozool 41:85–89

Kőhidai L, Vakkuri M, Keresztesi M, Leppaluoto J, Csaba G (2003) Induction of melatonin synthesis in Tetrahymena pyriformis by hormonal imprinting—a unicellular "factory" of the indoleamine. Cell Mol Biol 49:521–524

Kőhidai L, Lajkó E, Pállinger É, Csaba G (2012) Verification of epigenetic inheritance in a unicellular model system: multigenerational effects of hormonal imprinting. Cell Biol Int 36:951–959

Lajkó E, Pállinger É, Csaba G (2011) Investigations on the triiodothyronine (T3)-specificity of thyrotropic (TSH) and gonadotropic (HCG) hormone in the unicellular Tetrahymena. Acta Microbiol Immunol Hung 58:85–91

Lajkó E, Pállinger É, Csaba G (2012) Durable effect of heat-stress on the hormone production of Tetrahymena. Effect of insulin on the consequences of stress. Acta Microbiol Immunol Hung 59:249–256

Lamontagne NS, Johnson DF, Holmlund CE (1976) The transformation of testosterone by Tetrahymena. J Steroid Biochem 7:177–183

Lamontagne NS, Johnson DF, Holmlund CE (1977) The conversion of progesterone to pregnenolone by Tetrahymena pyriformis. J Steroid Biochem 8:329–334

Leclerq B, Exbrayat JM, Duyme F, De Coninck J (2002) New approach to model effects of darkness, melatonin and serotonin on Tetrahymena thermophila growth and production of hydrolytic enzymes. Biotechn Lett 34:769–774

Leick V, Bog-Hansen TC, Juhl HA (2001) Insulin/FGF-binding ciliary membrane glycoprotein from Tetrahymena. J Membr Biol 181:47–53

Lenard J (1992) Mammalian hormones in microbial cells. Trends Biochem Sci 17:147–150

Lenhoff HM (1968) Behavior, hormones and Hydra. Research on behavior of lower invertebrates may help elucidate some cellular actions of hormones. Science 161:434–472

Leonarditis G, Galanopoulos D (2011) Emerging roles of phosphoinositide-specific phospholipases C in the ciliates Tetrahymena and Paramecium. Comm Integrat Biol 45:576–578

LeRoith D, Shiloach J, Roth J, Lesniak MA (1980) Evolutionary origins of vertebrate hormones: substances similar to mammalian insulins are native to unicellular eukaryotes. Proc Natl Acad Sci USA 77:6184–6188

LeRoith D, Liotta AS, Roth J, Shiloach J, Lewis ME, Pert CB, Krieger DT (1982) Corticotrophin and beta-endorphin-like materials are native to unicellular organisms. Proc Natl Acad Sci USA 79:2086–2090

LeRoith D, Shiloach J, Berelowitz M, Frohman LA, Liotta AS, Krieger DT, Roth J (1983) Are messenger molecules in microbes the ancestors of the vertebrate hormones and tissue factors? Fed Proc 42:2602–2607

LeRoith D, Shiloach J, Heffron R, Rubinovitz C, Tanenbaum R, Roth J (1985) Insulin-related meterial in microbes: similarities and differences from mammalian insulin. Can J Biochem Cell Biol 63:839–849

LeRoith D, Roberts C Jr, Lesniak MA, Roth J (1986) Receptors for intercellular messenger molecules in microbes: Similarities to vertebrate receptors and possible implications for diseases in man. Experientia Suppl. 53:167–180

Linder JU, Schultz JE (2002) Guanylyl cyclases in unicellular organisms. Mol Cell Biochem 230:149–158

Linder JU, Engel P, Reimer A, Krüger T, Plattner H, Schultz A, Schultz JE (1999) Guanylyl cycleses with the topology of mammalian adenylyl cyclases and an N-terminal P-type ATP-ase-like domain in Paramecium, Tetrahymena and Plasmodium. EMBO J 18:4222–4232

Lyte M (2004) Microbial endocrinology and infectious disease in the 21st century. Trends Microbiol 12:14–20

Munaron L (2011) Shuffling the cards in signal transduction: calcium, arachidonic acid and mechanosensitivity. World J Biol Chem 2:59–66

Nandini-Kishore SG, Thompson GA Jr (1979) Increased levels of adenosine 3',5'-cyclic monophosphate in Tetrahymena stimulated by glucose and mediated by Ca2+ and epinephrine. Proc Natl Acad Sci USA 76:2708–2711

Nandini-Kishore SG, Thompson GA Jr (1980) Evidence for a pronounced secretion of cyclic AMP by Tetrahymena. J Cell Biol 85:910–915

Ng WL, Bassler BL (2009) Bacterial quorum-sensing network architectures. Annu Rev Genet 43:197–222

Pertseva M (1991) The evolution of hormonal signaling system. Comp Biochem Physiol A 100:775–787

Quinones-Maldonado V, Renaud FL (1987) Effect of biogenic amines on phagocytosis in Tetrahymena thermophila. J Protozool 34:435–438

Renaud FL, Colon I, Lebron J, Ortiz N, Rodriguez F, Cadilla C (1995) A novel opioid mechanism seems to modulate phagocytosis in Tetrahymena. J Eukaryot Microbiol 42:205–207

Renaud FL, Chiesa R, Rodriguez F, Tomassini N, Marino M (1996) Studies on the opioid mechanism in Tetrahymena. In: Csaba G, Müller WEG (eds) Signaling mechanisms in Protozoa and Invertebrates. Springer, Berlin, pp 29–39

Rodriguez N, Renaud FL (1980) On the possible role of serotonin in the regulation of regeneration of cilia. J Cell Biol 85:242–247

Rodriguez E, Lazaro MI, Renaud FL, Marino M (2004) Opioid activity of beta-endorphin-like proteins from Tetrahymena. J Eukaryot Microbiol 51:60–65

Ryals PE (2009) Inositols and phosphoinositides in Tetrahymena. Acta Protozool 48:191–202

Schultz JE, Schönborn C (1994) Cyclic AMP formation in Tetrahymena pyriformis is controlled by a K(+)-conductance. FEBS Lett 19:322–328

Schultz JE, Schönefeld U, Klumpp S (2005) Calcium/calmodulin-regulated guanylate cyclase and calcium-permeability in the ciliary membrane from Tetrahymena. Eur J Biochem 137:89–94

Schwabe C, LeRoith D, Thompson RP, Shiloach J, Roth J (1983) Relaxin extracted from protozoa (Tetrahymena pyriformis). Molecular and immunological properties. J Biol Chem 258:2778–2781

Selivanova GV, Shemarova IV, Vlasova TD (2002) Epidermal growth factor stimulates proliferation and DNA synthesis in ciliate cells. Ontogenez 33:457–460

Shemarova IV, Selivanova GV, Vlasova TD (2002) The influence of epidermal growth factor and insulin on proliferation and DNA synthesis in ciliates Tertrahymena pyriformis. Tsitologia 44:1097–1103

Shemarova IV, Selivanova GV, Vlasova TD (2007) The influence of the EGF on proliferative signal transduction in ciliate Tetrahymena pyriformis. Tsitologia 49:156–160

Shpakov AO, Derkach KV, Uspenskaja ZI, Shpakova EA, Kuznetsova LA, Plesneva SA, Pertseva MN (2004) Molecular mechanisms of regulatory action of adrenergic receptor agonists on functional activity of adenylyl cyclase signaling system of the ciliates Dileptus anser and Tetrahymena pyriformis. Tsitologia 46:317–325

Shpakov AO, Derkach KV, Uspenskaja ZI (2012) Glucose and cyclic adenosine monophosphate stimulate activities of adenylate cyclase and guanylate cyclase of Tetrahymena pyriformis infusoria. Bull Exp Biol Med 152:427–430

Stoka AM (1999) Phylogeny and evolution of chemical communication: an endocrine approach. J Mol Endocrinol 22:207–225

Suzuki Y, Nagao S, Abe K, Hirabayashi T, Watanabe Y (1981) Tetrahymena calcium-binding protein is indeed a calmodulin. J Biochem 89:333–336

Ud-Daula A, Pfister G, Schramm KW (2012) Identification of dopamine receptor in Tetrahymena thermophila by fluorescent ligands. Pak J Biol Sci 15:1133–1138

Umeki S, Nozawa Y (1996) Adenylate and guanylate cyclase in Tetrahymena. In: Csaba G, Müller WEG (eds) Signaling mechanisms in Protozoa and Invertebrates. Springer, Berlin, pp 40–60

Witzany G (2008) Biocommunication of unicellular and multicellular organisms. tripleC 6:24–53

Witzany G (2010) Uniform categorization of biocommunication in bacteria, fungi and plants. World J Biol Chem 26:160–180

Zhang S, Ling Z, Wong S, Nozawa Y, Umeki S (2010) Tetrahymena cell culture. Wiley Online Library 2010

Chapter 9
Signaling Through GPI-Anchored Surface Antigens in Ciliates

Yelena Bisharyan and Theodore Clark

Abstract Antibody-mediated cross-linking of GPI-linked surface membrane proteins known as immobilization antigens (i-antigens) causes rapid loss of forward swimming in motile ciliates. While the physiological relevance of this response is unclear, i-antigens are nevertheless capable of transducing signals involved in cellular defense in the context of host-pathogen and prey-predator interactions. Here we demonstrate an essential role for calcium in the immobilization response. Using *Tetrahymena thermophila* cells lines expressing the calcium-sensitive reporter construct, GCamp2, we find that antigen clustering and cessation of movement are accompanied by a rapid and marked increase in intracellular [Ca^{++}]. Elevation of cytosolic [Ca^{++}] and immobilization are blocked by the addition of calcium chelators (BAPTA-AM and EGTA) and inhibitors of mammalian L-type calcium channel blockers (verapamil and nifedipine) to the culture medium suggesting a role for extracellular calcium in triggering these events. Inhibition of forward swimming by the calcium ionophore, A23187, provides additional strong evidence of a direct role for calcium in the immobilization response. The association of i-antigens with so-called lipid rafts, and the massive reorganization of the plasma membrane in response to antibody binding, suggest a number of mechanisms that could regulate calcium homeostasis following lateral clustering of GPI-anchors at the cell surface.

9.1 Introduction

Cessation of forward swimming in the presence of specific antisera was among the earliest observed responses of ciliated protozoa to external stimuli. Described by Rössle over a century ago (Rössle 1905), the immobilization reaction has been

Y. Bisharyan
Tetragenetics Inc, 91 Mystic Street, 02474 Arlington, MA, USA

T. Clark (✉)
Department of Microbiology and Immunology, Cornell University, 14853 Ithaca, NY, USA
e-mail: tgc3@cornell.edu

G. Witzany and M. Nowacki (eds.), *Biocommunication of Ciliates*,
DOI 10.1007/978-3-319-32211-7_9

extensively investigated in both free-living and parasitic ciliates including *Paramecium spp.*, *Tetrahymena spp.*, and *Ichthyophthirius multifiliis* (Gerber et al. 2002; Clark and Forney 2003; Simon and Schmidt 2007). The target antigens involved in this response (immobilization antigens, or i-antigens) are abundant GPI-anchored proteins on ciliary and plasma membranes that are encoded by families of structurally related genes. By and large, i-antigen paralogs are expressed in a mutually exclusive fashion resulting in distinct serotypes identifiable with specific antibodies. In the case of *Paramecium* and *Tetrahymena*, serotype switching can occur in response to changes in environmental conditions involving, at least in some cases, RNAi-mediated control of gene expression (Marker et al. 2010; Baranasic et al. 2014). Antigenic shift has not been documented in *I. multifiliis*, although strains that stably express different antigens on their surface can easily be identified among geographic isolates, and whole genome sequencing has revealed at least 17 i-antigen genes along with 4 apparent pseudogenes in the most well characterized, G5, parasite isolate (Coyne et al. 2011).

While loss of motility in response to antibody binding is an easily observable phenotype, neither the mechanism underlying the phenomenon, nor its functional significance is fully understood. Free-living organisms such as *Paramecium* and *Tetrahymena* do not encounter antibodies in their natural environment, and src-family tyrosine kinases that transduce signals via GPI-anchored proteins in multicellular organisms, are absent from protists. By contrast, i-antigens are the primary targets of the host immune response to *Ichthyophthiriius*, and antibody-mediated cross-linking can induce a behavioral response leading to premature exit of parasites from the skin of infected fish (Clark et al. 1996; Clark and Dickerson 1997; Clark and Forney 2003). Along the same lines, *Paramecium* and other free-living ciliates serve as prey species for larger predatory organisms in ponds, and several lines of evidence suggest that i-antigens, along with related GPI-anchored proteins (known as “kairomones”), have communicative functions in prey-predator interactions (see Simon and Kusch 2013, for review). Such observations would argue that i-antigens play a general role in cellular defense (which, at its core, requires sensing of extracellular cues) and one could easily imagine that antibodies mimic and, perhaps, amplify the affects of natural ligands for these proteins, at least in free-living ciliates.

Although i-antigens are restricted to the outer leaflet of the lipid bilayer, there are numerous indications that they can transduce signals across the plasma membrane beginning with the fact that immobilization requires antigen clustering at the cell surface rather than antibody binding per se (Eisenbach et al. 1983; Clark et al. 1996), a hallmark of signal transduction phenomena in higher eukaryotes (Cebecauer et al. 2010). Consistent with this, we showed that cytosolic Ca^{++} levels become elevated in response to i-antigen cross-linking leading to a previously unrecognized phenomenon in which *Tetrahymena* and *Ichthyophthirius* jettison large numbers of intact mitochondria to the extracellular space (Bisharyan and Clark 2011). Here we demonstrate that elevated Ca^{++} is, in fact, a trigger for immobilization. Furthermore, we show that i-antigens are associated with detergent-resistant membrane fractions and that antigen clustering results in extensive reorganization of the plasma membrane, including lipid raft components, within a short time of antibody addition. This reorganization is manifested at the cellular level by extensive tubulation/vesiculation of the plasma membrane, migration of vesicles to the tips of cilia, and vesicle shedding to the surrounding medium. The role of Ca^{++} in initiating these phenomena, along with the effects of membrane reorganization on Ca^{++} homeostasis and adaptation to changes in the environment are discussed.

9.2 Results

9.2.1 *i-antigen Cross-Linking Stimulates Elevated Cytosolic [Ca^{++}] in Transgenic* Tetrahymena Thermophila

As demonstrated previously (Bisharyan and Clark 2011), transgenic cell lines of *T. thermophila* expressing the i-antigens of the parasitic ciliate, *Ichthyophthiriius multifiliis*, are rapidly immobilized following incubation in parasite-specific i-antigen antibodies, and can serve as useful tools for studying transmembrane signaling events linked to GPI-anchored proteins in ciliates. The distribution of the IAG52B (serotype D) i-antigen on the surface of infective theronts of *I. mutlifiliis* is shown in Fig. 9.1a, and the response of transgenic *Tetrahymena* expressing the same antigen to monoclonal antibody, G3-61, which recognizes conformational and immobilizing epitopes on the *IAG52B* gene product (Clark et al. 1996; Clark and Forney 2003) is shown in Fig. 9.1b, c. In the absence of antibody transgenic cells swim normally (Fig. 9.1b), while in the presence of high concentrations of antibody, cells cease forward swimming within 1–2 min of antibody treatment and settle to the bottom of the culture dish (Fig. 9.1c). Over time, cells aggregate into distinct foci and remain immotile for varying periods of time up to several hours (data not shown).

To determine whether intracellular [Ca^{++}] becomes elevated in response i-antigen cross-linking, we generated dually transformed *T. thermophila* cell lines harboring genes for both the *Ichthyophthirius* IAG52B i-antigen and a high-affinity calcium-sensitive reporter, G-CaMP2, under the control of a cadmium-inducible (MTT1) promoter (Bisharyan and Clark 2011). Elevation of cytosolic free [Ca^{++}] results in high signal-to-noise ratios of GFP fluorescence in a variety of cell types following expression of the recombinant G-CaMP2 fusion protein (Nakai et al. 2001). As shown previously (Bisharyan and Clark 2011), addition of mAb G3-61 to *T. thermophila* cell lines expressing both the G-CaMP2 probe and the IAG52B parasite surface antigen led to an obvious and sustained increase in green fluorescence detectable by confocal microscopy (Fig. 9.2a, b). Spectrofluorimetric measurements showed that GFP fluorescence was typically several fold above background in antibody-treated cells (Fig. 9.1c), comparable to control cell cultures treated with the calcium ionophore, A23187, in the absence of mAb G3-61 (Fig. 9.2d). Cytosolic Ca^{++} levels remained elevated for considerable periods of time lasting up to several hours after treatment with antibody.

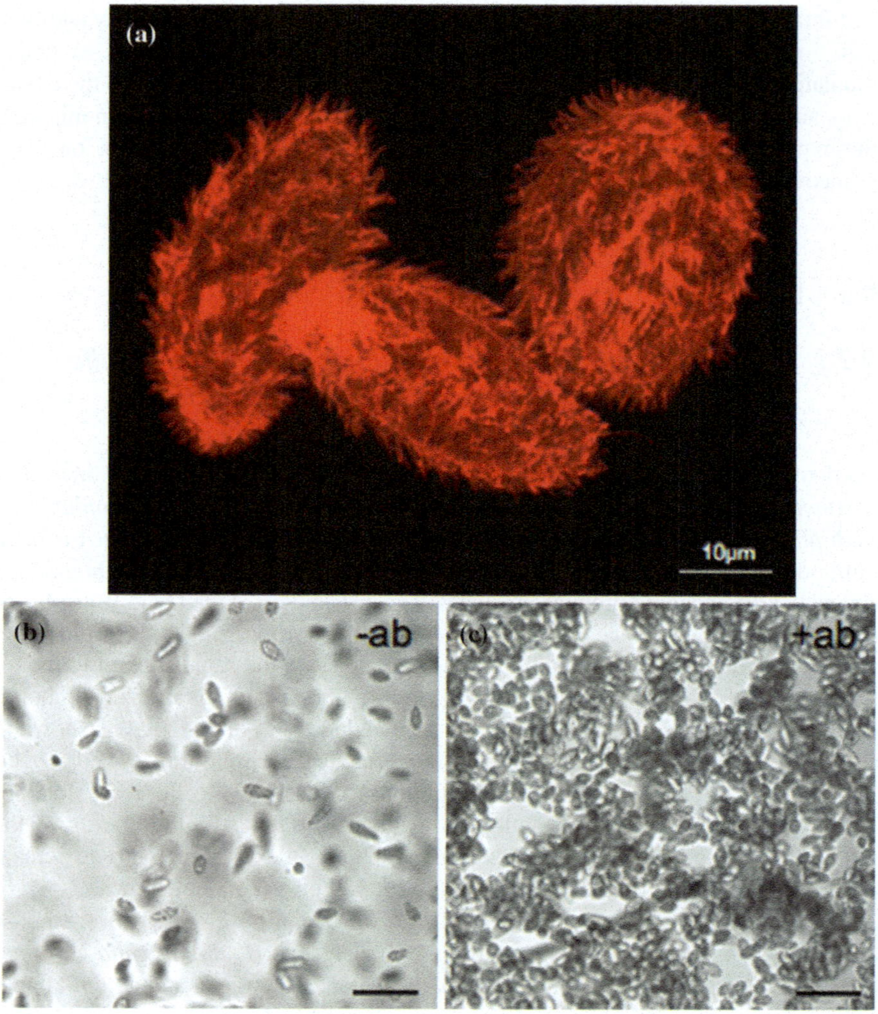

Fig. 9.1 Immobilization of transgenic *Tetrahymena*. Panel **a** Confocal immunofluorescence micrograph of infective theronts of serotype D *Ichthyophthirius multifiliis* labeled with monoclonal antibody G3-61 followed by goat anti-mouse Ig coupled to Texas Red as previously described (Bisharyan and Clark 2011). Note the uniform distribution of antigens (*red*) on ciliary and plasma membranes. Panels **b**, **c** Light microscopic images of *T. thermophila* cell cultures expressing the *IAG52B* parasite gene before (**b**) and after (**c**) treatment with mAb G3-61. Whereas cells are swimming freely in panel (**b**), cells treated with antibody are immobilized and aggregated at the bottom of the culture dish in panel (**c**). Magnification bars in (**b**) and (**c**) = 50 μm

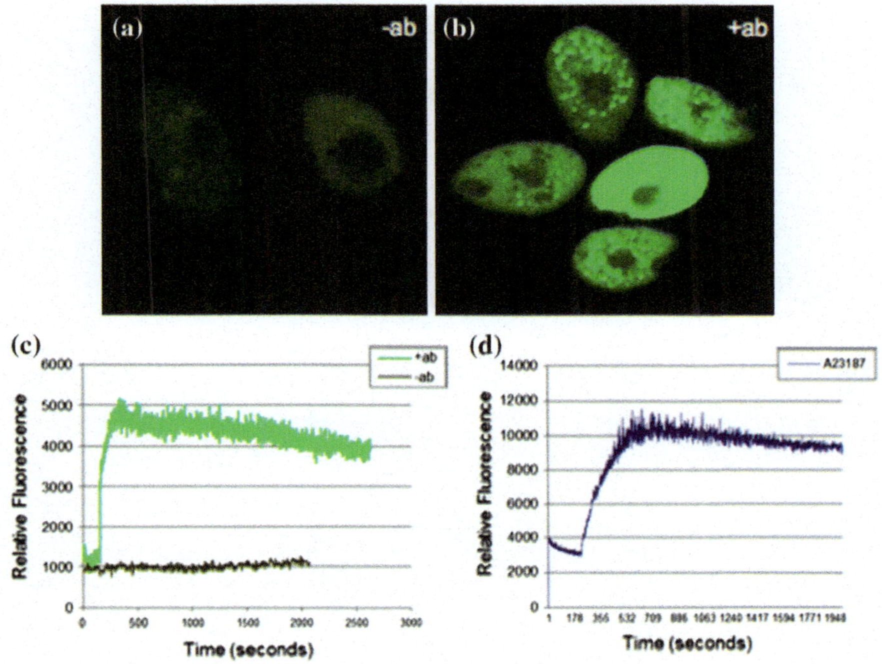

Fig. 9.2 Elevation of cytosolic [Ca^{++}] in response to i-antigen clustering. Panels **a** and **b** show confocal images of live *Tetrahymena* cultures expressing transgenes for the parasite IAG52B i-antigen and the GCamp2 reporter before (**a**) and after (**b**) treatment with mAb G3-61. Note the bright fluorescence in panel (**b**). Panel **c** shows spectrofluorimetric readings from the same cultures in the presence (*green trace*) and absence (*black trace*) of the i-antigen-specific mAb. Panel **d** shows spectrofluorimetric readings of a parallel culture in the presence of the calcium ionophore, A23187, at a concentration of 5 mM

9.2.2 Extracellular Calcium as a Trigger for Elevated Cytosolic [Ca^{++}] and Immobilization

A variety of pathways regulating calcium homeostasis in ciliates have been demonstrated (Plattner 2014, 2015) and the elevation of intracellular [Ca^{++}] described above could involve release from internal stores (alveolar sacs, endoplasmic reticulum, and/or mitochondria), or influx from the outside, or both. To begin to address this question, we examined the effects of calcium chelators added to the culture medium on cytosolic Ca^{++} levels in response to antibody treatment in transgenic *T. thermophila* cell lines. As shown in Fig. 9.3a, addition of BAPTA-AM or EGTA to the medium abrogated the increase in cytosolic [Ca^{++}] induced by antibody treatment. Importantly, this coincided with a dramatic change in swimming behavior in response to antibody binding including a marked decrease in the number of cells that were immobilized and a reduction of the overall duration of the immobilization response (cells that were initially immobilized began to swim

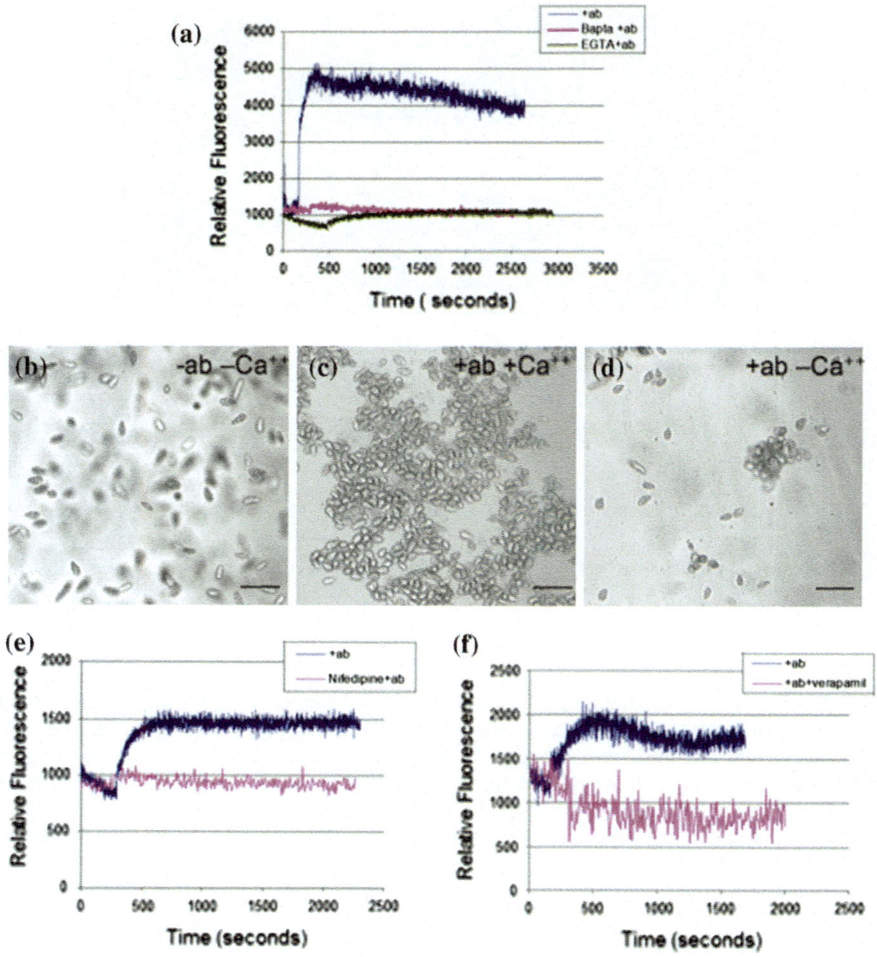

Fig. 9.3 Immobilization is a Ca^{++}-dependent phenomenon. Panel **a** Spectrofluorimetric readings of double transgenic *Tetrahymena* expressing the IAG52B i-antigen and GCamP2 reporter following addition of mAb G3-61 and pretreated with BAPTA-AM (*pink trace*) or EGTA (*black trace*) at concentrations of 20 and 15 mM, respectively. The *dark blue* trace shows fluorescence levels in the absence of chelators. Panel **b** Light micrograph showing double transgenic *Tetrahymena* pretreated with BAPTA-AM (20 μM) and in the absence of immobilizing antibody. Cells swam freely but with reduced velocity in the presence of chelator. Panel **c** The same cells as in **b** following addition of antibody and without pretreatment with chelator. As expected these cells were immobilized and remained clustered in multicellular aggregates even 2 h after antibody addition. Panel **d** The same cells as in (**c**) only pretreated with BAPTA-AM (20 μM). Note that the vast majority of these cells were swimming freely 30 min post-antibody addition. Panels **e**, **f** Spectrofluorimetric readings of double transgenic *Tetrahymena* with (*pink traces*) or without (*dark blue traces*) pretreatment with 50 μM nifedipine (**e**) or 30 μM verapamil(**f**) and following the addition of mAb G3-61

within 15–30 min following antibody treatment as compared to controls that remained immobilized for an hour or more) (Fig. 9.3b–d).

The effects described above clearly suggested a role for Ca^{++} in triggering the immobilization response. This was further indicated by experiments in which varying concentrations of the calcium ionophore, A23187, were added to wild type cells. While very low concentrations had no effect, concentrations in the 1–2 μM range caused backward swimming followed by aggregation of ~50 % of cells, and concentrations in the range of 5–10 μM led to rapid cessation of swimming by all cells in culture (data not shown).

The observations described above suggested a possible role for extracellular Ca^{++} in triggering both immobilization and the elevation of cytosolic [Ca^{++}] that accompanies antibody binding, and led to further questions regarding potential mechanisms of calcium entry into cells including the involvement of either mechanosensitive or voltage-sensitive channels that are known to regulate calcium influx in *Paramecium* and presumably other ciliates as well (Plattner 2014). In a previous study, we showed that verapamil, a known L-type calcium channel blocker in mammals, inhibited the elevation of cytosolic Ca^{++} that follows antibody treatment of transgenic *Tetrahymena* cell lines (Bisharyan and Clark 2011). We repeated that study using verapamil, and a second inhibitor of mammalian voltage-gated calcium channels, namely, nifedipine. As shown in Fig. 9.3e, f both drugs abrogated elevation of cytosolic [Ca^{++}] that normally occurs following the addition of antibody. Moreover, drug-treated cells continued to swim and mimicked the behavior of cells pretreated with calcium chelators before the addition of antibody (data not shown).

9.2.3 *Association of i-antigens with Lipid Rafts*

While the response to channel inhibitors suggested a role for voltage-gated channels in elevation of intracellular [Ca^{++}] in response to antibody binding, other mechanisms could account for this as well including release from internal stores with or without subsequent store-operated calcium entry (SOCE). In this case, second messengers such as inositol 1,4,5 trisphosphate ($InsP_3$) and/or nicotinic acid-adenine dinucleotidephosphate (NAADP) acting on $InsP_3$ and/or ryanodine-like receptors, respectively, could trigger release of Ca^{++} from storage compartments leading to elevation of cytosolic [Ca^{++}] and possibly entry via the equivalent of mammalian calcium release-activated (CRAC) channels. In mammalian cells, the clustering of GPI-anchored proteins in so-called lipid rafts (Rajendran and Simons 2005; Simons and Sampaio 2011) and the coordinate assembly of signaling complexes on the inner leaflet of the plasma membrane are associated with activation of phospholipase C and production of InsP3 from inositol 4,5 bisphosphate (Gamper and Shapiro 2007; Johnson and Rodgers 2008). With that in mind, we sought to characterize lipid

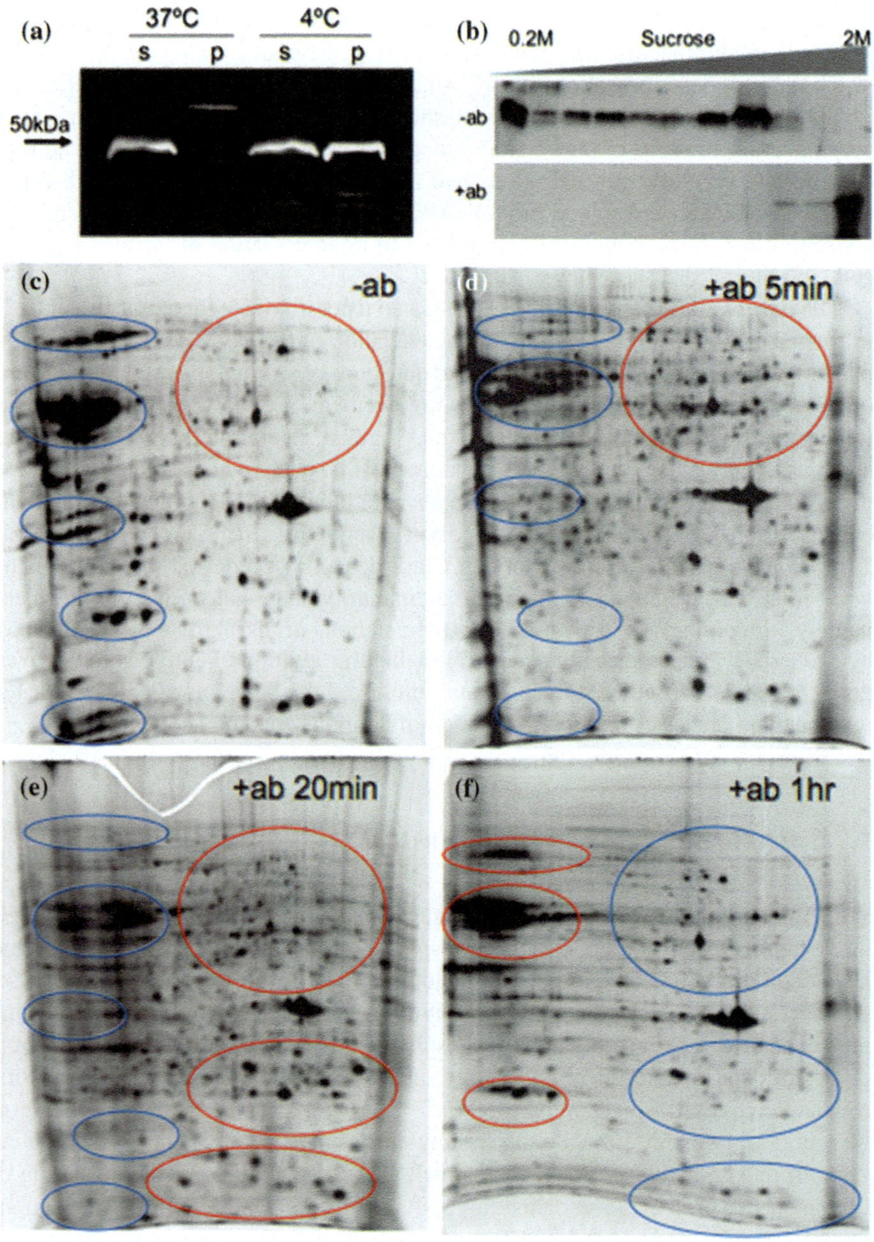
(a)
37°C
4°C
s p s p
50kDa
(b)
0.2M
Sucrose
2M
-ab
+ab
(c)
-ab
(d)
+ab 5min
(e)
+ab 20min
(f)
+ab 1hr

◂ **Fig. 9.4** Association of recombinant i-antigens with lipid rafts. Panel **a** *Tetrahymena* cell lines expressing the parasite IAG52B i-antigen were pelleted by low speed centrifugation, resuspended in buffer A (25 mM Tris pH 7.3, 150 mM NaCl, 5 mM EDTA, containing protease inhibitors E64 and PMSF), and lysed in an equal volume of the same buffer containing 2 % Triton-X100 at 37 or 4 °C. Samples were dounce-homogenized, incubated for 1 h on ice and then centrifuged at 10,000 X g for 20 min. Supernatants (s) and high-speed pellets (p) were subjected SDS-PAGE and Western blotting. Blots were probed with rabbit polyclonal antisera against affinity-purified IAG52B i-antigen followed by secondary goat-anti-rabbit IgG coupled to horseradish peroxidase, and signals developed as previously described (Bisharyan and Clark 2011). Note the differences in the levels of detergent insoluble protein in samples lysed at 37 °C versus 4 °C. Panel **b** Transgenic *T. thermophila* expressing the IAG52B parasite i-antigen were lysed as above in cold Triton X-100-containing buffer either before or 60 min after treatment with mAb G3-61. Cell lysates were then brought to 1 M sucrose and layered under 0.2–0.9 M sucrose gradients in 13.5 ml plastic centrifuge tubes. Gradients were centrifuged at 247,000 X g for 15–17 h at 4 °C using Beckman SW41Ti rotor. Fractions (1 ml) were collected starting from the top of the gradients, and aliquots subjected to Western blotting analysis. Fractions 1–6 represent low-density, lipid raft-containing fractions. Western blots were probed with monospecific rabbit antisera against the IAG52B i-antigen as in (**a**). Panels **c**–**f** Transgenic cells before and at varying times after treatment with mAb G3-61 were lysed and subjected to equilibrium density centrifugation on sucrose gradients as in (**b**). Fractions 1-6 were then harvested, subjected to 2-D PAGE and silver stained. The total lipid raft proteome is shown 0 min (**c**), 5 min (**d**), 20 min (**e**) and 60 min (**f**) after antibody treatment. Blue circles are examples of proteins that were lost and red circles proteins that were recruited to raft-containing fractions at different times after treatment

raft-containing fractions from transformed *T. thermophila* with the expectation that recombinant i-antigens associate with and become enriched in raft-containing fractions following antibody treatment.

Operationally, raft-associated proteins coalesce into insoluble aggregates (also referred to as detergent-resistant membranes) when cells are lysed in cold Triton X-100, allowing further purification by floatation on sucrose density gradients (Lingwood and Simons 2007). Consistent with previous findings in *Tetrahymena mimbres* (Zhang and Thompson 1997), prior to antibody treatment a substantial proportion of the recombinant IAG52B i-antigen expressed in *T. thermophila* was insoluble and appeared in the high speed pellet fractions of cells lysed in cold detergent (Fig. 9.4a). Moreover, when this detergent-resistant fraction was subjected to equilibrium sedimentation on sucrose density gradients (Fig. 9.4b), at least half the recombinant protein was detected in raft-containing fractions at or near the top of the gradients. While this clearly suggested an association of i-antigens with lipid rafts, when detergent-resistant membranes were prepared from the same cells after treatment with mAb G3-61, the IAG52B protein completely disappeared from low-density, raft-containing gradient fractions (Fig. 9.4b). The recombinant i-antigen was not the only protein that behaved this way. Indeed, analysis of the total lipid raft proteome over time showed that while the levels of many proteins were unchanged, others waxed and waned following antibody treatment (Fig. 9.4c–f).

9.2.4 Antibody-Mediated Cross-Linking Releases i-antigens into the Culture Medium

The dramatic loss of i-antigens from lipid raft-containing fractions following antibody treatment was somewhat surprising. Nevertheless, shedding of GPI-proteins into the culture medium is a well-known phenomenon in various protozoan species (Simon and Kusch 2013), and loss of i-antigens from the cell surface in response to antibody binding was described in ciliates as early as the 1950s (Beale and Kacser 1957). To determine whether disappearance of i-antigens from lipid raft fractions was the result of the protein being shed into the culture medium, we examined whole cells and cell-free culture supernatant fractions before and after treatment with mAb G3-61. Prior to antibody treatment, recombinant i-antigens were easily detected in whole cell pellets by Western blotting but not in culture supernatant fractions (Fig. 9.5a). By contrast, signals from whole cells dropped substantially following antibody addition, and large amounts of i-antigen appeared in the culture supernatant indicating that it was, in fact, being shed.

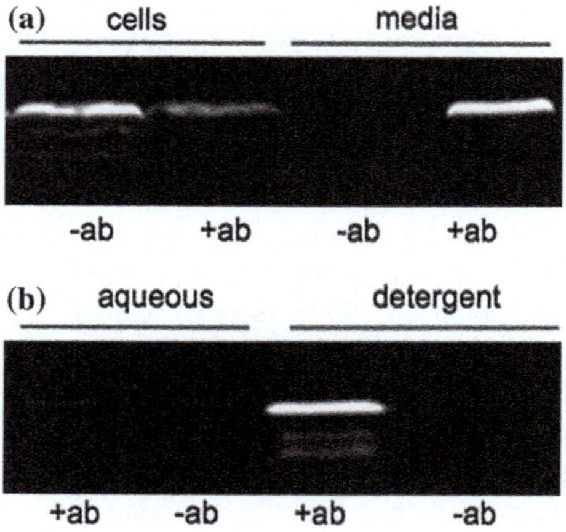

Fig. 9.5 i-antigens are shed from cells with their GPI-anhcors intact. Panel **a** Transformed *T. thermophila* were maintained in the presence (+ab) or absence (−ab) of mAb G3-61 for 30 min and the cells and culture supernatant fractions separated by centrifugation (10,000 X g for 10 min). Cells and media were then fractioned by SDS-PAGE and subjected to Western blotting using rabbit antisera against the IAG52B protein and goat anti-rabbit IgG coupled to horseradish peroxidase as previously described (Bisharyan and Clark 2011). Note the appearance of the recombinant antigen in the media following antibody treatment. Panel **b** Culture supernatant fractions prepared as in (**a**) were subjected to phase partitioning in Triton X-114 as previously described (Clark et al. 2001). Material in the aqueous and detergent phases were harvested, concentrated by acetone precipitation, fractionated by SDS-PAGE and subjected to Western blotting as above. Note that the detergent-soluble material from cell cultures treated with antibody (+ab) shows a strong signal for the recombinant IAG52B protein indicating that the shed protein retains its GPI-anchor

To determine whether release of i-antigens to the surrounding media was the result of enzymatic cleavage of either the protein or its GPI-anchor, we subjected culture supernatant fractions from antibody treated cells to phase partitioning in Triton X-114 (Bordier 1981; Clark et al. 2001). If the GPI-anchor or a large part of the N-terminal ectodomain of the protein were being cleaved prior to shedding, i-antigens would be expected to partition with the hydrophilic aqueous phase following extraction of culture supernatants with Triton-X114 at 30 °C (Clark et al. 2001). To the contrary, Western blots of material from detergent and aqueous phases showed that i-antigens partitioned exclusively with the hydrophobic (detergent) phase indicating that their GPI anchors were intact and that the shed protein was not being cleaved (Fig. 9.5b).

9.2.5 i-antigens Are Shed in Membrane Vesicles

Loss of antigens from the cell surface was clearly evident by confocal microscopy. As shown in Fig. 9.6a and d, cells fixed prior to antibody treatment showed uniform localization of i-antigens on ciliary and plasma membranes in both *I. multifiliiis* theronts, and transformed *T. thermophila* cell lines. When antibodies were added prior to fixation and then fixed at later time points, antigens rapidly redistributed to the tips of cilia (Fig. 9.6b, c, e, f), and, in most cases, were no longer visible within 30 min of treatment.

Because the majority of i-antigen present in the medium was recoverable in high-speed pellets of culture supernatant fractions (not shown), it seemed likely the protein was being shed in membrane vesicles. To determine whether this was the case, we analyzed cells by transmission electron microscopy (TEM) at varying times after treatment with monospecific polyclonal antibodies against the parasite i-antigen. Examination of thin sections revealed extensive tubulation/vesicle blebbing from plasma membranes within ~5 min of antibody treatment, followed by accumulation of membrane vesicles in large aggregates on plasma and ciliary membranes including the ciliary tips (Fig. 9.6g–j). As expected, TEM revealed large numbers of membrane vesicles in high-speed pellets prepared from culture supernatants of these cells (Fig. 9.7a), and analysis of frozen sections showed abundant labeling of vesicle aggregates at the cell surface when samples were reacted with gold-conjugated anti-rabbit secondary antibodies (Fig. 9.7b). Finally, cryosections of cells fixed prior to antibody addition showed no vesicle aggregates, and uniform distribution of gold along the ciliary and plasma membranes (Fig. 9.7c).

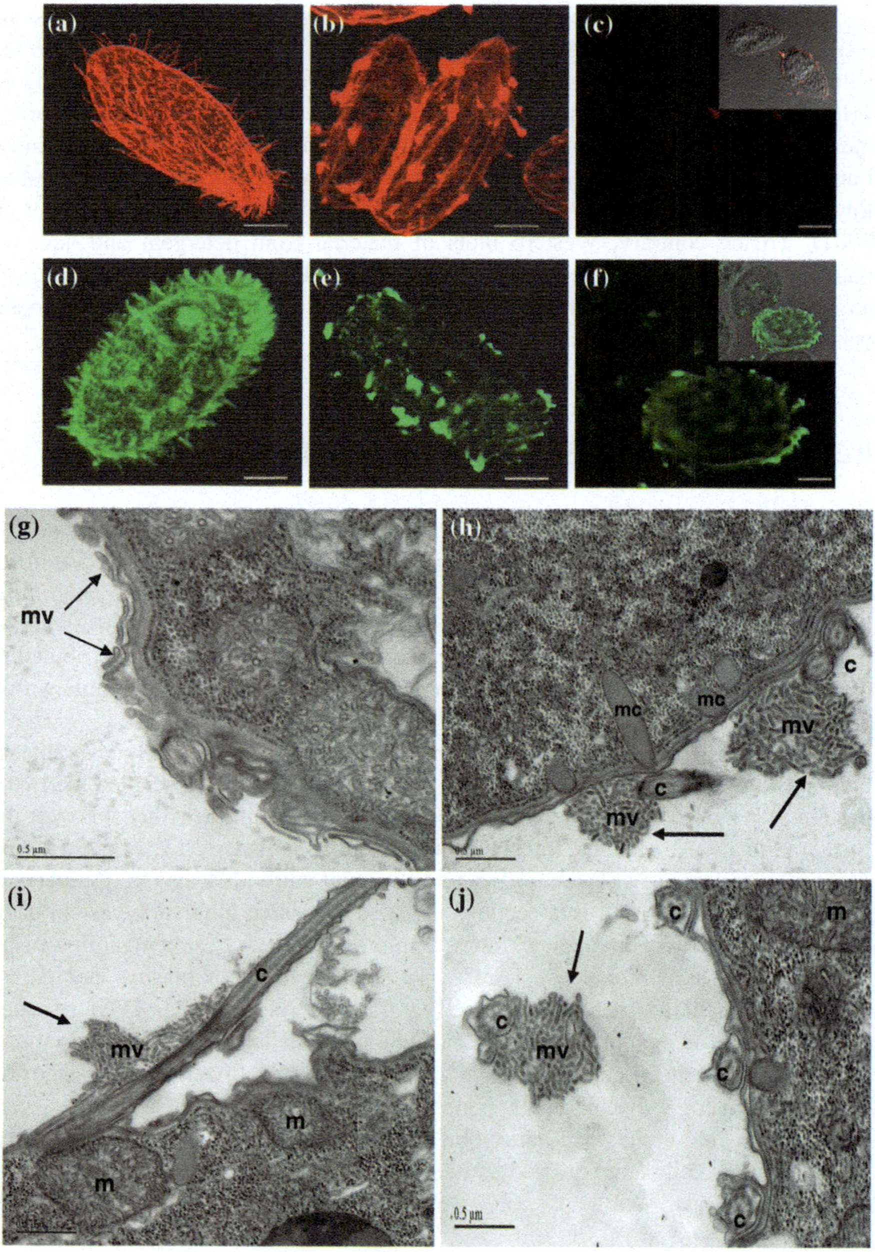
(a)
(b)
(c)
(d)
(e)
(f)
(g)
mv
0.5 μm
(h)
c
mc
mc
mv
c
mv
(i)
c
mv
m
m
(j)
c
m
c
mv
c
c
0.5 μm

◀ **Fig. 9.6** i-antigens are shed from the cell surface in membrane vesicles. Panels (**a**–**c**) Confocal microscopic images of recombinant *T. thermophila* expressing the IAG52B parasite i-antigen. In (**a**), cells were fixed in paraformaldehyde, washed, and labeled with mAb G3-61 (1:100 dilution of hybridoma culture supernatant in 1 % BSA/PBS buffer) and secondary rhodamine (TRITC) tagged goat anti-mouse IgG (1:500 dilution in 1 % BSA/PBS buffer). In (**b**) and (**c**), cells were incubated in mAb G3-61, then fixed at 15 min (**b**) or 1 h (**c**) following antibody treatment. Fixed cells were labeled with secondary TRITC-tagged anti-mouse IgG as above. Panels **d**–**f** Confocal images of serotype D *Ichthyophthirius* theronts labeled as in (**a**–**c**) above, only using fluorescein (FITC)-tagged anti-mouse antibody as the secondary probe. In (**d**), cells were fixed first before labeling with primary and secondary antibody, while in (**e**) and (**f**) cells were fixed at 15 (**e**) or 40 min (**f**) after treatment with mAb G3-61, and then labeled with the secondary FITC-tagged antibody probe. Insets in (**c**) and (**f**) show merged fluorescence and bright field images of cells within the field. Panels **g**–**j** Transmission electron micrographs of cells treated with antibody 5 min (**g**) and 30 min (**h**–**j**) after antibody treatment. Reorganization of the plasma membrane begins with the formation of tubules and vesicles (~50–100 nm diameter) along the plasma membrane, which eventually coalesce into large aggregates that appear on both plasma and ciliary membranes (arrows). Longitudinal sections along the length of the cilia often show regions of electron density at the margins of the axonemes in the same location as the aggregated membrane vesicles outside the cell (panel I). (c-cilia, mv-membrane vesicles, mc-mucocycts)

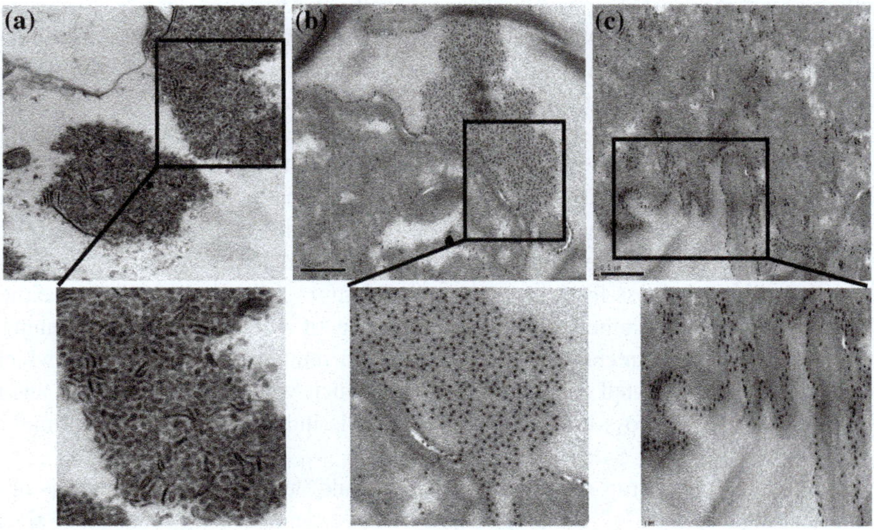

Fig. 9.7 i-antigens are highly enriched in shed vesicles. Panel **a** Low magnification transmission electron micrograph of vesicles harvested from culture supernatants of *T. thermophila* cells expressing the IAG52B parasite i-antigen and treated for 30 min with i-antigen-specific mAb G3-61. A higher magnification image of the vesicles in the inset is shown immediately below. Panel **b** Low magnification image of a cryosection through cells undergoing vesicle shedding 30 min following addition of monospecific rabbit antisera against the IAG52B protein (Clark and Forney 2003). The inset captures a region from a large vesicle aggregate where it borders the plasma membrane. The higher magnification image (*below*) shows heavy labeling of the vesicle aggregate with gold-conjugated secondary antibody. Panel **c** Low magnification cryoEM image from control cells reacted with primary and secondary antibodies after freezing and sectioning. Note the uniform distribution of gold on plasma and ciliary membranes and the absence of membrane vesicles (*inset below*)

9.3 Discussion

As elegantly described by (Plattner 2014), almost all the major pathways regulating calcium homeostasis in human cells had evolved prior to the emergence of metazoa, and ciliates provide exquisite models for studying both the evolution of calcium signaling pathways and the mechanisms that control Ca^{++} influx, storage and release in eukaryotes. Here we show that crosslinking of i-antigens at the surface of *Tetrahymena* induces a sustained increase in cytosolic [Ca^{++}] that we believe contributes to a number of physiological responses, most notably immobilization of cells. Using the GCamp2 reporter, elevated Ca^{++} levels could be detected within seconds of antibody addition, peaking between 3–5 min and remaining elevated for periods lasting up to several hours. Within the same time course, 95–100 % of cells were immobilized suggesting that increased levels of intracellular Ca^{++} and cessation of movement are linked. This was further indicated by the fact that pharmacological agents that blocked elevation of intracellular Ca^{++} in response to antibody (including Ca^{++} chelators and L-type calcium channel inhibitors) prevented immobilization, and that the calcium ionophore, A23187, had the same effect on swimming behavior as i-antigen specific mAbs.

A number of lines of evidence suggest that the trigger for elevated intracellular [Ca^{++}] (if not the source) is extracellular calcium. As described above, addition of moderate concentrations of Ca^{++} chelators to the culture media (sufficient to allow normal swimming, al beit at a slower rate), blocked the increase in intracellular [Ca^{++}] detectable with the GCamp2 reporter, as did the L-type calcium channel blockers, verapamil and nifedipine. With the caveat that many drugs designed for mammalian channels have off-target, or no effect in ciliates (Plattner et al. 2009), the results described here with calcium channel blockers suggest a model in which i-antigen cross-linking activates calcium selective channels, either voltage-dependent or otherwise. To test the role of voltage-dependent channels directly, it would be interesting to examine pawn mutants of *Paramecium*, which lack ciliary voltage-gated Ca^{++} channels, and which would be predicted to swim following i-antigen cross-linking if these channels indeed play a role in immobilization (Schein 1976; Satow and Kung 1976).

Issues relative to drug specificity aside, it would be difficult to ascribe the elevation of intracellular Ca^{++} observed here to voltage-dependent channels alone since these appear to be restricted to ciliary membranes (at least in *Paramecium*), and current evidence suggests that little if any calcium moves from the cilia to the cytosol under normal conditions (Plattner 2014). Still, experiments with *Paramecium* have demonstrated that spillover can occur when cells are over-stimulated (for example, by rapidly increasing the ratios of extracellular to intracellular [Ca^{++}]) (Plattner et al. 2006). Certainly, non-voltage dependent transporters could also mediate Ca^{++} entry. These include mechanosensitive Ca^{++} influx channels on somatic (non-ciliary) membranes, hypothetical CRAC channels, and others (Hofer and Brown 2003). If calcium-selective channels are indeed involved in Ca^{++} entry into cells, a number of important questions remain including (1) how these channels are activated in response to antibody binding; (2) whether Ca^{++} entry from the outside triggers Ca^{++} release from internal stores; and, (3) why cells stop swimming in response to elevated cytosolic [Ca^{++}].

With regard to activation of calcium selective channels in response to i-antigen cross-linking, one could image a number of mechanisms by which this could occur depending on the types of channels involved. In the case of voltage-dependent (e.g. L-type Ca^{++}) channels, membrane depolarization is one possibility, although electrophysiological studies in *Paramecium* suggest that cells maintain their voltage gradients following antibody binding (Ramanathan et al. 1983). Alternatively, lateral clustering of GPI-anchored proteins in plasma and ciliary membranes could influence channel activity either directly, through alterations in the biophysical properties of the lipid bilayer (Dai et al. 2009), or indirectly, through a variety of mechanisms including activation of membrane-associated kinases and/or phosphatases that might alter channel function (Dolphin 2009). Previous studies have shown that i-antigens comprise a substantial fraction of total membrane protein in *Paramecium* and *Ichthyophthirius* (Clark and Forney 2003), and massive loss of GPI-anchors associated with i-antigen shedding would be expected to alter membrane fluidity on a global scale. At the same time, preliminary studies in our laboratory have documented changes in the phosphorylation status of a wide range of proteins associated with lipid raft fractions following antibody treatment and i-antigen shedding (Y.B., data not shown). Activation of voltage-gated channels via post-translational modification could lead to calcium entry to cilia, and possibly spillover into the cytosol as described above. Alternatively, activation of non-voltage dependent channels on the plasma membrane could allow direct entry of calcium to the cytosol.

Superimposed on this is the potential for release of calcium from internal stores that could account for the sustained elevation of cytosolic [Ca^{++}] reported here and elsewhere (Bisharyan and Clark 2011). In cardiac muscle, voltage-dependent Ca^{++} influx through L-type calcium channels in the plasmalemma leads to activation of ryanodine receptors and release of stored Ca^{++} from the sarcoplasmic reticulum (SR) to the cytosol (the so-called, calcium-induced calcium release [CIRC] mechanism). In ciliates, the alveolar sacs, which lie just beneath the plasma membrane, can act as calcium storage compartments similar to the SR (Satir and Wissig 1982; Plattner 2014). Furthermore, both ryanodine-like and InsP3 receptors are present on alveolar membranes and could potentially regulate Ca^{++} release to the cytosol (Plattner 2014).

While evidence for a CICR-like mechanism in ciliates is lacking (Plattner 2014), release of calcium from internal stores could occur via phospholipase C activation and production of $InsP_3$ at the plasma membrane. Activation of phospholipase C has been linked to the assembly of signaling scaffolds on lipid rafts in a variety of cell types (Gamper and Shapiro 2007; Johnson and Rodgers 2008), and $InsP_3$ production in these cells in response to i-antigen clustering would not be surprising. Resulting depletion of Ca^{++} from internal stores could, in turn, lead to capacitative (store-operated) calcium entry through the interaction of proteins (equivalent to

Oria1 and Stim1 in mammals) on alveolar and plasma membranes to form active CRAC channels at the cell surface.

As shown here, immobilization of *Tetrahymena* in response to i-antigen clustering is accompanied by elevation of cytosolic [Ca^{++}], and can be elicited by the addition of a calcium ionophore to the culture medium suggesting a direct effect of calcium on forward movement. While the effects of Ca^{++} on the direction and frequency of ciliary beat have been documented in a variety of systems (Nakamura and Tamm 1985; Andrivon 1988; Iwadate and Nakaoka 2008; Schmid and Salathe 2011) modulation of these parameters, and motility overall, involves a complex interplay between internal Ca^{++} levels and other second messengers such as cGMP and cAMP, as well as ATP as an energy source. In this regard, we recently found that *Tetrahymena* and *Ichthyophthirius* jettison large numbers of mitochondria to the extracellular space following i-antigen clustering (Bisharyan and Clark 2011). Mitochondrial extrusion is also Ca^{++} dependent and occurs in response to heat shock as well as surface antigen cross-linking. Heat shock is known to cause rapid depletion of intracellular ATP (Findly et al. 1983) and immobilization in *Tetrahymena* (Y.B., personal communication), and the release of mitochondria to the extracellular space would likely to contribute to ATP loss. In virtually all ciliates, mitochondria are localized in the cortical cytoplasm subjacent to cilia and provide the ATP that drives dynein-based microtubule sliding. Thus, a scenario in which elevation of cytosolic [Ca^{++}] leading to a drop in intracellular ATP could easily explain the rapid loss of forward swimming in response antibody independent of a direct effect of Ca^{++} on microtubule sliding and/or coordinated ciliary beat.

While antibody-mediated loss of i-antigens from the surface of *Tetrahymena* and *Ichthyophthirius* has been previously reported (Bisharyan and Clark 2011), the TEM studies described here suggest a massive reorganization of plasma and ciliary membranes in response to i-antigen clustering. This begins with the formation of tubules and vesicles on the plasma membrane and culminates with the shedding of large assemblages of membrane aggregates from the tips of cilia. Interestingly, longitudinal sections cut through the long axis of the cilia show regions of electron density within the axoneme in the same locations of the aggregated membrane vesicles on the outside of the cell (Fig. 9.6i). These electron densities resemble intraflagellar transport particles (Kozminski et al. 1993), which drive membrane cargo (including secretory “ectosomes”) on the cilia and flagella of other cell types (Follit et al. 2006; Wood et al. 2013).

Shedding of membrane vesicles (ectosomes; exosomes; microvesicles; etc.) is a well described phenomenon in virtually all eukaryotic cells, and is associated with surface membrane traffic, horizontal transfer of proteins and mRNAs among neighboring cells, and intercellular signaling and cell-cell communication (Fevrier and Raposo 2004; Cocucci et al. 2009; Lotvall and Valadi 2007). It is worth noting in this regard that following addition of i-antigen-specific antibodies to transformed *Tetrahymena*, cells became immobilized first, and then aggregated into large clusters while at the same time shedding membrane vesicles. Thus, while aggregation of cells following antibody binding could result from a physical cross-linking of cells via bivalent antibodies, it may also reflect an active process involving cell-cell communication. Regardless, given the adaptive nature of ciliates and the ability of *Paramecium* and *Tetrahymena* to shift antigen expression in response to environmental stimuli, it is tempting to speculate that membrane shedding in response to i-antigen cross-linking reflects a normal process that cells use to adjust their surface membrane lipids and proteins in order to enhance fitness and overall survival.

Lastly, regardless of the functional significance of membrane shedding, the ability to induce membrane vesicle formation on such a large scale through addition of antibodies or heat shock may provide a useful tool to study how membrane nematogens alter curvature of the lipid bilayer to induce tubulation and/or membrane vesicle budding (Ramakrishnan et al. 2013).

Acknowledgements We wish to thank Dr. Feng-Xia Lian (New York University) and Mrs. Shannon Caldwell (Cornell University) for their excellent technical assistance with cryo-immunoEM and conventional transmission electron microscopy, respectively; Mr. Mozammal Hossein for maintenance of parasites (Cornell University); Dr. David Russell (Cornell University) for the use of his spectrofluorimeter; and Drs. Donna Cassidy-Hanley, David Holowka, and Mandayam Parthasarathy (Cornell University) for their helpful discussions during the course of the work. The research reported in this publication was supported in part by the Office of The Director, National Institutes of Health Office of Research Infrastructure Programs under Award Number P40OD010964 to TGC. The content is solely the responsibility of the authors and does not necessarily represent the official views of the National Institutes of Health.

References

Andrivon C (1988) Membrane control of ciliary movement in ciliates. Biol Cell 63:133–142

Baranasic D, Oppermann T, Cheaib M, Cullum J, Schmidt H, Simon M (2014) Genomic characterization of variable surface antigens reveals a telomere position effect as a prerequisite for RNA interference-mediated silencing in *Paramecium tetraurelia*. MBio 5(6):e01328

Beale GH, Kacser H (1957) Studies on the antigens of *Paramecium aurelia* with the aid of fluorescent antibodies. J Gen Microbiol 17:68–74

Bisharyan Y, Clark TG (2011) Calcium-dependent mitochondrial extrusion in ciliated protozoa. Mitochondrion 11:909–918

Bordier C (1981) Phase separation of integral membrane proteins in Triton X–114. J Biol Chem 256:1604–1607

Cebecauer M, Spitaler M, Sergé A, Magee AI (2010) Signalling complexes and clusters: functional advantages and methodological hurdles. J Cell Sci 123:309–320

Clark TG, Dickerson HW (1997) Antibody-mediated effects on parasite behavior: evidence of a novel mechanism of immunity against a parasitic protist. Parasitol Today 13:477–480

Clark TG, Forney JD (2003). Free-living and parasitic ciliates. In: Craig A, Scherf A (eds) Antigenic variation, Academic Press, London, pp 375–402

Clark TG, Lin TL, Dickerson HW (1996) Surface antigen cross-linking triggers forced exit of a protozoan parasite from its host. Proc Natl Sci USA 93:6825–6829

Clark TG, Gao Y, Gaertig J, Wang X, Cheng G (2001) The i-antigens of *Ichthyophthirius multifiliis* are GPI-anchored proteins. J Eukaryot Micro 48:332–337

Cocucci E, Racchetti G, Meldolesi J (2009) Shedding microvesicles: artefacts no more. Trends Cell Biol 19:43–51

Coyne RS, Hannick L, Shanmugam D, Hostetler JB, Brami D, Joardar VS, Johnson J, Radune D, Singh I, Badger JH, Kumar U, Saier M, Wang Y, Cai H, Gu J, Mather MW, Vaidya AB, Wilkes DE, Rajagopalan V, Asai DJ, Pearson CG, Findly RC, Dickerson HW, Wu M, Martens C, Van de Peer Y, Roos DS, Cassidy-Hanley DM, Clark TG (2011) Comparative genomics of the pathogenic ciliate *Ichthyophthirius multifiliis*, its free-living relatives and a host species provide insights into adoption of a parasitic lifestyle and prospects for disease control. Genome Biol 12:R100

Dai S, Hall DD, Hell JW (2009) Supramolecular assemblies and localized regulation of voltage-gated ion channels. Physiol Rev 89:411–452

Dolphin A (2009) Calcium channel diversity: multiple roles of calcium cahnnel subunits. Curr Opin Neurobiol 19:237–244

Eisenbach L, Ramanathan R, Nelson DL (1983) Biochemical studies of the excitable membrane of *Paramecium tetraurelia*. IX. Antibodies against ciliary membrane proteins. J Cell Biol 97:1412–1420

Fevrier B, Raposo G (2004) Exosomes: endosomal-derived vesicles shipping extracellular messages. Curr Opin Cell Biol 16:415–421

Findly RC, Gillies RJ, Shulman RG (1983) In vivo phosphorus-31 nuclear magnetic resonance reveals lowered ATP during heat shock of *Tetrahymena*. Science 219:1223–1225

Follit JA, Tuft RA, Fogarty KE, Pazour GJ (2006) The intraflagellar transport protein IFT20 is associated with the Golgi complex and is required for cilia assembl. Mol Biol Cell 17:3781–3792

Gamper N, Shapiro MS (2007) Target-specific PIP(2) signalling: how might it work? J Physiol 582:967–975

Gerber CA, Lopez AB, Shook SJ, Doerder FP (2002) Polymorphism and selection at the SerH immobilization antigen locus in natural populations of *Tetrahymena thermophila*. Genetics 160:1469–1479

Hofer AM, Brown EM (2003) Extracellular calcium sensing and signalling. Nat Rev Mol Cell Biol 4:111–119

Iwadate Y, Nakaoka Y (2008) Calcium regulates independently ciliary beat and cell contraction in *Paramecium* cells. Cell Calcium 44:169–179

Johnson CM, Rodgers W (2008) Spatial Segregation of phosphatidylinositol 4,5-bisphosphate (PIP(2)) Signaling in immune cell functions. Immunol Endocr Metab Agents Med Chem 8:349–357

Kozminski KG, Johnson KA, Forscher P, Rosenbaum JL (1993) A motility in the eukaryotic flagellum unrelated to flagellar beating. Proc Nat Acad Sci USA 90:5519–5523

Lingwood D, Simons K (2007) Detergent resitance as a tool in membrane research. Nat Protoc 2:2159–2165

Lotvall J, Valadi H (2007) Cell to cell signalling via exosomes through esRNA. Cell Adh Migr 1:156–158

Marker S, Le Mouël A, Meyer E, Simon M (2010) Distinct RNA-dependent RNA polymerases are required for RNAi triggered by double-stranded RNA versus truncated transgenes in *Paramecium tetraurelia*. Nucleic Acids Res 38:4092–4107

Nakai J, Ohkura M, Imoto K (2001) A high signal-to-noise Ca(2 +) probe composed of a single green fluorescent protein. Nat Biotechnol 19:137–141

Nakamura S, Tamm SL (1985) Calcium control of ciliary reversal in ionophore-treated and ATP-reactivated comb plates of ctenophores. J Cell Biol 100:1447–1454

Plattner H (2014) Calcium regulation in the protozoan model, Paramecium tetraurelia. J. Eukaryot Microbiol 61:95–114

Plattner H (2015) Signalling in ciliates: long- and short-range signals and molecular determinants for cellular dynamics. Bio Rev (In Press)

Plattner H, Diehl S, Husser MR, Hentschel J (2006) Sub-second calcium coupling between outside medium and subplasmalemmal stores during overstimulation/depolarisation-induced ciliary beat reversal in *Paramecium* cells. Cell Calcium 39:509–516

Plattner H, Sehring IM, Schilde C, Ladenburger EM (2009) Pharmacology of ciliated protozoa–drug (in)sensitivity and experimental drug (ab)use. Int Rev Cell Mol Biol 273:163–218
Rajendran L, Simons K (2005) Lipid rafts and membrane dynamics. J Cell Sci 118:1099–1102
Ramakrishnan N, Sunil Kumar PB, Ipsen JH (2013) Membrane-mediated aggregation of curvature-inducing nematogens and membrane tubulation. Biophys J 104:1018–1028
Ramanathan R, Saimi Y, Peterson JB, Nelson DL, Kung C (1983) Antibodies to the ciliary membrane of *Paramecium tetraurelia* alter membrane excitability. J Cell Biol 97:1421–1428
Rössle R (1905) Spezifische seren gegen infusorien. Arch Hyg Bakteriol 54:1–31
Satir BH, Wissig SL (1982) Alveolar sacs of *Tetrahymena*: ultrastructural characteristics and similarities to subsurface cisterns of muscle and nerve. J Cell Sci 55:13–33
Satow Y, Kung C (1976) Mutants with reduced Ca activation in *Paramecium aurelia*. J Membr Biol 28:277–294
Schein SJ (1976) Calcium channel stability measured by gradual loss of excitability in pawn mutants of *Paramecium aurelia*. J Exp Biol 65:725–736
Schmid A, Salathe M (2011) Ciliary beat co-ordination by calcium. Biol Cell 102:159–169
Simon MC, Kusch J (2013) Communicative functions of GPI-anchored surface proteins in unicellular eukaryotes. Crit Rev Microbiol 39:70–78
Simon MC, Schmidt HJ (2007) Antigenic variation in ciliates: antigen structure, function, expression. J Eukaryot Microbiol 54:1–7
Simons K, Sampaio JL (2011) Membrane organization and lipid rafts. Cold Spring Harb Perspect Biol 3:a004697
Wood CR, Huang K, Diener DR, Rosenbaum JL (2013) The cilium secretes bioactive ectosomes. Curr Biol 23:906–911
Zhang X, Thompson GA Jr (1997) An apparent association between glycosylphosphatidylinositol-anchored proteins and a sphingolipid in *Tetrahymena mimbres*. Biochem J 323:197–206

Chapter 10
Ciliate Communication via Water-Borne Pheromones

Pierangelo Luporini, Claudio Alimenti, Bill Pedrini and Adriana Vallesi

Abstract Ciliates communicate via diffusible signaling pheromones that have been identified in association with genetic mechanisms of mating types, either binary or multiple, that control the ciliate-specific sexual phenomenon of conjugation. Except for a glycoprotein and a tryptophan-related molecule acting as pheromones in *Blepharisma*, the other pheromones are cysteine-rich proteins varying in length among *Euplotes* species from 32 to 109 amino acid residues. These proteins are all structurally homologous sharing a common three-helix bundle fold, which allows them to bind cells in competition with one another in close analogy with the families of protein growth factors and cytokines of multicellular organisms. Although identified in the extracellular environment for their mating-induction activity, pheromones act also, and presumably primarily, as autocrine growth factors that bind to, and promote the reproduction of the same cells from which they are constitutively released.

10.1 Introduction

Ciliates, like many other organisms from bacteria to animals, communicate via water-borne chemical messages. Being secreted by an individual and received by other individuals of the same species in which they elicit a specific behavioral reaction, or a developmental process, they fully fit with the pheromone definition of Karlson and Luscher (1959). Ciliate pheromones are the effectors of an allo-recognition mechanism identified with genetically determined systems of mating types, which in ciliates may either be of binary and closed type and recall the dual nature of sex as typical in *Paramecium*, or of multiple, virtually 'open' type

P. Luporini (✉) · C. Alimenti · A. Vallesi
Laboratory of Eukaryotic Microbiology and Animal Biology, School of Biosciences and Veterinary Medicine, University of Camerino, 62032 Camerino, (MC), Italy
e-mail: piero.luporini@unicam.it

B. Pedrini
Paul Scherrer Institute, 5232 Villigen PSI, Switzerland

G. Witzany and M. Nowacki (eds.), *Biocommunication of Ciliates*,
DOI 10.1007/978-3-319-32211-7_10

and recall the incompatibility systems of flowering plants as is the case in many spirotrichs (Dini and Nyberg 1993; Phadke and Zufall 2009). They are the chemical markers that distinguish cells of one mating type from cells of another mating type, thus permitting them to discriminate between sibs (self) and genetically distinct (non-self) co-specifics. This discrimination becomes manifest when comparing the behavior of cells that are left to coexist with other cells of their same lineage (clone), with the behavior of cells that are mixed with others of a different clone. While in the former case cells maintain their individual vegetative stage by growing (reproducing) by binary fission, in the latter they are induced to socialize. They temporarily exit from the reproductive stage and engage in a sexual phenomenon of conjugation which is unique to ciliates. It involves the union in mating pairs of two vegetative cells which generate gamete-nuclei from their diploid germinal micronuclei only as a consequence of their mating union, and use these 'male' and 'female' gamete-nuclei to operate a mutual fertilization and exchange of genes, as occurs in simultaneous hermaphroditism in animals. Furthermore, these mating pairs may equally be 'heterotypic', i.e. formed between cells of different mating types, or 'homotypic', i.e. formed between cells of the same type, which are as fully fertile as the heterotytpic ones except for the unique case of *Blepharisma* (Miyake 1981). It is just in relation to this unique capacity of ciliates to form homotypic mating pairs, that nearly 75 years ago Kimball (1942) was able to pioneer the identification of ciliate pheromones in culture filtrates of a relatively common species of fresh water, *Euplotes patella*.

With regard to the binary system of mating types described by Sonneborn (1937) for the control of conjugation in *Paramecium aurelia*, Kimball (1939, 1942) observed two basic divergences in *E. patella*. First, *E. patella* was shown to control conjugation through a multiple system of mating types (Mt I, II, III and so forth) determined at a single mating-type locus through a multiple series of alleles regulated by relationships of co-dominance. Second, it was observed that while *P. aurelia* cells of the two different mating types start sticking into large clumps immediately after being mixed, *E. patella* cells of any two of the multiple mating types must interact for a discrete interval of time to become able to stick together in the formation of mating pairs. In inspecting the reasons for these divergences, Kimball assessed cell-free filtrates from an array of clonal cell cultures for any potential effect on the mating performances of clonal cell cultures of other mating types. By surprisingly observing a strong filtrate activity in inducing the formation of homotypic mating pairs in specific sets of cell cultures and not in others, he thus obtained clear evidence that *E. patella*, and ciliates in general can freely release their fully active 'mating-type factors' into the extracellular environment. We now usually call these water-borne factors 'pheromones', or 'gamones' or 'mating pheromones', and use the Kimball mating induction assay as the most practical procedure to identify pheromone-secreting ciliate species.

10.2 Pheromone-Secreting Species

Species of *Blepharisma* (Miyake 1981), *Euplotes* (Luporini et al. 1983), and *Dileptus anser* (Yudin et al. 1990; Uspenskaya and Yudin 2016) in general prove to be immediately prone to reveal their nature as pheromone-secreting species by responding positively to mating induction assays. Others, responding negatively, have usually been regarded as species that retain their pheromones bound to the cell surface, communicating only through direct physical cell–cell contacts. Nevertheless, although adverse to responding positively in mating induction assays, species of *Ephelota*, *Tokophrya*, and *Oxytricha* were revealed to communicate via pheromones. In the sessile and stalked suctorian *E. gemmipara* (Grell 1953), *T. lemnarum* and *T. infusionum* (Sonneborn 1978), cells of different mating types grown in close, but not physical contact with one another, have been observed to extend their bodies into pseudopodium-like projections directly oriented toward the prospective mate, thus clearly denoting the occurrence of an intercellular exchange of environmental mating inducing messages. In *O. bifaria* (Esposito et al. 1976), pheromone communication has been inferred by pre-incubating cultures of different mating types in chambers connected via micro-pore filters, and observing that mating pairs start forming in their mixtures within a significantly shorter gap of time ('waiting period') than in mixtures between non pre-incubated cell cultures.

The currently meager number of known pheromone-secreting species clearly cautions us from any generalization. However, the fact that these species lie in distinct clades of the ciliate phylogenetic tree is indicative that communication via pheromones in ciliates is likely much more common than so far suspected, and might represent an ancient evolutionary trait.

10.3 Pheromone Secretion

Ciliate pheromone secretion appears in general to follow a constitutive pathway. In *E. octocarinatus*, immunocytochemical analyses have provided evidence that, on the way to the extracellular environment, pheromones transit and are probably stored in saccule-like cortical structures ('ampules') surrounding the pits lodging the ciliary roots (Kush and Heckmann 1988a). In *B. japonicum* (Sugiura et al. 2005) and *E. octocarinatus* (Kusch and Heckmann 1988b), pheromones have been reported to be released only in temporal coincidence with developmental and physiological stages in which cells are competent to mate. In *E. raikovi*, instead, pheromone secretion starts from the very beginning of the life cycle when cells are still 'immature' for mating and continues throughout, regardless of whether cells are growing in presence of food and not competent to mate, or in early stationary phase and competent to mate (Vallesi et al. 1995).

Deep inter-specific variations exist for the amounts in which pheromones are secreted. In *Euplotes* species, only 5.6 μg of protein are purified from 10 liters of culture filtrates in *E. octocarinatus* (Kusch and Heckmann 1988b), and 3.30 mg in *E. raikovi* (Concetti et al. 1986). In this species, so far unrivalled for pheromone quantitative production, it has been roughly calculated a production (under optimized conditions) of 10–15 pg of pheromone/cell/day. Since the adoption of different procedures in extracting pheromones from culture filtrates cannot explain these inter-specific differences in pheromone production, it is likely that they reflect genetically established species-specific eco-genetic strategies of life.

Quantitative variations in pheromone secretion are also remarkable at intraspecific level between cells of different mating types (Raffioni et al. 1992), and appear to be due to inter-mating type variations in the degree of amplification of the pheromone coding genes, with the higher rates of pheromone production that correlate to the higher gene copy numbers (La Terza et al. 1995). They well account for the ample variations in the intensity of mating reactions that one cell type usually manifest in interacting with the other types of the system, as well as for the variable ratios of heterotypic to homotypic mating pairs that are commonly observed among different mating combinations (Luporini and Miceli 1986).

10.4 Pheromone Structures

Although ciliate pheromones were first isolated and characterized from *Blepharisma* (Miyake 1981), a wider structural characterization has been obtained in *Euplotes* which can be regarded as an immense source of pheromone structural polymorphism. This ciliate is unique for a diversification into dozens of species diversified into dozens of cell (mating) types, each distinct from all the others for its own structurally unique pheromone. The isolation of soluble pheromones also from *E. crassus* (Alimenti et al. 2011), that has usually been regarded as unable to secrete pheromones due to its unresponsiveness to mating induction assays, makes it evident that pheromone secretion is a general phenomenon within *Euplotes*. Taken together with the ubiquitous distribution of *Euplotes* species in nature, this presumes that *Euplotes* pheromones are a pervasive chemical component of every aquatic habitat of the globe, that as such might be exploited (for example, by creating gradients) as an environmental chemical signal also by other organisms.

Except for the *Blepharisma* pheromone known as 'gamome-2' or 'blepharismone', which is a cell-type non-specific tryptophan-derivative identified as 3-(2'-formylamino-5'-hydroxybenzoyl)lactate (Miyake 1981), all the other structurally known ciliate pheromones are cell-type specific proteins. As is usual for secreted proteins, they are synthesized in the form of 'immature' cytoplasmic precursors (prepropheromones) from which the soluble forms are released after co- and post-translational proteolytic cleavages of the signal-peptide and pro-segment. Only the 272-amino acid sequence of the *B. japonicum* pheromone, known as 'gamone-1' or 'blepharmone', carries six covalently bound sugars (Sugiura et al. 2001). *Euplotes* pheromones are not glycosylated, even if sugars and/or other compounds can associate non-covalently to the native protein conditioning its activity (Concetti et al. 1986).

Their structures are known from *E. raikovi*, *E. nobilii*, *E. octocarinatus*, *E. crassus* and *E. petzi*; not from *E. patella*, in spite of the pioneer role played by this species in the story of ciliate pheromones. However, *E. patella* pheromones, that have been also partially purified (Akada 1986), can safely be predicted to find close structural counterparts with those known from *E. octocarinatus* with which *E. patella* is phylogenetically and ecologically very closely related (Di Giuseppe et al. 2014).

Pheromones of *Euplotes* have rather acidic isoelectric points (3.7–4), and polypeptide chains that vary greatly in length from species to species, including only 32 amino acid residues in *E. petzi* (Pedrini et al. manuscript submitted) and 85 to 108 in *E. octocarinatus* (Schulze-Dieckhoff et al. 1987). In *E. crassus*, due to a phenomenon of gene duplication, they form two distinct subfamilies with sequences of 45 and 56 amino acids, respectively (Vallesi et al. 2014).

The large intra-specific variations that are usually observed in the percentages of pheromone sequence identity have no relationship with the degree of affinity/divergence that exists in the mating compatibility between any two cell types. To exemplify from *E. raikovi*, there is much stronger mating compatibility between the two cell types which secrete pheromones E*r*-1 and E*r*-2 that have only 25 % sequence identity, than between the cell types distinguished by pheromones E*r*-11 and E*r*-20 that have 59 % sequence identity. In general, a strict sequence conservation involves only the sites occupied by Cys residues and, to a lesser extent, the amino-terminal site where the conservation of the same, or a similar, residue clearly reflects a structural requirement for the processing enzyme that releases the mature protein from the precursor.

Independently of the extensive variations in their amino acid sequences, all *Euplotes* pheromones have molecular structures that, while clearly unique, closely mimic one another consistently with their multi-allelic determination at the same genetic locus. These structures are known for pheromones E*r*-1, E*r*-2, E*r*-10, E*r*-11, E*r*-22 and E*r*-23 of *E. raikovi* (Brown et al. 1993; Luginbühl et al. 1994; Mronga et al. 1994; Ottiger et al. 1994; Liu et al. 2001; Zahn et al. 2001), pheromones E*n*-1, E*n*-2, *En*-6 and E*n*-A1 of *E. nobilii* (Placzek et al. 2007; Pedrini et al. 2007; Di Giuseppe et al. 2011) and, most recently, for the pheromone E*p*-1 of *E. petzi* (Pedrini et al. manuscript submitted). Each has been determined by analyses of NMR spectroscopy and X-ray crystallography (limitedly to pheromone E*r*-1) of native protein preparations.

With reference to the more extensively analyzed pheromone families of *E. raikovi* (Fig. 10.1) and *E. nobilii* (Fig. 10.2), the common architectural core is provided by an up-down-up, anti-parallel bundle of three helices, that are held together by the conserved spatial arrangement of the disulfide bridges (three in *E. raikovi*, four in *E. nobilii*). The first and third helices take a regular α-conformation, while the second helix may shift from a regular to a more distorted shape. Each pheromone reveals its multiple structural specificities distributed across several districts of the molecule, but the most apparent hallmarks systematically reside in the greatly variable

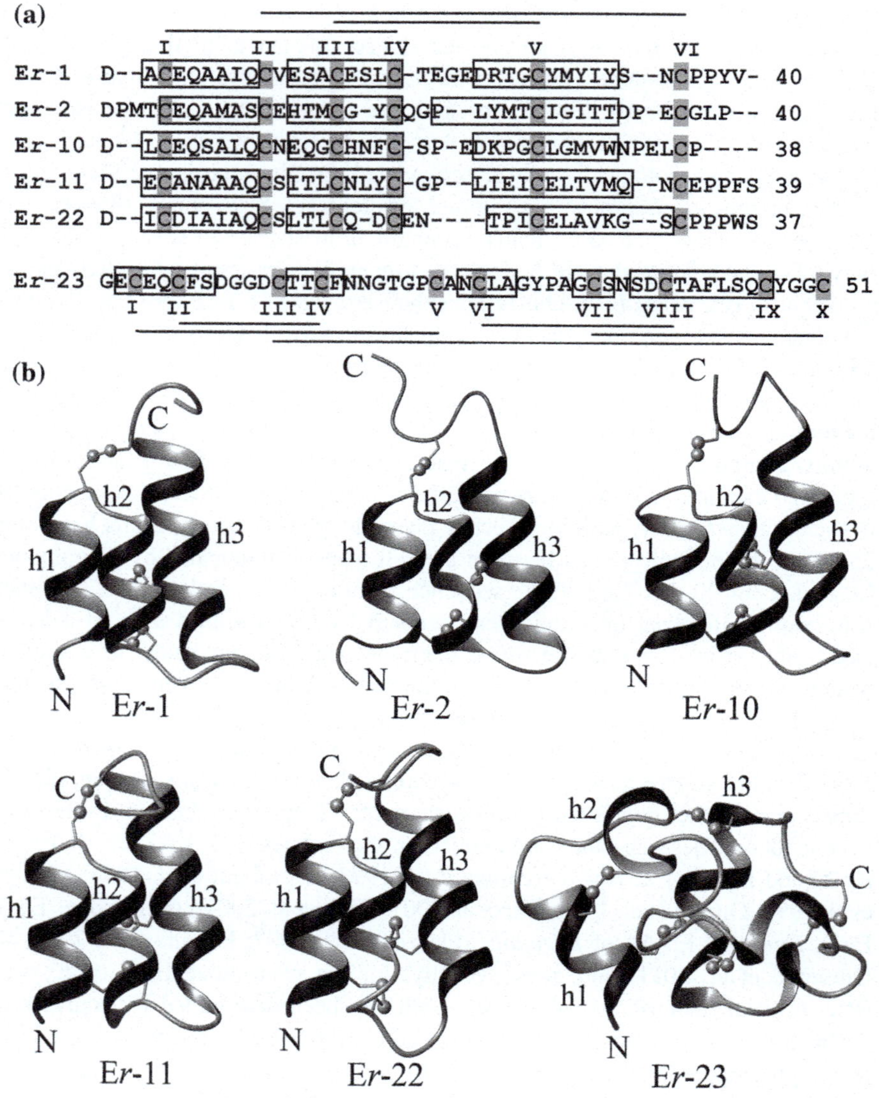

Fig. 10.1 The *E. raikovi* pheromone family. **a** Amino acid sequences of pheromones with known three-dimensional structures. The sequence of pheromone E*r*-23, a deviant member of the family, is not aligned with the other sequences. The cysteines are shadowed, indicated by progressive Roman numerals and connected by horizontal lines according to their pairings in disulfide bonds. Boxes identify the sequence segments arranged in helical structures. **b** Ribbon presentations of the molecular structures determined by NMR analysis of native protein preparations. The three α-helices (h) common to all six molecules are numbered progressively from the amino (N) to the carboxyl (C) chain end. The two additional helical motifs unique to the E*r*-23 are not numbered. The disulfide bonds are represented as sphere and stick diagrams. Protein data bank (PDB) entries: E*r*-1, 1ERC; E*r*-2, 1ERD; E*r*-10, 1ERP; E*r*-11, 1ERY; E*r*-22, 1HD6; E*r*-23, 1HA8

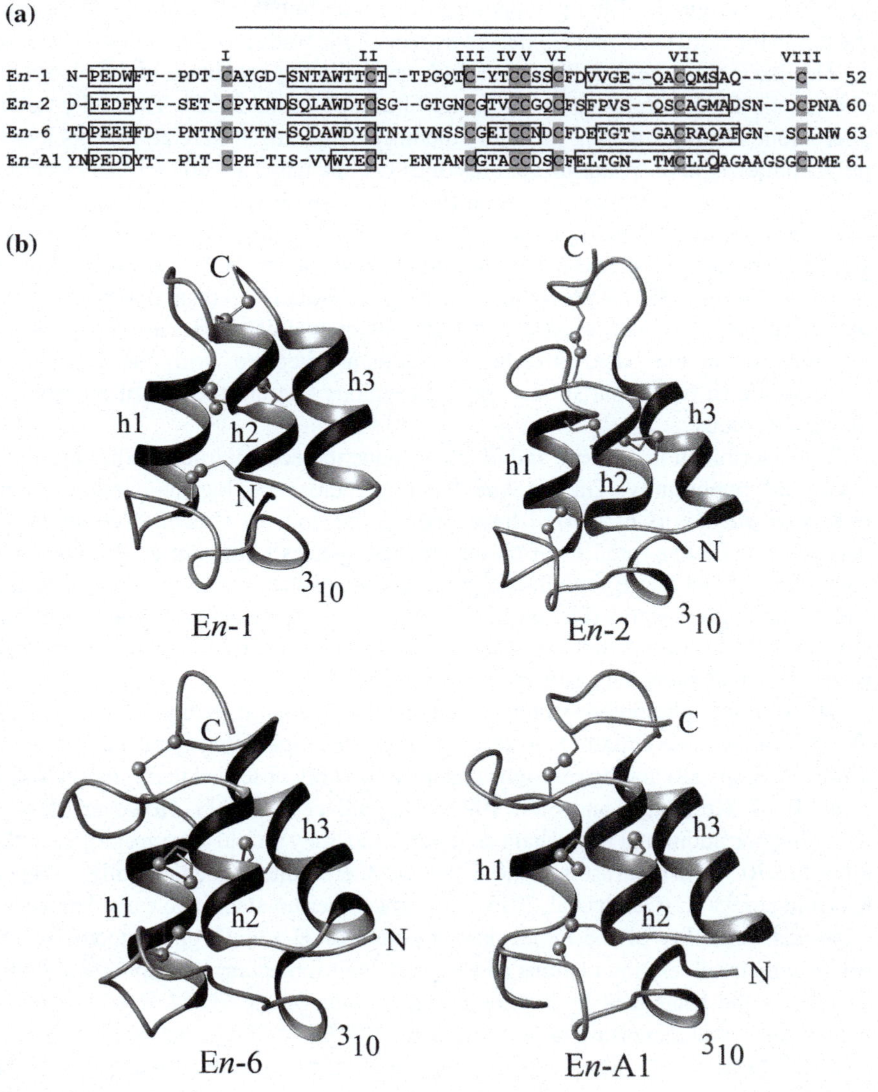

Fig. 10.2 The *E. nobilii* pheromone family. **a** Amino acid sequences of the four polar *E. nobilii* pheromones with known three-dimensional structures. Pheromones E*n*-1, E*n*-2 and E*n*-6 are of Antarctic origin, and E*n*-A1 of Arctic origin. Indications as in Fig. 10.1a. **b** Ribbon presentations of the molecular structures determined by NMR analysis of native protein preparations. The 3_{10} helical turn at the amino (N) terminal is indicated together with the three α-helices (h) numbered progressively from the N to the carboxyl (C) chain end. The disulfide bonds are represented as sphere and stick diagrams. PDB entries: E*n*-1, 2NSV; E*n*-2, 2NSW; E*n*-6, 2JMS; E*n*-A1, 2KK2

geometry and length of the polypeptide carboxy-terminal segment extending from the top of the third helix and connected by a disulfide bridge to the helix bundle core.

This basic three-helix fold pattern finds an intriguing exception in the *E. raikovi* pheromone E*r*-23 (Zahn et al. 2001; Di Giuseppe et al. 2002). This fully active pheromone is unique among its family members (Fig. 10.1). In relation to its longer polypeptide chain of 51 amino acid residues (vs. the basic 38–40) and ten cysteines (vs. the basic six), it possesses a molecular structure characterized by an eccentric five-disulfide bridge pattern and a five-helix molecular fold dictated by the necessity of accommodating two new additional helical motifs. However, in spite of its eccentric fold, the E*r*-23 pheromone structure is equally reconcilable with all the other structures. Its first, second and fifth helices maintain the same up-down-up geometry as in the basic three-helix bundle architecture, with the consequent conservation of functional surface-exposed regions that are the site of intermolecular associations.

The determination of the *E. nobilii* pheromone molecular structures has provided a twofold opportunity. On the one hand, the close phylogenetic relationships linking *E. nobilii* to *E. raikovi* have opened the door to structural comparisons between pheromone families of closely related species (Alimenti et al. 2009). On the other, the fact that *E. nobilii* is a polar species, isolated from both the Antarctic and Arctic marine waters (Di Giuseppe et al. 2011), has made it feasible to seek the structural modifications that psychrophilic proteins in general have evolved to cope with a thermodynamically adverse environment.

The *E. nobilii* pheromone structures extensively overlap with those of *E. raikovi* at the level of the three-helix bundle core, and this overlapping in general well accounts for the inter-species mating reactions that are frequently observed in *Euplotes* (Nobili et al. 1978; Kuhlmann and Sato 1993; Alimenti et al. 2011). However, despite including one additional intra-chain disulfide bond, they unfold at temperatures in the range 55–70 °C, in contrast with the *E. raikovi* pheromones which are fully stable up to heating to 95 °C (Geralt et al. 2013). This lower thermo-stability clearly denotes an improved flexibility of the molecular backbone of *E. nobilii* pheromones, which enhance their solvent interactions and weaken their molecular hydrophobic core by decreasing the content of hydrophobic amino acids (44 % vs. 54 % in *E. raikovi* pheromones) and increasing those of the polar (43 % vs. 33 %) and aromatic amino acids (11 % vs. 6 %). Most apparent is, however, the improved extension of the unstructured regions in *E. nobilii* pheromones. The three helices of these pheromones are all appreciably shorter than their counterparts in *E. raikovi*, and their polypeptide chains at the amino-terminus are all characterized by a segment of 15–17 residues that includes only a 3_{10}-helix turn as secondary rigid structure (Fig. 10.2).

The recent determination of the NMR solution structure of the *E. petzi* pheromone E*p*-1, one of the members of the *E. petzi* pheromone family, has brought to light a two-helix molecular fold stabilized by four disulfide bridges that is new among the *Euplotes* pheromone structures (Fig. 10.3). Considering the early divergence of *E. petzi* in the *Euplotes* phylogenetic tree (Di Giuseppe et al. 2014), this two-helix fold (that is likely distinctive of the whole *E. petzi* pheromone family) can be regarded as an evolutionary precursor of the three-helix fold of the

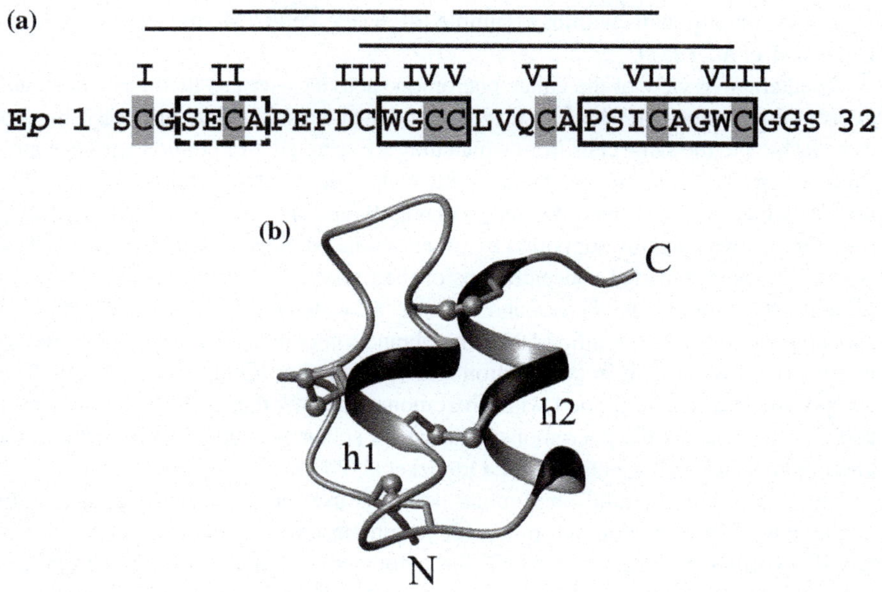

Fig. 10.3 Pheromone E*p*-1 of *E. petzi*. **a** Amino acid sequence with indications as in Fig. 10.1a. A dashed box has been added to enclose four residues involved in the formation of a helix-like turn recalling the first helix of the *E. raikovi* and *E. nobilii* pheromones. **b** Ribbon presentation of the molecular structure determined by NMR analysis of a native protein preparation. The two α-helices (h) are numbered from the amino (N) to the carboxyl (C) chain end. The disulfide bonds are represented as sphere and stick diagrams. PDB entry: 2N2S

E. raikovi and *E. nobilii* pheromones (Pedrini et al. manuscript submitted). Not only are its two helices spatially equivalent to the second and the third helix of these pheromones, but its amino-terminal segment also includes a four-residue turn that clearly presages a subsequent development into a well-defined helical secondary structure.

10.5 Pheromone Activity

The structural divergence between the glycoprotein gamone 1 and the tryptophan derivative gamone 2 of *B. japonicum* has for long time suggested that ciliate pheromones are to be viewed only as 'sexual signals' that cells, defined as 'sexually complementary', exchange and bind to accomplish mutual stimulation for mating and fertilizing (Miyake 1981, 1996). In this context, ciliate pheromones have been reported to function also as chemo-attractant (Afon'kin and Yudin 1987; Kosaka 1991; Sugiura et al. 2010), and to be active at 10^{-11}–10^{-12} M concentrations by measuring the minimum protein amount necessary to induce the formation of at least one mating pair within a sample of tester cells (Miyake 1981; Schulze Dieckhoff et al. 1987).

An alternative view of the ciliate pheromone activity warrants that these molecules evolved not as 'non-self', but as 'self' recognition signals that bind, primarily, in autocrine fashion to the same cells from which they are secreted, promoting their vegetative (mitotic) growth (Luporini and Miceli 1986; Vallesi et al. 1995). The mating induction activity would therefore be a derived, secondarily acquired property. This self recognition model was initially supported by observations on the ability of *Euplotes* cells to 'sense' the environmental concentrations of their secreted pheromones (Luporini and Miceli 1986; Kusch and Heckmann 1988b), as well as cross-linking experiments showing that cells use a specific 14 kDa membrane component to bind their pheromones in autocrine fashion and in competition with one another (Ortenzi et al. 1990). This component was then identified with pheromone isoforms that cells co-synthesize, in addition to their soluble pheromones, through a splicing mechanism from the same pheromone genes (Miceli et al. 1992; Ortenzi et al. 2000).

These membrane-bound pheromone isoforms use the uncleaved signal peptide of the cytoplasmic pheromone precursor to remain anchored to the cell surface as type-II membrane proteins, in which the cytoplasmic domain is formed by a new sequence at the amino terminus and the extracellular ligand-binding domain is structurally identical (because of the common origin from the same gene) to the secreted pheromone. Based on this structural equivalence between the two ligand and ligand-binding moieties, the pheromone-receptor interactions on the cell surface have been thought of as being mimicked by the protein-protein interactions that tightly pack molecules of the *E. raikovi* pheromone E*r*-1 into crystals, originally resolved at the limit of 1.6 Å (Weiss et al. 1995) and now of 0.7 Å (A. Finke and M. May, PSI, personal communication). As shown in Fig. 10.4, in the two dimensional array of the crystals, half of the pheromone molecules (resembling triangular pyramids with the three faces formed by two helices each) have their amino terminus pointing in one direction and can represent the membrane-bound pheromone isoforms; the other half point in the opposite direction thus representing the soluble molecules. Their cooperative association involves all three helices (and faces) of each molecule and arises from the formation of two distinct types of dimer. Dimer 1 is a symmetrical structure between two molecules that associate with their helices 1 and 2, while dimer 2 involves two molecules that associate by essentially stacking their helix 3 in anti-parallel fashion.

Differently from the instability that dimers and oligomers of higher order show in solution (Bradshaw et al. 1990), the energies of dimerization involved in the formation of both dimer types have been calculated to become much stronger in the crystal due to the partial immobilization of one molecule with respect to another (Weiss et al. 1995). In this case, each molecule can contribute approximately half (1510 Å^2) of its surface to oligomerization, which is much more than the required minimum of 600 Å^2. As happens in the crystal, oligomer stabilization may similarly take place on the cell surface between the soluble pheromone forms and the membrane-bound pheromone isoforms because these isoforms, being embedded in the membranes, have a much reduced rotational and translation energy compared to the soluble forms. And this

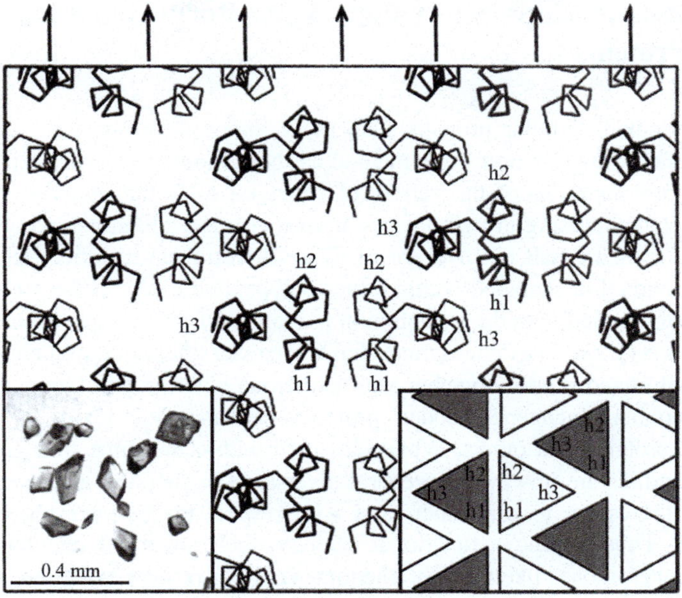

Fig. 10.4 Crystallographic *xy* plane of the *E. raikovi* pheromone E*r*-1 (top view) showing the intermolecular helix-helix interactions that cooperatively arrange molecules into two types of dimer. Interactions between two molecules that associate in dimer 1 are indicated by twofold rotation axes on the figure top, and interactions between two molecules that associate in dimer 2 are indicated by twofold screw axes. Half the molecules mimicking the receptor-binding moieties on the cell surface are in bold and sketched as filled triangles in the inset on the right. They have their amino chain end oriented toward the inside of the plane and their carboxyl chain end oriented toward the outside. The other half molecules mimicking soluble pheromone molecules are not in bold and sketched as light triangles in the inset. They have an upside-down orientation. The three helices of each molecule are identified as h1, h2 and h3. In the inset of the left, E*r*-1 crystals as visible on light microscopy. After Weiss et al. (1995). PDB entry: 1ERL

stabilization would be even stronger starting from two partially immobilized molecules, such as may be the case when two cells interact for mating, exposing their membrane-bound pheromone isoforms on the surface. While these isoforms undergo internalization through endocytotic vesicles upon autocrine pheromone binding, they remain blocked on the cell surface upon paracrine (heterologous) pheromone binding, implying that they can interact through the extracellular space, and function as adhesion molecules which cells use to unite in mating pairs (Vallesi et al. 2005).

10.6 Relationships with Other Water-Borne Signaling Proteins

To assess small globular proteins such as *Euplotes* pheromones for their evolutionary relationships, a more valid method has been thought of as being provided by a comparative analysis of the variety of structural folds that functionally diverse proteins (encompassing growth factors, toxins, enzyme inhibitors and others) have adopted for their small (an average of 57 ± 29 residues) disulfide-rich domains, either individual or included within larger polypeptide chains (Cheek et al. 2006). In this analysis (and with the exception of pheromone E*r*-23 unique for its five-helix globular array), *Euplotes* pheromones have been classified, due to their three-helix bundle right-handed core, together with numerous other protein families including anaphylotoxins, oncogene-encoded proteins, sea anemone toxins, cysteine-rich secretory proteins and others. While classified within the same fold group, these protein families are however in general diversified in function and mechanism of action with respect to one other, and with respect to *Euplotes* pheromones in particular. Thus lacking a functional analogy, their structural relationships with *Euplotes* pheromones likely reflect phenomena of convergent evolution, rather than homology.

Instead, more significant structure-function relationships have been detected between the *E. raikovi* pheromone family and another family of long-distance water-borne protein pheromones that various species of the common marine sea hare, *Aplysia*, freely release from their yolk gland into the environment, and use in phenomena of intra- and inter-specific communication and attraction of sexually mature individuals towards egg cordons (Painter et al. 1991). In their active form, these 56–58 amino acid proteins, designated as 'attractins' in relation to their most apparent activity, have a compact folded α-hairpin structure that is stabilized by three tightly conserved disulfide bonds and dominated by two anti-parallel helices (Schein et al. 2001). The second of these helices basically includes the heptapeptide Ile-Glu-Glu-Cys-Lys-Thr-Ser. This building block, which is common to all attractins of the different species of *Aplysia* and well accounts for the attractin inter-specific attractiveness, finds a close counterpart in the third helix of *E. raikovi* pheromones not only in terms of the conformation of its helical backbone, but also of the orientation of its side chains (Painter et al. 2004). Moreover, it likely functions as the key domain that attractins use in their receptor binding interactions, since their activity is abolished by mutating the charged residues that are exposed on the surface of their second helix. Similarly, *E. raikovi* pheromones may change their activity from autocrine to paracrine, determining the ensuing of intraclonal selfer (homotypic) mating pairs, in the case that their third helix is the site of as little as a single amino acid substitution (Raffioni et al. 1992), or of the oxidation of a Met residue (Alimenti et al. 2012).

10.7 Conclusions

The determination of a significant number of molecular structures of *Euplotes* pheromones has cast aside any doubts that, in full accord with their genetic determination at a single multi-allelic locus, these water-borne signaling molecules form species-specific families of small, disulfide-rich proteins. Each is similar to the others in the global fold and, at the same time, distinct from the others for structural specificities that evolution imposes on the common fold. This virtually unlimited intra- and inter-specific structural polymorphism, which is matched only by families of protein growth factors, chimokines and cytokines of pluricellular organisms, well explains the wide quantitative and qualitative variations, as well as the casual inter-specific mating interactions that are a common feature of ciliate high-multiple (open) mating systems. Ciliate pheromones have been discovered and are usually identified in culture filtrates in relation to their promptly assessed paracrine mating induction activity, which serves natural populations to reshuffle their gene pools and 'rejuvenate' by initiating new life cycles. However, an additional autocrine growth-promoting activity has been shown to be carried out by *Euplotes* pheromones. Although repeatedly postulated on theoretical grounds, this activity is much more troubling to assess experimentally. Yet, it is most likely primarily serving individuals to use their pheromones to promote their own reproduction and spread into the environment.

An autocrine activity requires that a cell secreting a signal also has a receptor for that signal. This requirement has been satisfied by the finding that *Euplotes* cells bind their secreted pheromones to the extracellular domain of longer membrane-bound pheromone isoforms that they co-synthesize with pheromones through a splicing mechanism of the same pheromone genes. In addition to providing a further link to growth factors such as EGF and TGFα that are also synthesized as membrane-bound forms for use in juxtacrine mechanisms (Bosenberg and Massagué 1993), this finding provides also a parsimonious explanation on the origin of new mating types in open mating systems such as those of *Euplotes* and other spirotrichous ciliates. Any mutation in the coding sequence of a pheromone gene in addition to resulting in the synthesis of a new pheromone, would also result in the synthesis of a new pheromone receptor thus ensuring the co-evolution of the two basic molecular units that are necessary to make functional a new mating type.

The three-helix fold that characterizes the molecular structure of *Euplotes* pheromones is an ancient and common motif in protein architecture (Cheek et al. 2006), which accounts for the capacity of these molecules to cross-react with other unrelated systems such as lymphocyte cell lines, conditioning their growth rate and expression of specific genes (Cervia et al. 2013). These cross-reactions, that in all probability reflect cases of chance structural convergence, stimulate research interest in ciliate pheromones also in view of a their potential applicative use.

Acknowledgments Financial support to research was principally provided by PRIN projects of the Ministero dell'Istruzione, dell'Università e della Ricerca (MIUR), and projects of the

Programma Nazionale di Ricerche in Antartide (PNRA). The expert advice of Dr. Gill Philips in editing the English usage of the text is gratefully acknowledged.

References

Afon'kin SY, Yudin AL (1987) Distribution of *Dileptus anser* (Ciliata, Holotricha, Gymnostomatida) of various genotypes for *mat* locus in the concentration gradient of their gamones. Acta Protozool 26:91–100

Akada R (1986) Partial characterization of the mating-inducing factors (gamones) produced by mating type VI cells in the ciliate *Euplotes patella* syngen 2. J Exp Zool 233:169–173

Alimenti C, Vallesi A, Pedrini B, Wüthrich K, Luporini P (2009) Molecular cold-adaptation, comparative analysis of two homologous families of psychrophilic and mesophilic signal proteins of the protozoan ciliate, *Euplotes*. IUBMB Life 61:838–884

Alimenti C, Vallesi A, Federici S, Di Giuseppe G, Dini F, Carratore V, Luporini P (2011) Isolation and structural characterization of two water-borne pheromones from *Euplotes crassus*, a ciliate commonly known to carry membrane-bound pheromones. J Eukaryot Microbiol 58:234–241

Alimenti C, Vallesi A, Luporini P, Buonanno F, Ortenzi C (2012) Cell aging-induced methionine oxidation causes an autocrine to paracrine shift of the pheromone activity in the protozoan ciliate, *Euplotes raikovi*. Exp Cell Res 318:144–151

Bosenberg MW, Massagué J (1993) Juxtacrine cell signaling molecules. Curr Opin Cell Biol 5:832–838

Bradshaw RA, Raffioni S, Luporini P, Chait BT, Lee T, Shively J (1990) Amino acid sequence-mass spectrometric analyses of mating pheromones of the ciliate *Euplotes raikovi*. In: Horvath C, Nikelly JG (eds) Analytical biotechnology. American Chemical Society, Washington DC, pp 154–161

Brown LR, Mronga S, Bradshaw RA, Ortenzi C, Luporini P, Wüthrich K (1993) Nuclear magnetic resonance solution structure of the pheromone E*r*-10 from the ciliated protozoan *Euplotes raikovi*. J Mol Biol 231:800–816

Cervia D, Catalani E, Belardinelli MC, Perrotta C, Picchietti S, Alimenti C et al (2013) The protein pheromone E*r*-1 of the ciliate *Euplotes raikovi* stimulates human T-cell activity: involvement of interleukin-2 system. Exp Cell Res 319:56–67

Cheek S, Krisna SS, Grishin NV (2006) Structural classification of small, disulfide-rich protein domains. J Mol Biol 359:215–237

Concetti A, Raffioni S, Miceli C, Barra D, Luporini P (1986) Purification to apparent homogeneity of the mating pheromone of *mat-1* homozygous *Euplotes raikovi*. J Biol Chem 261:10582–10586

Di Giuseppe G, Erra F, Dini F, Alimenti C, Vallesi A, Pedrini B, Wüthrich K, Luporini P (2011) Antarctic and Arctic populations of the ciliate *Euplotes nobilii* show common pheromone-mediated cell-cell signaling and cross-mating. Proc Natl Acad Sci USA 108:3181–3186

Di Giuseppe G, Erra F, Frontini FP, Dini F, Vallesi A, Luporini P (2014) Improved description of the bipolar ciliate, *Euplotes petzi*, and definition of its basal position in the *Euplotes* phylogenetic tree. Eur J Protistol 50:402–411

Di Giuseppe G, Miceli C, Zahn R, Damberger F, Wüthrich K, Luporini P (2002) A structurally deviant member of the *Euplotes raikovi* pheromone family, E*r*-23. J Eukaryot Microbiol 49:86–92

Dini F, Nyberg D (1993) Sex in ciliates. In: Jones JG (ed) Advances in microbial ecology, vol 13. Plenum Press, New York, pp 85–153

Esposito F, Ricci N, Nobili R (1976) Mating-type-specific soluble factors (gamones) in cell interaction of conjugation in the ciliate *Oxytricha bifaria*. J Exp Zool 197:275–281

Geralt M, Alimenti C, Vallesi A, Luporini P, Wüthrich K (2013) Thermodynamic stability of psychrophilic and mesophilic pheromones of the protozoan ciliate *Euplotes*. Biology 2:142–150

Grell KG (1953) Die Konjugation von *Ephelota gemmipara* R. Hertwig. Arch Protistenk 98:287–326

Karlson P, Luscher M (1959) 'Pheromones', a new term for a class of biologically active substances. Nature 183:55–56

Kimball RF (1939) Mating types in *Euplotes*. Am Nat 73:451–456

Kimball RF (1942) The nature and inheritance of mating types in *Euplotes patella*. Genetics 27:269–285

Kosaka T (1991) Mature cells attracting cells of the complementary mating type in *Euplotes woodruffi* syngen 3 (Ciliophora, Hypotrichida). Zool Sci 8:681–692

Kuhlmann HW, Sato K (1993) Interspecific mating reactions between *Euplotes octocarinatus* and *Euplotes patella* syngen 2. Eur J Protistol 29:24–31

Kusch J, Heckmann K (1988a) Gamones are secreted in *Euplotes octocarinatus* via the cortical ampules. Eur J Protistol 23:273–278

Kusch J, Heckmann K (1988b) Gamone production in large-scale cultures of *Euplotes octocarinatus*. J Protozool 35:337–340

La Terza A, Miceli C, Luporini P (1995) Differential amplification of pheromone genes of the ciliate *Euplotes raikovi*. Dev Genet 17:272–279

Liu A, Luginbühl P, Zerbe O, Ortenzi C, Luporini P, Wüthrich K (2001) NMR structure of the pheromone E*r*-22 from *Euplotes raikovi*. J Biomol NMR 19:75–78

Luginbühl P, Ottiger M, Mronga S, Wüthrich K (1994) Structure comparison of the NMR structures of the pheromones E*r*-1, E*r*-10, and E*r*-2 from *Euplotes raikovi*. Protein Sci 3:1537–1546

Luporini P, Miceli C (1986) Mating pheromones. In: Gall JG (ed) The molecular biology of ciliated Protozoa. Academic Press, New York, pp 263–299

Luporini P, Miceli C, Ortenzi C (1983) Evidence that the ciliate *Euplotes raikovi* releases mating inducing factors (gamones). J Exp Zool 226:1–9

Miceli C, La Terza A, Bradshaw RA, Luporini P (1992) Identification and structural characterization of a cDNA clone encoding a membrane-bound form of the polypeptide pheromone Er-1 in the ciliate protozoan *Euplotes raikovi*. Proc Natl Acad Sci USA 89:1988–1992

Miyake A (1981) Cell interaction by gamones in *Blepharisma*. In: O'Day DH, Horgen PA (eds) Sexual interaction in eukaryotic microbes. Academic Press, New York, pp 95–129

Miyake A (1996) Fertilization and sexuality in ciliates. In: Hausmann K, Bradbury PC (eds) Ciliates: cells as organisms. Fischer Verlag, Stuttgart, pp 243–290

Mronga S, Luginbühl P, Brown LR, Ortenzi C, Luporini P, Bradshaw RA, Wüthrich K (1994) The NMR solution structure of the pheromone E*r*-1 from the ciliated protozoan *Euplotes raikovi*. Protein Sci 3:1527–1536

Nobili R, Luporini P, Dini F (1978) Breeding systems, species relationships and evolutionary trends in some marine species of Euplotidae (Ciliata Hypotrichida). In: Battaglia B, Beardmore J (eds) Marine organisms: genetics, ecology and evolution. Plenum Press, New York, pp 591–616

Ortenzi C, Alimenti C, Vallesi A, Di Pretoro B, La Terza A, Luporini P (2000) The autocrine mitogenic loop of the ciliate *Euplotes raikovi*, the pheromone membarne-bound forms are the cell binding sites and potential signalling receptors of soluble pheromones. Mol Biol Cell 11:1445–1455

Ortenzi C, Miceli C, Bradshaw RA, Luporini P (1990) Identification and initial characterization of an autocrine pheromone receptor in the protozoan ciliate *Euplotes raikovi*. J Cell Biol 111:607–614

Ottiger M, Szyperski T, Luginbühl P, Ortenzi C, Luporini P, Bradshaw RA, Wüthrich K (1994) The NMR solution structure of the pheromone E*r*-2 from the ciliated protozoan *Euplotes raikovi*. Protein Sci 3:1515–1526

Painter SD, Chong MG, Wong MA, Gray A, Cormier JG, Nagle GT (1991) Relative contributions of the egg layer and egg cordon to pheromonal attraction and the induction of mating and egg-laying behavior in *Aplysia*. Biol Bull 181:81–94

Painter SD, Cummuns SF, Nichols AE, Akalal DBG, Schein CH, Braun W et al (2004) Structural and functional analysis of *Aplysia* attractins, a family of water-borne protein pheromones with interspecific attractiveness. Proc Natl Acad Sci USA 101:6929–6933

Pedrini B, Placzek WJ, Koculi E, Alimenti C, La Terza A, Luporini P, Wüthrich K (2007) Cold-adaptation in sea-water-borne signal proteins, sequence and NMR structure of the pheromone E*n*-6 from the Antarctic ciliate *Euplotes nobilii*. J Mol Biol 372:277–286

Phadke SS, Zufall RA (2009) Rapid diversification of mating systems in ciliates. Biol J Linn Soc 98:187–197

Placzek WJ, Etezady-Esfarjani T, Herrmann T, Pedrini B, Peti W, Alimenti C, Luporini P, Wüthrich K (2007) Cold-adapted signal proteins, NMR structures of pheromones from the Antarctic ciliate *Euplotes nobilii*. IUBMB Life 59:578–585

Raffioni S, Miceli C, Vallesi A, Chowdhury SK, Chait BT, Luporini P, Bradshaw RA (1992) Primary structure of *Euplotes raikovi* pheromones, comparison of five sequences of pheromones from cells with variable mating interactions. Proc Natl Acad Sci USA 89:071–2075

Schein CH, Nagle GT, Page JS, Sweedler JV, Xu Y, Painter SD et al (2001) *Aplysia* attractin: biophysical characterization and modeling of a water-borne pheromone. Biophys J 81:463–472

Schulze-Dieckhoff H, Freiburg M, Heckmann K (1987) The isolation of gamone 3 and 4 of *Euplotes octocarinatus*. Eur J Biochem 168:89–94

Sonneborn TM (1937) Sex, sex inheritance and sex determination in *Paramecium aurelia*. Proc Nat Acad Sci USA 23:378–395

Sonneborn TM (1978) Genetics of cell-cell interaction in ciliates. In: Lerner RA, Bergsma D (eds) The molecular basis of cell-cell interaction. Alan R Liss, New York, pp 417–427

Sugiura M, Harumoto T (2001) Identification, characterization, and complete amino acid sequence of the conjugation-inducing glycoprotein (blepharmone) in the ciliate *Blepharisma japonicum*. Proc Natl Acad Sci USA 98:14446–14451

Sugiura M, Kawahara S, Iio H, Harumoto T (2005) Developmentally and environmentally regulated expression of gamone 1, the trigger molecule for sexual reproduction in *Blepharisma japonicum*. J Cell Sci 118:2735–2741

Sugiura M, Shiotani H, Suzaki T, Harumoto T (2010) Behavioral changes induced by conjugation-inducing pheromones, gamomes 1 and 2, in the ciliate *Blepharisma japonicum*. Eur J Protistol 46:143–149

Vallesi A, Alimenti C, Federici S, Di Giuseppe G, Dini F, Guella G, Luporini P (2014) Evidence for gene duplication and allelic codominance (not hierarchical dominance) at the mating-type locus of the ciliate, *Euplotes crassus*. J Eukaryot Microbiol 61:620–629

Vallesi A, Ballarini P, Di Pretoro B, Alimenti C, Miceli C, Luporini P (2005) The autocrine, mitogenic pheromone-receptor loop of the ciliate *Euplotes raikovi*, pheromone-induced receptor internalization. Eukaryot Cell 4:1221–1227

Vallesi A, Giuli G, Bradshaw RA, Luporini P (1995) Autocrine mitogenic activity of pheromones produced by the protozoan ciliate *Euplotes raikovi*. Nature 376:522–524

Uspenkaya ZI, Yudin AL (2016) Fifty years of research on serotypes and mating types of *Dileptus anser*. A review. Eur J Protistol 53:31–44

Weiss MS, Anderson DH, Raffioni S, Bradshaw RA, Ortenzi C, Luporini P, Eisenberg D (1995) A cooperative model for ligand recognition and cell adhesion, evidence from the molecular packing in the 1.6 Å crystal structure of the pheromone E*r*-1 from the ciliate protozoan *Euplotes raikovi*. Proc Natl Acad Sci USA 92:10172–10176

Yudin AL, Afon'kin SYu, Parfenova EV (1990) Mating pheromones in the ciliate *Dileptus anser* Tsitologia 32:107–116 (In Russian with English summary)

Zahn R, Damberger F, Ortenzi C, Luporini P, Wüthrich K (2001) NMR structure of the *Euplotes raikovi* pheromone E*r*-23 and identification of its five disulfide bonds. J Mol Biol 313:923–931

Chapter 11
Communication in *Tetrahymena* Reproduction

Wendy Ashlock, Takahiko Akematsu and Ronald Pearlman

Abstract This chapter addresses genetic communication, the transfer of information from one generation to the next. We focus on the model ciliate, *Tetrahymena*. Communication in *Tetrahymena* involves communication between the organism and the environment, communication between organisms, and communication within the organism to distinguish self from non-self and germline from soma. Ciliates have a unique genomic structure, with two nuclei within the same cytoplasm, a large polyploid somatic nucleus and a smaller diploid germinal nucleus, both sharing the same genetic information. We will discuss how the structure of *Tetrahymena*'s genomes enables the expression of all possible combinations of alleles, and how it contributes to the silencing of genetic invaders (e.g. transposons). We will discuss how the genomic structure is maintained in a process that involves the coexistence of up to five nuclei in the same cytoplasm and the destruction of unneeded nuclei in a process called programmed nuclear death (PND) with analogies to autophagy and apoptosis.

11.1 Introduction

In this chapter, we discuss genetic communication, the transfer of information from one generation of *Tetrahymena* to another. The unique nuclear dimorphic genomic structure of *Tetrahymena* (and other ciliates) makes this a complex story. *Tetrahymena* carries its genetic information in two genomes, the somatic macronu-

---

W. Ashlock · R. Pearlman (✉)
Department of Biology, York University, 4700 Keele Street, Toronto, ON M3J 1P3, Canada
e-mail: ronp@yorku.ca

W. Ashlock
e-mail: washlock@yorku.ca

T. Akematsu
Department of Chromosome Biology, University of Vienna, Dr. Bohr-Gasse 9, 5. Stock, 1030 Vienna, Austria
e-mail: takahiko.akematsu@univie.ac.at

G. Witzany and M. Nowacki (eds.), *Biocommunication of Ciliates*,
DOI 10.1007/978-3-319-32211-7_11

cleus (MAC) and the germinal micronucleus (MIC). The MAC is physically larger than the MIC and is polyploid with about 45 genome copies. The smaller MIC is diploid, but carries more information. All information in the MAC is also contained in the MIC, but, in addition, the MIC contains additional information in the form of thousands of internal eliminated sequences (IES). These IES consist of repetitive elements, transposable elements (TEs) and remnants of TEs, both active and inactive. The MIC is arranged in five metacentric chromosomes, while the MAC is fragmented into ~225 autonomously replicating fragments, pseudochromosomes.

The story of *Tetrahymena* reproduction comes in two parts, because the organism reproduces in two different ways, asexually and sexually. Asexual reproduction occurs when environmental conditions are good and food is not limiting. Communication in asexual reproduction is directed by the polyploid structure of the MAC. This structure leads to a remarkable diversity of offspring, with all present alleles expressed in some of the offspring. The result leads to an extraordinary level of communication between the genetic information, represented in many different ways, and the environment.

Sexual reproduction, or conjugation, occurs in times of starvation. It is technically not reproduction, as no new cells are created, but involves the exchange of genetic information between two *Tetrahymena* cells. Conjugation is a complicated process, involving meiotic and mitotic nuclear divisions, the deletion of information from the MIC, the construction of a new MAC through the fragmentation of the MIC chromosomes, the deletion of IES, and the destruction of the old MAC (programmed nuclear death or PND).

An important part of the biocommunication involved in conjugation is the process of distinguishing self from non-self involving TEs. TEs are mobile portions of genomes that result from invasion of the genome by viruses. They make up a large portion of most eukaryotic genomes: nearly half of the human genome is composed of TEs; about 10 % of *Drosophila*'s genome is, and some plant genomes like maize and barley are over 85 % TEs. TEs often function mutualistically with their hosts. They can affect gene expression; they impact genome structure; they can drive evolution. But, they must be regulated, as uncontrolled multiplication of TEs can be harmful. Insertions can disrupt functional DNA and double-strand breaks induced on transposition can threaten genome integrity. Even inactive TEs are potentially harmful. Because they exist in many copies, they can create deletions and duplications during recombination. For more information about TEs, see (Craig 2002; Fedoroff 2012; Lisch and Slotkin 2011; Muñoz-López and García-Pérez 2010; Joly-Lopez and Bureau 2014). Most organisms regulate their TEs through silencing. Ciliates take a more radical approach, deleting the TEs entirely in the formation of the MAC from the MIC, creating a MAC genome that is virtually free of TEs. The language used to distinguish self from non-self consists of small RNAs, called scanRNAs.

PND involves another kind of communication, the communication between germline and soma. It occurs after genome rearrangement when the parental and progeny MACs coexist in the same cytoplasm. In order for the progeny genome to be expressed, the parental genome must be destroyed. We will discuss this process and the signals regulating it.

11.2 Asexual Reproduction: Communication with the Environment

The *Tetrahymena* life cycle is shown in Fig. 11.1. Stage A represents vegetative growth. During this stage, *Tetrahymena* reproduce amitotically through binary fission. This process has been described in detail (Cole and Sugai 2012; Orias et al.

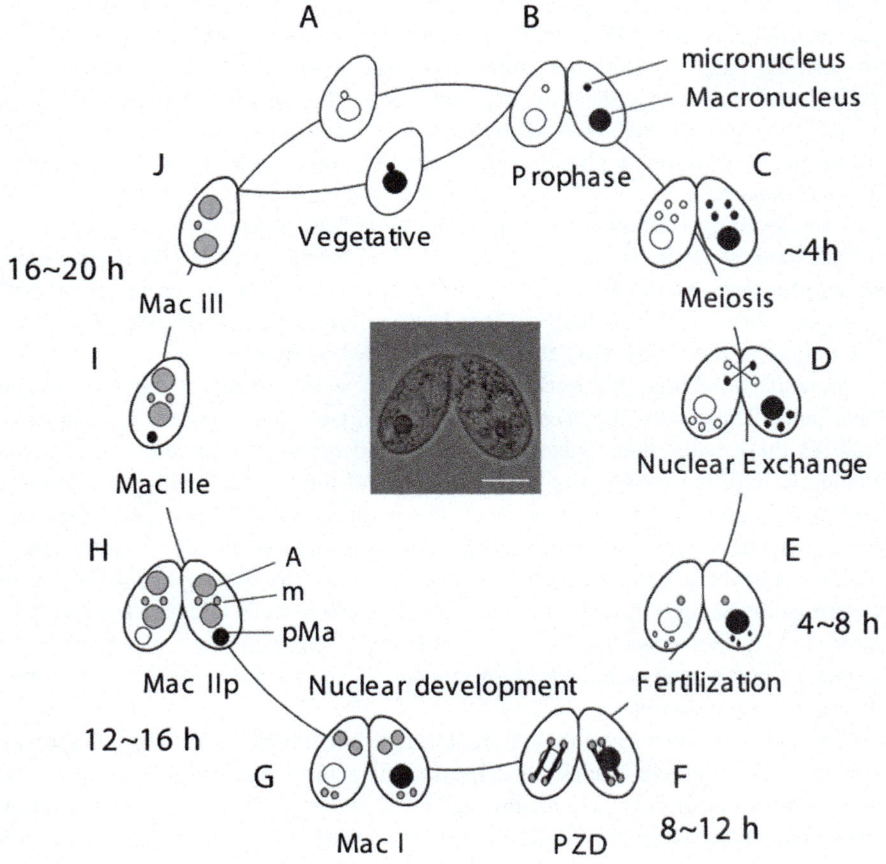

Fig. 11.1 Life Cycle of *Tetrahymena*. **a** Vegetative phase. **b** Meiotic prophase. **c** Meiosis. **d** Nuclear exchange. **e** Fertilization. **f** Postzygotic division. Fertilized nucleus divides twice. **g** MAC I. Anterior nuclei differentiate into new MAC, while posterior nuclei remain MIC. **h** MAC IIp. The parental MAC (pMa) begins to degenerate. The anlagen is designated by "A"; the MIC is designated by "m". **i** MAC IIe. Exconjugants separate. **j** MAC III. One of the two MICs is eliminated. Figure from Akematsu and Endoh (2010) licensed under http://creativecommons.org/licenses/by/2.0/)

2011). The MIC undergoes mitotic division, and the MAC divides amitotically with no involvement of nuclear envelope breakdown, no formation of a mitotic spindle or condensed chromosomes. The MAC and MIC have temporally distinct cell cycles and they divide at different times (Karrer 2012). The MIC mitosis happens first, and the MAC fission is coincident with the division of the cell. After 50–80 fissions, *Tetrahymena* become adolescents. This means they can mate with mature cells, but not with other adolescents. They become fully mature after 20–25 more fissions (Rogers and Karrer 1985).

Communication in this part of the life cycle can be thought of in terms of how genetic information is passed from a *Tetrahymena* cell to its progeny clones and how this allows the information to interact with the environment in diverse ways. The manner in which the genetic information is stored in the MAC directs this process. The MAC is polyploid with about 45 copies of its genetic material, with the exception of the rDNA, which has ~9000 copies. In addition, the MAC is fragmented into ~225 pseudochromosomes. These pseudochromosomes (also called autonomously replicating fragments or ARFs, subchromosomal pieces, or nanochromosomes) are flanked by telomeres but have no centromere. This arrangement leads to a phenotypically diverse set of descendants for a single *Tetrahymena* cell.

This diversity is manifest through phenotypic assortment (Merriam and Bruns 1988; Nanney 1980; Orias and Flacks 1975). During MAC amitosis, the pseudochromosomes are copied, and are distributed randomly between the two MAC genomes. After sufficient fissions, this leads to homozygous individuals pure for each allele in the MAC, but retaining both alleles in the MIC.

To understand how this works, imagine a box with an equal number of red and blue balls. These balls represent the copies of a gene with two alleles. Duplicate each of these balls, then shake up the box, and divide the contents equally and randomly into two boxes. It is likely that more of the red balls will go in one box and more of the blue in the other. After this is repeated many times, each box will have only red or blue balls. The mathematical formula for the rate of assortment is $1/(2N - 1)$, where N is the number of gene copies. So, for *Tetrahymena*, the rate of assortment is about 0.011, meaning it takes about 89 fissions to create homozygous clones. A further consequence of this process is co-assortment. Genes that are together on the same pseudochromosome sort in the same way most of the time, creating linkage groups.

The polyploid nature of the MAC, through phenotypic assortment, allows all alleles, even recessive alleles, to come to full expression. Fragmentation of the MAC allows for diverse combinations of these alleles. Thus, asexual reproduction communicates the genetic information in the organism to the environment in a multiplicity of combinations. This gives scope for natural selection to operate in the short term. Sexual recombination, on the other hand, creates an independent longer term source of diversity.

11.3 Conjugation: Communication Within the Cell

In ciliates reproduction and transfer of genetic information are separate. Reproduction occurs during vegetative growth. *Tetrahymena* can live indefinitely undergoing only vegetative growth, and only exchange genetic information when food is scarce. Transfer of genetic information is through a sexual process of mating and conjugation involving meiosis of the MIC, genetic exchange and genome rearrangements during the formation of a new MAC. This happens without fusion or disintegration of the parent cells. The parental nuclei and the progeny nuclei coexist in the same cell. Thus, communication in conjugation involves interactions between the different nuclei in the cell. During conjugation the number of nuclei in the cell varies from two to five.

Conjugation progresses through several stages, shown in Fig. 11.1. The diploid MIC undergoes meiosis (Stage C) creating four haploid meiotic products. One of the meiotic products is selected to mitotically divide and the other meiotic products are degraded. Gametes are exchanged between cells of different mating types through a temporary junction (Stage D "nuclear exchange"). The gametes fuse to make a zygotic nucleus (Stage E). The zygote divides twice (Stage F "postzygotic division"). The posterior pair become new MICs, and the anterior pair undergo genomic rearrangement to create new MACs (anlagen) using information from the parental MAC (Stage G "MAC 1"). The parental MAC undergoes programmed nuclear death (PND) (Stage H "MAC IIp"), and the pairs separate forming exconjugants (Stage I "MAC IIe"). Finally, one of the MICs in the exconjugants is eliminated and the cell undergoes its first fission into karyonides (Stage J "MAC III").

Meiosis in *Tetrahymena* is unusual in that during meiotic prophase, the MIC expands to about 50 times its usual diameter and organizes into a bouquet arrangement (Loidl et al. 2012). Meiosis does not involve a synaptonemal complex (SC), which is used by other organisms to stabilize nascent crossovers. Discussion of signals involved in meiosis can be found in Lukaszewicz et al. (2013), and Mochizuki et al. (2008).

11.3.1 Genome Rearrangement: Communication Between MIC and MAC

Genome rearrangement (Mochizuki 2010b, 2012) involves the transfer of information from the newly created MIC to the new MAC. This is shown in Stage G in Fig. 11.1, where the newly created zygotic anterior nuclei differentiate into MAC. Communication in genome rearrangement involves not just the transfer of information, but also the editing and rearranging of it. The five MIC chromosomes are fragmented into hundreds of MAC autonomously replicating fragments (ARFs or pseudochromosomes), and thousands of internal eliminated sequences (IES) are deleted. For reviews, see (Chalker and Yao 2011; Chalker et al. 2013). In some ciliates, such as *Oxytricha*, this process involves extensive scrambling of sequences (Chen et al. 2014), but in *Tetrahymena* the sequence arrangement in the MAC mostly matches that in the MIC. It was originally estimated that about a fifth of the genetic information in the MIC is deleted during MAC formation (Yao and Gorovsky 1974), but comparison of the MAC and MIC sequences shows that it is actually about a third (unpublished data Ashlock 2015). Deleted IES are immediately degraded. In *Tetrahymena*, IES rarely occur in genes, and, when they do, they usually occur in introns. Their deletion boundaries are imprecise (Vogt et al. 2013), unlike in other ciliates such as *Paramecium* where deletion boundaries are precise and many IES are within genes (Arnaiz et al. 2012; Gratias and Bétermier 2003; Kapusta et al. 2011). *Tetrahymena* IES range in size from tens of base pairs to tens of thousands.

Many TEs have been identified in IES. Tel-1 elements (Cherry and Blackburn 1985), REP non-LTR retrotransposons (Fillingham et al. 2004), a type of Polinton transposon unique to *Tetrahymena* called Tlr (Kapitonov and Jurka 2006; Krupovic et al. 2014; Wuitschick et al. 2002), and Tc1/mariner transposons (Eisen et al. 2006) have been described. Tc1/mariner transposons (Plasterk et al. 1999) are widespread throughout eukaryotes, found in plants, animals, and fungi. Tel-1, Tlr, and Tc1/mariner transposons are Class 2 TEs, transposons that transpose via a DNA intermediate in a cut-and-paste manner. Non-LTR retrotransposons are Class 1 TEs, transposons that transpose via an RNA intermediate in a copy-paste manner. For more information about TEs and how they affect genomes, see (Craig 2002; Slotkin and Martienssen 2007; Smit 1999). Now that both the MAC and the MIC genomes have been sequenced (Eisen et al. 2006; http://www.ciliate.org; http://www.broadinstitute.org/annotation/genome/Tetrahymena/MultiHome.html), it is possible to create a comprehensive catalogue of all types of TEs present in *Tetrahymena* nuclei (Fass et al. 2011). For a review of what is known about the composition of IES and their removal, see (Chalker and Yao 2011).

11.3.1.1 Cis-Acting Signals Near IES Boundaries

Communication of the exact location of excision boundaries comes from *cis*-signals from IES flanking sequence and is still not completely understood. There are no consistent sequence motifs, but specific sequences have been shown experimentally for specific IES to be both necessary and sufficient for specifying IES boundaries. The M IES has been extensively studied (Godiska and Yao 1990). A 10 bp *cis*-acting sequence 40–50 bp on either side of the M IES was identified. The M IES is known to have an alternate deletion boundary (Austerberry et al. 1989), and it was found that a copy of the *cis*-acting 10 bp sequence also existed 40–50 bp from that boundary. It was also shown that moving the sequence altered the deletion boundary. The R IES exhibits similar behavior, but with a different sequence (Chalker et al. 1999). *Cis*-acting 70 bp regions 30 bp on either side of the IES specify its boundaries. These sequences are different from each other and different from the *cis*-acting sequence for the M IES. They can substitute functionally for each other, but were found to have a "complex nature." Six point mutations did not alter their function. A necessary and sufficient signal for the deletion of the Tlr transposon was found 51 or more bp from its boundary (Patil and Karrer 2000). An IES found within an intron of the gene *ARP*1, called mse2.9, was found to have a complex sequence element 47–81 bp from its right side that was necessary and sufficient for excision (Fillingham et al. 2001). Substituting the flanking sequences from the M and the R IES for this sequence resulted in correct excision. It has been shown (Carle 2013) that the protein Lia3 was necessary for correct excision of the M IES and other IES with similar *cis*-acting boundary sequences. They propose that it and its three Lia homologs control the excision boundaries of different subsets of IES. For IES in the M IES subset, this is done by binding to a RNA/DNA hybrid quadruplex formed between the ends of the IES during the transcription that occurs before IES excision.

11.3.1.2 ScanRNA Model

The *cis*-acting sequences only explain how, once a region is targeted, the deletion boundaries are determined. Communication of which specific regions to target is done using small RNAs. This is analogous to the way TEs in other organisms are targeted for silencing (Chalker et al. 2013; Kataoka and Mochizuki 2011; Noto et al. 2015; Schoeberl and Mochizuki 2011). TEs are silenced in many organisms using an RNAi pathway (Girard and Hannon 2008; Malone and Hannon 2009). Small RNAs target sequences where, together with an Argonaute protein, they form an RNA-induced silencing complex (RISC). The RISC silences the TEs in a variety of different ways; it can target the cleavage of RNA transcripts; it can inhibit protein synthesis, and it can direct chromatin modifications that inhibit transcription (Levin and Moran 2011; Malone and Hannon 2009; Slotkin and Martienssen 2007). The silencing is reversible because of the potential usefulness of TEs to the organism. In *Tetrahymena*, after a region in the anlagen is targeted by the scanRNAs for deletion, it is marked as heterochromatin and excised with the involvement of the domesticated PiggyBac transposase Tpb2 (Vogt and Mochizuki 2013). Because *Tetrahymena* retains a copy of the sequence in the MIC, it can afford to delete the marked sequences, and this is an example of irreversible gene silencing (Mochizuki and Gorovsky 2004).

TEs are silenced in eukaryotes using a variety of different small RNAs. Mammals and flies use piwi-interacting RNAs (piRNAs) (Guzzardo et al. 2013; Senti and Brennecke 2010; Siomi et al. 2011) which are small non-coding RNAs that form RNA-protein complexes through interactions with piwi proteins. Plants and yeast use small interfering RNAs (siRNAs), which are double-stranded RNA molecules that interfere with the expression of DNA with complementary sequence by breaking down the mRNA after transcription (Agrawal et al. 2003).

A model for how DNA excision is specified in *Tetrahymena* has been developed (Mochizuki and Gorovsky 2004; Mochizuki 2010, 2012; Mochizuki et al. 2002; Yao et al. 2003). The IES are identified in a unique manner. Small RNAs, called scanRNAs are bidirectionally transcribed from the entire MIC genome and diced into 28 nt molecules. These scanRNAs are transferred to the parental MAC. Any scanRNAs homologous to sequences in the parental MAC are degraded. The remaining scanRNAs, which are homologous to IES in the parental genome, are transferred through the cytoplasm to the new MAC where they target the IES to which they are homologous for elimination.

It has recently been shown (Schoeberl et al. 2012) that the transcription of the scanRNAs is not uniform across the MIC. IES are preferentially, though not exclusively, transcribed, and there are large (>100 Kbp) regions that are scanRNA deserts. This means that TE insertions in non-IES sequences are less likely to be eliminated. The scanRNA model was recently updated (Noto et al. 2015). The refinement postulates two types of scanRNAs, early and late, for two types of IES, Type-A IES and Type-B IES. This refinement increases the robustness of the model, protecting against the possibility that errors in DNA elimination will be inherited, and provides an explanation for the uneven transcription of the MIC.

Type A IES are young IES that contain potentially active TEs. They occur in non-gene-rich regions of the MIC. Type B IES occur in gene-rich regions and are more likely to contain degenerate TEs (higher AT-content, less sequence homology to known TEs). The Type B IES have short sequences with similarity to Type A IES. The updated model postulates that early scanRNAs are transcribed from Type A IES. As well as targeting the Type A IES for elimination, these early scanRNAs recognize Type B IES and create late scanRNAs that target their elimination. Some of these late scanRNAs recognize other Type B IES, creating a "chain reaction" of late scanRNA production. It is suggested (Noto et al. 2015) that the *cis*-acting sequences described in the preceding section could be a check on the production of late scanRNAs, making sure they stay within the boundaries of IES and avoid deleting essential genomic regions.

RNA involvement in gene silencing in ciliates will be described in more detail in another chapter of this volume (Nowaki, Chap. 3).

11.3.1.3 Sequence Fragmentation

Besides editing the MIC genome through the deletion of IES, genome rearrangement involves sequence fragmentation. Where to do this is communicated through short sequence patterns called chromosome breakage sequences (CBS). Five metacentric MIC chromosomes are converted into hundreds of MAC autonomously replicating fragments (pseudochromosomes), which are then replicated to about 45 copies each. In addition, a ribosomal DNA pseudochromosome is created. In most eukaryotes the sequence encoding ribosomal DNA, rDNA, consists of tandem repeats. In *Tetrahymena* there is only one copy in the MIC, but to create the MAC rDNA, this sequence is excised, converted to a 21 Kbp palindromic sequence (inverted dimer), and amplified to about 9000 copies (Challoner et al. 1985; Engberg and Nielsen 1990; Kiss and Pearlman 1981; Yao et al. 1985; Yao 1986).

Sequence fragmentation was first studied in the context of rDNA excision (Yao 1981). At each breakage site there is a conserved 15 bp CBS. CBS are surrounded by AT-rich sequence. Fragmentation, unlike IES excision and unlike in other ciliates such as *Paramecium* (Le Mouel et al. 2003), is precise (Fan and Yao 1996). Only the CBS and a few bases on either side are deleted (at most 75 bases total). De novo telomeres are added to the ends of the fragments to form MAC pseudochromosomes.

A genome-wide study of CBS has been done (Hamilton et al. 2006). Thirteen CBS variants of the classic CBS sequence, TAAACCAACCTCTTT, were found. They all fit this form, - AAACCAACC - C - - -, where the hyphens represent variable bases. The variable locations are not unconstrained, however. For the sequence to be functional, only two base substitutions from the classic sequence can be made.

Recent analysis of the MIC sequence (E. Orias personal communication) has identified 221 CBS, accounting for all the MAC chromosome ends. Analysis of these sequences support a model in which CBS are derived from a transposon whose endonuclease was domesticated. Comparison of CBS across *Tetrahymena* species (P. Huvos personal communication), *T. thermophila, T. malaccensis, T. elliotti, and T. borealis*, shows that CBS sites are conserved, meaning that MAC chromosome lengths are conserved across species.

11.3.1.4 Mechanism of IES Excision

The mechanism of IES excision in *Tetrahymena*, i.e., the way in which the deletion information is communicated, resembles transposition of TEs. A ciliate model was proposed based on detailed study of a few IES (Klobutcher and Herrick 1997). Deletion is initiated by a double-strand break at one end of the IES. Four bp 5′ overhangs with 3′ A residues are created on both sides of the break. The 3′ hydroxyl group of the A residue on the MAC-destined side of the break executes a nucleophilic attack on the other side of the IES, forming a junction that excludes the IES (Saveliev and Cox 1996). This is similar to the excision process for cut-and-paste TEs recently described (Muñoz-López and García-Pérez 2010).

This resemblance to cut-and-paste TEs suggests that some sort of transposase is involved. An important function of TEs in genomes is to provide a source of new genes. This process is called exaptation or domestication. Coding sequences from TEs acquire a new function that is beneficial to the host organism. Examples include: telomerase, *RAG*1 (involved in V(J)D recombination), and a gene found in flowering plants called Mustang (reviewed in Volff 2006). The mechanism of IES excision motivated researchers to look for a domesticated transposase.

A transposase from the piggyBac family was recently identified in *Tetrahymena* (Cheng et al. 2010; Vogt and Mochizuki 2013). The gene encoding this transposon-derived protein is called *TPB*2 (*Tetrahymena* piggyBac-like transposase 2), and its product has been shown to be necessary for IES elimination. It localizes to heterochromatin created in the scanRNA process and shows some preference for the piggyBac recognition sequence, 5′-TTAA-3′.

A recent review (Vogt et al. 2013) contrasts the IES excision mechanisms in *Tetrahymena*, *Paramecium*, and *Oxytricha*. *Paramecium* uses a process similar to *Tetrahymena* with a domesticated piggyBac protein called Pgm (piggyMac) (Baudry et al. 2009; Dubois et al. 2012). Its excision boundaries are precise, however, and flanked by 5′-TA-3′. *Oxytricha*, on the other hand, does not have a domesticated transposase. Instead, thousands of TBE (telomere-bearing element) transposons in the MIC are expressed during conjugation, participating in the elimination of IES. This mutualistic arrangement leads to purifying selection of the TBE transposons in the MIC, though they are eliminated in the MAC.

11.3.1.5 TEs that Escape into the MAC

In many organisms, TEs function mutualistically with their hosts. The information in the TEs is often useful. But, in *Tetrahymena*, if the TEs exist only in the transcriptionally inactive MIC, how can their information be shared? The answer lies in imperfections in the model for IES elimination mediated by small RNAs. The robustness of this model was estimated by simulating accidental loss of scanRNA expression from Type A IES (Noto et al. 2015). This simulation predicted that scanRNAs expressed from 1 % of nonredundant Type A IES could identify up to 84 % of total IES, and scanRNAs expressed from 10 % of nonredundant Type A IES could identify up to 91 % of total IES. This estimate of a highly robust model is consistent with the virtual non-existence of TEs in the MAC.

However, the existence of the *TPB*2 gene in *Tetrahymena* is evidence that the TE elimination process is imperfect. At least one piggyBac TE was able to get into the MAC to be domesticated. Systems for TE silencing in other eukaryotes are similarly imperfect. This is desirable as TEs are occasionally beneficial to the organism. A BLASTN search of *Tetrahymena* genes against IES demonstrates that 2 % of genes contain perfect matches to portions of IES of length the size of a scanRNA or longer and 4 % of genes have matches with e-values less than 1e-100 (unpublished data, Ashlock 2015). These could be domesticated TEs, or evidence that IES elimination from within genes is less robust than IES elimination in intergenic regions.

Repeated elements in IES are found in multiple degraded copies in the MAC. An example is the R-indel (Huvos 2004). The R-indel was the first IES to be detected within another IES. It was found to exist in one of the well-studied IES, a Type B IES called the R IES, in some strains but not others, suggesting active transposition. Many copies of it exist in MIC-limited sequences, both as complete IES and within other IES. There are more than 2000 copies of the R-indel in MIC-limited sequence, and about 200 copies in MAC sequence (unpublished data, Ashlock 2015). The REP non-LTR retrotransposon (Fillingham et al. 2004), is a *Tetrahymena*-specific TE. The original copies of it were found to be flanked by a complex direct repeat. This complex direct repeat occurs more than 2000 times in the MAC (unpublished data, Ashlock 2015). This repeat contains a protein domain that is part of many genes, sometimes in multiple copies.

Because of the imprecise deletion boundaries of IES in *Tetrahymena*, remnants of IES exist in the MAC. About half of the IES deletion sites in the MAC have sequence with similarity to IES (unpublished data, Ashlock 2015) on one or both sides of the deletion. There is also documented microheterogeneity of just a few base pairs (Austerberry et al. 1989). BLASTN search for sequence with similarity to known boundaries of the REP non-LTR transposon and the Tlr DNA transposon shows that deletion boundaries can vary by as much as 1200 bp (unpublished data, Ashlock 2015). This is consistent with the experimental result of inserting a chimeric construct for which the elimination boundaries occurred within 500–1000 bp of the target sequence (Noto et al. 2015). This suggests that the boundary determining *cis*-acting sequences are not present for every IES.

Many IES in *Tetrahymena* consist of groups of TE-derived sequence, some nested within each other. Tlr DNA transposons (Wuitschick et al. 2002) are non-autonomous derivatives of Polington DNA transposons (Kapitonov and Jurka 2006), a TE that is known to prefer to insert within other TEs. BLASTN search reveals more than 3000 fragments of Tlrs in the MIC, and 120 fragments in the MAC, suggesting 96 % insert into pre-existing IES in the MIC (unpublished data, Ashlock 2015).

11.3.2 Programmed Nuclear Death: Communication Between Germline and Soma

An important component of the conjugation process is the degradation of the parental MAC after DNA rearrangement. This begins when the Twi1-scanRNA complex retained in the parental MAC moves to the anlagen (Akematsu et al. 2014). Only the parental MAC degrades via chromatin changes induced by apoptosis proteins and is selectively eliminated from the cytoplasm by a highly elaborated autophagic/lysosomal process, while the other co-existing nuclei, the developing progeny MAC and MIC, are unaffected. This process is called Programmed Nuclear Death (PND) and plays an important role in the lifecycle of ciliates through not only completing the parental lifespan but also ensuring that the progeny genotype is expressed after conjugation. PND in *Tetrahymena* starts simultaneously with initiation of the progeny nuclear development in anlagen, implying a possible existence of molecular crosstalk among the nuclei to manage the fate of the different types of genomes in response to commitment to developmental programming. Much remains unclear about the mechanisms of PND, but *Tetrahymena* with the co-operating PND and nuclear development is a useful model system that allows rapid experimental readout for a better understanding of communication between germline and soma, which directly relates to universal issues on aging, senescence and rejuvenation (reprogramming) of eukaryotic cells, without involving the cell membrane boundary.

11.3.2.1 Autophagic/Lysosomal Process to Eliminate the Parental MAC

Autophagy is a remarkable cellular catabolism or self-eating process conserved in all eukaryotes and plays an essential role in cellular integrity through not only removal of undesirable organelles and proteins from the cytoplasm but also management of cell stress and nutrient homeostasis under starvation conditions. It is a form of communication of a cell with itself. In macroautophagy, which is the main type of autophagy, sequestration of targeted components with a double-membrane structure (isolation membrane) is critical to protect other organelles and cytoplasm from harmful digestive enzymes and is called autophagosome formation. The source of isolation membrane originates from the endoplasmic reticulum-mitochondria contact site in mammalian cells (Hamasaki et al. 2013). Seventeen autophagy-related (*ATG*) genes have been identified necessary to complete autophagosome formation. The role of class III phosphatidylinositol 3-kinase (otherwise known as Vps34) is also necessary to expand the isolation membrane through recruitment of Atg proteins on the nucleation site. The autophagosome then fuses with lysosomes to form an autolysosome, and the sequestered components are digested in an acidic environment and recycled. For reviews, see (He and Klionsky 2009; Mariño et al. 2014; Ohsumi 2014).

PND appears controlled by a specialized type of macroautophagy. When the anlagen starts to develop, the envelope of the parental MAC becomes stainable with a fluorescent autophagosome indicator monodansylcadaverine (MDC) (Akematsu et al. 2010). The time of appearance of the MDC stainability is consistent with the specific localization of Atg8s (Liu and Yao 2012), the most well characterized autophagosome component, on the surface of the parental MAC, strongly suggesting involvement of autophagosome formation or an equivalent structure on the surface of the parental MAC as the initial step of PND to sequester the nucleus destined to die from the other co-existing anlagen. The role of the Vps34 ortholog in *Tetrahymena* as well as Atg8s is essential for biogenesis of the autophagosomal property as in yeast and mammalian macroautophagy (Akematsu et al. 2014; Liu and Yao 2012). However, the membrane dynamics remains unclear, whether the process is achieved through engulfment of the nucleus by an isolation membrane in a similar manner to the typical macroautophagy or aggregation and the subsequent fusion of small autophagosomes on the nucleus. Since only nuclear membrane was observed around the degenerating parental MAC under transmission electron microscopy (Akematsu et al. 2010), the envelope of the parental MAC might change its nature as if it is an autophagosome using a common mechanism but without other membrane dynamics different from the typical macroautophagy seen in other systems.

Once PND starts with formation of the autophagosomal property on the surface of the parental MAC, the nucleoplasm begins to condense and reduces in size (Kobayashi and Endoh 2003). In this period, the parental MAC is surrounded by a large number of digestive vesicles, which consist of at least two different types of vesicles that are small autophagosomes in which mitochondria that have undergone loss of membrane potential are incorporated, as well as lysosomes (Akematsu et al.

2010; Kobayashi and Endoh 2005). These vesicles fuse with the periphery and release their contents into the inside of the nucleus in a stepwise fashion. First, the periphery of the nucleus fuses with the small autophagosomes from which mitochondrial apoptotic molecules, such as apoptosis-inducing factor (AIF) and DNases, are released into the nucleus (Akematsu and Endoh 2010). AIF is a mitochondrial flavoprotein and promotes the activities of mitochondrial DNases represented by endonuclease G (EndoG) in animal apoptosis. At least twelve DNase activities have been detected in *Tetrahymena* mitochondrial extracts (Osada et al. 2014), and they seem to have a pH optimum at near neutral as does mammalian EndoG and are responsible for the first wave of nuclear degradation, including the condensation of the parental MAC, kilobase-size DNA fragmentation, and the subsequent oligonucleosome-sized ladder formation in collaboration with AIF (Akematsu and Endoh 2010; Kobayashi and Endoh 2005). These phenomena are similar to the apoptotic events in Programmed Cell Death (PCD) in multicellular organisms, so PND is sometimes referred to as nuclear apoptosis (Endoh and Kobayashi 2006). On the other hand, lysosomes remain at the periphery of the parental MAC during the early stage of PND and release their contents into the nucleus in the next step. The nucleus undergoes further degradation in an acidic environment and is eventually reabsorbed. According to genetic analysis of PND using a *Tetrahymena* mutant pair nullisomic for chromosome 3 of the MIC (NULLI 3), the final resorption process is controlled by zygotic gene expression (Davis et al. 1992), while the early stage of PND mediated by the collaboration of AIF and mitochondrial DNases depends on parental genotype. Unlike the early stage of PND, acidic enzymes similar to DNase II, which is involved in apoptosis in Chinese hamster ovary cells, or DNase β and γ, both of which are involved in apoptosis, are likely involved in the later stage (Mpoke and Wolfe 1996).

11.3.2.2 Possible Molecular Signals to Ensure the Selective Nuclear Macroautophagy

The exact mechanism by which the selective nuclear elimination is communicated is still unclear. However, it has been suggested that the macronuclear envelope with autophagosomal property might exhibit some molecules on the surface as guidance of the disused nucleus, since abolished autophagosome formation on the parental macronuclear envelope by knocking out of *ATG*8s or *VPS*34 prevents approach of the digestive vesicles to the nucleus. In mitochondria-specific autophagy called mitophagy in mammals for example, the loss of membrane potential on aged or damaged mitochondria are specifically targeted by the autophagosome (Kim et al. 2007). By analogy, in mammals, cells in apoptosis are known to expose some molecules on the cell surface as "eat-me" signals, toward activated macrophages. The most common molecules used as eat-me signals are phosphatydylserine (PS) and a variety of glycocalyx (glycoprotein) compounds, both of which are restricted to the inner leaflet of the bi-layer plasma membrane by active transport from the outer to the inner leaflet (Depraetere 2000; Eda et al. 2004; Savill and

Gregory 2007). In *Tetrahymena* PND, as eat-me signals in the apoptotic cell, exposure of PS and sugars such as N-acetyl- D-glucosamine and sialic acid are found on the degenerating parental macronuclear envelope, while they are not found on other nuclear envelopes (Akematsu et al. 2010). These molecules are referred to as possible "attack-me" signals toward the autophagic/lysosomal process and could play roles as ligands to receptors on the surface of the digestive vesicles.

11.3.2.3 Commitment to PND and Its Relation to Progeny Nuclear Development

In *Tetrahymena*, PND occurs simultaneously with the initiation of progeny nuclear differentiation in the anlagen. Once PND starts with the formation of the autophagosomal property on the parental macronuclear envelope, transcriptional activity of the parental MAC is replaced by the developing macronuclear anlagen as represented by translocation of RNA polymerase II (RNApol II), a central enzyme for gene transcription (Mochizuki and Gorovsky 2004). The language of recruitment of RNApol II to targeted promoters, as well as enhancing the level of transcription, is transmitted through chromatin structure. At the early stage of PND, the chromatin in the parental MAC becomes heterochromatic while in the developing macronuclear anlagen, it becomes euchromatic. For instance, dimethylation of histone H3 at lysine 4 (H3K4), which is an epigenetic mark for active transcription, occurs in the developing macronuclear anlagen concomitant with the loss of the methyl mark from the parental MAC (Akematsu et al. 2010, 2014). The opposite chromatin remodeling in the two generations of the MAC might be mediated by the macronuclear-specific nuclear pore complex (NPC), which characterizes somatic features (Iwamoto et al. 2009) and disappears from the parental macronuclear membrane when the progeny macronuclear anlagen starts to develop (Iwamoto et al. 2015). Disappearance of NPC from the parental macronuclear membrane is likely coincident with the time of appearance of the autophagosomal property on the surface of the parental MAC, suggesting that PND and development act in close co-operation. Perhaps the two distinct programmings are co-regulated by upstream signaling that might exist earlier than the stage of karyogamy formation, which corresponds to fertilization in metazoa, since a karyogamy deficient pair affects neither activation of nuclear development nor the PND process (Fukuda et al. 2015).

11.4 Conclusion

There are several types of bio-communication; communication with the environment, communication among organisms, communication within organisms, and communication used to differentiate self from non-self. *Tetrahymena* reproduction uses all of these. Asexual reproduction is a way of communicating with the environment, with phenotypic assortment allowing the expression of diverse combinations of alleles. The exchange of genetic material during conjugation is a type of communication between two *Tetrahymena* organisms. The process of conjugation involves communication within the cell involving the genome rearrangements needed to build the new MAC from the new MIC, and involving the process of PND, ridding the cell of the old MAC. The deletion of thousands of IES during genome rearrangement is a type of communication involving the differentiation of self from non-self as well as a communication between parent and offspring, with early scanRNAs recognizing sequences existing in the offspring but not the parent, and late scanRNAs recognizing non-self, TE-like sequences.

Acknowledgments Work presented from REP's laboratory was supported by grants from the Canadian Institutes of Health Research (CIHR) and the Natural Sciences and Engineering Research Council (NSERC) of Canada. WA was supported by a postdoctoral fellowship from NSERC and TA was supported in the Pearlman lab by a Banting Fellowship and is presently supported by an INDICAR Fellowship. We thank students and postdoctoral fellows, particularly Dr. Jyoti Garg, in the REP lab for discussion and our collaborator Dr. J.S. Fillingham for discussion.

References

Agrawal N, Dasaradhi PVN, Mohmmed A, Malhotra P, Bhatnagar RK, Mukherjee SK (2003) RNA interference: biology, mechanism, and applications. Microbiol Mol Biol Rev MMBR 67 (4):657–685

Akematsu T, Endoh H (2010) Role of apoptosis-inducing factor (AIF) in programmed nuclear death during conjugation in Tetrahymena thermophila. BMC Cell Biol 11:13

Akematsu T, Fukuda Y, Attiq R, Pearlman RE (2014) Role of class III phosphatidylinositol 3-kinase during programmed nuclear death of Tetrahymena thermophila. Autophagy 10 (2):209–225

Akematsu T, Pearlman RE, Endoh H (2010) Gigantic macroautophagy in programmed nuclear death of Tetrahymena thermophila. Autophagy 6(7):901–911

Arnaiz O, Mathy N, Baudry C, Malinsky S, Aury JM, Denby Wilkes C, … Sperling L (2012) The Paramecium germline genome provides a niche for intragenic parasitic DNA: evolutionary dynamics of internal eliminated sequences. PLoS Genet 8(10):e1002984

Austerberry CF, Snyder RO, Yao MC (1989) Sequence microheterogeneity is generated at junctions of programmed DNA deletions in Tetrahymena thermophila. Nucleic Acids Res 17 (18):7263–7272

Baudry C, Malinsky S, Restituito M, Kapusta A, Rosa S, Meyer E, Betermier M (2009) PiggyMac, a domesticated piggyBac transposase involved in programmed genome rearrangements in the ciliate Paramecium tetraurelia. Genes Dev 23(21):2478–2483

Carle CM (2013) A Novel G-Quadruplex Binding Protein Determines the Boundaries of Genome Rearrangements in Tetrahymena thermophila. Washington University in St. Louis. http://openscholarship.wustl.edu/art_sci_etds/115

Chalker DL, Meyer E, Mochizuki K (2013) Epigenetics of ciliates. Cold Spring Harb Perspect Biol 5(12):a017764

Chalker DL, Terza A La, Wilson A, Kroenke CD, Yao M (1999) Flanking regulatory sequences of the Tetrahymena R Deletion Element Determine the Boundaries of the DNA rearrangement. Mol Cell Biol 19(8):5631–5641

Chalker DL, Yao M-C (2011) DNA Elimination in ciliates: transposon domestication and genome surveillance. Annu Rev Genet 45(1):227–246

Challoner PB, Amin AA, Pearlman RE, Blackburn EH (1985) Conserved arrangements of repeated DNA sequences in nontranscribed spacers of ciliate ribosomal RNA genes: evidence for molecular coevolution. Nucleic Acids Res 13(7):2661–2680

Chen X, Bracht JR, Goldman AD, Dolzhenko E, Clay DM, Swart EC, Landweber LF (2014) The architecture of a scrambled genome reveals massive levels of genomic rearrangement during development. Cell 158(5):1187–1198

Cheng C, Vogt A, Machizuki K, Yao M (2010) A domesticated piggybac transposase plays key roles in heterochromatin dynamics and DNA cleavage during programmed DNA deletion in Tetrahymena thermophila. Mol Biol Cell 21:1753–1762

Cherry JM, Blackburn EH (1985) The internally located telomeric sequences in the germ-line chromosomes of Tetrahymena are at the ends of transposon-like elements. Cell 43(3 Pt 2):747–758

Cole E, Sugai T (2012) Developmental progression of Tetrahymena through the cell cycle and conjugation. Methods Cell Biol 109:177–236

Craig NL (2002) Mobile DNA II. ASM Press, Washington, DC

Davis MC, Ward JG, Herrick G, Allis CD (1992) Programmed nuclear death: apoptotic-like degradation of specific nuclei in conjugating Tetrahymena. Dev Biol 154(2):419–432

Depraetere V (2000) "Eat me" signals of apoptotic bodies. Nat Cell Biol 2(6):E104

Dubois E, Bischerour J, Marmignon A, Mathy N, Régnier V, Bétermier M (2012) Transposon invasion of the Paramecium germline genome countered by a domesticated PiggyBac transposase and the NHEJ pathway. Int J Evol Biol 2012:436196

Eda S, Yamanaka M, Beppu M (2004) Carbohydrate-mediated phagocytic recognition of early apoptotic cells undergoing transient capping of CD43 glycoprotein. J Biol Chem 279(7):5967–5974

Eisen JA, Coyne RS, Wu M, Wu D, Thiagarajan M, Wortman JR, … Orias E (2006) Macronuclear genome sequence of the ciliate Tetrahymena thermophila, a model eukaryote. PLoS Biol 4(9): e286

Endoh H, Kobayashi T (2006) Death harmony played by nucleus and mitochondria: nuclear apoptosis during conjugation of Tetrahymena. Autophagy 2(2):129–131

Engberg J, Nielsen H (1990) Complete sequence of the extrachromosomal rDNA molecule from the ciliate Tetrahymena thermophila strain B1868VII. Nucleic Acids Res 18(23):6915–6919

Fan Q, Yao M (1996) New telomere formation coupled with site-specific chromosome breakage in Tetrahymena thermophila. Mol Cell Biol 16(3):1267–1274

Fass JN, Joshi NA, Couvillion MT, Bowen J, Gorovsky MA, Hamilton EP, … Collins K (2011) Genome-scale analysis of programmed DNA elimination sites in Tetrahymena thermophila. G3: Genes Genomes Genet 1:515–522

Fedoroff NV (2012) Transposable elements, epigenetics, and genome evolution. Science 338:758–767

Fillingham JS, Bruno D, Pearlman RE (2001) Cis-acting requirements in flanking DNA for the programmed elimination of mse2.9: a common mechanism for deletion of internal eliminated sequences from the developing macronucleus of Tetrahymena thermophila. Nucleic Acids Res 29(2):488–498

Fillingham JS, Thing TA, Vythilingum N, Keuroghlian A, Bruno D, Golding GB, Pearlman RE (2004) A non-long terminal repeat retrotransposon family is restricted to the germ line micronucleus of the ciliated protozoan Tetrahymena thermophila. Eukaryot Cell 3(1):157–169

Fukuda Y, Akematsu T, Attiq R, Tada C, Nakai Y, Pearlman RE (2015) Role of the cytosolic heat shock protein 70 Ssa5 in the ciliate protozoan Tetrahymena thermophila. J Eukaryot Microbiol 62(4):481–493

Girard A, Hannon GJ (2008) Conserved themes in small-RNA-mediated transposon control. Trends Cell Biol 18(3):136–148

Godiska R, Yao M (1990) A programmed site-specific DNA rearrangement in Tetrahymena thermophila requires flanking polypurine tracks. Cell 61:1237–1246

Gratias A, Bétermier M (2003) Processing of double-strand breaks is involved in the precise excision of Paramecium internal eliminated sequences. Mol Cell Biol 23(20):7152–7162

Guzzardo PM, Muerdter F, Hannon GJ (2013) The piRNA pathway in flies: highlights and future directions. Curr Opin Genet Dev 23(1):44–52

Hamasaki M, Furuta N, Matsuda A, Nezu A, Yamamoto A, Fujita N, Yoshimori T (2013) Autophagosomes form at ER-mitochondria contact sites. Nature 495(7441):389–393

Hamilton EP, Williamson S, Dunn S, Merriam V, Lin C, Vong L, … Orias E (2006) The highly conserved family of Tetrahymena thermophila chromosome breakage elements contains an invariant 10-base-pair core. Eukaryot Cell 5(4):771–80

He C, Klionsky DJ (2009) Regulation mechanisms and signaling pathways of autophagy. Annu Rev Genet 43:67–93

Huvos P (2004) A member of a repeat family is the source of an insertion-deletion polymorphism inside a developmentally eliminated sequence of Tetrahymena thermophila. J Mol Biol 336 (5):1061–1073

Iwamoto M, Koujin T, Osakada H, Mori C, Kojidani T, Matsuda A, … Haraguchi T (2015) Biased assembly of the nuclear pore complex is required for somatic and germline nuclear differentiation in Tetrahymena. J Cell Sci 128(9):1812–23

Iwamoto M, Mori C, Kojidani T, Bunai F, Hori T, Fukagawa T, … Haraguchi T (2009) Two distinct repeat sequences of Nup98 nucleoporins characterize dual nuclei in the binucleated ciliate tetrahymena. Curr Biol: CB 19(10):843–847

Joly-Lopez Z, Bureau TE (2014) Diversity and evolution of transposable elements in Arabidopsis. Chromosome Res: Int J Mol Supramol Evol Aspects Chromosome Biol 22(2):203–216

Kapitonov VV, Jurka J (2006) Self-synthesizing DNA transposons in eukaryotes. Proc Natl Acad Sci USA 103(12):4540–4545

Kapusta A, Matsuda A, Marmignon A, Ku M, Silve A, Meyer E, Bétermier M (2011) Highly precise and developmentally programmed genome assembly in Paramecium requires ligase IV-dependent end joining. PLoS Genet 7(4):e1002049

Karrer KM (2012) Nuclear dualism. Methods Cell Biol 109:29–52

Kataoka K, Mochizuki K (2011) Programmed DNA elimination in Tetrahymena: a small RNA-mediated genome surveillance mechanism. Adv Exp Med Biol 722:156–173

Kim I, Rodriguez-Enriquez S, Lemasters JJ (2007) Selective degradation of mitochondria by mitophagy. Arch Biochem Biophys 462(2):245–253

Kiss GB, Pearlman RE (1981) Extrachromosomal rDNA of Tetrahymena thermophila is not a perfect palindrome. Gene 13(3):281–287

Klobutcher LA, Herrick G (1997) Developmental genome reorganization in ciliated protozoa: the transposon link. Prog Nucleic Acid Res Mol Biol 56:1–62

Kobayashi T, Endoh H (2003) Caspase-like activity in programmed nuclear death during conjugation of Tetrahymena thermophila. Cell Death Differ 10(6):634–640

Kobayashi T, Endoh H (2005) A possible role of mitochondria in the apoptotic-like programmed nuclear death of Tetrahymena thermophila. FEBS J 272(20):5378–5387

Krupovic M, Bamford DH, Koonin EV (2014) Conservation of major and minor jelly-roll capsid proteins in Polinton (Maverick) transposons suggests that they are bona fide viruses. Biol Direct 9(1):6

Le Mouel A, Butler A, Caron F, Meyer E (2003) Developmentally regulated chromosome fragmentation linked to imprecise elimination of repeated sequences in Paramecia. Eukaryot Cell 2(5):1076–1090

Levin HL, Moran JV (2011) Dynamic interactions between transposable elements and their hosts. Nat Rev Genet 12(9):615–627

Lisch D, Slotkin RK (2011) Strategies for silencing and escape: the ancient struggle between transposable elements and their hosts. Int Rev Cell Mol Biol 292:119–152

Liu M-L, Yao M-C (2012) Role of ATG8 and autophagy in programmed nuclear degradation in Tetrahymena thermophila. Eukaryot Cell 11(4):494–506

Loidl J, Lukaszewicz A, Howard-Till RA, Koestler T (2012) The Tetrahymena meiotic chromosome bouquet is organized by centromeres and promotes interhomolog recombination. J Cell Sci 125(Pt 23):5873–5880

Lukaszewicz A, Howard-Till RA, Loidl J (2013) Mus81 nuclease and Sgs1 helicase are essential for meiotic recombination in a protist lacking a synaptonemal complex. Nucleic Acids Res 40:9296–9309

Malone CD, Hannon GJ (2009) Small RNAs as guardians of the genome. Cell 136(4):656–668

Mariño G, Niso-Santano M, Baehrecke EH, Kroemer G (2014) Self-consumption: the interplay of autophagy and apoptosis. Nat Rev Mol Cell Biol 15(2):81–94

Merriam EV, Bruns PJ (1988) Phenotypic assortment in Tetrahymena thermophila: assortment kinetics of antibiotic-resistance markers, tsA, death, and the highly amplified rDNA locus. Genetics 120(2):389–395

Mochizuki K (2010a) DNA rearrangements directed by non-coding RNAs in ciliates. Wiley Interdisc Rev RNA 1(3):376–387

Mochizuki K (2010b) RNA-directed epigenetic regulation of DNA rearrangements. Essays Biochem 48(1):89–100

Mochizuki K (2012) Developmentally programmed, RNA-directed genome rearrangement in Tetrahymena. Dev Growth Differ 54(1):108–119

Mochizuki K, Fine NA, Fujisawa T, Gorovsky MA (2002) Analysis of a piwi-related gene implicates small RNAs in genome rearrangement in Tetrahymena. Cell 110(6):689–699

Mochizuki K, Gorovsky MA (2004) Small RNAs in genome rearrangement in Tetrahymena. Curr Opin Genet Dev 14:181–187

Mochizuki K, Novatchkova M, Loidl J (2008) DNA double-strand breaks, but not crossovers, are required for the reorganization of meiotic nuclei in Tetrahymena. J Cell Sci 121(Pt 13):2148–2158

Mpoke S, Wolfe J (1996) DNA digestion and chromatin condensation during nuclear death in Tetrahymena. Exp Cell Res 225(2):357–365

Muñoz-López M, García-Pérez JL (2010) DNA transposons: nature and applications in genomics. Curr Genomics 11(2):115–128

Nanney DL (1980) Experimental ciliatology: an introduction to genetic and developmental analysis in ciliates. John Wiley & Sons Inc

Noto T, Kataoka K, Suhren JH, Hayashi A, Woolcock KJ, Gorovsky MA, Mochizuki K (2015) Small-RNA-mediated genome-wide trans-recognition network in Tetrahymena DNA elimination. Mol Cell 59(2):229–242

Ohsumi Y (2014) Historical landmarks of autophagy research. Cell Res 24(1):9–23

Orias E, Cervantes MD, Hamilton EP (2011) Tetrahymena thermophila, a unicellular eukaryote with separate germline and somatic genomes. Res Microbiol 162(6):578–586

Orias E, Flacks M (1975) Macronuclear genetics of Tetrahymena. I. Random distribution of macronuclear genecopies in T. pyriformis, syngen 1. Genetics 79(2):187–206

Osada E, Akematsu T, Asano T, Endoh H (2014) A novel mitochondrial nuclease-associated protein: a major executor of the programmed nuclear death in Tetrahymena thermophila. Biol Cell 106(3):97–109

Patil NS, Karrer KM (2000) A developmentally regulated deletion element with long terminal repeats has cis-acting sequences in the flanking DNA. Nucleic Acids Res 28(6):1465–1472

Plasterk RH, Izsvak Z, Ivics Z, Izsvák Z, Ivics Z (1999) Resident aliens: the Tc1/mariner superfamily of transposable elements. Trends Genet 15(8):326–332

Rogers MB, Karrer KM (1985) Adolescence in Tetrahymena thermophila. Proc Natl Acad Sci USA 82(2):436–439

Saveliev SV, Cox MM (1996) Developmentally programmed DNA deletion in Tetrahymena thermophila by a transposition-like reaction pathway. The EMBO Journal 15(11):2858–2869

Savill J, Gregory C (2007) Apoptotic PS to phagocyte TIM-4: eat me. Immunity 27(6):830–832

Schoeberl UE, Kurth HM, Noto T, Mochizuki K (2012) Biased transcription and selective degradation of small RNAs shape the pattern of DNA elimination in Tetrahymena. Genes Dev 26(15):1729–1742

Schoeberl UE, Mochizuki K (2011) Keeping the soma free of transposons: programmed DNA elimination in ciliates. J Biol Chem 286(43):37045–37052

Senti K-A, Brennecke J (2010) The piRNA pathway: a fly's perspective on the guardian of the genome. Trends Genet: TIG 26(12):499–509

Siomi MC, Sato K, Pezic D, Aravin AA (2011) PIWI-interacting small RNAs: the vanguard of genome defence. Nat Rev Mol Cell Biol 12(4):246–258

Slotkin RK, Martienssen R (2007) Transposable elements and the epigenetic regulation of the genome. Nat Rev Genet 8(4):272–285

Smit AFA (1999) Interspersed repeats and other mementos of transposable elements in mammalian genomes. Curr Opin Genet Dev 9:657–663

Vogt A, Goldman AD, Mochizuki K, Landweber LF (2013) Transposon domestication versus mutualism in ciliate genome rearrangements. PLoS Genet 9(8):e1003659

Vogt A, Mochizuki K (2013) A domesticated PiggyBac transposase interacts with heterochromatin and catalyzes reproducible DNA elimination in Tetrahymena. PLoS Genet 9(12):e1004032

Volff J-N (2006) Turning junk into gold: domestication of transposable elements and the creation of new genes in eukaryotes. BioEssays: News Rev Mol Cell Dev Biol 28(9):913–922

Wuitschick JD, Gershan JA, Lochowicz AJ, Li S, Karrer KM (2002) A novel family of mobile genetic elements is limited to the germline genome in Tetrahymena thermophila. Nucleic Acids Res 30(11):2524–2537

Yao MC, Zhu SG, Yao CH (1985) Gene amplification in Tetrahymena thermophila: formation of extrachromosomal palindromic genes coding for rRNA. Mol Cell Biol 5(6):1260–1267

Yao M-C (1981) Ribosomal RNA gene amplification in Tetrahymena may be associated with chromosome breakage and DNA elimination. Cell 24(3):765–774

Yao M-C (1986) Amplification of ribosomal RNA genes. In: The molecular biology of ciliated protozoa. Elsevier, pp 179–201

Yao M-C, Fuller P, Xi X (2003) Programmed DNA deletion as an RNA-guided system of genome defense. Science (New York, N.Y.) 300(5625):1581–1584

Yao M-C, Gorovsky M (1974) Comparison of the sequences of macro- and micronuclear DNA of Tetrahymena pyriformis. Chromosoma 48(1):1–18

Chapter 12
Cell-Cell Interactions Leading to Establishment of a Mating Junction in *Tetrahymena* and *Paramecium*, Two "Contact-Mediated" Mating Systems

Eric S. Cole

Abstract *Tetrahymena* and *Paramecium* initiate mating in response to direct physical contact between cells of complementary mating types. Genes that determine mating-type have been cloned and sequenced for both organisms. What remains unclear is how physical interactions between complementary mating type proteins trigger cell adhesion, and assembly of the transient, specialized mating junction through which gametic nuclei are exchanged. It has proven challenging to tease apart the mechanism of cell signaling from that of cell adhesion. Ultimately, mating cells establish cytoplasmic continuity through a process of membrane fusion mediated (at least in *Tetrahymena*) by a universal gamete fusogen homologous to the Hap2/GCS1 protein first identified in plants, algae, protists and many of the lower animals.

12.1 Introduction

The unicellular condition is the archetype for life on earth. Every problem of physiology, reproduction, motility and homeostasis was first solved by molecular mechanisms operating within single cells. Relatively late in evolutionary history these mechanisms, originally adapted for life within a unicellular context, were co-opted to serve multicellular tissues and organs. Cell signaling at the unicellular level is important in two broad contexts: recognizing the presence of predators or prey within the environment (allospecific interactions), and recognizing and responding to members of your own species (conspecific interactions).

Ciliates respond to contact with members of their own species in several situations. "Quorum sensing" behaviors can lead to changes in ciliate behavior (Jacob, et al., this volume). Elevated cell densities increase the frequency of conspecific interactions triggering encystment in some ciliates, possibly serving as a form of population control at high densities (See Bracht, this volume). Historically, *Tetrahymena* cultures

E.S. Cole (✉)
Biology Department, St. Olaf College, Northfield, MN 55057, USA
e-mail: colee@stolaf.edu

G. Witzany and M. Nowacki (eds.), *Biocommunication of Ciliates*,
DOI 10.1007/978-3-319-32211-7_12

at low cell densities have been difficult to establish. This dependence upon "neighbors" to establish a proliferative culture is common in a variety of tissue-culture situations, and in ciliates has been associated with some form of autocrine signaling (Rasmussen and Wheatley 2007), (See also György and Kőhidai, this volume). These observations suggest a form of cooperativity between cells as proliferative cultures become established. Finally sexual reproduction requires sophisticated communication between cells. Cell-cell interactions between complementary sexes or "mating types" trigger a complex developmental program leading to nuclear exchange. It is the signaling events associated with mating and more specifically with the construction of a mating junction that this chapter will focus on.

12.2 Ciliate Mating: A Distinct Form of Gamete Fusion

Cell-cell interactions leading to conjugation (mating) in ciliates have been likened to the interaction between egg and sperm in multicellular organisms (Austin 1965; Hiwatashi 1969; Metz 1954; Miyake 1996). One significant difference is that, whereas most mating systems involve anisogamy (eggs and sperm differing dramatically in size and structure), ciliate mating is an example of the isogametic condition in which both gametes are identical. *Vorticella* and *Ephelota* represent exceptions to this rule exhibiting microgametic and macrogametic forms, (Finley 1939; Grell 1967). Isogamy is also seen in yeasts and other fungi, as well as flagellate and filamentous algae. Another significant difference is that, in virtually all other mating systems, gamete fusion is complete and irreversible. Once the membrane of the male gamete fuses with the membrane of the female gamete, the two become integrated irreversibly in the resulting zygote. Most ciliates, on the other hand, meet and establish cytoplasmic continuity through a specialized region of the plasma membrane that is both spatially limited and transient, the "mating junction" (Cole 2006; Cole et al. 2015; Wolfe 1982, 1985). Haploid nuclei are exchanged across this cytoplasmic bridge that then seals over, restoring cytoplasmic autonomy to each of the mating partners. Such transient, reversible mating junctions are unique in the world of eukaryotic organisms.

Despite these differences, ciliates must solve many of the same problems facing the egg and sperm. One such problem involves cell-cell recognition. There appear to be two models for the cell signaling that constitute sexual recognition and that launch conjugal behavior and sexual reproduction in ciliates. In one model, complementary mating types secrete sex-specific, diffusible molecules (pheromones) that trigger pre-conjugal behavior in the mating partner. Ciliates that deploy diffusible pheromones [sometimes referred to as "gamones", (Miyake 1974)], are reviewed by Luporini, [(Luporini et al. 2015), and in this volume]. The other model for cell-cell communication leading to the launch of conjugal development involves direct physical contact between cells of complementary mating type, and invokes surface molecules rather than secreted signals. Curiously, the choice to deploy membrane-bound vs secreted mating signals does not seem to follow any obvious or consistent phylogenetic trends. Though *Euplotes* represents a classic example of secreted sex-pheromones, there are species within the *Euplotes* complex that deploy contact-mediated signaling rather than secreted molecules, (Dini 1984; Gates 1990). Contact-mediated conjugal behavior is more consistently represented in two oligohymenophorean ciliate genera: *Paramecium* and *Tetrahymena,* and the molecular players are just beginning to emerge.

12.3 Cell-Signaling Versus Cell Adhesion: The Central Problem

Historically the language of cell biology categorizes protein-protein interactions between cells as either serving to bring about cell adhesion or cell signaling. The difference appears to be the "output" of such interactions. When surface proteins bind to their complements on a neighboring cell and the principal result is adhesion, we classify such proteins as cell-adhesion molecules and designate them as having a structural role. If protein/receptor interactions trigger physiological responses, we tend to ignore the structural nature of the adhesion that occurs between ligand and receptor, and focus on their role in cell-cell communication. What such semantic categories tend to obscure is that cell interactions of an adhesive nature often have physiological consequences. Cell adhesion molecules communicate signals from the extracellular environment to the intracellular environment (Bhatt et al. 2013; Geiger and Yamada 2011; Hatzfeld et al. 2014; Shen et al. 2012). The challenge of disentangling cell adhesion from cell signaling is especially prominent when exploring the communication that precedes sexual union in ciliates. In examples involving secreted pheromones the situation at first seems less confusing as signaling (via secreted mating-type pheromones) and cell adhesion (a contact-mediated response) would appear to be discrete phenomena (See Luporini, this volume). However, Weiss et al. (1995), in solving the 3-D crystal structure of the *Euplotes* ER1 pheromone, proposed a fascinating model in which the secreted pheromones could cause aggregation of surface receptors on the mating partner. The receptors, by virtue of their aggregate increase in adhesivity, in turn could mediate the subsequent cell adhesion with the complementary receptors on the mating partner. This attractive, yet hypothetical model highlights our difficulty.

The problem of distinguishing sexual signaling from cell adhesion in contact-mediated systems is made more complicated by the possibility that cell adhesion could involve the mating-type determinants themselves, bypassing the need for receptors. Alternatively, cell-signaling via mating-type determinants might trigger adhesivity as a secondary cellular response. In short, signaling and adhesion may or may not be mediated by the same molecules. Critical to discerning the separate or conflated nature of cell-signaling and cell-adhesion in the contact-mediated species will be discovering whether or not cell-adhesion is generic or restricted to cells of complementary mating types. In short, is the mechanism of cell-adhesion dependent or independent of the mechanism of mating type recognition?

12.4 *Tetrahymena* Versus *Paramecium* (a Matter of Timing?)

In *Tetrahymena*, mixing complementary mating types results in an hour to an hour and a half delay before cell-adhesion and pair formation. Presumably, this hour-long waiting period ["co-stimulation": (Bruns and Palestine 1975)] involves repetitive contacts between cells of complementary mating types, and a prolonged intracellular response that readies the cells for pairing. *Paramecium,* on the other hand, exhibit cell adhesion immediately upon mixing. This "mating reaction" is manifest as a massive aggregation from which committed pairs emerge an hour to an hour and a half later. It is tempting to infer parallels between the two models. Perhaps the necessary physical contacts that occur between *Tetrahymena* cells, mediated through brief and perhaps random collisions, result in the same sort of signaling interactions that occur within the more highly stabilized mating aggregates of *Paramecium*. If true, then the mating reaction observed in *Paramecium* (that has lent itself to a rich history of experimental manipulation) may yield insight into the events that occur during co-stimulation mediated through the brief, cell-cell collisions of *Tetrahymena*. At the very least, details from the *Paramecium* mating reaction can help us design testable predictions as we explore the more transient interactions between *Tetrahymena* cells as they prepare for mating.

12.5 The *Paramecium* Model

Paramecium offers the archetype of contact-mediated cell signaling leading to conjugal behavior among ciliates. The phenomenon of mating in *Paramecium* was first recognized by O.F. Müller in a paper published posthumously in 1786. He was the first to discern that this was a sexual process rather than an example of asexual reproduction through "longitudinal fission". Tracy Sonneborn isolated the first mating types (named odd and even, or "O" and "E") in 1937, and opened the door to a genetic analysis of the maternal pattern of mating type inheritance. Mating involves three steps. First, cells of complementary mating type adhere through contacts on **cilia** located on the ventral-anterior surface of the cell (Byrne 1973; Cohen and Siegel 1963; Hiwatashi 1961); (Fig. 12.1). This happens within seconds of mixing two mating types together: reviewed by Miyake (1996), and Hiwatashi (1981). Next, cells form a region of cell adhesion located at the anterior tip. This region, called the **anterior holdfast** (Jennings 1911), involves a region of membrane that is first denuded of cilia 30 min after mixing, (Hiwatashi 1955), and serves as a site of cell adhesion (45–60 min after mixing). "Holdfast pairs" emerge from the mating aggregates after 60–90 min, are relatively easy to dissociate by mechanical agitation, and are commonly referred to as "loose pairs". The region of cell adhesion appears to spread from anterior to posterior along the ventral surface preceded by a wave of local deciliation. After two hours, a region on the ventral

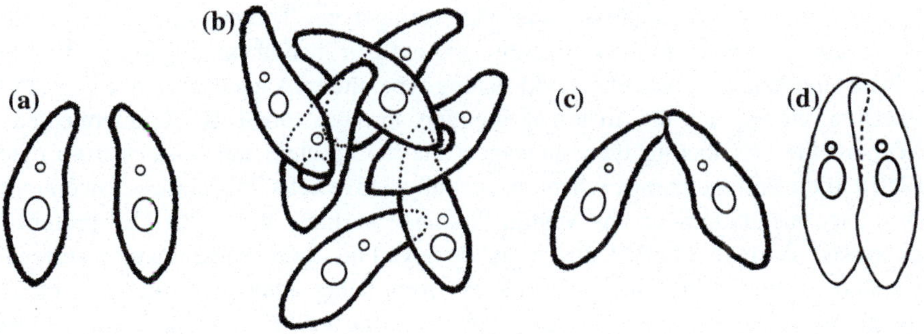

Fig. 12.1 Diagram depicting the stages of mating in *Paramecium*. **a** Cells prior to mixing. **b** Pairs shortly after mixing undergoing aggregation during the "mating reaction". **c** One hour later, individual pairs emerge from aggregates, joined at the anterior "holdfast". **d** Later pairs become joined at both the anterior holdfast and the par-oral cone

surface located within the oral region everts to form the "**fertilization cone**" (Diller 1936) (Fig. 12.1). This specialized region of membrane and cytoplasm houses the pronuclei prior to exchange, and ultimately becomes the region of membrane fusion through which pronuclei are exchanged (a "**nuclear exchange junction**"). Once membrane fusion is established, these "tight pairs" are thoroughly committed to conjugal development though as we shall see, nuclear events of meiosis and macronuclear reorganization can be triggered even without membrane fusion. The process of aggregation that initiates this mating sequence requires contact between cells of complementary mating type within normal, natural conditions.

In 2014, Singh et al. working in the laboratory of Eric Meyer, identified the *P. tetraurelia* gene (**mtA**) encoding the "even" mating type protein, and demonstrated the protein's localization and requirement for determining mating-type as well as species specificity (Singh et al. 2014). Furthermore, they unveiled a molecular mechanism by which the maternal genome directs gene rearrangement leading to expression of the "E" mating type protein, a fascinating mechanism involving RNA-guided DNA excision. The "E" mating type protein is a trans-membrane protein that localizes to cilia on the anterior-ventral surface only in cells expressing the "even" mating type. When this mating-type protein is not expressed, cells adopt the "default" option and express the "O" mating type.

In a separate species, (*P. caudatum*), Xu and Takahashi (2001) generated a monoclonal antibody that seems to recognize the "O" mating-type antigen. Curiously, this antigen is expressed in both "O" and "E" mating types, but only becomes localized at the cell surface in ciliary membranes within cells expressing the odd mating type. (The "O" antigen remains cytoplasmic in the "even" cells). It should be acknowledged that different (though related) species were involved in these two discoveries, and there are likely to be differences in the details of their respective mechanisms. For example, Singh et al. (2014) demonstrated inter-species differences in how the mtA gene is silenced within "O" cells. That said, expression of the *P. septaurelia* mtA gene in the *P. tetraurelia* species supports a cross-species mating reaction and suppresses the odd mating type. The emerging view is that *Paramecium* express the “O” mating type antigen in both mating types. In cells whose “mothers” expressed the odd-mating type, the mtA gene is silenced by DNA excision guided by the maternal genome through small RNA intermediates. Without the “E” mating type substance, the “O” protein translocates from cytoplasm to the cell membrane where, in sexually active cells, it becomes concentrated in ciliary membranes on the ventral “mating” surface. When the “E” protein is expressed from the mtA locus, it is deployed to these same mating surfaces, somehow suppressing the “O” protein from being integrated into the ciliary membranes. Hence, “even” mating types express the “E” protein from the mtA locus on their ventral cilia, and “odd” mating types express the “O” protein (gene unknown) on their ventral cilia. That both mating type substances are proteins is supported by experiments in which aggregate-inducing cilia (see below) of either mating type are exposed to trypsin, with a consequent loss of aggregate-inducing potential (Kitamura and Hiwatashi 1978).

The question we return to is what role do “O/E” molecular interactions play during co-stimulation leading to pair formation and mating? The answer goes to the question: how do we dissect *cell adhesion* from *cell-signaling*? In *Paramecium*, there appears to be one immediate event that requires odd/even mating type interactions: the cell adhesion phenomenon, or “mating reaction”. Mature, starved cells of complementary mating types are competent to form mating aggregates, but the response appears dependent upon physical interaction of the “O” and “E” surface proteins. It is tempting to imagine that the “O” and “E” surface proteins act as the immediate ligand and receptor molecules that generate cell adhesion. These “O/E” molecular interactions would then stimulate cells to assemble a mating junction (the paroral cone) preparing for membrane fusion and initiating meiosis. A problem with this intuitive model is the observation that, under certain circumstances, cells can form homotypic unions. If cell adhesion is mediated through heterotypic protein interactions, how can cells expressing identical mating types form adhesive pairs? Regardless of whether or not mating type proteins serve both signaling and adhesive roles, the heterotypic interaction is a pre-requisite for the immediate expression of cell-adhesivity resulting in cell aggregation under normal mating conditions.

A second possible role for “O/E” molecular interactions is controversial. This comes from an observation by Metz who noticed that heterotypic pairs become mechanically stabilized after about 2 h whereas homotypic pairs are more easily dissociated by drawing them in and out of a glass pipette in a viscous medium (saturated methyl cellulose) (Metz 1947). This observation suggests that homotypic pairs may not form the “tight pairs” characterized by fusion at the paroral cone, though they do appear to adhere at this membrane locus. This lone observation begs further investigation in that it bears significantly on the role of the mating type proteins. We will see that, in *Tetrahymena*, a second role for mating type proteins in stabilizing heterotypic pairs over homotypic pairs is a more well-established phenomenon.

12.5.1 Homotypic Versus Heterotypic Pairing

In a series of classic experiments, the conditions necessary to provoke a "mating reaction" (cell aggregation via ciliary adhesion) were explored. Remarkably, dead cells of one mating type when mixed with starved (mating-competent) live cells of a complementary mating type, can provoke cell-cell adhesion in the living cells [(homotypic cell adhesion) (Metz 1946, 1947) reviewed by Metz (1954), and Hiwatashi (1969)]. Such necrophilic stimulation drives aggregation as well as "pair formation" among the living, homotypic cells. Homotypic pairs even initiate the nuclear events associated with conjugation: meiosis and nuclear reorganization, resulting in a type of self-mating. Subsequently, it was shown that cilia isolated from one mating type could trigger the mating reaction and the above-mentioned homotypic pairing events (Fukushi and Hiwatashi 1970; Miyake 1964). This result has been obtained with numerous species of *Paramecium*. Taking this reductionist approach one step further, mating type-specific membrane vesicles isolated from ciliary preparations were shown to trigger homotypic pairing events in cells expressing the complementary mating type (Kitamura and Hiwatashi 1976, 1980; Watanabe 1977). Curiously, these isolated vesicles seem to have two biological effects with different specific activities. The lower (less stable) activity seems to provoke the "mating reaction" mediated by ciliary adhesion (aggregation). The more potent (and more stable) vesicle-mediated activity triggers homotypic pairing without the aggregation phenomenon (Hiwatashi 1981; Kitamura and Hiwatashi 1980). This raises the possibility that contact between heterotypic mating-type proteins play two significant roles, one occurring early (ciliary adhesion leading to cell clumping) and one later (during membrane adhesion between the ventral cell-surfaces beginning at the anterior holdfast and extending to the developing paroral cone).

This series of elegant investigations provides clarification of the role of the mating type proteins. The Odd/Even mating proteins appear to be necessary for catalyzing the mating reaction through a generalized, and non-specific cell-adhesion mechanism. That is to say, "O"/"E" molecular interactions operating at the cell-surface, seem to activate a general cell-adhesion mechanism that is not mating-type specific. Once this cell-adhesion mechanism is triggered, it is relatively indiscriminant, permitting both homotypic and heterotypic cell adhesion. One implication of this is that there may be another cell-surface molecule that mediates cell adhesion, one that is expressed on both mating types, and who's adhesive interactions are only activated after being stimulated by neighboring heterotypic, mating-type protein interactions. (A more complicated possibility, is that heterotypic mating type protein interactions activate a non-specific adhesivity within the mating type proteins themselves). A candidate for a second, generic cell-adhesion molecule has been identified (Ognibene et al. 2008). A 38 kD protein was identified (by virtue of its glycosylation state, and pattern of expression), and an antibody raised against it. The antigen is located on the cell surface and ciliary membranes of mating-competent cells in both mating types, and was absent from immature or non-competent cells. Exposure to the antiserum, (but not to pre-immune serum) delays the mating reaction (aggregation) and inhibits pair-formation in a dose-dependent fashion. Hence, this protein seems to participate in both the rapid, cilia-mediated aggregation reaction and in the subsequent establishment of loose-pairing via adhesion at the anterior holdfast.

A second implication of these findings is that cell adhesion (observed even in homotypic pairs) can trigger downstream events including formation of the paroral cone and the nuclear events associated with meiosis and nuclear reorganization. This is tricky! It remains ambiguous whether the downstream events observed in homotypic pairs are triggered by the generic cell adhesion mechanism or directly by the instigating heterotypic (mating type) interactions that provoke generic cell adhesion, since all these events are initially triggered by heterotypic mating type substances presented as dead cells, cilia or isolated ciliary membranes.

12.5.2 Chemical Induction of Homotypic Pairing

Homotypic partners of Paramecium can be stimulated to form pairs in the complete absence of substances derived from cells of the complementary mating type. This can be achieved though a number of chemical agents [(Hiwatashi 1969; Miyake 1961, 1981) reviewed by Miyake (1996)]. Typically, chemical induction is performed by exposing cells to a medium with low calcium concentrations (typically below 0.1 mM, Miyake 1958), and elevated levels of some other inorganic cation creating a “cation imbalance” (Miyake 1958). Hi-K^+/low-Ca^{++} is a classic exposure. Chemical induction appears to act by hacking the cell’s signaling machinery rather than by triggering a natural response. To begin with, chemical-induction initiates homotypic pair formation in a way that bypasses the normal, cilia-mediated cell aggregation reaction (pairs form directly after hour-long exposure without forming the typical aggregates or clumps). Furthermore, 1 uM $LaCl_3$ (a known inhibitor of calcium influx) suppresses the high K^+ trigger of homotypic pairing but has no effect on natural, heterotypic pairing (Cronkite 1979). This suggests that ion-imbalance triggers an artificial influx of calcium across the plasma membrane from the external medium, which in turn signals events leading to pair formation. Cation-imbalance suitable for triggering homotypic pairing can also trigger ciliary reversal, or an “avoidance reaction” (Kamada and Kinosita 1940). Ciliary reversal, in turn, has also been linked to elevations of intracellular calcium (Naitoh and Eckert 1974), and more specifically to the activation of K^+-sensitive calcium channels. Behavioral mutants that block calcium-induced-ciliary reversal [pawn B and pawn C^{ts} (Kung and Naitoh 1973)], also fail to respond to chemical induction of pairing [(Cronkite 1976), reviewed by Hiwatashi and Kitamura (1985)]. An attractive working hypothesis is that chemical induction of mating (at least through the manipulation of ion-imbalance) may be mediated through artificial stimulation of K^+-sensitive calcium channels. Though chemical induction of homotypic pairing can be brought on by an artificially generated calcium signal, it suggests that during normal heterotypic pair formation (following the mating reaction), there may be a form of signaling that triggers natural, internal calcium release. The artificial signal mimics natural signaling in the host of downstream consequences that it triggers. These include decreased cell-motility, ciliary denudation along the ventral mating surface, adhesion along this surface beginning at the anterior holdfast and extending to the oral apparatus, development of the paroral cone (the region of future nuclear exchange), and initiation of the nuclear events associated with conjugation. It is significant that application of A23187 (a calcium ionophore) can trigger early micronuclear behaviors associated with mating, (Cronkite 1979) but appears insufficient to trigger other events associated with homotypic pairing. A calcium-signal serving as a second messenger during mating is an attractive hypothesis with some compelling and supportive evidence, but it is likely that the details of natural signaling are subtle and complex.

Alternatively, nuclear stains that bind AT-rich regions of the DNA (acridine dyes, DAPI, Hoechst), have also been shown to trigger pairing in *Paramecium* (Miyake 1968). At first, this result seems enigmatic. However, aromatic cations (such as DAPI, Hoechst and Acridine orange) can intercalate into the DNA resulting in charge neutralization and chromatin condensation (Kapuscinski 1990). One can imagine that induced changes in chromatin condensation might lead to activation of some (ill-defined) signaling mechanism that triggers mating in *Paramecium*. A more interesting possibility follows from the observation that binding of chromatin by aromatic cations can lead to displacement of endogenous, chromatin-bound calcium ions. A paper by Przywara et al. (1991), demonstrates that acridine orange blocks intra-nuclear (but not cytoplasmic) elevation in free calcium in response to neural stimulation. These authors suggest that the chromatin itself houses a large pool of bound calcium, and that AO acts by displacing this intra-nuclear pool of calcium. In such a model, bathing *Paramecium* cells in one of these aromatic cations could lead to a sudden displacement of this chromatin-bound calcium, and the resulting transient elevation of free intra-nuclear calcium could trigger gene expression leading to conjugal pair formation. Hence there may be a connection between mechanisms that trigger homotypic pairing through manipulation of Ca^{++} and the curious case of chromatin binding agents doing the same. Such a purely speculative model obviously requires testing.

12.5.3 Mating-Type Signaling: One Role or Two?

There is some ambiguity in the literature regarding whether homotypic *Paramecium* pairs undergo membrane fusion at the paroral cone and subsequent cross-fertilization, or undergo a form of self-fertilization or "autogamy". Metz carefully observed homotypic pairs arising from mixtures of live and dead *P. aurelia* cells expressing complementary mating types (Metz 1948). By subjecting such pairs to repeated expulsion through a pipette, Metz noticed that heterotypic pairs exhibited a point of persistent adhesion at the paroral cone as well as at the anterior holdfast. He anticipated (correctly) that the paroral cone was where membrane fusion was initiated, preparing the way for nuclear exchange and creating a tight bond between mating partners. This tight paroral adhesion was (he claimed) absent in homotypic pairs induced by exposure to dead partners of opposite mating type. In describing such homotypic pairs he writes: "*Union involves only the anterior or holdfast regions of the pair members… Paroral cones form but these structures do not fuse*." In contrast, Miyake (1968) studying genetically marked *P. aurelia* cells of identical mating type, demonstrated that chemically-induced homotypic pairs *were* capable of exchanging nuclei (though at reduced efficiency). This is an important point in that it raises the question: are heterotypic mating-type interactions involved in facilitating membrane fusion at the paroral cone? Granted, the genetic data is more persuasive than microscopic observation, but one cannot help remarking that Metz was examining homotypic pairs induced by whole cell interactions, whereas Miyake was experimenting with a more artifactual, chemical-induction of pairing. In this author's mind, it is an open question whether or not mating-type proteins perform two signaling functions: first to trigger agglutination by "activating" adhesive proteins that mediate ciliary contacts, and second, to stabilize (heterotypic) mating pairs by promoting membrane fusion at the future nuclear exchange junction.

12.5.4 The Developmental Response to Cell-Cell Signaling

There are numerous responses to the cellular contact that initiates mating in *Paramecium* [reviewed by Mikami (2000)]. Within minutes, there is an increase in cytoplasmic streaming (Kobayashi et al. 1999). This is followed by migration of the germinal micronucleus from its docking site alongside the macronucleus (Fujishima and Hiwatashi 1977). Subsequently, cellular motility (ciliary action) is diminished. After 30 min, a wave of ventral cilia degeneration begins at the anterior end, and extends posteriorly. There is suggestion that the early micronuclear migration (emm) is triggered by fluxes in internal calcium levels (Yang and Haga 1993). After 45–60 min, cells form a region of membrane adhesion, the anterior holdfast. After two hours, cells are firmly bonded by membrane fusion at the paroral cone. Subsequently, the program of meiotic divisions commences leading to pro-nuclear exchange and fusion.

12.6 The *Tetrahymena* Model

Conjugation in *Tetrahymena thermophila* represents a second model in which mating is provoked by contact-mediated cell-signaling. Like *Paramecium*, mating in *Tetrahymena* requires cells to be mature, nutritionally starved and allowed to physically interact with starved cells of a complementary mating type. The

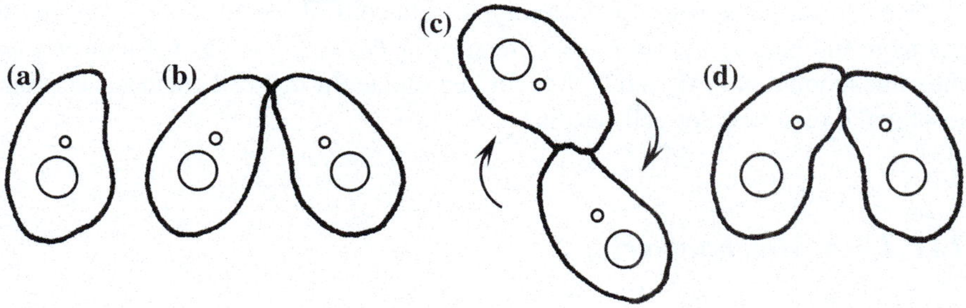

Fig. 12.2 Diagram depicting stages of mating in *Tetraymena*. **a** Un-paired cell. **b** One hour after mixing, cells form "loose pairs" joined at the anterior tips. **c** "Tumbling" early during pair-formation, as pairs adopt a "pinwheel" configuration, and spin. **d** Tight pairing as membranes form hundreds of focal fusion sites within the mating junction

phenomena of maturity/immaturity have been well studied in both organisms (Bleyman 1971; Haga and Hiwatashi 1981; Jennings 1944; Rogers and Karrer 1985). In *Tetrahymena*, the requisite period of nutritional deprivation (two or more hours) is referred to as "initiation" (Bruns and Brussard 1974). The period of interaction involving direct physical contact between cells of complementary mating type is referred to as "co-stimulation" (Bruns and Palestine 1975; Finley and Bruns 1980). Unlike *Paramecium*, there is no conspicuous clumping or aggregation of cells immediately upon mixing. Pairs form after approximately one hour of co-stimulation (Fig. 12.2), roughly the same waiting period required for binary pairs to emerge from cell aggregates in *Paramecium*. It is this period of co-stimulation that will be reviewed.

The requirement for physical contact between cells of complementary mating type was demonstrated by separating complementary mating types (of which there are seven) by a Millipore filter with 8 μm diameter pores and demonstrating that fluid exchange via diffusion was insufficient to shorten the one-hour waiting period associated with co-stimulation (Takahashi 1973). Without the distraction of a "Mating Reaction" (mixed cell-aggregates mediated by ciliary contact in *Paramecium*), attention has focused on the cell-cell interactions that lead to actual pair formation in *Tetrahymena*. The model that has emerged suggests that co-stimulation is achieved by multiple, repetitive collisions between cells of mixed mating type (Allewell et al. 1976; Brown et al. 1993). This obligate period of heterotypic collision has two components. In the early stage, (first 20 min), pairing can be blocked by physical agitation (rapid shaking), or by exposure to Actinomycin D (blocking RNA transcription), (Allewell and Wolfe 1977). After this early period, shaking or adding ActD no longer prevents pair formation. These results suggest that an early stage of reciprocal cell signaling requires 20-min of uninterrupted, heterotypic cell-collisions and that these repetitive contacts trigger transcription of gene products necessary for pair formation. These "stimulated" cells require another 40 min to develop the capacity for pairing through cell adhesion at a specific cellular location similar to that seen in *Paramecium:* along the ventral surface, anterior to the oral apparatus.

One can liken the period of heterotypic collisions observed in *Tetrahymena* to the more intimate aggregation response seen in *Paramecium*, in that both occupy the same amount of time, and both seem to promote the cell-cell contacts necessary to provoke readiness for cell fusion.

12.6.1 A Matchmaker?

Though this form of cellular communication appears to be contact-mediated, there is surprisingly strong evidence supporting the role of a "third party", some soluble, secreted material that mediates cell-cell contact (Adair et al. 1978). This material is *not* mating-type specific, and is secreted both during starvation (initiation) and during co-stimulation. It appears in the medium within 15 min of nutritional shift-down, accumulates to critical concentration within one hour of starvation, and its synthesis is blocked by cycloheximide implying a requirement for protein synthesis (Wolfe et al. 1980). Furthermore, this substance (modestly heat-stable and un-dialyzable) appears to be necessary for both co-stimulation and the cell-adhesion that follows co-stimulation (Wolfe et al. 1979). When *Tetrahymena* cells are washed, they must "condition" their medium (presumably by secreting this factor) for three hours before they are capable of undergoing co-stimulation and pairing. This "matchmaker" function is immediately restored to washed, co-stimulating cell cultures if their own conditioned medium (the supernatant) is added back (Adair et al. 1978; Wolfe et al. 1979). These findings increase our cast of *Tetrahymena* molecules in search of identity to include mating-type factors necessary for co-stimulation, soluble factors that promote co-stimulation leading to cell adhesion and a membrane fusogen. *Tetrahymena* might also have cell adhesion molecules associated with the membrane that are activated by heterotypic mating-type interactions.

12.6.2 Dead Cells, Cilia and Ciliary Vesicles: Membrane Proxies for Cell-Cell Contact

Many of the classic experiments performed in *Paramecium* were replicated in *Tetrahymena* with some modification. Unfortunately, many of these results have never been published. Toshiro Sugai repeated the "dead-cell" co-stimulation experiments in the following way: Glutaraldehyde-treated dead mating type "A" cells were washed thoroughly and added to live, starved mating type "B" cells for 1 h. Similarly, glutaraldehyde-treated mating type "B" cells were added to live mating type "A" cells for 1 h. Then these two macabre cultures were mixed. Pairs involving live cells appeared within a very short time compared with live cells undergoing normal co-stimulation time (5 min vs. 25 min at 26 °C). The most straightforward interpretation is that in *Tetrahymena* as in *Paramecium*, surface molecules (even from dead cells) are capable of mediating the signaling that brings about co-stimulation (Sugai, personal communication).

Attempts to provoke co-stimulation in *Tetrahymena* by exposure to detached cilia, have failed (Wolfe et al. 1993). It has been suggested that this is because the cells eat the cilia, thereby breaking the requisite nutritional deprivation. The question arises: why did this experiment ever work in *Paramecium*? The answer might be, that conjugation in *Tetrahymena* has a more stringent requirement for nutritional deprivation than it does in *Paramecium*. Alternatively, ciliary binding is so agressive in *Paramecium*, that loose-cilia quickly adhere to the cells' own cilia, and are removed from the feeding current.

There is preliminary evidence that membrane vesicles derived from cilia are also competent to promote pair formation in *Tetrahymena* (Love and Rotheim 1984). This is not as dramatic as the response seen in *Paramecium*. To begin with, the vesicles deployed do not seem to provoke homotypic pairing in *Tetrahymena* as they do in *Paramecium*. The response in *Tetrahymena* is more subtle. The percent of cells engaged in pairing appears to be boosted by pre-treating starved cells with ciliary vesicles isolated from cells expressing a complementary mating type. It is significant that pairing does not appear to be accelerated in most experiments. The effect is, however, mating type-specific. Curiously the vesicles were isolated from vegetative cells (not starved, initiated cells). It would be more useful to see how vesicles from "initiated" cells would behave in a similar assay. Paradoxically, once co-stimulation is achieved, the same ciliary vesicles appear to block pair formation. The author's of that study interpret these results by suggesting that ciliary vesicles bind to initiated cell membranes early on to promote co-stimulation, and later, the self-same binding serves to sterically inhibit interaction with a mating partner. As a preliminary study, these observations are provocative, and reinforce earlier findings suggesting that co-stimulation requires physical contact between cells.

12.6.3 Co-stimulation and Homotypic Pairing in **Tetrahymena**

One difference between *Tetrahymena* and *Paramecium*, is the readiness of the latter to form adhesive interactions between cells of the same mating type (homotypic pairing). Indeed, many experiments were made possible in *Paramecium* through the exploitation of homotypic pairing as a convenient bio-assay: namely the use of membrane-proxies (dead cells, cilia, and ciliary vesicles) as well as chemical induction of co-stimulation. The absence of conspicuous homotypic pairing in *Tetrahymena* led many to conclude that it does not occur in this model organism. One report defies this assumption by deploying cells with conspicuous nuclear markers (Kitamura et al. 1986). This study demonstrated that, early during *Tetrahymena* pair-formation, as many as 45 % of the observed pairs are homotypic in nature and these early pairs of identical mating type quickly dissociate leading to a greater and greater prevalence of heterotypic pairs. This finding has significance. It strengthens the hypothesis, first hinted at in the *Paramecium* studies, that mating type proteins perform two functions. First, they trigger non-specific adhesivity between cells of complementary mating type (and in a species-specific way). Second, they may mediate stabilization of heterotypic mating pairs, possibly facilitating the membrane fusion event that promotes transition from weak to strong pairing. The observation of early, transient homotypic pairs in *Tetrahymena* (as well as *Paramecium*) reinforces the need to search for and identify non-specific cell adhesion molecules whose normal activity must first be stimulated by interaction of mating-type proteins.

Another study explored co-stimulation in cell mixtures involving disparate numbers of mating types (Suganuma et al. 1984). A 9:1 ratio of cells was allowed to interact for 30–60 min before mixing with a similar 1:9 ratio mixture. Mixing a 9:1 with a 1:9 mixture yields a resultant 1:1 ratio of mixed mating types. Suganuma reported a very rapid rise in pairing following mixing of the two disparate (co-stimulated) mixtures. This result demonstrates that co-stimulation can occur even when one mating type is in a minority, and that co-stimulation is necessary but not sufficient for stable pair formation. It seems that continued stimulation by a cell of complementary mating type is required for tight-pairing to occur. Toshiro Sugai (who supplied the mating type cultures for the Suganuma study), performed follow-up experiments on skewed sex ratio cultures. With highly mating-reactive cells, he observed that in 1:9 cultures, pairing exceeded 10 % by quite a lot, reaching as high as 80 % in some cultures (indistinguishable from 1:1 cultures). These numbers then plummeted to 10–20 % as pair stabilization occurred. These unpublished results again suggest that homotypic pairing does occur, and that heterotypic interactions are necessary for "stabilization" of persistent pairs.

12.6.4 Cloning the Mating Type Locus

Though studies on *Tetrahymena* co-stimulation events got off to a slow start (in contrast to the vigorous efforts launched in the *Paramecium* field), it was the *Tetrahymena* community (and the lab of Eduardo Orias in particular) that first identified and unraveled the mating type gene locus and the molecular mechanism of mating type determination (Cervantes et al. 2013). This was made possible by a magnificent, community-led effort to sequence the entire *Tetrahymena* genome (Eisen et al. 2006). It should be noted that both *Tetrahymena* and *Paramecium* lagged far behind the secreted-pheromone/gamone signaling systems observed in *Euplotes* and *Blepharisma*, [see Luporini this volume, and (Luporini et al. 2015)]. Mating-type molecules from these organisms were identified in the pre-genomics era by brute-force biochemistry.

Six *Tetrahymena* mating type gene pairs were found to reside at a single locus within the germ-line micronucleus. Each pair of incomplete genes is arranged in a head to head configuration (one pair for each mating type possibility). During conjugation, genome reorganization results in deletion of all but one pair of head-to-head genes, and joining the mating-type specific sequence to a membrane-spanning domain. The resultant recombinant gene pair is expressed from the somatic macronucleus, and determines which of seven mating types will be expressed. (To account for the seven observed mating types with only six genes, one must factor in that there is an allelic variant within the greater *Tetrahymena* population). Each mating type gene encodes a mating-type specific region and a common, trans-membrane domain. The predicted protein structure supports a role for the resulting mating type protein at the cell surface, most likely interacting with the compatible proteins expressed from cells of complementary mating types (Cervantes et al. 2013). Though the genetic basis of mating type determination has now been firmly established, the actual role of the mating type proteins remains unsolved.

12.6.5 Cloning the Membrane Fusogen

With the advent of bio-informatics and proteomics, rapid progress has been made in identifying proteins that are involved in cell-cell interactions at the mating junction. An ancestral male-specific gamete fusogen Hap2/GCS1 (Wong and Johnson 2010) has been implicated in gamete fusion in plants, protists, and some of the lower animals. Through bio-informatics, a *Tetrahymena* homolog was identified, cloned, GFP-tagged and knocked out (Cole et al. 2014). Knockouts were shown to undergo normal cell adhesion, but failed to form the trans-membrane pores that are necessary for nuclear exchange. Without the ability to establish cytoplasmic continuity, cell pairs remained fragile (weak pairing), yet adhesion alone was sufficient to provoke all the nuclear events associated with conjugation [meiosis and nuclear reorganization reviewed in Cole and Sugai (2012)]. This result reinforces work from *Paramecium* suggesting that actual pair-fusion is not necessary in order to trigger the nuclear events associated with conjugation (Miyake 1964; Fujishima and Hiwatashi 1970). Furthermore, experiments in *Tetrahymena* demonstrating that disrupting pairs very early in mating (1.5 h) does not necessarily prevent them from completing the program of nuclear events associated with conjugation (Kiersnowska et al. 2000; Virtue and Cole 1999).

12.6.6 ConA Receptors: A Potential Cell-Adhesion System

Researchers have flanked the early events of cell-signaling in *Tetrahymena*, by first identifying and cloning genes encoding the mating type determinants, and then identifying the membrane fusogen, Hap2, that brings about cell union. What remain to be found are the adhesion molecules that mediate loose pairing and the soluble "matchmaking" factor that may facilitate co-stimulation and loose pairing. Our best candidates for the former are surface glycoproteins that interact with a plant lectin, concanavalin A. Lectins are carbohydrate-binding proteins known for their ability to adhere to integral membrane glycoproteins. Concanavalin A (conA) is a tetramer derived from the jack bean that specifically recognizes terminal alpha-D-mannosyl or alpha-D-glucosyl residues exposed on surface glycoproteins. ConA has been used in two different ways to explore co-stimulation in *Tetrahymena*. First, conA can block pair formation presumably by steric interference with the cell adhesion molecules themselves (Frisch and Loyter 1977; Pagliaro and Wolfe 1987). Second, fluorescently tagged conA ligands decorate the developing adhesion zone in cells undergoing co-stimulation [(Suganuma and Yamamoto 1988; Watanabe et al. 1981; Wolfe and Feng 1988; Wolfe et al. 1986), reviewed by Cole (2006)]. This fluorescent probe has been useful as a marker for co-stimulation, and as a probe used in assays to purify proteins associated with the mating junction (Cole et al. 2008). Despite identification of numerous conA binding proteins through SDS-PAGE techniques (Dentler 1992; Driscoll and Hufnagel 1999; Wolfe et al. 1993), the specific conA receptors on the *Tetrahymena* cell membrane responsible for cell-adhesion remain to be identified. One of the most interesting findings is that when conA binds to surface proteins in *Tetrahymena*, it causes them to become anchored to the cytoskeleton (Pagliaro and Wolfe 1987). It isn't clear that natural cell-signaling associated with natural co-stimulation results in a similar cytoskeletal anchoring of membrane receptors, but it is a provocative lead.

ConA-binding experiments suggest that normal cell-adhesion (possibly mediated by conA receptor proteins during co-stimulation in *Tetrahymena*) might trigger cortical reorganization in the mating cells via cytoskeletal interactions. Cytoskeletal changes associated with establishment of the mating junction have been documented including changes in microtubule dynamics (Gaertig and Fleury 1992; Kushida et al. 2015; Orias et al. 1983), as well as more unique elements of the ciliate cytoskeleton including filament-forming proteins comprising the non-microtubule cytoskeleton known as the "epiplasm" [fenestrin: (Cole et al. 2008; Nelsen et al. 1994), and Cit1 (Numata et al. 1985)].

The model that has emerged suggests that mating-type interactions trigger a global redistribution of conA-receptors from a dispersed pattern, to a pattern concentrated over the developing adhesion zone in a process known as "conA receptor tipping" (Wolfe and Feng 1988). Here they mediate cell adhesion in an (initially) mating-type-independent fashion creating loose pairs among co-stimulated cells. Re-localization of the conA receptors may also initiate reorganization of the cytoskeleton as cells prepare to form a mating junction. Continuing interactions (again involving the mating-type proteins) provoke transformation from loose pairing to tight pairing as the Hap2 fusogen comes into play creating hundreds of independent membrane fusion events (Cole 2006; Cole et al. 2014; Orias 1986; Wolfe 1982).

The role of the "soluble protein" in *Tetrahymena* co-stimulation and pair formation remains a mystery. It is conceivable, that it serves as a co-factor, stabilizing structural interactions between the mating-type proteins, or between mating type proteins and the cell-adhesion molecules themselves. Alternatively, it could play some sort of signaling role, helping cells to maintain the co-stimulated state until tight pairing is achieved.

12.6.7 Membrane Trafficking and Cell Signaling

At the time of this writing, nothing is known of how the newly identified mating type proteins physically interact to initiate the cellular response that leads to adhesion and tight bonding through membrane fusion, and launch the program of cortical remodeling and nuclear activities associated with conjugation. There are, however, some intriguing leads that involve membrane trafficking. Drugs that interfere with endocytosis (Rotheim and Love 1982), also block pair formation in *Tetrahymena*. These authors suggest that ligand-receptor complexes may be internalized by endocytosis as part of the signaling process. CDA12p, a protein associated with recycling endosomes, decorates vesicles that swarm the mating junction during conjugation (Zweifel et al. 2009). ("Signals" involved in vesicle traffic within ciliates are reviewed by Plattner, this volume). Recent electron tomography has revealed extracellular micro-vesicles (EMVs) that are secreted into the space between mating partners early during pair formation. These appear to be subsequently engulfed into large, vesicle-filled membrane compartments through a process of bulk endocytosis (Cole et al. 2015). It is intriguing to speculate that *Tetrahymena* may signal one another via a process involving secretion and absorption of extracellular, membrane-bound vesicles. The potential signaling role of shed micro-vesicles has received a great deal of attention in the recent literature [See Review: (Cocucci and Meldolesi 2015; Cocucci et al. 2009)]. It is noteworthy that, in *Chlamydomonas,* membrane vesicles are shed constitutively from the flagella during all stages of the life cycle. During preparatory periods of nitrogen starvation, (a pre-requisite for mating), vesicles shed from the flagella are capable of provoking homotypic cell adhesion. Mating type "+" vesicles induce homotypic agglutination in mating type "−" cells and vice versa (McLean et al. 1974; Van Den Ende 1981; Wiese 1965). Vesicles harvested from vegetative cells failed to induce agglutination. One distinction worth making, extracellular micro-vesicles (EMVs) observed at the *Tetrahymena* mating junction appear to be shed from the plasma membrane itself, and not from cilia. (There are no cilia within the mating junction, yet vesicle shedding is clearly occurring from the plasma membrane.) Furthermore, it is not yet clear whether EMV shedding is confined to the mating junction and to the life history stage of conjugation, or whether shedding is a constitutive process as seen in *Chlamydomonas*. It is also noteworthy that in the ciliate, *Euplotes raikovi*, the secreted mating-type pheromone that binds to a homotypic receptor is subsequently internalized by endocytosis, and the resulting "signaling endosomes" appear to trigger a mitogenic response (Vallesi et al. 2005). Curiously, in the *Euplotes* example, pheromone binding to a heterotypic receptor appears to block endocytosis, and triggers a mating response. In light of these findings, the signaling potential of ciliary-vesicles derived from mating ciliates as well as studies involving "conditioned medium" should be re-evaluated.

12.7 Mating Type Proteins as Cell-Adhesion Ligands

Figure 12.3 illustrates the essential elements of this model. Mating type protein "A" (dispersed over the cell surface or ventral cilia) binds directly to a dispersed population of mating type "B" proteins. In *Tetrahymena*, such interactions are brief (no single ligand-ligand interaction is sufficient to anchor vigorously swimming cells). In response to these transient ligand interactions, a "tipping phenomenon" is triggered in which mating type receptors migrate to the anterior cortex that will become the holdfast (in *Paramecium*) and the mating junction (in *Tetrahymena*). Fluorescent ConA labeling reveals just such a "tipping" phenomenon in *Tetrahymena* (Wolfe and Feng 1988). As the density of mating type proteins increases, the same ligand-ligand interactions that served a signaling function, acquire sufficient aggregate binding force to mediate cell adhesion. This could be favored by an (observed) decrease in cell motility. For *Tetrahymena*, this model does a reasonably good job of accounting for the observed features of pre-conjugal behavior. One problem for this model is the observation of transient homotypic pairing early during the loose-pairing episode. One must modify the model to accommodate same-sex mating affinities between "like" mating type proteins. At first, this exception appears deadly to the model. If identical mating-type proteins can bind one another (mediating homotypic pair formation) then one should observe pair formation even in starved, same-sex cultures, which one does not. A gentle rescue of the model may be possible. It is not clear how each mating type protein (of which there are seven in *Tetrahymena*) is capable of recognizing and binding to the other six proteins that comprise the mating type spectrum, without also binding to itself! If one imagines that self-binding does occur (though perhaps with lower affinity), then our model predicts that homotypic protein interactions will fail to achieve the threshold necessary for triggering a signaling event. However, if heterotypic signaling achieves the "tipping" phenomenon, (concentrating the mating type receptors) then even low-affinity homotypic associations might, in aggregate, be sufficient to bring about cell-adhesion. One would predict that homotypic adhesions would be less stable, giving way to the more stable heterotypic pairs (as is seen). It is remarkable, and perhaps noteworthy, that many elements of the emerging model for establishing a cell-cell junction during co-stimulation and mating resemble those associated with the formation of the neural synapse. See (Bemben et al. 2015).

In *Paramecium*, adhesion is instantaneous and mediated by proteins on the ventral cilia. Later, these same cilia are resorbed and a secondary form of adhesion is mediated by the denuded plasma membrane at the anterior holdfast. Homotypic pairing is more vigorous than in *Tetrahymena*, and can be induced by chemical signals that bypass the aggregation phenomenon. To accommodate these observations, the model needs help. The immediacy of cell adhesion argues against a

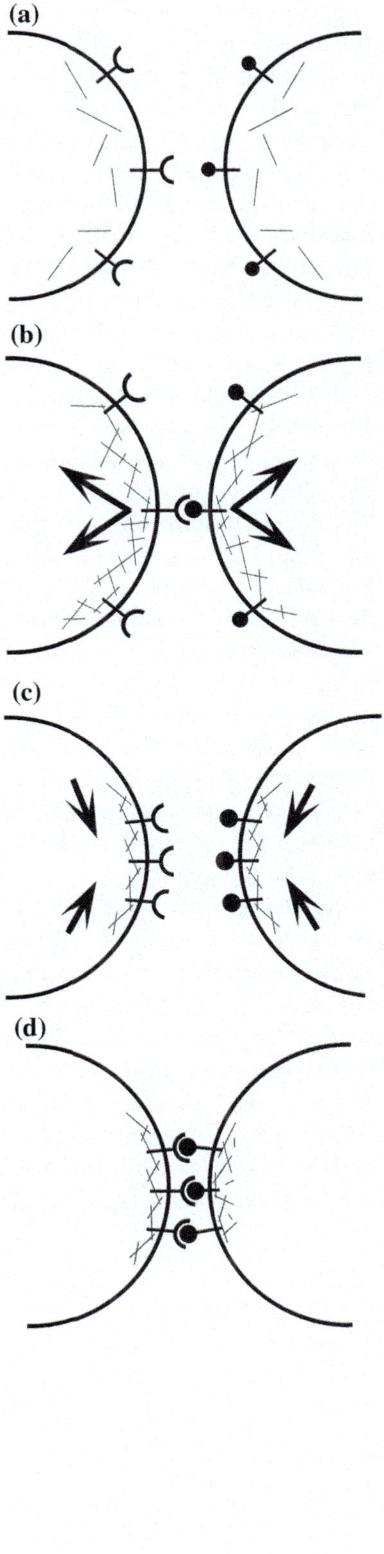

Fig. 12.3 A simple model depicting events that accompany "co-stimulation" in *Tetrahymena*. In this model, the mating-type proteins serve as both signal molecules and cell-adhesion molecules. (**a**) Contact between mating type proteins leads to (**b**) a cell-signal that triggers association of the mating type receptors with the cortical cytoskeleton. Subsequently, mating type receptors migrate together (**c**), where they can now serve a role in cell adhesion (**d**).

"tipping" phenomenon (at least in the cilia where initial adhesion takes place). Clearly the requisite cell-adhesion ligands are already present and sufficiently concentrated in the ventral cilia to support adhesion resulting in aggregation into mating clumps, and yet homotypic clumping does not occur without exposure to materials (dead cells, cilia, ciliary vesicles) isolated from the other mating type! Can we still invoke the "simple" model in which signaling and adhesion are mediated by the same surface molecules? Possibly. In all cases where immediate cell clumping associated with the "mating reaction" can be triggered, heterotypic membrane proxies are involved. One could suggest that these membrane proxies serve as multivalent ligands, bringing cells of the same mating type together and triggering changes that prepare them for holdfast formation. In this version of the model, homotypic ciliary adhesion (associated with immediate clumping) still requires the high affinity interactions afforded by heterotypic mating type protein associations. Membrane surrogates would act as multivalent cross-linkers bringing homotypic pairs together for subsequent membrane associations. The formation of loose pairs via formation of the anterior holdfast, resembles the scenario painted for *Tetrahymena*. Some form of "tipping" could concentrate the mating type proteins into aggregates capable of creating a holdfast, even among homotypic unions. Significantly, resorption of the cilia (not shedding) might offer a mechanism for this. As ciliary axonemes disassemble, the corresponding ciliary membrane presumably spreads and is incorporated into the surrounding somatic membrane of the future holdfast. This could result in the generation of localized concentrates of the mating type receptors. Heterotypic membrane stimulation, or chemical signals that result in resorption of anterior ventral cilia, could generate patches of concentrated mating type proteins sufficient to bring about loose pairing. (Curiously, even in *Tetrahymena*, the anterior holdfast membrane is devoid of cilia. Ciliary resorption has never been reported, but this is a much smaller membrane domain, and it might have been missed).

An alternate model, (Fig. 12.4), must involve more players. Heterotypic interactions between mating type proteins could "activate" localized adhesion molecules that subsequently mediate aggregation (in *Paramecium*) and later pair formation. In this model, the cell adhesion molecules are not mating-type specific, and membrane proxies would trigger a generalized adhesivity. To account for the preponderance of heterotypic pairs (and their selective stabilization over homotypic pairs in *Tetrahymena*), secondary roles must be invoked for the mating type proteins in favoring heterotypic unions. Even more complex models can be envisioned. Neither of these models accounts for the synergistic effect of the soluble factor observed in *Tetrahymena* cultures (Adair et al. 1978), though these are not hard to envision.

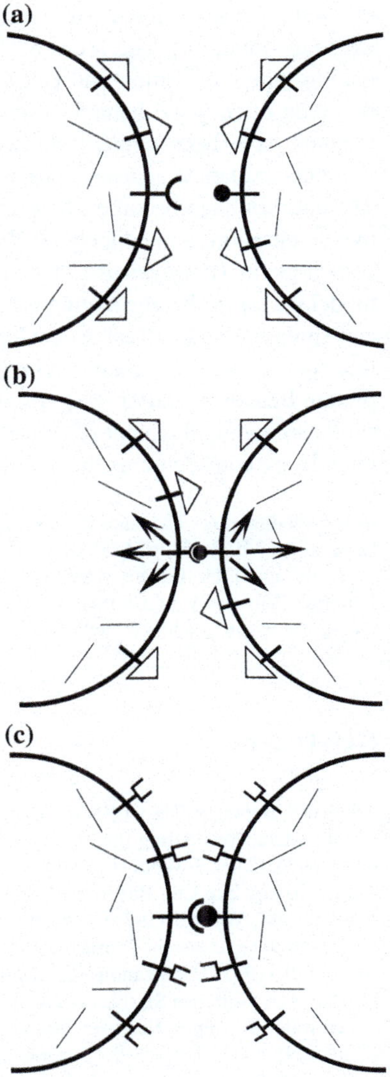

Fig. 12.4 A more complex, alternative model. In this model, binding of complementary mating type proteins (**a**, **b**) generate a signal that alters the conformation of a common surface glycoprotein, resulting in cell adhesion (**c**). Presumably, the mating type ligands help stabilize heterotypic pairs over homotypic pairs in *Tetrahymena*

12.8 Conclusions

Pre-conjugal events that trigger mating in ciliates take two forms. Species represented by *Euplotes* and *Blepharisma*, secrete their signaling molecules. Species such as *Tetrahymena* and *Paramecium* require physical contact of surface ligands. The latter two contact-mediated models differ from one another in that *Paramecium* undergoes mass adhesion immediately upon mixing, with "loose pairs" emerging an hour later, whereas *Tetrahymena* undergo an hour of random collisions prior to "loose pair" formation. In both species, loose pairing is followed by the development of "tight pairing" correlated with membrane fusion at their respective nuclear exchange junctions. The focus of this review has been to explore the link between mating type recognition and cell adhesion associated with pair formation and the initiation of conjugal behavior. In both species, genes encoding mating type determinants have been cloned and sequenced. These genes encode trans-membrane proteins and hence appear appropriate for deployment in the role of cell signaling. The question then becomes, how do interactions between mating type proteins lead to cell adhesion, immediately (in the case of *Paramecium*) and after prolonged, cell collisions (in *Tetrahymena*)? Two models seem ready for consideration. In the first model we assert that the mating type proteins are the sole actors, bringing about both cell-signaling and cell-adhesion. In the second model, cell-signaling (mediated by interactions between mating type proteins) triggers adhesivity by activating other protein ligands already present on the cell surface in inactive form, or by triggering synthesis and deployment of cell adhesion proteins. The relative merits of these two models are considered as they could apply to *Paramecium* and *Tetrahymena*.

Acknowledgments The author was supported by an RUI grant from the National Science Foundation, **MCB:1517624**. The author would especially like to thank Dr. Toshiro Sugai (Ibaraki U., Mito Japan), Dr. Eduardo Orias (University of California, Santa Barbara), and Dr. Pierangelo Luporini (University of Camerino, Italy) for formative discussions and constructive feedback during the writing of this review.

References

Adair WS, Barker R, Turner RS Jr, Wolfe J (1978) Demonstration of a cell-free factor involved in cell interactions during mating in *Tetrahymena*. Nature (London) 274:54–55

Allewell JM, Oles J, Wolfe J (1976) A physicochemical analysis of conjugation in *Tetrahymena pyriformis*. Exp Cell Res 97:394–405

Allewell NM, Wolfe J (1977) A kinetic analysis of the memory of a developmental interaction. Mating interactions in *Tetrahymena pyriformis*. Exp Cell Res 109(1):15–24

Austin CR (1965) Fertilization. Prentice-Hall, Engelwood Cliffs, NJ

Bemben MA, Shipman SL, Nicoll RA, Roche KW (2015) The cellular and molecular landscape of neuroligins. Trends Neurosci 38:496–505

Bhatt T, Rizvi A, Batta SPR, Kataria S, Jamora C (2013) Signaling and mechanical roles of E-cadherin. Cell Commun Adhes 20:189–199

Bleyman LK (1971) Aging in early maturing mutants of *Tetrahymena pyriformis*. Genetics 68:S6

Brown F, Tirone S, Wolfe J (1993) Early encounters of the repetitive kind: a prelude to cell adhesion in conjugating *Tetrahymena thermophila*. Dev Dyn 196:195–204

Bruns PJ, Palestine RF (1975) Costimulation in *Tetrahymena pyriformis*: a developmental interaction between specially prepared cells. Dev Biol 42:75–83

Bruns PJ, Brussard TB (1974) Pair formation in *Tetrahymena pyriformis*, an inducible developmental system. J Exp Zool 188:337–44

Byrne BC (1973) Mutagenic analysis of mating type and isolation of reactive cilia of both mating types in the ciliated protozoan. Paramecium aurelia syngen 4(33B):3492

Cervantes MD, Hamilton EP, Xiong J, Lawson MJ, Yuan D, Hadjithomas M, Miao W, Orias E (2013) Selecting one of several mating types through gene segment joining and deletion in *Tetrahymena thermophila*. PLoS Biol 11:e1001518

Cocucci E, Meldolesi J (2015) Ectosomes and exosomes: shedding the confusion between extracellular vesicles. Trends Cell Biol 25:364–372

Cocucci E, Racchetti G, Meldolesi J (2009) Shedding microvesicles: artefacts no more. Trends Cell Biol 19:43–51

Cohen LW, Siegel RW (1963) Mating-type substances of *Paramecium bursaria*. Genet Res 4:143–160

Cole E, Sugai T (2012) Developmental progression of *Tetrahymena* through the cell cycle and conjugation. Methods Cell Biol 109:177–236

Cole ES (2006) The *Tetrahymena* conjugation junction. In: Baluska F, Volkmann D, and Barlow PW eds. Cell-Cell Channels. Springer, 233 Spring Street, New York, Ny 10013, United States pp 39–62

Cole ES, Anderson PC, Fulton RB, Majerus ME, Rooney MG, Savage JM, Chalker D, Honts J, Welch ME, Wentland AL et al (2008) A proteomics approach to cloning fenestrin from the nuclear exchange junction of *Tetrahymena*. J Euk Microbiol 55:245–256

Cole ES, Cassidy-Hanley D, Pinello JF, Zeng H, Hsueh M, Kolbin D, Ozzello C, Giddings T, Winey M, Clark TG (2014) Function of the male-gamete-specific fusion protein HAP2 in a seven-sexed ciliate. Curr Biol 24:2168–2173

Cole ES, Giddings TH, Ozzello C, Winey M, O'Toole E, Orias J, Hamilton E, Guerrier S, Ballard A, Aronstein T (2015) Membrane dynamics at the nuclear exchange junction during early mating (one to four hours) in the ciliate *Tetrahymena thermophila*. Eukaryot Cell 14:116–127

Cronkite DL (1976) A role of calcium ions in chemical induction of mating in *Paramecium* tetraurelia. J Protozool 23:431–433

Cronkite DL (1979) An analysis of the mechanism of activation of rapid micronuclear migration, a very early event in conjugation of Paramecium caudatum. Proc Intl Congr Protozool 5:284–289

Dentler WL (1992) Identification of *Tetrahymena* ciliary surface proteins labeled with a sulfosuccinimidyl 6-(botinamido) hexanoate and concanavalin A and fractionated with Triton-X114. J Euk Microbiol 39:368–378

Diller WF (1936) Nuclear reorganization processes in Paramecium aurelia, with descriptions of autogamy and 'hemixis'. J Morphol 59:11–67

Dini F (1984) On the evolutionary significance of autogamy in the marine *Euplotes* (Ciliophora, Hypotrichida). Am Nat 123:151–162

Driscoll C, Hufnagel LA (1999) Affinity-purification of concanavalin A-binding ciliary glycoconjugates of starved and feeding *Tetrahymena thermophila*. J Euk Microbiol 46:142–146

Eisen JA, Coyne RS, Wu M, Wu D, Thiagarajan M, Wortman JR, Badger JH, Ren Q, Amedeo P, Jones KM et al (2006) Macronuclear genome sequence of the ciliate *Tetrahymena thermophila*, a model eukaryote. PLoS Biol 4:e286

Finley HE (1939) Sexual differentiation in Vorticella microstoma. J Exp Zool 81:209–229

Finley MJ, Bruns PJ (1980) Costimulation in *Tetrahymena*. II. A non-specific response to heterotypic cell-cell interactions. Dev Biol 79:81–94

Frisch A, Loyter A (1977) Inhibition of conjugation in *Tetrahymena pyriformis* by ConA- ConA localization of conA binding sites. Exp Cell Res 110:337–346

Fujishima M, Hiwatashi K (1977) Early step in initiation of fertilization in *Paramecium*- early micronuclear migration. J Exp Zool 201:127

Fukushi T, Hiwatashi K (1970) Preparation of mating reactive cilia from Paramecium caudatum by MnCl2. J Protozool 17:1

Gaertig J, Fleury A (1992) Spatiotemporal reorganization of intracytoplasmic microtubules is associated with nuclear selection and differentiation during developmental process in the ciliate *Tetrahymena thermophila*. Protoplasma 167:74–87

Gates MA (1990) Morphological drift accompanying nascent population differentiation in the ciliate *Euplotes vannus*. J Protozool 37:78–86

Geiger B, Yamada KM (2011) Molecular architecture and function of matrix adhesions. Cold Spring Harbor Perspect Biol 3:a005033

Grell KG (1967) Sexual reproduction in Protozoa. Pergamon Press, Oxford, London and New York

Haga N, Hiwatashi K (1981) A protein Called immaturin controlling sexual immaturity in Paramecium. Nature 289:177–178

Hatzfeld M, Wolf A, Keil R (2014) Plakophilins in desmosomal adhesion and signaling. Cell Commun Adhes 21:25–42

Hiwatashi K (1955) Studies on the conjugation of Paramecium caudatum. VI. On the nature of the union of conjugation. Sci Repts Tohoku Univ Ser 4 Biol 21:207–218

Hiwatashi K (1961) Locality of mating reactivity on the surface of *Paramecium* caudatum. Sci Repts Tohoku Univ Fourth Ser Biol 27:93–99

Hiwatashi K (1969) *Paramecium.* In: Fertilization, 2. Academic Press, New York, pp 255–293

Hiwatashi K (1981) Sexual interactions in *Paramecium* In: O'day DH, and Horgen PA (eds) Sexual interactions in eukaryotic microbes, pp 351–378

Hiwatashi K, Kitamura A (1985) Fertilization in *Paramecium.* In: Metz C, Monroy A (eds) Biology of fertilization, 1. Academic Press, New York, pp 57–85

Jennings HS (1911) Assortative mating, variability and inheritance of size, in the conjugation of paramecium. J Exp Zool 11:1–134

Jennings HS (1944) Paramecium bursaria: Life history. III. Repeated conjugations in the same stock at different ages, with and without inbreeding, in relation to mortality at conjugation. J Exp Zool 96:243–273

Kamada T, Kinosita H (1940) Calcium-potassium factor in ciliary reversal of Paramecium. Proc Imp Acad (Tokyo) 16:125–130

Kapuscinski J (1990) Interactions of nucleic-acids with fluorescent dyes—spectral properties of condensed complexes. J Histochem Cytochem 38:1323–1329

Kiersnowska M, Kaczanowski A, Morga J (2000) Macronuclear development in conjugants of Tetrahymena thermophila, which were artificially separated at meiotic prophase. The Journal of eukaryotic microbiology 47:139–147

Kitamura A, Hiwatashi K (1976) Mating-reactive membrane-vesicles from cilia of *Paramecium caudatum.* J Cell Biol 69:736–740

Kitamura A, Hiwatashi K (1978) Are sugar residues involved in specific cell recognition of mating in Paramecium J Exp Zool 203:99–107

Kitamura A, Hiwatashi K (1980) Reconstitution of mating active membrane-vesicles in *Paramecium.* Exp Cell Res 125:486–489

Kitamura A, Sugai T, Kitamura Y (1986) Homotypic pair formation during conjugation in *Tetrahymena thermophila.* J Cell Sci 82:223–234

Kobayashi K, Yoshimura C, Mikami K (1999) Protoplasmic streaming and nuclear differentiation during conjugation of Paramecium caudatum. Zool Sci 16:885–892

Kung C, Naitoh Y (1973) Calcium-induced ciliary reversal in extracted models of pawn, a behavioral mutant of *Paramecium.* Science 179:195–196

Kushida Y, Takaine M, Nakano K, Sugai T, Numata O (2015) Dynamic change of cellular localization of microtubule-organizing center during conjugation of ciliate Tetrahymena thermophila. Zool Sci 32:25–32

Love B, Rotheim MB (1984) Cell surface interactions in conjugation—*Tetrahymena* ciliary membrane vesicles. Mol Cell Biol 4:681–687

Luporini P, Alimenti C, Vallesi A (2015) Ciliate pheromone structures and activity: a review. Ital J Zool 82:3–14

McLean RJ, Laurendi CJ, Brown RM (1974) Relationship of gamone to mating reaction in *Chlamydomonas moewusii.* Proc Natl Acad Sci USA 71:2610–2613

Metz CB (1946) effects of various agents on the mating type substance of *Paramecium aurelia* variety 4. Anat Rec 94:347

Metz CB (1947) Induction of pseudo selfing and meiosis in *Paramecium aurelia* by formalin killed animals of opposite mating type. J Exp Zool 105:115–139

Metz CB (1948) The nature and mode of action of the mating type substances. Am Nat 82:85–95

Metz CB (1954) Mating substances and the physiology of fertilization in ciliates. In: Wenrich DH (ed) Sex in microorganisms, pp 284–334
Mikami K (2000) Fertilization in protozoa. In: Tarin JJ, Cano A (eds) Fertilization in protozoa and metazoan animals. Springer, Berlin, pp 1–25
Miyake A (1958) Induction of conjugation by chemical agents in Paramecium caudatum. J Inst Polytech Osaka City Univ Ser D Biol 9:251–296
Miyake A (1961) Artificial induction of conjugation by chemical agents in Paramecium of the aurelia group and some of its applications to genetic work. Am Zool 1:373–374
Miyake A (1964) Induction of conjugation by cell-free preparations in Paramecium multimicronucleatum. Science 146:1583–1585
Miyake A (1968) Induction of conjugation by chemical agents in Paramecium. J Exp Zool 167:359
Miyake A (1974) Cell interaction in conjugation of ciliates. Curr Top Microbiol Immunol 64:49–77
Miyake A (1981) Physiology and biochemistry of conjugation in ciliates. Biochemistry and physiology of protozoa 4:125–198
Miyake A (1996) Fertilization and sexuality in ciliates. In: Hausmann K, Bradbury PC (eds) Ciliates as organisms. Fischer, Stuttgart, pp 243–290
Naitoh Y, Eckert R (1974) The control of ciliary activity in Protozoa. In: Sleigh MA (ed) Cilia and flagella 1-500. Academic Press, London, New York., pp 305–352
Nelsen EM, Williams NE, Yi H, Knaak J, Frankel J (1994) "Fenestrin" and conjugation in *Tetrahymena thermophila*. J Euk Microbiol 41:483–495
Numata O, Sugai T, Watanabe Y (1985) Control of germ cell nuclear behavior at fertilization by Tetrahymena intermediate filament protein 314:192–193
Ognibene M, Della Giovampaola C, Trielli F, Focarelli R, Rosati F, Corrado MUD (2008) Identification and characterization of a 38 kDa glycoprotein functionally associated with mating activity of *Paramecium* primaurelia. Eur J Protistol 44:81–90
Orias E (1986) Ciliate conjugation. In: Gall JG (ed) The molecular biology of ciliated Protozoa, pp 45–84
Orias JD, Hamilton EP, Orias E (1983) A microtubular meshwork associated with gametic pronuclear transfer across a cell-cell junction. Science 222:181–184
Pagliaro L, Wolfe J (1987) Concanavalin-A binding induces association of possible mating type receptors with the cytoskeleton of *Tetrahymena*. Exp Cell Res 168:138–152
Przywara DA, Bhave SV, Bhave A, Wakade TD, Wakade AR (1991) Stimulated rise in neuronal calcium is faster and greater in the nucleus than the cytosol. Faseb J 5:217–222
Rasmussen MI, Wheatley DN (2007) Purification and characterisation of cell survival factor 1 (TCSF1) from *Tetrahymena thermophila*. J Cell Commun Signal 1:185–193
Rogers MB, Karrer KM (1985) Adolescence in *Tetrahymena thermophila*. Proc Natl Acad Sci USA 82:436–439
Rotheim MB, Love B (1982) Role for endocytosis in conjugation in Tetrahymena. Mol Cell Biol 2:633–637
Shen B, Delaney MK, Du XP (2012) Inside-out, outside-in, and inside-outside-in: G protein signaling in integrin-mediated cell adhesion, spreading, and retraction. Curr Opin Cell Biol 24:600–606
Singh DP, Saudemont B, Guglielmi G, Arnaiz O, Gout JF, Prajer M, Potekhin A, Przybos E, Aubusson-Fleury A, Bhullar S et al (2014) Genome-defence small RNAs exapted for epigenetic mating-type inheritance. Nature 509:447–453
Suganuma Y, Shimode C, Yamamoto H (1984) Conjugation in Tetrahymena: formation of a special junction area for conjugation during the co-stimulation period 33:10–18
Suganuma Y, Yamamoto H (1988) Conjugation in *Tetrahymena*: Its relation to concanavalin A receptor distribution on the cell surface. Zool Sci 5:323–330
Takahashi M (1973) Does fluid have any function for mating in Tetrahymena pyriformis. Sci Rep Tohoku Univ Fourth Ser (Biol) 36:223–229

Vallesi A, Ballarini P, Di Pretoro B, Alimenti C, Miceli C, Luporini P (2005) Autocrine, mitogenic pheromone receptor loop of the ciliate Euplotes raikovi: pheromone-induced receptor internalization. Eukaryot Cell 4:1221–1227

Van Den Ende H (1981) Sexual interactions in the green algae *Chlamydomonas eugametos*

Virtue MA, Cole ES (1999) A cytogenetic study of devolopment in mechanically disrupted pairs of *Tetrahymena thermophila*. J Euk Microbiol 46:597–605

Watanabe S, Toyohara A, Suzaki T (1981) The relation of concanavalin A receptor distribution to the conjugation process in *Tetrahymena thermophila*. J Protozool 28:171–175

Watanabe T (1977) Ciliary membranes and mating substances in *Paramecium caudatum*. J Protozool 24:426–429

Weiss MS, Anderson DH, Raffioni S, Bradshaw RA, Ortenzi C, Luporini P, Eisenberg D (1995) A cooperative model for receptor recognition and cell-adhesion—evidence from the molecular packing in the 1.6 Angstrom crystal structure of the pheromone ER-1 from the ciliated protozoan *Euplotes raikovi*. Proc Natl Acad Sci USA 92:10172–10176

Wiese L (1965) On sexual agglutination and mating-type substances (gamones) in isogamous heterothallic Chlamydomonads, I. Evidence of identity of gamones with surface components responsible for sexual flagellary contact. J Phycol 1:46

Wolfe J (1982) The conjugation junction of *Tetrahymena*: its structure and development. J Morph 172:159–178

Wolfe J (1985) Cytoskeletal reorganization and plasma membrane fusion in conjugating *Tetrahymena*. J Cell Sci 73:69–85

Wolfe J, Feng S (1988) Concanavalin A receptor "tipping" in *Tetrahymena* and its relationship to cell adhesion during conjugation. Development 102:699–708

Wolfe J, Meal KJ, Soiffer R (1980) Limiting conditions for conjugation in Tetrahymena: cellular development and factor active in conjugation. J Exp Zool 212:37–46

Wolfe J, Mpoke S, Tirone SF (1993) Cilia, ciliary concanavalin A-binding and mating recognition in *Tetrahymena thermophila*. Exp Cell Res 209:342–349

Wolfe J, Pagliaro L, Fortune H (1986) Coordination of concanavalin-A-receptor distribution and surface differentiation in *Tetrahymena*. Differentiation 31:1–9

Wolfe J, Turner R Jr, Barker R, Adair WS (1979) The need for an extracellular component for cell pairing in *Tetrahymena*. Exp Cell Res 121:27–30

Wong JL, Johnson MA (2010) Is HAP2-GCS1 an ancestral gamete fusogen?. Trends Cell Biol 20.3:134–141

Xu X, Takahashi M (2001) Molecular nature of mating-type substances in Paramecium caudatum. Dev Growth Differ 43:S135

Yang XU, Haga N (1993) Initiation of the earliest nuclear evenet in fertilization of *Paramecium* by the microinjection of calcium buffer. Zool Sci 10:859–862

Zweifel E, Smith J, Romero D, Giddings TH, Winey M, Honts J, Dahlseid J, Schneider B, Cole ES (2009) Nested genes CDA12 and CDA13 encode proteins associated with membrane trafficking in the ciliate *Tetrahymena thermophila*. Eukaryot Cell 8:899–912

Chapter 13
Mating Systems and Reproductive Strategies in *Tetrahymena*

Rebecca A. Zufall

Abstract Understanding the evolution of mating systems and reproductive strategies has long been of interest to biologists. When, with whom, and how often organisms engage in sexual reproduction are critical in determining the genetic composition and future evolutionary potential of populations. Studies of these features in *Tetrahymena* have provided a variety of insights into the evolutionary forces driving the evolution of sex and sexual systems. This chapter explores the evolutionary causes and consequences of *T. thermophila's* unusual mechanism of sex determination, the choice of when and with whom to mate, and mechanisms that may allow *Tetrahymena* to go for millions of years with no sexual reproduction.

13.1 Introduction

The costs of sexual reproduction are well known (Maynard Smith 1978; Bell 1982). Nonetheless, sexual reproduction appears to be the most successful evolutionary strategy across eukaryotes (Weismann 1887; Maynard Smith 1978; Bell 1982; Rice 2002). This is most likely due to the various benefits provided by sex, including facilitating the purging of deleterious mutations (Muller 1964; Kondrashov 1988), accelerating the accumulation of beneficial mutations (Fisher 1930; Muller 1932), and increasing the genetic variance in fitness, which makes natural selection more efficient (Weismann 1887; Kondrashov 1993; Burt 2000). The evolutionary significance of sexual reproduction, thus, makes an understanding of reproductive strategies and biocommunication among mating partners of central importance.

Despite, or possibly because of, the near ubiquity of sex, there is extensive diversity in every aspect of mating and reproduction across eukaryotes. For example, sexual reproduction may be obligate, as in most mammals, or facultative, as in some animals and plants and probably most protists (Jackson et al. 1985;

R.A. Zufall (✉)
Department of Biology and Biochemistry, University of Houston, Houston, TX 77204, USA
e-mail: rzufall@uh.edu

G. Witzany and M. Nowacki (eds.), *Biocommunication of Ciliates*,
DOI 10.1007/978-3-319-32211-7_13

Speijer et al. 2015). Within ciliates, extensive variation is found in all aspects of mating systems, including the number of mating types in a species, the ability to undergo autogamy, the molecular determinants of mating type, and the mechanism of mating type inheritance (Dini and Nyberg 1993; Miyake 1996; Phadke and Zufall 2009). The evolutionary forces responsible for the evolution of this wide diversity of reproductive traits remain elusive (Bachtrog et al. 2014).

This chapter focuses primarily on the reproductive strategies and mating behavior in the ciliate *Tetrahymena thermophila*. Reproduction in *T. thermophila* is of particular interest for a variety of reasons. First, the mechanism of sex determination is unusual and has unexpected effects on population processes, such as sex ratio. Studying mating behavior in this species is additionally valuable given the extensive research on cell-cell interactions in *Tetrahymena* (Cole 2016). Finally, *T. thermophila* and other species in this genus may provide novel insight into the evolution of sex by providing instances of successful asexual lineages.

13.2 Life Cycle

Like all ciliates, *Tetrahymena* are facultatively sexual. When food is abundant, cells divide by binary fission. In the absence of sufficient food and in the presence of mature cells of complementary mating type, conjugation is induced, micronuclei undergo meiosis, and meiotic products are reciprocally exchanged. Zygotic nuclei divide to produce new micronuclei and macronuclei in each progeny cell (Prescott 1994). The details of this process result in four genetically identical progeny (Nanney and Caughey 1953; Orias 1986), however these cells are not necessarily identical phenotypically due to cortical inheritance (Beisson and Sonneborn 1965; Nanney 1966), karyonidal inheritance of mating type (see Sect. 13.3.2), and other epigenetic phenomena (Fig. 13.1a). Following conjugation, cells are immature for a variable number of generations—40–60 generations in inbred strains and longer in natural isolates (Lynn and Doerder 2012). Upon maturity, individuals can mate with cells carrying any complementary mating type.

13.3 Sex Determination

13.3.1 Mating Types

Ciliates vary widely in the number of mating types found in a species. *T. thermophila* has seven mating types, called I-VII. Mating types are self-incompatible, but can mate with any other non-self mating type. Mating type switching does not

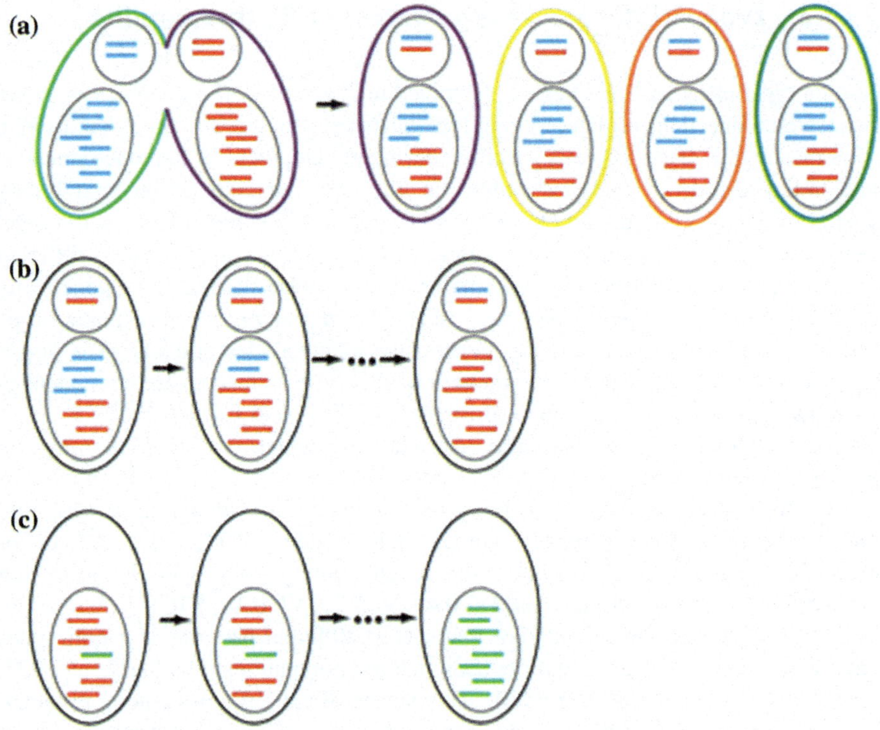

Fig. 13.1 Sexual conjugation and asexual division in *Tetrahymena*. **a** When sexually competent cells of compatible mating types encounter one another, they can form mating pairs. Following meiosis and mitosis of the micronucleus, pronuclei are reciprocally exchanged and fuse. The parental macronucleus is degraded and new macronuclei develop from mitotic products of the zygotic nucleus. Exconjugants separate and the first fission produces four genetically identical progeny, that may, when mature, differ in mating type. Different mating types are indicated by different cell outline colors. *Blue* and *red bars* in the nuclei represent different alleles of homologous chromosomes. Parents are shown as homozygotes and progeny are heterozygous. (See Orias et al. 2011 for more details.) **b** Amitosis of the macronucleus during asexual division results in phenotypic assortment, where one copy of each homologous chromosome eventually becomes fixed in the macronucleus and the other lost. **c** Phenotypic assortment also occurs with new mutations, shown in *green* (Color figure online)

occur in this species, so mating type is fixed in a clonal lineage until conjugation (Nanney and Caughey 1953). In natural populations, all seven mating types are found consistently (Doerder et al. 1995; Arslanyolu and Doerder 2000). Mating type is determined by alleles at the mating type (*mat*) locus, which can each specify four to six of the possible seven mating types (Nanney et al. 1955; Arslanyolu and Doerder 2000).

13.3.2 Probabilistic Sex Determination in T. thermophila

Mating type inheritance has been well studied in *T. thermophila*; nonetheless, many unanswered questions remain. Mating type inheritance in this species is karyonidal. This means that mating type is specified during development of the macronucleus, and independently in each karyonide, i.e. the daughter cells of the first binary fission following conjugation (Nanney and Caughey 1953; Fig. 13.1a). Karyonidal inheritance produces progeny of a single mating pair that may develop different mating types and therefore be able to mate with each other despite being genetically identical. Thus, despite the fact that *T. thermophila* are mating-type self-incompatible, matings among genetically identical descendants of a single pair can occur. The genetic consequences of this type of mating are equivalent to autogamy, the most extreme form of inbreeding (Dini and Corliss 2001).

The particular form of kayonidal inheritance of mating type in *T. thermophila* has been called "probabilistic sex determination" (Paixão et al. 2011). Each allele at the *mat* locus determines the probability with which an individual will develop into one of the seven mating types (Nanney and Caughey 1953; Orias 1981). For example, the *mat-2* allele specifies the seven mating types with the following probabilities: I, 0; II, 0.15; III, 0.09; IV, 0.47; V, 0.05; VI, 0.14; VII, 0.1 (N = 1090; Orias 1981). At least 14 *mat* alleles, that specify different patterns of mating types among progeny, have been identified from natural populations (Doerder et al. 1995, 1996; Arslanyolu and Doerder 2000). The pattern of mating types among progeny, however, can also be influenced by genetic background and environmental conditions (Nanney 1959; Arslanyolu and Doerder 2000; Phadke et al. 2014). Two types of *mat* alleles have been found in natural populations of *T. thermophila*: alleles that code for all mating types except I (e.g. *mat-2*; B-type alleles) and alleles that code for all mating types except IV and VII (e.g. *mat-3*; A-type alleles; Nanney 1959; Doerder et al. 1995; Arslanyolu and Doerder 2000).

The *mat-2* allele was recently sequenced from the micronuclear genome of an inbred strain, and experimentally verified to control both non-self recognition and successful mating (Cervantes et al. 2013; Cole 2016). In the micronucleus, this allele contains six pairs of incomplete genes that correspond to the six mating types that it specifies. During development of the macronuclear genome, all but one gene pair is deleted and the remaining genes are each joined to parts of the transmembrane domain that are shared across all mating type gene pairs. This results in cells that, when mature, express a single mating type (Cervantes et al. 2013).

The micronuclear organization and macronuclear development of this locus therefore explain how a single allele can encode multiple mating types (Cervantes et al. 2013). It does not, however, explain how the probability of expressing each mating type is determined; this remains an open question.

13.3.3 Why Probabilistic Sex Determination?

A peculiar consequence of probabilistic sex determination is that it allows for mating among genetically identical individuals, despite self-incompatibility of mating types. Self-incompatibility is hypothesized to have evolved because it prevents close inbreeding (Barrett 1988). However, self-incompatibility also frequently breaks down under small population sizes because it reduces the availability of compatible mates (Barrett 1988; Busch and Schoen 2008).

Probabilistic sex determination likely provides two benefits under small effective population sizes. First, because all seven sexes can be encoded by just two different alleles, a population founded by two individuals that differ at the *mat* locus, or a single heterozygote, can generate all seven sexes in a population. This is not possible with other types of genetic sex determination mechanisms, e.g. synclonal or cytoplasmic inheritance found in other ciliates (Lynn and Doerder 2012). Given this mechanism, is it perhaps not surprising that all populations where sufficient numbers of *T. thermophila* have been studied contain all seven mating types (Doerder et al. 1995; Arslanyolu and Doerder 2000). Why this number is fixed at seven within *T. thermophila*, but varies among species, remains a mystery.

The second benefit of probabilistic sex determination in small populations comes from the fact that the progeny of a single mated pair can mate among themselves. This increases the probability in a sparse population that an individual will encounter another compatible individual. There is a potential cost of such matings, however, due to the fact that the resulting progeny will be highly inbred. If a population harbors recessive deleterious mutations, then inbreeding will result in a reduction in fitness of the inbred offspring, relative to outbred individuals, a phenomenon known as inbreeding depression (Charlesworth and Charlesworth 1987). Work by Nanney (1957) showed evidence of inbreeding depression in lab strains of *T. thermophila.* However, recent analysis of inbreeding effects on natural isolates revealed no evidence of inbreeding depression (Dimond and Zufall in press). If probabilistic sex determination does indeed lead to frequent inbreeding, then it is likely that most recessive deleterious alleles would be purged from the population and little cost to inbreeding would remain (Crow 1970). However, data on heterozygosity at the *mat* and *serH* loci suggest that outcrossing may actually be frequent (Nanney 1980; Arslanyolu and Doerder 2000; Gerber et al. 2002). Further measures of heterozygosity in natural populations could provide insight into the relative frequencies of inbreeding and outcrossing.

13.3.4 Probabilistic Sex Determination and Sex Ratio

Population sex ratio refers to the relative abundance of each sex, or mating type, in a population. Fisher's sex ratio theory predicts that negative frequency-dependent selection will lead to the establishment of even, or equal, sex ratios in populations

(Fisher 1930; Shaw and Mohler 1953). When a sex is rare, it will have more access to potential mates, and thus will tend to have more offspring. This implies that genotypes that produce more offspring of the rare type will have more grandoffspring and those genotypes will spread in the population. Biased, or uneven, sex ratios are often explained in terms of deviations from Fisher's assumptions (West 2009).

Natural populations of *T. thermophila* exhibit uneven sex ratios (Doerder et al. 1995). To determine whether these uneven sex ratios could be explained by the mechanism of sex determination, Paixão et al. (2011) modeled the effects of probabilistic sex determination on sex ratios. They found that, indeed, probabilistic sex determination is sufficient to explain the observed deviations from even sex ratios in natural populations of *T. thermophila* (Paixão et al. 2011). Probabilistic sex determination can result in uneven population sex ratios because the frequencies of different mating types cannot be independently regulated since they are encoded by the same allele. Thus, unless there are a very large number of *mat* alleles in a population, natural selection can be constrained from moving a population to an even sex ratio (Paixão et al. 2011).

13.4 Mating Behavior

Much effort has been focused on studying the mating behaviors of animals, however comparatively little is known about similar behaviors among microbes (but see, e.g., Jackson and Hartwell 1990; Luporini et al. 2005).

13.4.1 The Stages of Mating

Mating in *Tetrahymena* is initiated by starvation. Following initiation, co-stimulation occurs when cells of compatible mating types come into contact (Cole 2016). Mating pairs start to form about an hour following co-stimulation. The interacting cells involved in costimulation are not necessarily the same as those involved in pairing, and the extent of costimulation by one compatible sex does not affect the efficiency of pairing with another compatible sex (Finley and Bruns 1980).

13.4.2 Mate Choice

Mating is random when two individuals in a population are just as likely as any other two individuals to mate. Random mating is thus expected when there is little variance in the perceived quality of potential mates. In natural populations, mating is rarely random (Partridge 1983).

Mate choice has primarily been examined in regard to choice among various genotypes. However, given the large number of mating types in *T. thermophila*, as in many microbial eukaryotes, it is possible that mating preferences may also evolve with respect to mating type. In other words, individuals of one mating type may choose to mate preferentially with individuals with one or more other mating types, regardless of their genotype.

There are two stages of mating where such preferences could be manifest: costimulation and pair formation. Current data indicate that mating types can functionally substitute for one another during costimulation (Finley and Bruns 1980) and that there is no preference among mating types during pair formation (Phadke et al. 2012). These results indicate that the choice among mating types, in this species, is limited to self-/non-self discrimination.

13.4.3 Frequency of Mating

Sexual reproduction is thought to provide both direct and indirect benefits. For example, in ciliates sexual reproduction initiates "resetting" of the macronucleus. This can directly benefit a cell that may have undergone otherwise irreparable damage, e.g. loss of chromosome arms or whole chromosomes (Bell 1988). Indirect benefits are generally the result of increased genetic variation among the offspring population, allowing more efficient purging of deleterious mutations and increased rates of adaptation (Fisher 1930; Muller 1932, 1964; Kondrashov 1988, 1993; Burt 2000).

However, sexual reproduction is also costly. While isogametic species, such as ciliates, do not experience the famous twofold cost of males like anisogametic species (Bell 1982), they may experience other costs. These costs could include genome dilution, the costs of searching for and choosing mates, undergoing meiosis, and recombination breaking up successful gene combinations (reviewed in Lehtonen et al. 2012).

The question then arises, for species with facultative sexuality, given these costs and benefits, how frequently will sexual reproduction occur in a population? In ciliates, the upper limit of this frequency is determined by a period of sexual immaturity. Following mating, *T. thermophila* undergo a period of immaturity during which they are incapable of conjugation for ~40–100 rounds of asexual division. Thus, *T. thermophila* can mate at most every 40–100 generations. Doerder et al. (1995) found that in natural populations immature cells are common. This indicates that sexual reproduction occurs frequently. The number of immature cells also varies with the season. Since sexual reproduction is signaled by starvation, the seasonal variation in mating frequency likely corresponds with the seasonal variation in bacterial food supply (Doerder et al. 1995).

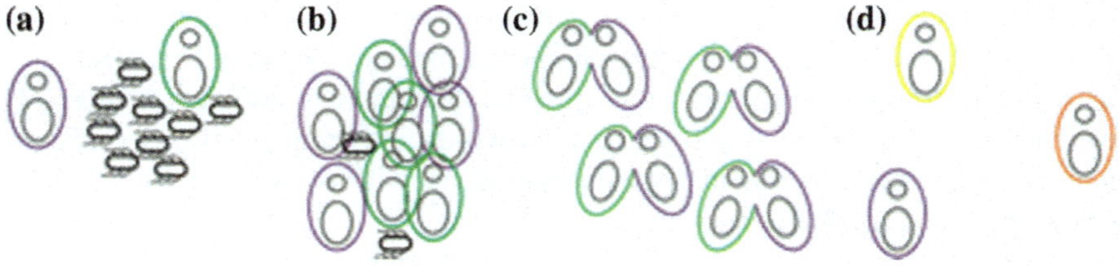

Fig. 13.2 Model of mating behavior. **a** Cells are attracted to food sources, shown here as a cluster of bacteria. **b** Cells feed on the bacteria and multiply, increasing the local density. **c** Once the food supply is exhausted, mating pairs form among complementary mating types. (Colors as in Fig. 13.1.) **d** Progeny then disperse to search for a new food source (Doerder et al. 1995)

13.4.4 Model of Mating Behavior

Doerder et al. (1995) propose the following model of mating behavior in *T. thermophila* (Fig. 13.2). Cells are attracted to food sources, where they feed and undergo asexual division, increasing local population density. When the food supply is exhausted, cells starve and become sexually competent. Due to the high density of cells present, costimulation, which is density dependent, can occur (Finlay and Bruns 1980). So long as at least two mating types are present, mating pairs will then form. Finally, the resulting progeny will disperse to find a new food source.

This model also helps explain the large number of mating types found in *T. thermophila*. Greater than two mating types in a population increases the chances of finding a compatible mate when there is limited time in which to find a mating partner (Iwasa and Sasaki 1987). In the Doerder et al. (1995) model, mating occurs in a limited time window between the onset of sexual competence and dispersal, both of which are likely triggered by the exhaustion of food resources.

Under this model, we might expect frequent mating among close relatives since population expansion occurs by asexual division of clonal lineages. Probabilistic sex determination may further increase the level of inbreeding by allowing mating among the asexual progeny of genetically identical siblings. The observed absence of inbreeding depression found in natural populations (see Sect. 13.3.3) supports this model.

13.5 Evolution of Asexuality

13.5.1 Asexuality Is Rare

Despite its many costs, most eukaryotic species undergo at least occasional sexual reproduction, indicating that sex is an evolutionarily successful strategy with benefits outweighing the costs (see Sect. 13.1; Weismann 1887; Maynard Smith 1978; Bell 1982; Rice 2002). By forgoing the benefits of sex, asexual populations are expected to be evolutionary "dead ends" (Stebbins 1957; Maynard Smith 1978), due to genetic deterioration or an inability to adapt to environmental change. Consistent with this prediction, many asexual lineages show signs of accelerated accumulation of deleterious mutations compared to their sexual relatives (Paland and Lynch 2006; Johnson and Howard 2007; Neiman et al. 2010; Henry et al. 2012; Tucker et al. 2013; Hollister et al. 2015).

13.5.2 Asexuality Is Common in Tetrahymena

Many microbial eukaryotes have been reported to be asexual, but upon further inspection are found to be "secretively sexual" (Dunthorn and Katz 2010). For example, species that have never been observed to undergo conjugation may nonetheless maintain meiosis-specific genes (Schurko and Logsdon 2008; Malik et al. 2008) or show evidence of recombination (Cooper et al. 2007).

Nuclear dualism in ciliates provides a largely unambiguous marker for asexuality that avoids the problem of secretive sex: when the germline micronucleus is absent, cells are either incapable of sexual reproduction (Kaney and Speare 1983) or acquire a new micronucleus from their mating partner (Allen et al. 1967). While amicronucleate lab strains have been generated, few ciliates are ever found in nature without a micronucleus, except in the genus *Tetrahymena* (Doerder 2014).

In stark contrast to other ciliates, Doerder (2014) found that fully one quarter of *Tetrahymena*-like wild isolates were amicronucleate. Phylogenetic analysis of these isolates reveals that there have been many independent origins of amicronucleates in *Tetrahymena* and the closely related genus *Glaucoma*. Some *Tetrahymena* species have only been found as amicronucleates, e.g. *T. pyriformis* and *T. furgasoni*, and others contain both micronucleate and amicronucleate isolates, including *T. thermophila* (Doerder 2014).

13.5.3 Tetrahymena *Genome Architecture May Facilitate Asexuality*

Perhaps the most surprising finding regarding asexuality in *Tetrahymena* is the estimated ages of some of the asexual lineages. Based on a molecular clock, Doerder (2014) estimated that some asexual lineages of *Tetrahymena* may be up to millions of years old. If these estimates are accurate, they indicate that *Tetramymena* is one of a rare group of organisms that can be evolutionarily successful in the absence of sexual reproduction.

The success of asexuality in *Tetrahymena* is likely at least partially a consequence of their genome architecture, in particular the structure of the macronuclear genome (Doerder 2014). The *Tetrahymena* macronuclear genome contains $\sim$45 copies of each of the 225 acentromeric chromosomes (Orias et al. 2011; Eisen et al. 2006).

During asexual division, the macronucleus divides by amitosis, an unusual form of nuclear division where homologous chromosomes are randomly segregated among daughter cells (Allen and Nanney 1958; Orias and Flacks 1975). This process results in a phenomenon known as phenotypic assortment, in which heterozygous macronuclear chromosomes can become completely homozygous within ~200 rounds of asexual division following conjugation (Doerder et al. 1992; Fig. 13.1b). New mutations similarly have been shown to assort in both micronucleate and amicronucleate *Tetrahymena* (Orias and Newby 1975; Byrne 1978; Fig. 13.1c).

The process of phenotypic assortment during amitosis may allow *Tetrahymena* to experience some of the benefits of sexual reproduction under asexuality. Under mitosis, asexual reproduction results in the irreversible accumulation of deleterious alleles, a phenomenon known as Muller's ratchet (Muller 1964; Felsenstein 1974; Haigh 1978). In contrast, amitosis may approximate the effects of meiotic recombination and allow the purging of deleterious alleles and fixation of beneficial alleles (Doerder 2014). For example, in a population with two segregating alleles that confer different fitnesses, amitosis will produce asexual progeny that differ in the number of copies of each allele. Selection will favor the progeny with more copies of the beneficial allele, eventually resulting in the elimination of the deleterious allele and fixation of the beneficial allele.

In some species of ciliates, amitosis results in unequal numbers of chromosomes among daughter cells, and is thought to be responsible for senescence of cell lineages that are grown asexually (Bell 1988). In contrast, *T. thermophila* appear to have a mechanism that controls macronuclear chromosome copy number, and therefore do not senesce when grown asexually (Preer and Preer 1979; Orias et al. 2011). Thus, chromosome copy number control, in combination with amitosis, is likely an essential feature of successful asexuality in ciliates.

13.6 Conclusion

We have learned a great deal about mating and reproductive strategies in *Tetrahymena* over the last 60 years. Nonetheless, many unanswered questions remain. Modern experimental approaches are likely to yield additional insight into these issues. For example, population genomic sequencing can shed light on mating system parameters, such as the frequency of sexual reproduction and the rates of inbreeding/outcrossing. Sequencing of the micronuclear genome may elucidate the mechanism of probabilistic sex determination. Experimental evolution will allow tests of sex ratio evolution and adaptation. And more powerful computation will allow exploration of the roles of various genomic features on the success of asexuals.

Tetrahymena continues to be a powerful model system in cell and molecular biology and genetics. The work described in this chapter indicates that this genus is also an important model system in studying a variety of outstanding questions in evolutionary biology.

Acknowledgments I am grateful to Paul Doerder for many helpful discussions and insights into these topics and to Sujal Phadke, Ricardo Azevedo, Kristen Dimond, Hongan Long, and Joe West. This work was supported in part by the National Institute of General Medical Science of the National Institutes of Health (award no. R01GM101352).

References

Allen SL, Nanney DL (1958) An analysis of nuclear differentiation in the selfers of *Tetrahymena*. Am Nat 92:139–160

Allen SL, File SK, Koch SL (1967) Genomic exclusion in *Tetrahymena*. Genetics 55:823–837

Arslanyolu M, Doerder FP (2000) Genetic and environmental factors affecting mating type frequency in natural isolates of *Tetrahymena thermophila*. J Eukaryotic Microbiol 47:412–418

Bachtrog D, Mank JE, Peichel CL, Kirkpatrick M, Otto SP, Ashman T-L, Hahn MW, Kitano J, Mayrose I, Ming R, Perrin N, Ross L, Valenzuela N, Vamosi JC, The Tree of Sex Consortium (2014) Sex determination: why so many ways of doing it? PLoS Biol 12:e1001899

Barrett SCH (1988) The evolution, maintenance, and loss of self-incompatibility systems. In: Lovett Doust J, Lovett Doust L (eds) Plant reproductive ecology: patterns and strategies. Oxford University Press, Oxford, pp 98–124

Beisson J, Sonneborn TM (1965) Cytoplasmic inheritance of the organization of the cell cortex in *Paramecium aurelia*. Proc Natl Acad Sci USA 53:275–282

Bell G (1982) The masterpiece of nature: the evolution and genetics of sexuality. University of California Press, Berkeley

Bell G (1988) Sex and death in protozoa. Cambridge University Press, Cambridge

Burt A (2000) Perspective: sex, recombination, and the efficacy of selection—was Weismann right? Evolution 54:337–351

Busch JW, Schoen DJ (2008) The evolution of self-incompatibility when mates are limiting. Trends Plant Sci 13:128–136

Byrne BC (1978) Induced resistance to 6-methylpurine and cycloheximide in *Tetrahymena*. II. Variation within vegetative cultures of micronucleate *T. thermophila* and amicronucleate *T. pyriformis*. Genetics 89:703–710

Cervantes MD, Hamilton EP, Xiong J, Lawson MJ, Yuan D, Hadjithomas M, Miao W, Orias E (2013) Selecting one of several mating types through gene segment joining and deletion in *Tetrahymena thermophila*. PLoS Biol 11:e1001518

Charlesworth D, Charlesworth B (1987) Inbreeding depression and its evolutionary consequences. Annu Rev Ecol Syst 18:237–268

Cole ES (2016) Cell-Cell interactions leading to establishment of a mating junction in *Tetrahymena* and *Paramecium*, two "Contact-Mediated" mating systems. In: Witzany G, Nowacki M (eds) Biocommunication of Ciliates. doi:10.1007/978-3-319-32211-7_12

Cooper MA, Adam RD, Worobey M, Sterling CR (2007) Population genetics provides evidence for recombination in *Giardia*. Curr Biol 17:1984–1988

Crow JF (1970) Genetic loads and the cost of natural selection. In: Kojima K (ed) Mathematical topics in population genetics. Springer, Berlin Heidelberg, pp 128–177

Dimond KL, Zufall RA (in press) Hidden genetic variation in the germline genome of *Tetrahymena thermophila*. J Evol Biol.

Dini F, Corliss JO (2001) Protozoan sexuality. In: eLS, Wiley, Chichester

Dini F, Nyberg D (1993) Sex in ciliates. In: Jones JG (ed) Advances in microbial ecology, vol 13. Springer, US, pp 85–153

Doerder FP (2014) Abandoning sex: multiple origins of asexuality in the ciliate *Tetrahymena*. BMC Evol Biol 14:112

Doerder FP, Deak JC, Lief JH (1992) Rate of phenotypic assortment in *Tetrahymena thermophila*. Dev Genet 13:126–132

Doerder FP, Gates MA, Eberhardt FP, Arslanyolu M (1995) High-frequency of sex and equal frequencies of mating types in natural-populations of the ciliate *Tetrahymena thermophila*. Proc Natl Acad Sci USA 92:8715–8718

Doerder FP, Arslanyolu M, Saad Y, Kaczmarek M, Mendoza M, Mita B (1996) Ecological genetics of *Tetrahymena thermophila:* mating types, i-antigens, multiple alleles and epistasis. J Eurkaryotic Microbiol 43:95–100

Dunthorn M, Katz LA (2010) Secretive ciliates and putative asexuality in microbial eukaryotes. Trends Microbiol 18:183–188

Eisen JA, Coyne RS, Wu M, Wu DY, Thiagarajan M, Wortman JR, Badger JH, Ren QH, Amedeo P, Jones KM et al (2006) Macronuclear genome sequence of the ciliate *Tetrahymena thermophila*, a model eukaryote. PLoS Biol 4:1620–1642

Felsenstein J (1974) The evolutionary advantage of recombination. Genetics 78:737–756

Finley MJ, Bruns PJ (1980) Costimulation in *Tetrahymena*. II. A nonspecific response to heterotypic cell–cell interactions. Dev Biol 79:81–94

Fisher RA (1930) The genetical theory of natural selection. Clarendon Press, Oxford

Gerber CA, Lopez AB, Shook SJ, Doerder FP (2002) Polymorphism and selection at the SerH immobilization antigen locus in natural populations of *Tetrahymena thermophila*. Genetics 160:1469–1479

Haigh J (1978) The accumulation of deleterious genes in a population–Muller's Ratchet. Theor Popul Biol 14:251–267

Henry L, Schwander T, Crespi BJ (2012) Deleterious mutation accumulation in asexual *Timema* stick insects. Mol Biol Evol 29:401–408

Hollister JD, Greiner S, Wang W, Wang J, Zhang Y, Wong GK-S, Wright SI, Johnson MTJ (2015) Recurrent loss of sex is associated with accumulation of deleterious mutations in Oenothera. Mol Biol Evol 32:896–905

Iwasa Y, Sasaki A (1987) Evolution of the number of sexes. Evolution 41:49–65

Jackson CL, Hartwell LH (1990) Courtship in *S. cerevisiae*: both cell types choose mating partners by responding to the strongest pheromone signal. Cell 63:1039–1051

Jackson JBC, Buss LW, Cook RE (1985) Population biology and evolution of clonal organisms. Yale University Press

Johnson SG, Howard RS (2007) Contrasting patterns of synonymous and nonsynonymous sequence evolution in asexual and sexual freshwater snail lineages. Evolution 61:2728–2735

Kaney AR, Speare VJ (1983) An amicronucleate mutant of *Tetrahymena thermophila*. Exp Cell Res 143:461–467

Kondrashov AS (1988) Deleterious mutations and the evolution of sexual reproduction. Nature 336:435–440

Kondrashov AS (1993) Classification of hypotheses on the advantage of amphimixis. J Hered 84:372–387

Lehtonen J, Jennions MD, Kokko H (2012) The many costs of sex. Trends Ecol Evol 27:172–178

Luporini P, Alimenti C, Ortenzi C, Vallesi A (2005) Ciliate mating types and their specific protein pheromones. Acta Protozool 44:89–101

Lynn DH, Doerder FP (2012) The life and times of *Tetrahymena*. In: Collins K (ed) Methods in cell biology. Academic Press, pp. 9–27

Malik S-B, Pightling AW, Stefaniak LM, Schurko AM, Logsdon JM (2008) An expanded inventory of conserved meiotic genes provides evidence for sex in *Trichomonas vaginalis*. PLoS ONE 3:e2879

Maynard Smith J (1978) The evolution of sex. Cambridge University Press, Cambridge New York

Miyake A (1996) Fertilization and sexuality in ciliates. In: Hausmann K, Bradbury PC (eds) Ciliates: cells as organisms. Gustav Fischer, Stuttgart, pp 243–290

Muller HJ (1932) Some genetic aspects of sex. Am Nat 66:118–138

Muller HJ (1964) The relation of recombination to mutational advance. Mutat Res 106:2–9

Nanney DL (1957) Inbreeding degeneration in *Tetrahymena*. Genetics 42:137–146

Nanney DL (1959) Genetic factors affecting mating type frequencies in variety 1 of *Tetrahymena pyriformis*. Genetics 44:1173–1184

Nanney DL (1966) Corticotypes in *Tetrahymena pyriformis*. Am Nat 100:303–318
Nanney DL (1980) Experimental ciliatology: an introduction to genetic and developmental analysis in ciliates. Wiley, New York
Nanney DL, Caughey PA (1953) Mating type determination in *Tetrahymena pyriformis*. Proc Natl Acad Sci USA 39:1057–1063
Nanney DL, Caughey PA, Tefankjian A (1955) The genetic control of mating type potentialities in *Tetrahymena pyriformis*. Genetics 40:668–680
Neiman M, Hehman G, Miller JT, Logsdon JM, Taylor DR (2010) Accelerated mutation accumulation in asexual lineages of a freshwater snail. Mol Biol Evol 27:954–963
Orias E (1981) Probable somatic DNA rearrangements in mating type determination in *Tetrahymena thermophila*: a review and a model. Dev Genet 2:185–202
Orias E (1986) Ciliate Conjugation. In: Gall JG (ed) The molecular biology of ciliated protozoa. Academic Press, Orlando, pp 45–84
Orias E, Flacks M (1975) Macronuclear genetics of *Tetrahymena* I. Random distribution of macronuclear gene copies in *T. pyriformis*, syngen 1. Genetics 79:187–206
Orias E, Newby CJ (1975) Macronuclear genetics of *Tetrahymena*. II. Macronuclear location of somatic mutations to cycloheximide resistance. Genetics 80:251–262
Orias E, Cervantes MD, Hamilton EP (2011) *Tetrahymena thermophila*, a unicellular eukaryote with separate germline and somatic genomes. Res Microbiol 162:578–586
Paixão T, Phadke SS, Azevedo RBR, Zufall RA (2011) Sex ratio evolution under probabilistic sex determination. Evolution 65:2050–2060
Paland S, Lynch M (2006) Transitions to asexuality result in excess amino acid substitutions. Science 311:990–992
Partridge L (1983) Non-random mating and offspring fitness. In: Bateson P (ed) Mate choice. Cambridge University Press, Cambridge, pp 227–256
Phadke SS, Zufall RA (2009) Rapid diversification of mating systems in ciliates. Biol J Linnean Soc 98:187–197
Phadke SS, Cooper L, Zufall RA (2012) Is evolution of mating preferences inevitable? Random mating in the multisex system of *Tetrahymena thermophila*. Int J Evol Biol 2012:e201921
Phadke SS, Paixão T, Pham T, Pham S, Zufall RA (2014) Genetic background alters dominance relationships between *mat* alleles in the ciliate *Tetrahymena thermophila*. J Hered 105:130–135
Preer JR, Preer LB (1979) The size of macronuclear DNA and its relationship to models for maintaining genic balance. J Protozool 26:14–18
Prescott DM (1994) The DNA of ciliated protozoa. Microbiol Rev 58:233–267
Rice WR (2002) Experimental tests of the adaptive significance of sexual recombination. Nat Rev Genet 3:241–251
Schurko AM, Logsdon JM (2008) Using a meiosis detection toolkit to investigate ancient asexual "scandals" and the evolution of sex. BioEssays 30:579–589
Shaw RF, Mohler JD (1953) The selective significance of the sex ratio. Am Nat 87:337–342
Speijer D, Lukeš J, Eliáš M (2015) Sex is a ubiquitous, ancient, and inherent attribute of eukaryotic life. Proc Natl Acad Sci USA 112:8827–8834
Stebbins GL (1957) Self fertilization and population variability in the higher plants. Am Nat 91:337–354
Tucker AE, Ackerman MS, Eads BD, Xu S, Lynch M (2013) Population-genomic insights into the evolutionary origin and fate of obligately asexual *Daphnia pulex*. Proc Natl Acad Sci USA 110:15740–15745
Weismann A (1887) On the signification of the polar globules. Nature 36:607–609
West S (2009) Sex allocation. Princeton University Press

Chapter 14
Social Information in Cooperation and Dispersal in *Tetrahymena*

Staffan Jacob, Jean Clobert, Delphine Legrand, Nicolas Schtickzelle and Alexis S. Chaine

Abstract Most organisms are able to use social information to adjust key behaviours in their lifecycle including dispersal and cooperation. Ciliate microcosms provide a highly powerful tool to study the role of biocommunication in ecology and evolution because they allow simultaneous study of the chemical carriers of information and their eco-evolutionary consequences within short spatial and temporal scales. Here, we review what is known about the existence of social information use and its consequences on dispersal and cooperation in *Tetrahymena*, and point out the usefulness of *Tetrahymena* microcosms to further develop our understanding of the mechanisms and implications of information use in eco-evolutionary dynamics.

14.1 Introduction

Throughout their lifetime, organisms can acquire and use information to maximize their reproduction and survival. Finding food, mates, habitats, avoiding predators and parasites, allocating energy and time to reproduction or self-maintenance are all situations for which taking into account information about the environmental

S. Jacob (✉) · J. Clobert · D. Legrand · A.S. Chaine
Evolutionary Ecology Group, Station d'Ecologie Théorique et Expérimentale du CNRS, UMR5321, 2 route du CNRS, 09200 Saint-Girons, France
e-mail: jacobstaffan@gmail.com

J. Clobert
e-mail: jean.clobert@EcoEx-Moulis.cnrs.fr

D. Legrand
e-mail: delphine.legrand@EcoEx-Moulis.cnrs.fr

A.S. Chaine
e-mail: alexis.chaine@cnrs.fr

S. Jacob · N. Schtickzelle
Biodiversity Research Centre, Université Catholique de Louvain, Earth and Life Institute, Croix du Sud 4, L7-07-04, 1348 Louvain-la-Neuve, Belgium
e-mail: nicolas.schtickzelle@uclouvain.be

G. Witzany and M. Nowacki (eds.), *Biocommunication of Ciliates*,
DOI 10.1007/978-3-319-32211-7_14

context is crucial to maximize benefits and reduce associated costs. Organisms should thus continuously gather information about their surroundings and make optimal decisions based on this information (Dall et al. 2005; Schmidt et al. 2010).

Over the past few decades, behavioural ecologists have done an excellent job defining and classifying the different sorts of information gathering processes and the use of this information. Information can be defined as any feature gathered from the abiotic environment (physical cues such as temperature or humidity), the biotic environment (the presence of a competitor, a predator, a parasite or a member of its own species) or perception of one's own internal state (e.g. condition, health; Dall et al. 2005; Schmidt et al. 2010). When information is released by individuals, either from the same or another species, we can distinguish between two types of information transfer. Information can be passively released, meaning that the individual providing the information makes it inadvertently (cues), or the individuals can actively release information, so as to benefit from the reaction of other individuals gathering that information (signals; Danchin 2004; Dall et al. 2005; Seppänen et al. 2007; Schmidt et al. 2010).

The impact of information release and use on different eco-evolutionary processes such as population or community dynamics and local adaptation have been widely studied and reviewed during the past decades (e.g. Danchin 2004; Dall et al. 2005; Seppänen et al. 2007; Schmidt et al. 2010). In particular, studying the evolution of information use is central for our understanding of how individuals deal with their environment by adjusting their phenotype to the local environment, or moving toward another environment. While movements of individuals in a landscape (i.e. dispersal) is a major driver of ecological and evolutionary dynamics, this behaviour has long been considered as fixed or random, thus assuming that individuals do not use information to choose whether to stay in a habitat or leave and join another (Hamilton and May 1977; Patterson et al. 2008; Clobert et al. 2009; Chaine and Clobert 2012). Conversely, a large body of empirical work shows that dispersal can be a plastic trait that depends on the internal state of individuals and their environmental context (*reviewed in* Bowler and Benton 2005; Clobert et al. 2009). This means that individuals are not blind, but able to use information in a decision-making process during dispersal (Reed et al. 1999; Bowler and Benton 2005; Edelaar et al. 2008; Clobert et al. 2009; Cox and Kesler 2012; Jacob et al. 2015a).

Individuals can use a large variety of abiotic and biotic factors as information sources to adjust their dispersal decisions. For instance, temperature, resources and the presence and phenotype of conspecifics have all been demonstrated as major drivers of dispersal decisions (e.g. Doligez et al. 2004; Danchin 2004; Clobert et al. 2009; Chaine and Clobert 2012; Legrand et al. 2015). While effects of temperature and resources are relatively straightforward (individuals usually prefer habitats with optimal temperature and high resource availability), the effects of the presence and type of conspecifics on dispersal decisions are more complex. At first sight, organisms are expected to avoid crowded habitats and thus to leave habitats with high competition to find less crowded ones (positive density-dependent dispersal; Gandon and Michalakis 1999; Bowler and Benton 2005; Poethke et al. 2007). However, staying with congeners can be beneficial in various ways, by for instance increasing the probability to find a mate and reproduce, diluting predation risk, or if high density provides a cue indicating a good-quality environment (negative density-dependent dispersal; Kim et al. 2009; Baguette et al. 2011; Fellous et al. 2012). Moreover, the effects of density on dispersal decisions are particularly intriguing in species where staying with congeners allows benefitting from social interactions, i.e. in cooperative species.

Cooperation between individuals is widespread in nature, having evolved in almost all biological systems from microorganisms to plants and metazoans, and relies on the emission of behaviours that provide a benefit to another individual (West et al. 2007; Williams et al. 2007; Darch et al. 2012). For cooperation to evolve and be maintained, individuals investing in these costly cooperative behaviours should obtain some benefits in return, i.e. they should have indirect fitness benefits (Hamilton 1964; West et al. 2007). Kin selection theory predicts that cooperative behaviours should be directed toward kin, making the increased reproductive success of the recipient of the cooperative behaviour beneficial for the emitter: by helping a kin to reproduce, an individual indirectly favours the transmission of its own genes (Hamilton 1964; West et al. 2007). As a consequence, individuals should tend to stay with congeners in the presence of kin cooperation, leading to the evolution of low dispersal rates and negative density-dependent dispersal (Hamilton 1964; Griffin and West 2002). However, while staying with kin provides benefits through kin cooperation, it also increases kin competition (i.e. the reduced indirect fitness resulting from competition between relatives), which can negate the benefits of kin cooperation (Hamilton 1964; Taylor 1992; Queller 1994). Information on the kin relationships within a population can therefore be used to modulate cooperative behaviour as well as dispersal.

The relationships between information, dispersal and cooperation have been particularly studied in laboratory cultures of protists, and especially in the ciliate *Tetrahymena*, for two reasons. First, protists are ideal systems to experimentally manipulate biotic and abiotic information, the spatial dimension and configuration of the landscape, and to measure cooperative behaviours (Schtickzelle et al. 2009; Chaine et al. 2010; Jacob et al. 2015b; Altermatt et al. 2015). Therefore, they provide an ideal system (i) to test theoretical predictions on the stability of cooperative strategies (West et al. 2007), the evolution of dispersal phenotypes (Pennekamp et al. 2014; Altermatt et al. 2015; Fronhofer et al. 2015), and metapopulation or metacommunity dynamics (Altermatt et al. 2015; Fronhofer and Altermatt 2015); (ii) to test for the role of biocommunication in dispersal decisions (Chaine et al. 2010; Jacob et al. 2015b) and in the expression of cooperation behaviours (Schtickzelle et al. 2009; Chaine et al. 2010); (iii) or to determine the respective roles of dispersal and cooperation on each other (Schtickzelle et al. 2009; Chaine et al. 2010). Secondly, *Tetrahymena* sp. are actively moving ciliates that present variation in dispersal-related traits distinguishing resident individuals, those that stay in their 'natal' patch, from disperser individuals, those that leave their 'natal' patch to join another (Fjerdingstad et al. 2007; Pennekamp et al. 2014). In addition, a specialized "dispersal morph" produces a flagellum under particular environmental conditions (Nelsen and Debault 1978). Fitness associated with each movement strategy can be estimated by contrasting behaviours between clonal lines and measuring the consequences for spatial population dynamics. Furthermore, individuals can cooperate under harsh conditions in order to improve their survival, by forming cell aggregates that facilitate the exchange of growth factors (Christensen et al. 1996, 2003; Schtickzelle et al. 2009; Chaine et al. 2010), and variation among clonal lines in this cooperative behaviour has been found (Schtickzelle et al. 2009). The existence of both dispersal and cooperation variation at the intra-specific level makes *Tetrahymena* is an ideal model organism to study the evolutionary interplay between cooperation and dispersal and the role of bio-communication in these processes.

In this chapter, we review the current knowledge about the existence of information use and its consequences on dispersal and cooperation in *Tetrahymena*. We first briefly introduce the effects of non-social information on dispersal decisions and cooperation to demonstrate the context-dependence of these behaviours. Then, we review the effects of social information on dispersal and cooperation, focusing on the density of congeners. Finally, because all individuals are not equal in a population, we point out that the condition and quality of congeners are important sources of information modulating the interaction between dispersal and cooperative decisions.

14.2 Non-social Environment Affects Dispersal

Living without information about the environment entails major risks. Among the large variety of environmental characteristics individuals should take into account in their behavioural decisions, the non-social environmental characteristics such as temperature and resources are of central importance in determining reproduction and survival. In Tetrahymena, the most obvious environmental characteristic likely to influence behavioural decisions is resource availability. For instance, while *Tetrahymena* sp. is mainly a clonally reproducing organism, a lack of food that induces starvation stimulates the expression of mating types and thus sexual reproduction in *Tetrahymena* sp. (Collins 2012). Resources are one of the major causes of dispersal as individuals benefit from moving from a poor quality to a high quality patch and therefore information about patch quality should therefore be critical to dispersal decisions.

Indeed, long distance dispersal is strongly dependent on resources, since the production of flagellated dispersal morphs occurs primarily when individuals are starved (Nelsen and Debault 1978). Further, *Tetrahymena* sp. are able to actively detect a distant food resource, and orientate towards it (Hellung-Larsen et al. 1990). In the presence of nutrients such as proteose peptone, platelet extract and fibroblast growth factor, cells show chemotaxis, orienting their movement trajectory towards the source of nutrients. Furthermore, some chemoattractants also increase swim velocity because they cause cell membrane hyperpolarization, in turn causing an increase in ciliary beat frequency, while other chemoattractants are effective without hyperpolarization and change in swim velocity, affecting only the directionality of movement (Collins 2012). Using *T. pyriformis*, Fronhofer et al. (2015) found significant effects of local resource abundance (bacterial density) on cell movements: velocity increased at higher resource concentrations but linearity of the movement trajectory decreased. As a result, despite changes in movement properties, the overall displacement rate was unaffected by bacterial abundance. This effect might emerge as a consequence of a more systematic foraging behaviour when resources are more abundant (Fronhofer et al. 2013).

Besides chemotaxis, food availability affects cell secretion of chemicals, by drastically reducing the diversity and changing the nature of compounds secreted by cells (Madinger et al. 2010). This means that *Tetrahymena* might obtain information about resource availability not only through personally acquired information, but also through intercellular chemical biocommunication. Furthermore, intercellular chemical signals are of major importance for cooperative behaviours in these species (Christensen et al. 1996, 2003; Schtickzelle et al. 2009; Chaine et al. 2010). We might thus expect cooperation in this species to be a plastic behaviour for which cells adjust their investment depending upon the balance between personal assessment of resource availability and intercellular chemical biocommunication. Since the chemicals involved in cooperation in *Tetrahymena thermophila* have been well known for decades (Christensen and Rasmussen 1992; Christensen et al. 1995, 1996, 2003; Rasmussen et al. 1996; Straarup et al. 1997), we argue that this species provides an excellent model for experimental investigation of theoretical predictions on the stability of cooperative strategies (e.g. Hamilton 1964; West et al. 2007).

Another environmental factor affecting dispersal in *Tetrahymena* is temperature. Swimming speed of *T. pyriformis* increases with temperature within the suitable physiological range of temperatures, which is also thought to be due to cell membrane hyperpolarization (Connolly et al. 1985). However, the same study also found that there were precipitous declines of swimming velocity at the hot (c. >38–40 °C) and cold (c. <10–12 °C) extremes, probably associated with a loss of membrane integrity. Interestingly, depending on the temperature at which the cells were acclimated, differences were observed in cell optimal temperatures (32 °C for 20 °C acclimated cells, 40 °C for 38 °C acclimated cells) and breaking points. This temperature effect on swimming velocity of *Tetrahymena* is in agreement with the observation of a faster and straighter movement of *T. pyriformis* (Koutna et al. 2004) and *T. thermophila* (Shiurba et al. 2006) when submitted to infrared light, known to induce heating. Infrared radiation also increased differentiation of cells into dispersal morphs (Shiurba et al. 2006).

14.3 Social Environment: Effects of Population Density on Dispersal and Cooperation

While habitat characteristics such as food availability and temperature are major determinants of an organism' survival and reproduction, and are known to influence *Tetrahymena* behaviour as pointed out above, the value of a given environment is also modulated by the density of congeners in this environment. Information about the social environment is therefore of central importance to decision making processes.

14.3.1 Competition and Dispersal Decisions

In *Tetrahymena*, growth becomes limited when density reaches a habitat's carrying capacity because of competition for resources, leading to classical logarithmic population growth (e.g. Fjerdingstad et al. 2007). But besides this negative effect of high density, very low density is also known to limit growth and to increase mortality (Christensen and Rasmussen 1992; Christensen et al. 1995). In *Tetrahymena thermophila* cells cultured in a synthetic nutrient medium, the 2 h doubling time at 37 °C can turn to 2 h half-life when density decreases below 500 cells/ml (Christensen and Rasmussen 1992; Christensen et al. 1995). Given these major effects of density on growth and survival, we expect *Tetrahymena* cells to adjust their dispersal decisions depending on information about the density of congeners in a habitat. Abundant empirical and theoretical evidence in a large variety of organisms points out the major role of population density in driving dispersal decisions (reviewed in Bowler and Benton 2005), and evidence supporting such density-dependent dispersal also exists for *Tetrahymena* (Pennekamp et al. 2014). In this context, a major advantage of clonal organisms cultured under controlled conditions is that the relative contribution of genetic and environmental factors and their interaction (GxE reaction norm approaches) in the acquisition and use of environmentally derived information can be easily disentangled (Pennekamp et al. 2014; Jacob et al. 2015b; Fig. 14.1). Genetic effects (G) signify the differences in dispersal among genotypes that are independent from the environmental context, environmental effects (E) are the extent of response to the environmental context that is common between genotypes, and their interaction (GxE) indicates genetic variation in context-dependent dispersal. Such an approach has been fruitfully used in *Tetrahymena*.

Using 44 genetic clonal strains of *Tetrahymena thermophila*, Pennekamp et al. (2014) found that 43 % of the variation in dispersal propensity is explained by the genotype, 13 % by the environment itself, and 2 % by the interaction between genotype and environment (Fig. 14.1). This demonstrates that dispersal in *Tetrahymena* is both genetically based and dependent upon information gathered from the environment. Importantly, the small but significant interaction between genotype and environment points out the existence of genetic variability in context-dependent dispersal. In other words, different genotypes use information in different ways and consequently there is some potential for plasticity in informed dispersal decisions to evolve (Clobert et al. 2009; Pennekamp et al. 2014).

As for density effects on growth, some evidence suggests that density-dependent dispersal in *Tetrahymena* might be mediated by intercellular chemical signalling. Fronhofer et al. (2015) provided experimental evidence that chemicals released in the media drives effects of density on mobility in *Tetrahymena pyriformis*. By separately manipulating the number of cells and the associated chemicals (by media

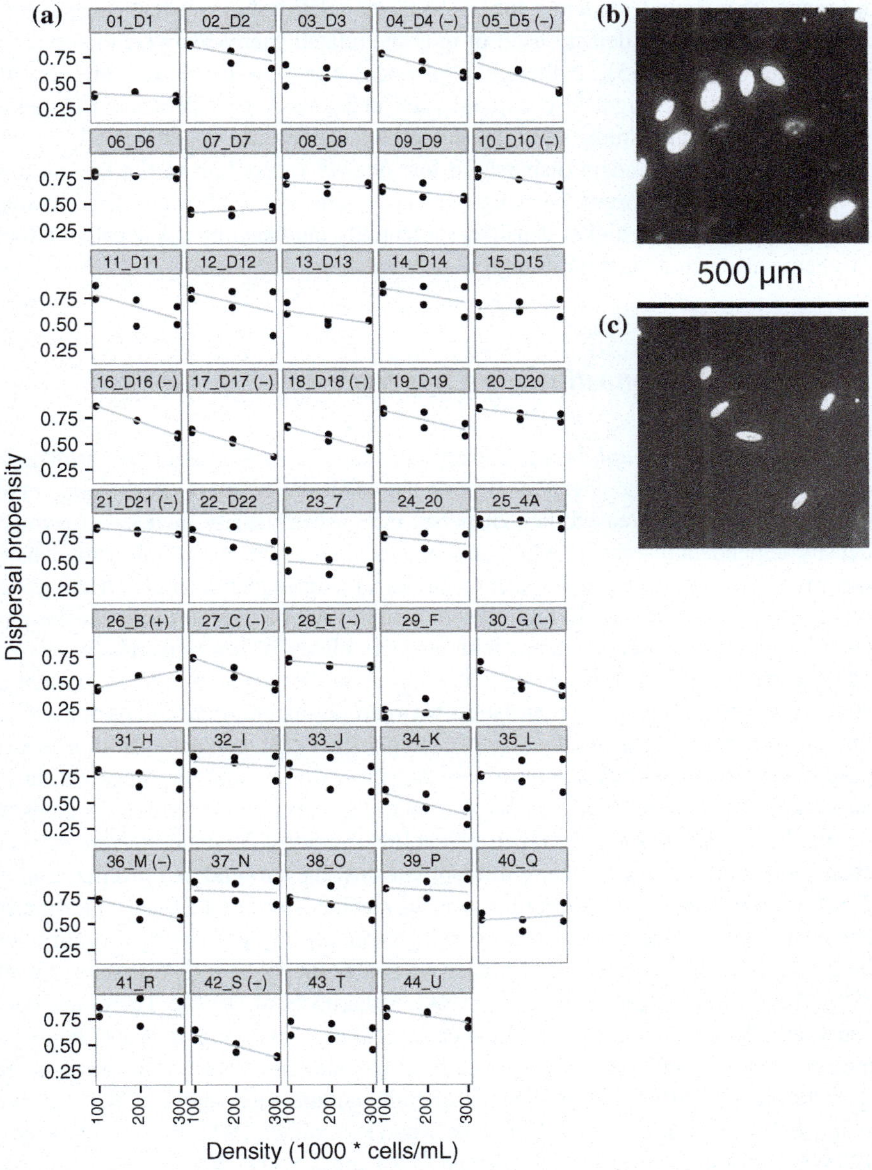

Fig. 14.1 Population density and dispersal in *Tetrahymena thermophila*. **a** Density-dependent dispersal decisions in 44 genetic clonal strains ("(-)" and "(+)" indicates significant negative or positive density-dependent dispersal rate). **b** Classical large and round cell shape under normal growth conditions. **c** Typical smaller and elongated shape of the fast-swimming cells (**a** from Pennekamp et al. 2014; **b**, **c** from Fjerdingstad et al. 2007)

filtering), they found evidence for a U-shaped relationship between density and mobility, regardless if density itself or only the related chemicals were manipulated (Fronhofer et al. 2015). Although a physical effect of crowding on reducing movement abilities might also explain such negative density-dependent dispersal reaction norms, this study strongly suggests the existence of chemically driven negative density-dependent dispersal at low density (according to the known positive effects of quorum sensing at low density), followed by positive density-dependence at higher densities (potentially mediated by the negative effects of chemicals such as insulin at high concentration).

14.3.2 From Competition to Cooperation

In *Tetrahymena thermophila*, the density-dependent dispersal reaction norms have mostly been found to be negative. Among the 44 strains tested by Pennekamp et al. (2014), 13 strains decreased their dispersal rates when density increased (negative density-dependent dispersal), while only one increased its dispersal rate with density (positive density-dependent dispersal; Fig. 14.1). Consequently, most of the genotypes tested tended to stay in high-density patches, while they left low-density patches. This pattern is in accordance with negative effects of low densities for growth and survival shown in this species (Christensen and Rasmussen 1992; Christensen et al. 1995). Such a positive correlation between population density and individual fitness, known as the Allee effect (Courchamp et al. 2008), can result from a lack of possibilities for social interactions at low densities, while higher densities will facilitate social interactions. This can be particularly important in unicellular organisms such as *Tetrahymena* sp in which growth at low density is dependent on among-cell chemical signalling. As for many microorganisms, the survival and proliferation of *Tetrahymena* depends on a quorum sensing mechanism: extracellular emission of chemical signals drives population growth, especially by preventing morality and favouring growth at low density (Christensen and Rasmussen 1992; Christensen et al. 1995, 1996, 2003; Rasmussen et al. 1996; Straarup et al. 1997). Therefore, cells cooperate through intercellular chemical signalling involving a broad range of molecules from hormones such as insulin, neurotransmitters such as endorphins, to 'growth factors' (TPAF) resembling cell signalling molecules such as cGMP (Leroith et al. 1982; Le Roith et al. 1983; Christensen and Rasmussen 1992; Köhidai and Csaba 1995; Csaba and Kovács 1999; reviewed in Rasmussen and Wheatley 2007; Csaba 2012). Interestingly, *T. thermophila* has been found to form aggregative groups, a behaviour hypothesized to favour the exchange of these specific chemicals (Schtickzelle et al. 2009; Chaine et al. 2010). Extensive genetic variation in this aggregative behaviour has been found in *T. thermophila*, which means that low, medium and high levels of cooperation exists at the intra-specific level in *T. thermophila* (Schtickzelle et al. 2009; Chaine et al. 2010). Furthermore, highly aggregative cell lines live longer on a given resource, suggesting benefits of aggregation (Chaine et al. 2010) (Fig. 14.2).

(a)

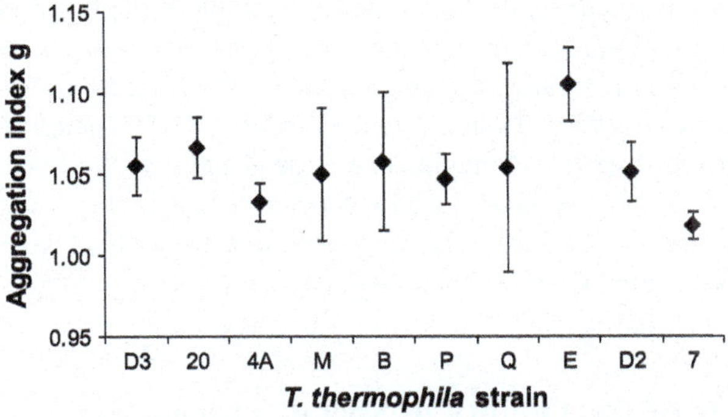

(b)

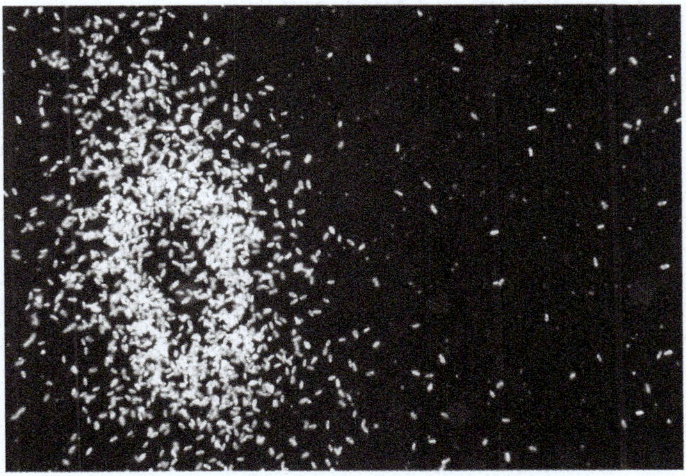

Fig. 14.2 Cooperative behaviour of *Tetrahymena thermophila*: **a** among strains variability in cooperation propensity (figure from Schtickzelle et al. 2009); **b** example of aggregation behaviour

The main hypothesis explaining the emergence and maintenance of cooperation is kin selection, which suggests that for cooperation to evolve, costly cooperative behaviours should be directed toward kin, thus indirectly increasing gene transmission (Hamilton 1964; West et al. 2007). The maintenance of kin-based groups allowing cooperative behaviours to be directed toward kin provides a simple mechanism to control invasion by cheaters. Dispersal, the ability of individuals to move from one place to another during their life, is expected to decrease relatedness in populations (Clobert et al. 2012). Since kin selection theory predicts that stable groups of kin should stabilize cooperation, cooperation should favour low dispersal (Hamilton 1964; Griffin and West 2002). As a corollary, dispersal has long been seen as an opposing force to cooperation. Indeed, *T. thermophila* genetic lines with low aggregation levels showed high dispersal rates, while highly aggregative lines disperse at lower rates, but with some cells specialized for long-distance dispersal (Schtickzelle et al. 2009).

However, while staying with kin provides benefits through kin cooperation, it also increases kin competition (i.e. the reduced indirect fitness resulting from competition between relatives), which can negate the benefits of kin cooperation (Hamilton and May 1977; Taylor 1992; Queller 1994). Therefore, individuals should avoid fixed dispersal strategies that depend only on the overall population density, and instead they should adjust dispersal as a function of information about the identity of others. For instance, a cooperative individual will benefit from staying with kin, while it should leave habitats composed of unrelated individuals.

14.4 Effects of the Quality of Social Information on Cooperation and Dispersal

In addition to their presence, individuals often provide additional information linked to their identity and this information can in turn influence survival, cooperation, and dispersal. For example, individuals differ in their genetic makeup, their current condition, and their recent or past experiences, which can all influence their phenotype and potentially some aspects of biocommunication. This sort of information can be presented inadvertently and still be used by conspecifics to modulate their behaviour, or it can be expressly produced to specifically encourage a response in receivers. In this section, we focus on two types of biocommunication, local and indirect, that depend on the identity and condition of individuals (i.e. their *quality*), and describe how these two types of information influence the other individuals in a population.

14.4.1 Quality/Condition of Individuals in a Population

The current health or condition of an individual is generally thought to have an important influence on its behaviour in most organisms, and we expect the same to be true in ciliates. Indeed, a well-known example of this in *Tetrahymena* is that expression of mating type and sexual reproduction can be stimulated under starvation (Collins 2012). This should have the practical effect of increasing gregariousness and cooperation among individuals under such conditions even if reduced food should also increase competition, although this possibility has not yet been tested. Likewise, dispersal is dependent upon cell current condition since starvation is known to induce the production of flagellated dispersal morphs as explained before in this chapter (Nelsen and Debault 1978). Indeed, we might predict that condition should influence most decisions a cell makes including cooperation, foraging, asexual reproduction and dispersal.

While the effect of an individual's condition on its behaviour seems obvious, the condition of other individuals in a population is less straightforward, and yet could similarly influence behaviour. Indeed, the social environment is known to influence cooperation (Hatchwell 2009) and dispersal behaviour (Léna et al. 1998) in a broad

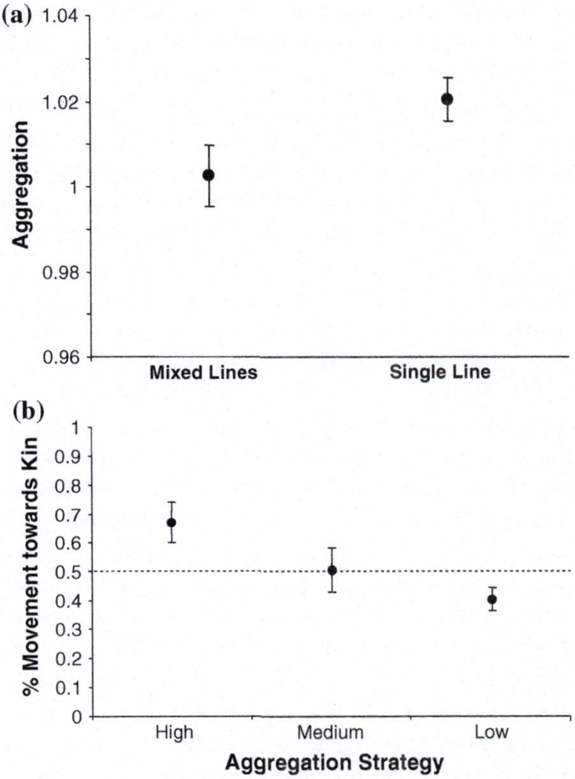

Fig. 14.3 Kinship, cooperation and dispersal in *Tetrahymena thermophila*. **a** Aggregation level decreases with relatedness, i.e. when clonal lines are mixed. **b** Cooperative clonal lines preferentially move toward kin, while non-cooperative lines are repulsed by kin (figures from Chaine et al. 2010)

variety of species, and we might expect that this should occur in ciliates as well. For example, cooperative aggregation behaviour in *Tetrahymena thermophila* is modulated depending on the average genetic relatedness of the population (Chaine et al. 2010; Fig. 14.3). Because cooperation is linked to kinship in this species (Chaine et al. 2010), we might predict that kinship within a population may also influence dispersal decisions (Hamilton and May 1977, see below). Most natural populations probably consist of a mix of individuals with different genotypes (Doerder et al. 1995; Zufall et al. 2013) and different current condition, and this variation could influence cooperation and dispersal behaviour. Critical to this prediction, though, is that some form of biocommunication occurs, so that individuals can detect the quality or character of individuals around them.

There is extensive evidence from the experimental literature that ciliates both produce and recognize substances that could permit biocommunication regarding individual quality (reviewed in Csaba 2012), and that such communication modulates behaviour. Moreover, the production of these molecules can vary according to the quality of individuals. For example, the production of insulin increases when cells are starved (Csaba et al. 2007) and it is likely that some insulin ends up in the environment either passively across the membrane (Weide et al. 2006; Madinger et al. 2010) or when cells die (Saitoh and Asai 1980) and therefore insulin concentration could provide information about the relative quality of other individuals in the population. Furthermore, *Tetrahymena* have receptors to identify most of the above molecules (Christopher and Sundermann 1992; Csaba 1994, 2012; Christensen et al. 2003). In other words, cells produce signal molecules (Csaba 2012; see also Fels 2009 for an interesting example with photons in *Paramecium*) and can use that information that is distributed in the environment, whether it comes from the same species or other sources (Wheatley et al. 1994).

The use of biological information from the environment can have important fitness consequences and so should influence behaviour and movement of organisms. For example, in a classic set of studies, Saitoh and Asai (1980) and Rassmussen and colleagues (Wheatley et al. 1993b; Schousboe and Rasmussen 1994; Christensen et al. 1995, 1996; Rasmussen et al. 1996) showed that exogenous molecules produced by *Tetrahymena* cells can help stabilize or increase growth in other stressed cells, or conversely reduce cell growth at higher densities (Christensen et al. 1996; reviewed in Wheatley et al. 1993a). At a finer scale, average relatedness of a population influences growth rates of cooperative genotypes, but have no effect on non-cooperative genotypes (Chaine et al. 2010). If individuals of different quality release different concentrations of these molecules, then we might expect cells to react to signal molecules in some way to maximize their fitness. Indeed, there is evidence that some of these exogenous signal molecules influence the behaviour of receiving cells: insulin in the environment decreases velocity in *Tetrahymena* (Kovács et al. 1994) and this shift in behaviour could aid in growth and survival of these cells in such an insulin-rich environment (Christensen et al. 1996). Furthermore, *Tetrahymena* shows chemotactic behaviour to a very broad range of molecules (Almagor et al. 1981; Hellung-Larsen et al. 1990; Leick et al. 1994) and this behaviour can be chemo-attractive (Hellung-Larsen et al. 1986; Wheatley et al. 1994; Leick et al. 1996) or chemo-repulsive (Francis and Hennessey 1995; Köhidai and Csaba 1996). In other words, cells likely move towards environments with beneficial exogenous molecules and move away from environments containing harmful molecules. Indeed, we showed that *Tetrahymena* cells can use molecules providing information about average genotype or social character of an environment to orient movement: cooperative cell lines preferentially moved towards habitat patches that had contained a kin population, whereas uncooperative lines moved away from kin (Chaine et al. 2010; Fig. 14.3). Therefore *Tetrahymena* appears to use intercellular chemical signals, produced either passively or actively, representing the quality of individuals in a population to alter cooperative behaviour and dispersal to improve fitness.

14.4.2 Social Information from the Landscape

If individuals exude information about their quality, then recent immigrants should provide information about surrounding populations (Greene 1987; Cote and Clobert

2007; Chaine et al. 2013). This notion has received very little experimental attention with one example in a lizard (Cote and Clobert 2007) and one test in *Tetrahymena* ciliates (Jacob et al. 2015b). Despite the limited number of tests, indirect information could be an important form of biocommunication in ciliates. *Tetrahymena* cells provide information about their genetic character or current condition which could be useful to other cells (see above) and if immigrants are recognizable (e.g. through flagella, Nelsen and Debault 1978; or movement behaviour, Schtickzelle et al. 2009), then the quality of individuals could provide information about surrounding populations. This information would be especially useful in species that live in patchy environments or have a sharp quality gradient and/or ephemeral environments, where movement to new, unknown patches might be beneficial.

In our work, we tested a number of different sources of information that could be carried by immigrants and that we hypothesized might influence dispersal behaviour. Of the five possible sources of information (density, number, food abundance, difficulty of dispersing, and cooperation propensity) that immigrants could have provided in our experiments, three were used to alter dispersal behaviour of residents (Jacob et al. 2015b). Interestingly, the information that was used (food abundance, social strategy, and difficulty of dispersing) seem to be more complex information than those that were not (population density, number of immigrants). This pattern strongly suggests that *Tetrahymena* has access to biological information in most cases, but only pays attention to information that might be of value to dispersal decisions. In our experiments, we also detected among strain variation (i.e. genetic variation, G) for how indirect information from immigrants was used, suggesting that some genetic lines might either be better at detecting and using information or that the benefit of using information differs for some genotypes relative to others (Jacob et al. 2015b). This demonstration of indirect information use in *Tetrahymena*, combined with the above argument for the abundance and importance of information transmitted between cells, suggests that the use of indirect information could be widespread in ciliates and could provide fruitful avenues for future research.

14.5 Conclusions and Future Directions

Information use in decision-making processes has major implications for a large variety of ecological and evolutionary processes in a large variety of organisms (Danchin 2004; Dall et al. 2005; Seppänen et al. 2007; Schmidt et al. 2010). As we reviewed here, *Tetrahymena* ciliates do not bypass this rule, since chemical biocommunication is for instance involved in the induction of sexual reproduction, population survival and growth, movement behaviours and investment in cooperation. Such evidence paves the way for experimental research about the ecological and evolutionary significance of informed behavioural processes. *Tetrahymena* microcosms indeed provide the opportunity to study all aspects of information use, from the chemical mechanisms carrying information, its consequences on metapopulation and metacommunity dynamics, and to the eco-evolutionary dynamics of information use. Indeed, protist microcosms offer the possibility to study variability in individual behaviours, their underlying mechanisms, how these behaviours turn out into fitness, and their consequences for population dynamics and evolutionary changes within relatively short spatial and temporal scales (Altermatt et al. 2015).

We have argued that, in addition to information gathered about the environment, organisms also adjust their behaviours based on their own condition or quality, the quality of others in their neighbourhood, and information provided by others about quality of distant populations. Very few direct tests of these notions exist in the ciliate literature, which essentially provides support for the general principles. Clearly, formal validations are now needed on these topics, and we highlight three main points here that we feel of particular interest. The first point is to understand how variation in the quality and condition of other individuals in the local environment influences behaviour. Most researchers work with homogenous, single-strained populations, but real populations probably contain mixes of individuals that differ in quality. Second, we need to improve our understanding of how individuals use direct and/or indirect information, e.g. from immigrants, to make decisions. Indirect information can come from many different origins, and could be of poor predictive power for example if non-local conditions are temporally or spatially variable. Ciliates provide an excellent model to push this field forward. Finally, information gathered from the quality of other individuals is context-specific, so interpretation of experiments could critically depends upon the natural environment in which strains evolved before laboratory use. As a result, we need to be aware that although ciliates provide a tractable model for exploring questions about information use in decision-making, a nearly complete lack of knowledge on their natural populations may biased our interpretation. We think that improving our knowledge on natural distributions, genetic structuring, and stability of populations (Doerder et al. 1995; Zufall et al. 2013) would greatly enhance the value of experiments in ciliate biocommunication.

Tetrahymena microcosms provide an excellent opportunity to develop our understanding of the implications of information use in eco-evolutionary dynamics. Manipulating the variability in cell condition and information use strategies and following their fitness in experimental systems with highly controlled environmental conditions will help testing theoretically derived predictions and therefore understanding the impact of information on eco-evolutionary processes. For instance, such an experimental setup can allow identification of which type of information is used (i.e. social or non-social), the effect of own-cell condition, the mechanisms behind decision-making processes, and how much the evolution of information use itself and the type of information taken into account are affected by the environmental context (e.g. quality of habitats, habitat loss, global warming, fragmentation). We therefore encourage evolutionary ecologists to use ciliate models in multidisciplinary studies covering all different levels of complexity related to information use, from the genetic and chemical mechanisms behind information to its consequences for population dynamics and evolution.

References

Almagor M, Ron A, Bar-Tana J (1981) Chemotaxis in *Tetrahymena thermophila*. Cell Motil Cytoskelet 1:261–268

Altermatt F, Fronhofer EA, Garnier A et al (2015) Big answers from small worlds: a user's guide for protist microcosms as a model system in ecology and evolution. Methods Ecol Evol 6:218–231. doi:10.1111/2041-210X.12312

Baguette M, Clobert J, Schtickzelle N (2011) Metapopulation dynamics of the bog fritillary butterfly: experimental changes in habitat quality induced negative density-dependent dispersal. Ecography 34:170–176. doi:10.1111/j.1600-0587.2010.06212.x

Bowler DE, Benton TG (2005) Causes and consequences of animal dispersal strategies: relating individual behaviour to spatial dynamics. Biol Rev 80:205–225. doi:10.1017/S1464793104006645

Chaine AS, Clobert J (2012) Dispersal in a changing world. In: Wong BBM, Candolin U (eds) Behavioural responses to a changing world. Oxford University Press, Oxford, UK, pp 63–79

Chaine AS, Schtickzelle N, Polard T et al (2010) Kin-based recognition and social aggregation in a ciliate. Evolution 64:1290–1300

Chaine AS, Legendre S, Clobert J (2013) The co-evolution of multiply-informed dispersal: information transfer across landscapes from neighbors and immigrants. PeerJ 1:e44

Christensen S, Rasmussen L (1992) Evidence for growth factors which control cell multiplication in *Tetrahymena thermophila*. Acta Protozool 31:215–219

Christensen ST, Wheatley DN, Rasmussen MI, Rasmussen L (1995) Mechanisms controlling death, survival and proliferation in a model unicellular eukaryote *Tetrahymena thermophila*. Cell Death Differ 2:301–308

Christensen ST, Kemp K, Quie H, Rasmussen L (1996) Cell death, survival and proliferation in *Tetrahymena thermophila*. effects of insulin, sodium nitroprusside, 8-bromo cyclic GMP, NG-methyl-L-arginine and methylene blue. Cell Biol Int 20:653–666

Christensen ST, Guerra CF, Awan A et al (2003) Insulin receptor-like proteins in *Tetrahymena thermophila* ciliary membranes. Curr Biol CB 13:R50–R52

Christopher GK, Sundermann CA (1992) Conventional and confocal microscopic studies of insulin receptor induction in *Tetrahymena pyriformis*. Exp Cell Res 201:477–484

Clobert J, Le Galliard J-F, Cote J et al (2009) Informed dispersal, heterogeneity in animal dispersal syndromes and the dynamics of spatially structured populations. Ecol Lett 12:197–209. doi:10.1111/j.1461-0248.2008.01267.x

Clobert J, Baguette M, Benton TG, Bullock J (2012) Dispersal ecology and evolution. Oxford University Press

Collins K (2012) *Tetrahymena thermophila*. Elsevier

Connolly JG, Brown ID, Lee AG, Kerkut GA (1985) Temperature-dependent changes in the swimming behaviour of Tetrahymena pyriformis-NT1 and their interrelationships with electrophysiology and the state of membrane lipids. Comp Biochem Physiol A Comp Physiol 81:303–310

Cote J, Clobert J (2007) Social information and emigration: lessons from immigrants. Ecol Lett 10:411–417

Courchamp F, Berec L, Gascoigne J (2008) Allee effects in ecology and conservation. Environ Conserv 36:80–85

Cox AS, Kesler DC (2012) Prospecting behavior and the influence of forest cover on natal dispersal in aresident bird. Behav Ecol 23:1068–1077. doi:10.1093/beheco/ars076

Csaba G (1994) Phylogeny and ontogeny of chemical signaling: origin and development of hormone receptors. Int Rev Cytol 155:1–48

Csaba G (2012) The hormonal system of the unicellular *Tetrahymena*: a review with evolutionary aspects. Acta Microbiol Immunol Hung 59:131–156. doi:10.1556/AMicr.59.2012.2.1

Csaba G, Kovács P (1999) Localization of beta-endorphin in tetrahymena by confocal microscopy. Induction of the prolonged production of the hormone by hormonal imprinting. Cell Biol Int 23:695–702. doi:10.1006/cbir.1999.0437

Csaba G, Kovács P, Pállinger É (2007) Effect of starvation on insulin production and insulin binding in *Tetrahymena*. Cell Biochem Funct 25:473–477. doi:10.1002/cbf.1333

Dall S, Giraldeau L, Olsson O et al (2005) Information and its use by animals in evolutionary ecology. Trends Ecol Evol 20:187–193. doi:10.1016/j.tree.2005.01.010

Danchin E (2004) Public information: from nosy neighbors to cultural evolution. Science 305:487–491. doi:10.1126/science.1098254

Darch SE, West SA, Winzer K, Diggle SP (2012) Density-dependent fitness benefits in quorum-sensing bacterial populations. Proc Natl Acad Sci 109:8259–8263. doi:10.1073/pnas.1118131109

Doerder FP, Gates MA, Eberhardt FP, Arslanyolu M (1995) High frequency of sex and equal frequencies of mating types in natural populations of the ciliate *Tetrahymena thermophila*. Proc Natl Acad Sci 92:8715–8718

Doligez B, Pärt T, Danchin E et al (2004) Availability and use of public information and conspecific density for settlement decisions in the collared flycatcher. J Anim Ecol 73:75–87

Edelaar P, Siepieski AM, Clobert J (2008) Matching habitta choice causes directed gene flow: a neglected dimension in evolution and ecology. Evolution 62:2462–2472

Fellous S, Duncan A, Coulon A, Kaltz O (2012) Quorum sensing and density-dependent dispersal in an aquatic model system. PLoS ONE 7:e48436. doi:10.1371/journal.pone.0048436

Fels D (2009) Cellular communication through light. PLoS ONE 4:e5086. doi:10.1371/journal.pone.0005086

Fjerdingstad EJ, Schtickzelle N, Manhes P et al (2007) Evolution of dispersal and life history strategies—*Tetrahymena* ciliates. BMC Evol Biol 7:133. doi:10.1186/1471-2148-7-133

Francis JT, Hennessey TM (1995) Chemorepellents in *Paramecium* and *Tetrahymena*. J Eukaryot Microbiol 42:78–83

Fronhofer EA, Hovestadt T, Poethke HJ (2013) From random walks to informed movement. Oikos 122:857–866

Fronhofer EA, Altermatt F (2015) Eco-evolutionary feedbacks during experimental range expansions. Nat Commun 6:6844. doi:10.1038/ncomms7844

Fronhofer EA, Kropf T, Altermatt F (2015) Density-dependent movement and the consequences of the Allee effect in the model organism *Tetrahymena*. J Anim Ecol 84:712–722. doi:10.1111/1365-2656.12315

Gandon S, Michalakis Y (1999) Evolutionarily stable dispersal rate in a metapopulation with extinctions and kin competition. J Theor Biol 199:275–290

Greene E (1987) Individuals in an osprey colony discriminate between high and low quality information. Nature 329:239–241

Griffin AS, West SA (2002) Kin selection: fact and fiction. Trends Ecol Evol 17:15–21

Hamilton WD (1964) The genetical evolution of social behaviour. I. J Theor Biol 7:1–16

Hamilton WD, May RM (1977) Dispersal in stable habitats. Nature 269:578–581

Hatchwell BJ (2009) The evolution of cooperative breeding in birds: kinship, dispersal and life history. Philos Trans R Soc B Biol Sci 364:3217–3227. doi:10.1098/rstb.2009.0109

Hellung-Larsen P, Leick V, Tommerup N (1986) Chemoattraction in *Tetrahymena*: on the role of chemokinesis. Biol Bull 170:357–367

Hellung-Larsen P, Leick V, Tommerup N, Kronborg D (1990) Chemotaxis in *tetrahymena*. Eur J Protistol 25:229–233. doi:10.1016/S0932-4739(11)80174-4

Jacob S, Bestion E, Legrand D, Clobert J, Cote J (2015a) Habitat matching and spatial heterogeneity of phenotypes: implications for metapopulation and metacommunity functioning. Evol Ecol 29:851–871

Jacob S, Chaine AS, Schtickzelle N et al (2015b) Social information from immigrants: multiple immigrant-based sources of information for dispersal decisions in a ciliate. J Anim Ecol 84:1373–1383. doi:10.1111/1365-2656.12380

Kim S-Y, Torres R, Drummond H (2009) Simultaneous positive and negative density-dependent dispersal in a colonial bird species. Ecology 90:230–239

Köhidai L, Csaba G (1995) Effects of the mammalian vasoconstrictor peptide, endothelin-1, on Tetrahymena pyriformis GL, and the immunocytological detection of endogenous endothelin-like activity. Comp Biochem Physiol C: Pharmacol Toxicol Endocrinol 111:311–316

Köhidai L, Csaba G (1996) Different and selective chemotactic responses of *Tetrahymena pyriformis* to two families of signal molecules: lectins and peptide hormones. Acta Microbiol Immunol Hung 43:83–91

Koutna M, Janisch R, Unucka M, Svobodnik A, Mornstein V (2004) Effects of Low-power Laser Irradiation on Cell Locomotion in Protozoa. Photochem Photobiol 80:531–534

Kovács P, Lovas G, Csaba G (1994) Influence of insulin on the movement of *Tetrahymena pyriformis*. Hormonal imprinting alters the velocity. Comp Biochem Physiol A Physiol 107:375–379. doi:10.1016/0300-9629(94)90395-6

Le Roith D, Shiloach J, Berelowitz M et al (1983) Are messenger molecules in microbes the ancestors of the vertebrate hormones and tissue factors? Fed Proc 42:2602–2607

Legrand D, Trochet A, Moulherat S et al (2015) Ranking the ecological causes of dispersal in a butterfly. Ecography 38:822–831

Leick V, Koppelhus U, Rosenberg J (1994) Cilia-mediated oriented chemokinesis in *Tetrahymena thermophila*. J Eukaryot Microbiol 41:546–553

Leick V, Grave M, Hellung-Larsen P (1996) Signal peptide-induced sensory behavior in free ciliates: bioassays and cellular mechanisms. In: Csaba PDG, Müller PDWEG (eds) Signaling mechanisms in protozoa and invertebrates. Springer, Berlin Heidelberg, pp 61–79

Léna J-P, Clobert J, de Fraipont M et al (1998) The relative influence of density and kinship on dispersal in the common lizard. Behav Ecol 9:500–507

Leroith D, Liotta AS, Roth J et al (1982) Corticotropin and beta-endorphin-like materials are native to unicellular organisms. Proc Natl Acad Sci USA 79:2086–2090

Madinger CL, Collins K, Fields LG et al (2010) Constitutive secretion in *Tetrahymena thermophila*. Eukaryot Cell 9:674–681. doi:10.1128/EC.00024-10

Nelsen E, Debault L (1978) Transformation in *Tetrahymena pyriformis*: description of an inducible phenotype. J Protozool 25:113–119

Patterson T, Thomas L, Wilcox C et al (2008) State–space models of individual animal movement. Trends Ecol Evol 23:87–94. doi:10.1016/j.tree.2007.10.009

Pennekamp F, Mitchell KA, Chaine A, Schtickzelle N (2014) Dispersal propensity in *Tetrahymena thermophila* ciliates-a reaction norm perspective: dispersal reaction norms of *T. thermophila* ciliates. Evolution n/a–n/a. doi:10.1111/evo.12428

Poethke HJ, Pfenning B, Hovestadt T et al (2007) The relative contribution of individual and kin selection to the evolution of density-dependent dispersal rates. Evol Ecol Res 9:41–50

Queller DC (1994) Genetic relatedness in viscous populations. Evol Ecol 8:70–73

Rasmussen MI, Wheatley DN (2007) Purification and characterisation of cell survival factor 1 (TCSF1) from *Tetrahymena thermophila*. J Cell Commun Signal 1:185–193. doi:10.1007/s12079-007-0016-9

Rasmussen L, Christensen ST, Schousboe P, Wheatley DN (1996) Cell survival and multiplication. The overriding need for signals: from unicellular to multicellular systems. FEMS Microbiol Lett 137:123–128

Reed JM, Boulinier T, Danchin E, Oring LW (1999) Informed dispersal: prospecting by birds for breeding sites. In: Nolan V, Ketterson ED, Thompson CF (eds) Current ornithology. Springer, US, pp 189–259

Saitoh T, Asai H (1980) Growth-controlling factors produced during the logarithmic and stationary phases of *Tetrahymena pyriformis*. Experientia 36:1375–1376

Schmidt KA, Dall SRX, van Gils JA (2010) The ecology of information: an overview on the ecological significance of making informed decisions. Oikos 119:304–316. doi:10.1111/j.1600-0706.2009.17573.x

Schousboe P, Rasmussen L (1994) Survival of *Tetrahymena thermophila* at low initial cell densities. Effects of lipids and long-chain alcohols. J Eukaryot Microbiol 41:195–199

Schtickzelle N, Fjerdingstad E, Chaine AS, Clobert J (2009) Cooperative social clusters are not destroyed by dispersal in a ciliate. BMC Evol Biol 9:251

Seppänen J-T, Forsman JT, Mönkkönen M, Thomson RL (2007) Social information use is a process across time, space, and ecology, reaching heterospecifics. Ecology 88:1622–1633

Shiurba R, Hirabayashi T, Masuda M, Kawamura A, Komoike Y, Klitz W, Kinowaki K, Funatsu T, Kondo S, Kiyokawa S, Sugai T, Kawamura K, Namiki H, Higashinakagawa T (2006) Cellular responses of the ciliate, Tetrahymena thermophila, to far infrared irradiation. Photochem Photobiol Sci 5:799–807

Straarup E, Schousboe P, Quie H et al (1997) Effects of protein kinase C activators and staurosporine on cell survival, proliferation and protein kinase activity in *Tetrahymena thermophila*. Microbios 91:181–190

Taylor PD (1992) Altruism in viscous populations—an inclusive fitness model. Evol Ecol 6:352–356

Weide T, Herrmann L, Bockau U et al (2006) Secretion of functional human enzymes by *Tetrahymena thermophila*. BMC Biotechnol 6:19. doi:10.1186/1472-6750-6-19

West SA, Griffin AS, Gardner A (2007) Evolutionary explanations for cooperation. Curr Biol 17: R661–R672

Wheatley DN, Christensen S, Schousboe P, Rasmussen L (1993a) Signaling in growth and death: adequate nutrition alone may not be sufficient for ciliates, a minireview. Cell Biol Internatl 17:817–823

Wheatley DN, Christensen ST, Schousboe P, Rasmussen L (1993b) Signalling in cell growth and death: adequate nutrition alone may not be sufficient for ciliates. A minireview. Cell Biol Int 17:817–823. doi:10.1006/cbir.1993.1143

Wheatley DN, Rasmussen L, Tiedtke A (1994) My favourite cell: *Tetrahymena*: a model for growth, cell cycle and nutritional studies, with biotechnological potential. BioEssays 16:367–372. doi:10.1002/bies.950160512

Williams P, Winzer K, Chan W, Camara M (2007) Look who's talking: communication and quorum sensing in the bacterial world. Philos Trans R Soc B Biol Sci 362:1119–1134

Zufall RA, Dimond KL, Doerder FP (2013) Restricted distribution and limited gene flow in the model ciliate *Tetrahymena thermophila*. Mol Ecol 22:1081–1091

Chapter 15
Symbiotic Associations in Ciliates: Ecological and Evolutionary Perspectives

Arno Germond and Toshiyuki Nakajima

Abstract Ciliates can develop facultative or permanent symbiotic associations with other species. The ecological success of mixotrophic symbiosis in ciliates lead us to question the physiological aspects and the evolutionary processes by which such associations can emerge, be maintained, and evolve. We highlight the symbiotic interactions between ciliated protozoan and other organisms, and discuss the general and essential aspects of these associations from the ecological and evolutionary perspectives. We first overview the ecological role of mixotrophic protists and the major types of symbiotic associations in aquatic environments (Sects. 15.2 and 15.3). Subsequently, we review the mechanisms employed for partner recognition and selective acquisition, for the infection process of symbiotic cells, and the physiological regulation used by the ciliate hosts to control or exploit these symbionts (Sect. 15.4). Lastly, we introduce a non-reductionist approach, by using an experimental synthetic ecosystem, which focuses on how ecosystem processes affect the emergence and evolution of symbiotic associations (Sect. 15.5).

15.1 Introduction

Protists occupy a wide range of environmental niches on earth and have high dispersability. This ecological success is partially explained by their capacity to develop facultative or permanent symbiotic associations with other species. These associations are characterized by their mixotrophic nature. Mixotrophy is a mix of

A. Germond
RIKEN, Laboratory for Comprehensive Bioimaging,
Quantitative Biology Center (QBiC), Osaka, Japan

T. Nakajima (✉)
Department of Biology, Ehime University, Ehime Prefecture, Japan
e-mail: nakajima@sci.ehime-u.ac.jp

G. Witzany and M. Nowacki (eds.), *Biocommunication of Ciliates*,
DOI 10.1007/978-3-319-32211-7_15

the heterotrophic and auto-(photo-)trophic nutrition, and the contribution of each nutritional mode can vary, depending on the associated organisms as well as the environmental factors (Jones 1994). Through mixotrophy, protist symbioses are considered to play a major role in biogeochemical cycling and trophic transfers in all aquatic environments (Esteban et al. 2010; Caron et al. 2012).

Although the existence of mixotrophic protists has been known for a century or so, the interest in them has drawn much attention only in the past couple of decades. As they can be easily maintained in laboratory cultures, model organisms of ciliate species have been used as a tool to test the ecological theories, such as prey-predator relationships. Later, it was demonstrated that mixotrophic associations have evolved independently in a large variety of taxonomic groups and species through secondary or tertiary symbiogenesis. Mixotrophic protists are a good model of symbiotic associations, through which plastids originated in eukaryotic organisms. Thus, ciliated organisms provide useful tools to analyze the physiological, ecological, and evolutionary processes of symbiotic associations.

In this chapter, we highlight the general and essential aspects of symbiotic interactions of ciliated protozoan with other organisms such as archaea, bacteria, and algae. In the first section, we briefly review the ecological role and importance of mixotrophic protists as trophic links in the aquatic food webs. In the second section, we present and provide examples of the major types of associations involving mixotrophic ciliates. For instance, ciliated organisms are known to develop mutualistic relationships with bacteria and photosynthetic microorganisms such as unicellular green algae (zoochlorellae), dinoflagellates (zooxanthellae), and cryptomonads. Ciliates can also sequester chloroplasts from ingested algal prey. Finally, ciliates are known as symbionts in the complex microflora of the gastrointestinal tracts of metazoans. In the third section, we describe the mechanisms for partner recognition and specificity, the mechanisms for the infection success of symbiotic cells, and the physiological regulatory mechanisms used by ciliate hosts to control or exploit symbionts. In the last section, we focus on the evolutionary ecological aspect of symbiogenesis. Here, we introduce an experimental approach to the algal-ciliate symbiotic associations by employing an experimental synthetic ecosystem. Along the text, we propose some perspectives on the studies for future investigations, in order to give the biologist the opportunity to make significant contributions to the understanding of mixotrophic symbioses in ciliates.

15.2 Ecological Importance of Aquatic Mixotrophic Ciliates

Mixotrophic ciliates are common in all aquatic habitats around the world and their distribution is controlled by dispersion and adaptation to the environmental heterogeneity. They can be found from equatorial to polar regions, in freshwater and seawater, in coastal or pelagic waters, in hydrothermal vents, anoxic sediments, hyporheic zones, and oxic as well as anoxic parts in eutrophic or euryhaline waters

(Andrushchyshyn et al. 2007; Burkholder et al. 2008; Stoecker 1999; Stoecker et al. 2009). In particular, the phagotrophic pelagic organisms are quantitatively as well as qualitatively dominated by flagellates and ciliates.

By endosymbiosis, protists organisms acquire novel metabolic functions. In the past 20 years, increasing evidence have suggested that mixotrophic protists play a crucial role in the recycling and flux exchanges of carbon, nitrogen, methane, sulfur, and other chemical compounds in aquatic natural ecosystems (Capone 2000; Caron et al. 2012). Specifically, inorganic carbon is obtained from the photosynthetic activities of protist symbionts (Jonsson 1987). Ciliated protists are also involved in the transfer of nitrogen in aquatic environments through their symbiotic interactions with nitrogen-fixing prokaryotes (Kneip et al. 2007). In addition, ciliates can also associate with methanogenic and sulfur-oxidizing symbionts.

Participating as both producers and consumers, mixotrophic ciliates are important intermediates between the decomposers (bacteria), phototrophic organisms (phytoplankton, algae), and metazoans. In particular, aquatic ciliates are important grazers of algae, bacteria, and other microorganisms. Thus, they are an important ecological link in aquatic ecosystems (Caron et al. 2012). For these reasons, aquatic ciliates and other protists are seen as crucial contributors of the biodiversity of the aquatic environment. Surveys are being conducted in situ in the water column across diverse locations of the world ocean to measure the biomass and productivity of the protistan communities (e.g., Carvalho and Granéli 2006; Ferhman 2009). However, it is difficult to obtain realistic estimates of the true contribution of mixotrophic associations, considering their diversity and the wide range of environments they inhabit.

A challenge for future studies will be to model and assess more precisely how the genetic composition and dynamics of mixotrophic communities affect the biogeochemical cycling and other ecological processes. To this end, new investigation methods that include both observational and experimental approaches can help in the assessment of predictable patterns (e.g., Carvalho and Granéli 2006; Ferhman 2009). In particular, low-cost, high-throughput sequencing can greatly advance the analysis of marine microbial community structure (Ferhman 2009). Other approaches are based on the use of mathematical models. In this context, a recent study by Mitra et al. (2014) suggested that in order to prevent a misinterpretation of the effects of climate change on biogeochemical cycling and the functioning of the biological pump, the inclusion of multi-nutrient mixotroph models within ecosystem studies is strongly desirable.

15.3 Diversity of Mixotrophic Associations in Ciliates

With more than 250 species of ciliates harboring symbionts, and the possibility for hosts to be inhabited by one or more genotypes, the diversity of mixotrophic protists is more complex than initially thought (Stoecker et al. 2009). Mixotrophic ciliates can harbor intracellular archaea, bacteria, or algae. They can also retain functional chloroplasts, or act as endosymbionts in the intestinal tracts of animals.

Overall, these associations of ciliates encompass three major types of nutrition, namely: phototrophic, chemosynthetic, and heterotrophic (Dziallas et al. 2012). Here, we review the different types of associations known in symbiotic ciliates. Although these interactions may appear to be highly specific and conserved (i.e., stable), recent studies demonstrated that the symbiogenesis may rather be a highly dynamic process (Dziallas et al. 2012; Pierce and Turner 1992). In Table 15.1, we present the major symbiotic associations involving ciliates hosts. The readers are requested to refer to the cited literature for further information.

Table 15.1 Major symbioses in ciliates (Ciliophora) living in Marine (Ma), Freshwater (Fr) environments, or obtained in experimental cultures (EC)

Host	Habitat	Symbionts or plastids (and their produced compounds, if known)	References
Acaryophyra sp.	Fr	*Chlorella*-like	Berninger et al. (1986), Finlay et al. (2006)
Climacostomum virens	Fr	*Chlorella*-like (Glucose, Fructose, Xylose)	Reisser and Kurmeier (1984), Reisser (1986)
Coleps hirtus	Fr	*Chlorella*-like (Xylose)	Reisser and Kurmeier (1984), Reisser (1986), Christopher and Patterson (1983), Klaveness (1984)
Cyclidium porcatum	Ma	Methanogenic archaea	Fenchel and Finlay (1991)
Disematostoma butschlii	Fr	*Chlorella*-like	Berninger et al. (1986)
Euplotes daidaleos	Fr	*Chlorella*-like (Fructose, Xylose)	Reisser and Kurmeier (1984), Reisser (1986)
Euplotes sp.		*Polynucleobacter necessarius*	
Frontonia leucas	Fr	*Chlorella*-like	Berninger et al. (1986), Sud (1968)
Holosticha viridis	Fr	*Chlorella*-like	Sud (1968)
Malacophrys sphagni	Fr	*Chlorella*-like	Kawakami and Kawakami (1985)
Maristentor dinoferus	Ma	Dinoflagellate *Symbiodinium*	Lobban et al. (2002)
Mesodinium rubrum		Dinophlagellates, or chloroplast from Cryptophytes	Bakker (1967), cited in Crawford (1989)
Ophrydium versatile	Fr	*Chlorella*-like	Kahl (1930), Seckbach (1994)
Paramecium bursaria	Fr	*Chlorella vulgaris* (Glucose, Maltose) *Micractinium reisseiri*	Reisser and Kurmeier (1984), Reisser (1986), Hoshina et al. (2005)

(continued)

Table 15.1 (continued)

Host	Habitat	Symbionts or plastids (and their produced compounds, if known)	References
Paramecium sp.	Fr	*Holospora*, *Caedibacter*	
Prorodon viridis	Fr	*Chlorella*-like	Sud (1968)
Prorodon ovum	Fr	*Chlorella*-like	Sud (1968)
Psilotricha viridis			Kahl (1930)
Spirostomim viridis	Fr	*Chlorella*-like	Seckbach (1994)
Stentor niger	Fr	*Chlorella*-like	Kawakami (1984)
Stentor polymorphus	Fr	*Chlorella*-like (Maltose)	Reisser and Kurmeier (1984), Reisser (1986)
Stentor roeseli		*Chlorella*-like	Kahl (1930)
Strombidium sp.	Ma,Fe	*Chloroplast*	
Strombidium acutum (S. Rassoulzadegani)	Ma	*Chloroplast* from Chlorophycea	Laval-Peuto and Rassoulzadegan (1988), Stoecker et al. (1988–1989)
Tutorials tricycle	Fr	*Chlorella*-like	Seckbach (1994)
Tetrahymena thermophila	EC	*Micractinium* sp. (Sucrose, Glycerol)	Nakajima et al. (2009), Germond et al. (2013a, b)
Tontonia spp.	Ma	Chloroplast from Chromophyte alga	Laval-Peuto and Febvre (1986), Laval-Peuto and Rassoulzadegan (1988)
Trimyema sp.	Ma	Methanogenic archaea, *Methanobacterium formicicum*	Wagener et al. (1990), Shinzato and Kamagata (2010)
Vorticella chlorellata	Fr	*Chlorella*-like	Graham and Graham (1978)
Vorticella sp.	Fr	*Chlorella*-like	Graham and Graham (1978)
Zoothamnium niveum	Ma	Chemoautotrophic bacteria	Hemprich and Ehrenberg (1831), cited in Rinke et al. (2006)

15.3.1 Anaerobic Ciliates as Host and Symbiont in the Gut Microbiota

Parasitic or mutualistic ciliates are found worldwide in anoxic aquatic sediments, and also as symbiotic species in the gut of many animals. An example of this is *Balantidium coli*, which is a parasitic species and the only member of the ciliate phylum known to be pathogenic to humans, causing the disease Balantidiasis. Parasitic species, like *B. coli*, often have complex life-cycles, which includes a free-living stage. To transfer from one host to another, ciliate parasites use several modes of transmission: direct transmission, feces-transmitted cysts, or vector-borne transmission through other organisms (through an infected insect, for example).

In addition to parasitic ciliates, mutualistic anaerobic ciliates are also known to inhabit the intestinal tracts of animal hosts, including insects, amphibian, and ruminants. Ruminants, in particular, possess obligate archaea, bacterial, and protozoal symbionts, of which anaerobic ciliated protozoa represent about 40 % of the total rumen biomass (Williams and Coleman 1991; Tannock 1995). Specifically, two major groups of mutualistic ciliates inhabit the rumen, the holotrichs and the entodiniomorphs. Within the intestinal tracts, ciliated protozoa ingest bacteria and small plant fragments. They digest starch and proteins in their food vacuoles and release volatile fatty acids, ammonia, hydrogen, amino acids, and sugars. Holotrichs can convert sugars into a starch-like compound. In return, these chemicals can be beneficial for various kinds of extracellular bacteria (methanogens, nitrogen-fixing or cellulose-degrading bacteria). In addition, anaerobic ciliates of the rumen can develop endosymbiotic associations, such as with methanogens, as detailed below. For example, *Dasytricha ruminantium* and *Entodinium* spp., found in the rumen of sheep, harbor methanogenic bacteria living in their cytoplasm (Finlay et al. 1994). Therefore, the dietary requirements of the host are dependent not only upon the ciliate population size, the outflow of protist biomass from the rumen, the availability of the cellular constituents in the lower digestive tract, but also on the co-interactions evolving between ciliates and their symbionts. Since they are specifically adapted to the host environment, parasitic and mutualistic ciliates that live as both host and symbionts in animal hosts are often hard to grow in experimental cultures. As a result, little is known about the effects of microbial community on the animal's physiology and/or immune system.

15.3.2 Marine Anaerobic Ciliates and Their Symbionts

Anaerobic ciliates adapt to the anoxic environments by relying on specific hydrogen-producing organelles called hydrogenosomes. The hydrogenosome is thought to have a unique symbiotic origin from mitochondria-derived organelles (Hackstein et al. 2004). It evolved repeatedly in 7 out of 22 ciliate taxa (Fenchel and Finlay 1995). The hydrogen wastes produced by the anaerobic host through the hydrogenosomes can be used as a metabolic resource by lithoautotrophic endosymbionts (e.g., methanogens) or by extracellular microorganisms, such as nitrogen-fixing prokaryotes (Fenchel and Finlay 1991).

Metopus contortus, *Trimyema* sp., and *Cyclidium porcatum* are well-known examples of ciliate hosts harboring symbiotic methanogenic archaea. In these symbiotic associations, it was suggested that the host benefited from its symbiotic partners in terms of growth rate and survival, and in turn controlled the life cycle of the symbionts. The symbiotic methanogens are transmitted vertically during the mitosis of the hosts (Fenchel and Finlay 1991; van Hoek et al. 2000), although there is a close relationship between free-living methanogenic archaea and endosymbiont, suggesting that horizontal transmission also happened several times in the evolutionary history of anaerobic ciliates (van Hoek et al. 2000; Hackstein 2010). For example, it was shown that *Methanobacterium formicicum*, symbiotic methanogens of *Trimyema compressum* could be lost in experimental culture conditions, can then be used for re-infection by symbiont-free ciliates (Wagener et al. 1990; Shinzato and Kamagata 2010). *Trimyema* does not require specific species of symbionts and contains different methanogenic species (Final et al. 1993). The biology and physiology of anaerobic ciliates were largely described by the work of Fenchel and Finlay and readers may refer to their work for further information (Fenchel and Finlay 1991, 1995, 2010).

15.3.3 Ciliates with Endosymbiotic Photo-Autotrophs

There exists a broad range of symbiotic interactions of ciliates with unicellular algae in freshwater as well as marine environments (Esteban et al. 2010). Endosymbionts are mostly eukaryotic and provide nutrients to the host through photosynthesis. In many cases, the host still relies on food uptake to live, however, when starved, the photosynthetic activity of the symbiotic cells supports and prolongs the survival of the hosts. This partially explains the ecological success of such mixotrophic associations in environments with limited food resources. In marine environments, endosymbiont phototrophs are largely represented by zooxanthellae algae (in fact, different genotypes of *Symbiodinium*) which inhabit the intracellular environment of invertebrates, radiolarians, foraminifera, and ciliates. However, there are only three ciliate species reported to harbor zooxanthellae: *Maristentor dinoferus*, *Paraeuplotes tortugensis*, and *Euplotes uncinatus* (Lobban et al. 2005). *Maristentor* is a large ciliate, which develops mutualistic associations with corals, and harbor 500–800 *Symbiodinium* symbionts per cell (Lobban et al. 2002). On the contrary, ciliates such as *Euplotes* can feed on coral tissues and acquire the coral's endosymbiotic dinoflagellates. The dinoflagellate symbionts usually living in corals were shown to be also photosynthetically active in ciliate organisms. Other symbionts of marine ciliates have also been reported, such as a sulfur purple bacterium capable of anoxygenic photosynthesis in *Strombidium purpureum* or a cyanobacterium species in *Codonella* (Esteban et al. 2010).

In freshwater, the dominant phototrophic type of symbiosis is represented by ciliate hosts harboring unicellular *Chlorella*-like green algae, or zoochlorellae. Recent molecular analyses lead to major modifications in the classification of green algae, in particular in the *Chlorella* genus (Hoshina et al. 2005; Pröschold et al. 2011). Consequently, symbiotic cells previously thought to be *Chlorella* algae may now belong to other genera (Hoshina et al. 2005; Germond et al. 2013a). *Chlorella*-like symbionts are found not only in ciliates, but also in freshwater invertebrates, such as *Chlorohydra* (e.g., *Hydra viridis*), and sponges. Symbiotic zoochlorellae likely evolved from free-living species, and can be transmitted from the host by horizontal or vertical transmission, as discussed in the last section of this chapter.

Algal-ciliate associations range from facultative to obligate and all phototrophic endosymbionts were shown to benefit the host through their photosynthetic activity. Moreover, depending on the environmental conditions or physiological needs, it is thought that hosts can optimize the nutrient production through behavioral (e.g., phototaxis) or physiological regulatory systems, as discussed in the next section.

15.3.4 Protists with Photosynthetic Organelles

Protist organisms may have functional chloroplasts of their own. Kleptochloroplastidy, also called 'kleptoplasty', or organelle retention, is the ability of a heterotrophic organism to sequester functional chloroplasts from algal prey (e.g., Gustafson et al. 2000). In many cases, the functional chloroplasts originate from cryptophytes and haplophytes organisms (Jones 1994), and must be constantly renewed. The frequency of renewal varies among different species and environmental conditions, and chloroplast may last from a few hours to a few days within the host (Dolan 1992). Often, the chloroplast-bearing protists are still dependent of another source of food besides light, for growth. Therefore, the mixotrophic regime varies from phototrophy to heterotrophy, depending on how much the organism relies on its chloroplasts (Jones 1994). Kleptochloroplastidy is widespread among planktonic marine ciliates such as in *Strombidium*, *Prorodon*, and *Mesodinium rubrum*. *Mesodinium rubrum*, for example, is a ciliate known to form coastal red tides. It acquires chloroplast from cryptophyte preys, which can be retained throughout the year (Gustafson et al. 2000). Kleptochloroplastidy was also shown to occur in freshwater ciliates such as *Histiobalantium natans* (Esteban et al. 2010), but was more commonly found in several marine invertebrates, phytoflagellates, and dinoflagellates (Venn et al. 2008).

15.4 Important Traits of Ciliate Symbioses

Although the nutritional aspect of symbiosis is often the main aspect that characterizes the symbiotic associations, the onset of symbiosis for horizontally transmitted symbionts includes a complex series of biological steps, which were referred to, by Nyholm and McFall-Ngai (2004) as 'the winnowing'. These steps could be listed as follows: (1) the encounter of the partners (2) the infection/ingestion and survival of the symbiont in the host until its durable establishment within the host cell, and (3) the transfer of symbiotic cells to the next generation. Each of these different steps may involve complex bio-communication processes such as extracellular sensoring (response toward environment), intracellular signaling, and interorganismic signaling. In this section, we review the processes that characterize symbiotic interactions. Here, we emphasize on ciliate symbioses, and consider the ciliate as hosts (containing partner) and voluntarily exclude the particular case of ciliate organisms as symbionts of intestinal microflora.

15.4.1 Horizontal Transfer of Symbiont and Selective Feeding

In many cases, the acquisition of symbiotic cells by horizontal transmission (i.e. ingestion or infection) is a necessity in mutualistic symbiosis. Although being a crucial aspect of mutualistic associations, little is known about the selective process (i.e., the recognition and the acquisition mechanisms) of new symbiotic cells by prospective hosts (for review, see Montagnes et al. 2008).

The acquisition of symbiont may be facilitated by behavioral and ecological factors. For example, the grazing behavior (phagotrophic activity) of ciliate organisms, and the physical proximity between prospective partners encourage symbiotic associations to emerge and evolve, as demonstrated previously in environmental (Summerer et al. 2008a, b) or experimental studies (Nakajima et al. 2009) (see Sect. 15.4). The cell size of symbionts, their cell defenses (e.g., the presence of spike or bristles on the algal cell surface), or the ability for the symbiont to form cell-aggregates to avoid predation, are other important factors that may increase or decrease the probability for ingestion.

In addition to these ecological factors, it was proposed that the prospective hosts rely on molecular communication signaling to seek and acquire symbiotic partners. While the process of inter-partner signaling and recognition mechanisms during the onset of mutualistic symbioses were particularly well-demonstrated in plant or animal symbiosis (e.g., squid-bacteria symbiosis, Nyholm and McFall-Ngai 2004), little is known about the communication processes occurring in ciliate symbioses.

One mechanism involved in partner recognition in ciliates includes the direct recognition of symbiont cell-surfaces. Such a process could explain the success of re-infection experiments performed for anaerobic ciliates and their methanogens (Shinzato and Kamagata 2010), or in phototrophic host such as in *Paramecium bursaria* and their algal symbiont (Kodama and Fujishima 2009). Although the factors involved in partners' recognition is still unclear, the possible involvement of a lectin/glycan system for direct recognition of the cell-surface was proposed in other symbiosis, such as in the corals/dinoflagellates (Wood et al. 2006) and in legume/bacteria symbiosis. Lectins can be free in solutions or located on the extracellular side of the organism's cell-wall. They bind specifically to glycans that compose the surface of other cells. For example, the dinoflagellate *Oxyrrhis marina* uses a selection mechanism based on Ca^{2+}-dependent mannose-binding lectin to capture and phagocytize algal prey (Wootton et al. 2007). However, the implication of lectin/glycan mechanisms in ciliate symbiosis remains to be demonstrated.

Mechanisms other than direct-cell surface recognition may favor the partner choice in local environments. For instance, it is known that bacteria and algal cells release a large array of chemical compounds in the surrounding environment, such as carbohydrates, water-soluble glycoconjugates, and proteins. Hypothetically, chemosensory attraction may facilitate protist hosts to find suitable partners. However, on the contrary, it was shown that some metabolites such as dimethylsulfide, amino acids, or algal toxins can deter the grazing behavior of protists, as demonstrated in the ciliate *Favella* sp. (e.g., Strom et al. 2003, 2007). Therefore, further research is needed to evaluate if ciliated organisms are able to perceive suitable partners from their secretome.

15.4.2 Infection of Symbiotic Within Ciliate Hosts

Once acquired by horizontal transfer, the chloroplast, the algal cells, or the bacterial symbiont must survive the digestive process and immune-defenses of the host in order to establish durably within the host, generally in its cytoplasm. Particles or organisms ingested by phagotrophy are rapidly localized within a digestive vacuole (DV) of the ciliate host. Newly formed DV is quickly acidified by acidosomal fusion, resulting in a decrease of the vacuole's pH (Fok and Allen 1983). If the acidosomal fusion occurs, then hydrolases enzymes (lysosomal enzymes) are released into the vacuole to break down whatever is inside (Allen and Fok 1988), a process during which the ingested algae should be partially or completely digested.

It was demonstrated that symbiotic cells are able to counter the digestion processes to successfully establish within the host. To briefly summarize, ingested bacteria or algal cells can (i) prevent of the host's vacuole acidification, (ii) prevent the fusion of lysosomal enzymes to the vacuole, (iii) resist the action of lysosomal enzymes, or (iv) escape from the DV by budding into the cytoplasm, enwrapped with a perivacuolar membrane which gives protection from the host's lysosomal fusion (O'Brien 1982; Fitt and Trench 1983; Kodama and Fujishima 2005). For example, in the hydra-algal symbiosis, algae are acid tolerant and may prevent the lysosomal digestion by the release of maltose (Huss et al. 1993/1994). In the ciliate *Paramecium bursaria*, after lysosomal fusion occurs, several cells of the symbiotic algae located within the same digestive vacuole temporarily resist the action of lysosomes, then they can escape by budding into the host cytoplasm (Kodama and Fujishima 2009).

Other molecular mechanisms may also participate in the non-recognition of symbiotic cells during infection. For example, it was shown that symbiotic algae treated with cellulase and pectinase or having been coated with specific antibodies or with lectins (Concanavalin A or *Ricinus communis* agglutinin) are usually not recognized as suitable for the host (Reisser et al. 1982). As a result, they are sequestered into food vacuoles or excreted by the cytoproct. The importance of the integrity of algal surface for partner recognition was also demonstrated not only in ciliates, but also in other protist species. For example, the incubation of *Symbiodinium* with Concanavalin A, *Limulus polyphemus* agglutinin (LPA), *Phaseolus vulgaris* erythroagglutinin (PHA-E), or wheat germ agglutinin (WGA) significantly reduced reinfection success in adult *A. pulchella* hosts (Lin et al. 2000). In *H. viridis*, Concanavalin A completely inhibited reinfection by *Chlorella* spp. (Meints and Pardy 1980).

Overall, these studies suggest that the algal cell-wall structure is potentially the basis for initiation and establishment of the symbiosis (Lee et al. 1985). Yet, more investigations are required to determine how the symbiotic cells successfully infect the cytoplasm of their hosts, and to what extent they are protected from immune responses or other regulatory mechanisms.

15.4.3 Benefits of Symbiosis

As seen in Sect. 15.1, ciliate hosts benefit from the nutrients produced by their symbiotic cells. For example, in *P. bursaria*, symbiotic cells can provide up to 85 % of the photosynthetically fixed carbon to the host (Muscatine et al. 1967). To maximize the nutritional benefits from the photosynthetic activity of their symbionts, ciliate host cells respond to light gradient and swim to brighter environment (positive phototaxis) (Dolan 1992). This behavioral phototaxis is mediated by the photoreceptors present in ciliate host or the photosynthetic symbiont, or both. Extended investigations in this regard have been carried out in the ciliate *Blespharisma japonica* and *Stentor coeruleus* (Lenci et al. 2001).

The nutritional benefit obtained from mixotrophy is often measured in terms of fitness or increased survival of the host, and is especially notable when resources are scarce in the environment. For example, experimental cultures containing *T. thermophila* hosts and symbiotic *Micractinium* green algae were performed to test the survival of *T. thermophila* in the absence of bacteria (see Sect. 15.4). When cultured without green alga, *T. thermophila* could survive about 1 week in a mineral salt medium. However, when cultured with symbiotic clones of *Micractinium* alga, *T. thermophila* could survive longer, for up to 130 additional days. In this association, *T. thermophila* benefits from the carbohydrates as well as the glycerol released by the algal cells (Nakajima et al. 2009, 2015; Germond et al. 2013b).

Symbiotic cells may also offer other benefits to their host besides the nutritional one. For example, it is likely that the symbiont of *Maristentor dinoferus* produce mycosporine-like amino acids, protecting the host against UV irradiation occurring in transparent and oligotrophic waters (Sommaruga et al. 2006). Symbiotic cells may also provide protection against parasites or participate in detoxification processes in their hosts (Dziallas et al. 2012).

On the other hand, it is worth saying that while the symbionts clearly benefits the hosts, the benefits of such symbiotic association for the symbiont are sometimes less obvious. Often, it was suggested that the symbiotic cells are protected against environmental changes or benefit from some nutrients procured by the host (e.g., McAuley 1988; McAuley et al. 1996). An example of symbiont benefit is discussed from an evolutionary perspective in Sect. 15.5.3.

15.4.4 Regulation of the Number of Symbiotic Cells by Host

It is not, however obvious that whether the symbiotic algal cells are always advantageous to their hosts, when the environmental variables (such as light, season, temperature, and other host-related physiological factors) are unfavorable (e.g., Achuthankutty 2006). For example, a large number of intracellular algal cells can lead to an accumulation of photosynthetically-derived reactive oxygen species (ROS) known to damage the host cell through oxidative stress. Because symbiont-bearing lifestyle may be associated with several possible costs, one may expect that ciliate hosts regulate the number of symbiotic cells in relation to their dietary needs and environmental changes to survive better in changing environment. Several regulation systems were proposed: (i) a controlled expulsion of the symbiont, (ii) the digestion of the symbiont or (iii) a physiological control of the symbiont division (e.g., through nutrient availability or pH). Furthermore, we cannot exclude the possibility that some intracellular symbionts act as cheaters, and do not participate in the host's survival, but keep benefiting from it.

Voluntary excretion and cell division were proposed as the two mechanisms used by ciliate hosts to regulate the intracellular population of symbiotic cells (McAuley 1986; Reisser et al. 1983; Reisser 1986). Specifically, symbionts of ciliate hosts can be transmitted vertically when passed directly from parent to daughter cells by binary division of the host cells, or released in the surrounding medium by voluntary excretion. Voluntary excretion, however, suggests the existence of some selective criteria. In this sense, some authors suggested that the selective criteria leading to the algal expulsion could be based on the photosynthetic activities of the symbiont or the production of ROS molecules. For example, Gerashchenko and colleagues, who studied the optical characterization of both symbiotic and free-living algal strains with respect to their ability to establish symbiosis with *P. bursaria*, showed that chlorophyll content per cell volume seems to be a valuable factor for predicting a favorable symbiotic relationship (Gerashchenko et al. 2000). Furthermore, it was also suggested that protistan hosts can control the size of the endosymbiont population by digesting the algae (Reisser et al. 1983), although little is known about the ability of protist hosts to digest or kill their symbiont (e.g., through immune-system-like responses) once they are established within the cell.

On the other hand, other studies conducted in the protist *H. viridis* suggested that some other factors, and not voluntary expulsion or digestion of symbiotic algae, are responsible for the limitation of the algal growth within the host and provide a basis for symbiont regulation (Muscatine and Pool 1979). Later, it was suggested that such a regulatory mechanism could imply a limitation of the nutrient provided by the host (Douglas 1994). One of the most attractive hypotheses so far is the apparent relation between pH in vacuoles and the regulation of the symbiotic algal cycle (Stabell et al. 2002). Specifically, it was demonstrated that the symbiotic *Chlorella* cells of *P. bursaria* and *H. viridis* hosts synthesize and release more carbohydrates when grown in experimental cultures at acidic pH, but that acidic pH also reduces the uptake of ammonium, which prevent the algal cell division (McAuley et al. 1996; Cernichiari et al. 1969; Douglas and Smith 1984; Douglas 1994; Dorling et al. 1997). However, later studies showed that the symbiotic cells of *P. bursaria* were established in the cytoplasm of hosts, in which they are probably maintained at neutral pH (Kodama and Fujishima 2005, 2009). Yet, it has not been excluded that the host might regulate its symbiotic population through temporal changes of the vacuole's pH (Dorling et al. 1997).

Further studies are desirable to identify and quantify the regulatory mechanisms existing in protistan symbioses. In addition, it may be interesting to determine whether similar mechanisms are being employed across these associations, or if there exist some other variations, depending on the interacting partners and ecological factors.

15.5 Experimental Approach for the Evolutionary Study of Symbiotic Associations Between Ciliates and Algae

In this section, we discuss the ecological and evolutionary mechanisms responsible for the emergence and evolution of mixotrophic symbiosis (i.e., symbiogenesis). Particularly, we hypothesize that symbioses tend to evolve at a mature stage of the ecosystem development. Moreover, as a perspective for future studies, we report that the use of synthetic ecosystems can be effective to analyze the evolution of symbiotic associations.

15.5.1 Evolutionary-Ecological Perspective of the Evolution of Symbiosis

Some authors suggested that symbiogenesis (i.e., the evolution of adaptation by symbiosis) may challenge Darwin's theory as the sole explanatory framework for adaptive evolution, since the adaptive traits of partners could be acquired by a non-gradual process (Margulis 1993; Roossinck 2005; Watson 2006). However, the endosymbiosis between auto-(photo-) and heterotrophic organisms exhibits multistage associations, suggesting that there exists a continuity between various stages. These stages include (i) a tentative or unstable stage, (ii) a stable stage, in which partners live together through their entire lifecycles and reproduce synchronously at a high rate of vertical transmission of symbionts (Weis 1977), and (iii) a genetically unified stage, often characterized by a reduction of the genome size of the symbionts and/or horizontal gene transfer (McFadden 2001; Howe et al. 2008). Furthermore, symbiotic associations at early evolutionary stages can be lost, replaced, or strengthened (Saldarriaga et al. 2001; Ishida and Green 2002; Horiguchi 2006). This dynamic aspect of multistage associations is considered to reflect potential conflicts between partners, depending on the cost and benefit of symbiosis under given ecological conditions. Therefore, it is suggested that symbiogenesis is driven gradually by natural selection operating on heritable phenotypes of participating organisms (Maynard Smith 1991).

What is the ecological condition in which endosymbiotic associations are favored? Generally, it is believed that the endosymbiotic association can evolve under resource-limited conditions and/or low dissolved oxygen concentrations (Reisser and Kurmeier 1984; Berninger et al. 1986; Reisser 1986; Finlay et al. 1996). How, then, can such resource-limited conditions be generated ecologically? Resource availability is strongly dependent on ecosystem dynamics. Ecosystem ecology suggested that ecosystems undergo a directional, predictable change in the community structure, energetics, nutrient cycling, and life-history characteristics of component species (e.g. Margalef 1963; Cooke 1967; Odum 1969; Christensen 1995; Faith et al. 2004). According to this view, ecosystems develop toward a mature stage, which is characterized by an equality between gross production and community respiration, a low ratio of gross production to standing biomass, a high ratio of supported biomass to unit energy flow, a detritus-dependent nutrient cycling, a high species diversity, web-like food chains, and a well-organized spatial heterogeneity. It is quite likely that the stages in ecosystem development have crucial impacts on the evolution of component species. In a mature stage of development, in which the available nutrients for producers are detritus-dependent, and consumers' survival and reproduction is highly density-dependent, organisms with higher reproductive and survival efficiency can evolve (Margalef 1963; Odum 1969). In such a situation, symbioses are expected to prevail due to their efficient growth and survival. In other words, the mature-stage conditions may function as a major selective force for symbiotic associations. Therefore, we can hypothesize that auto- and heterotrophic symbioses evolve in a mature stage of ecosystem development.

15.5.2 Synthetic Ecosystem as a New Tool for Analyzing the Evolution of Algal-Ciliate Associations

Testing hypothesis about biological evolution is usually limited by methodological difficulties. However, the use of experimental models, such as synthetic microbial ecosystem may be a useful approach to test evolutionary hypotheses concerning symbiogenesis, and may also provide the opportunity for creating new testable hypotheses. Algae, bacteria, and protists such as ciliates and flagellates constitute major components of the microbial loop in natural aquatic ecosystems (Azam et al. 1983; Ferhman 2009; Caron et al. 2012; Mitra et al. 2014). In microbial loop, the dissolved organic carbon released from algae and organic detritus, from all component species is mainly consumed by bacteria, which are then consumed by protists. Nakajima et al. (2009) developed an experimental semi-closed ecosystem model (called the CET microcosm), which simulates a microbial loop system, in order to directly observe an early evolutionary process of endosymbiosis emerging from organisms that were previously unassociated. The CET microcosm contains a green alga [*Micractinium* sp. Ehime, formerly described as *Chlorella vulgaris* (Germond et al. 2013a)], a bacterium (*Escherichia coli*), and a ciliate (*Tetrahymena thermophila*) in a glass bottle using a medium containing mineral salts, incubated statically without transfer on a 12L:12D cycle at 30 °C. *Micractinium* grows in the light using minerals in the medium and excretes photosynthetic organic compounds into the medium. *E. coli* grows by using the excreted organic compounds or detritus. In natural habitats, *T. thermophila* is a bacteria-feeder that commonly lives on decaying vegetation at the mud-water interface (Doerder et al. 1995). In the microcosm, *T. thermophila* grows by eating *E. coli*, although accidental ingestions of algal cells may occur. However, the ciliate cannot grow on *Micractinium* (Fig. 15.1a).

An initial 3 year culture of the microcosm without transfer revealed the emergence of symbiotic associations between *Micractinium* and *Tetrahymena*, and between *Micractinium* and *E. coli*, which evolved during a mature stage of ecosystem development of the CET microcosm, under a stoichiometric restriction. According to microscopic observations, *Tetrahymena* cells that contained *Micractinium* cells started to increase in number on around day 100, reaching approximately 80–90 % of the *Tetrahymena* population, and then maintained at a high relative density during the rest of the culture. Moreover, cell aggregates containing *Micractinium* and *E. coli* cells appeared and increased in abundance in the culture, suggesting that the association may act as a selective advantage in the resource-limited environment of the microcosm. In fact, it is likely that these cell-aggregates functioned as an ectosymbiotic association in which the partners benefit from resource exchanges facilitated by their close physical proximity (Nakajima et al. 2009, 2015). Moreover, the cell aggregates are considered to serve as refuges for *E. coli* from predation by *T. thermophila*, causing starvation of *T. thermophila*. Under this condition, the ciliate cells can benefit more by developing symbiosis with algal cells (Fig. 15.1b).

15.5.3 Selection Process for the Algal-Ciliate Endosymbiosis in CET Microcosm

An experimental analysis of the above CET microcosm was conducted by using several algal isolates obtained from within the *Tetrahymena* cells in the CET microcosm after it was cultured for 1819–1837 days. The analysis revealed that non-aggregate-forming algal isolates prolonged the longevity of both the ancestral and derived *Tetrahymena* clones (isolated from the microcosm after it had been cultured for 2668 days) under bacteria-free conditions, than that in the co-cultures with the ancestral *Micractinium* strain. Non-aggregate-forming algal isolates

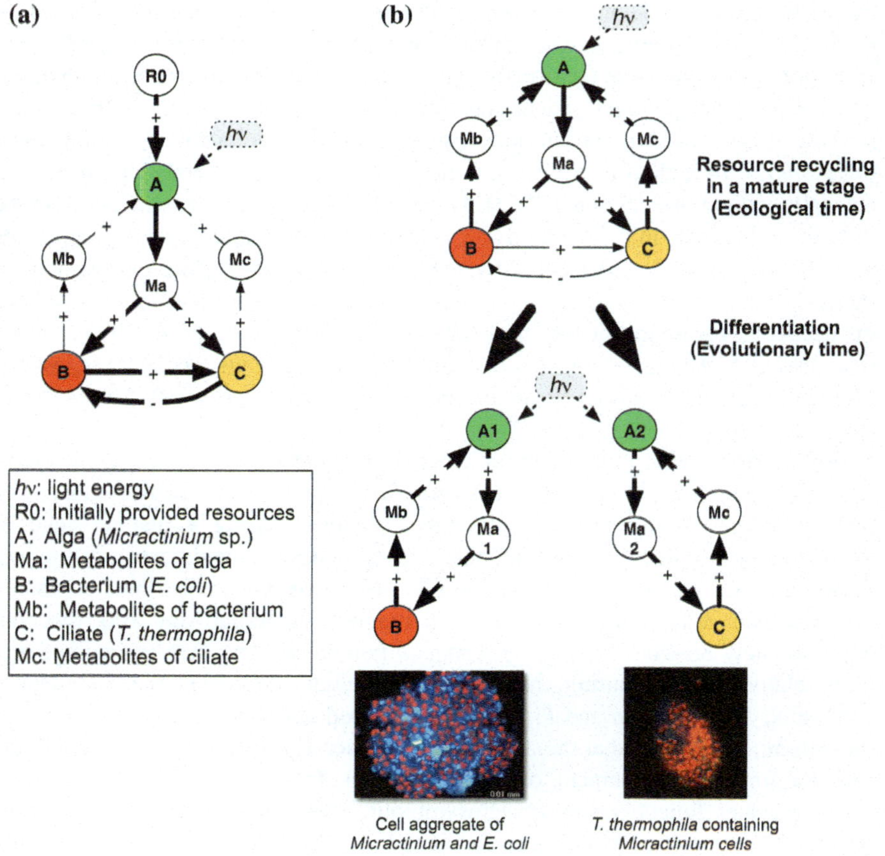

Fig. 15.1 **a** Mater/energy flows in CET microcosm at the beginning of culture. Thick arrows indicate major flows with the positive (+) or negative (−) interaction between species. **b** Mater/energy flows in the mature stage of CET microcosm culture. The *Micractinium* population differentiated into aggregate-forming ("A1") and non-aggregate-forming phenotypes ("A2"), although intermediate types existed (Nakajima et al. 2015). The former type developed an ectosymbiotic association with *E. coli*, whereas the latter type developed an endosymbiotic association with *T. thermophila*

released a large amount of photosynthetic products than the ancestral algal strain. In contrast, aggregate-forming algal isolates did not prolong the longevity of either ancestral or isolated *Tetrahymena* clones. However, the aggregate-forming isolates increased the survivability when cocultured with *E. coli*, than that with the non-aggregate forming algal types (Nakajima et al. 2015). This study demonstrates that the dispersive or the aggregative ability of algal cells affects how they interact with the other species, which in turn affects the evolution of symbiotic associations with other organisms such as ciliates and bacteria.

Moreover, our result suggested that competition between the different algal types, when present in the extra-host environment, favors the development of the host-benefiting symbiotic type. In fact, a competition experiment using a coculture with *E. coli* in the absence of *T. thermophila* revealed that the non-aggregate-forming type of *Micractinium* was less competitive in this extra-host environment than the aggregate-forming type. For this non-aggregate-forming type of algal cells, the host environment can allow a higher survival. From the perspective of the game theory, the non- or less- benefiting endoalgal cells would be passively affected in a negative manner by the death of the host; therefore, cooperative symbiont genotypes are favored, because the fitness of the symbiont is linked to host survival and reproduction. It is predicted that the vertical transmission of symbionts favors cooperative traits in the symbiont, because in that way the symbiont can increase its own fitness by benefiting the host (Trivers 1971; Bull and Rice 1991). In contrast, horizontal transmission of symbiont was reported to favor exploitative traits (i.e., parasitic lifestyle) (Bull 1994; Sachs et al. 2004; Sachs and Wilcox 2006). Therefore, during the algal diversification process, severe competition with a sister line that is adapted to the extra-host environment plays an important role in producing a selective force that favors host-benefiting traits in the endosymbiotic lineage.

The result of the CET microcosm experiment suggests a new perspective for understanding the dynamics of free-living and endosymbiotic life styles in nature. Recent studies revisited the phylogeny of *Chlorella*-like algae by using DNA sequencing to examine the phylogenetic relationship among various physiological and ecological forms of algae adapted to free-living (planktonic), edaphic, and endosymbiotic niches. These studies reconstructed phylogenetic relationships between species/strains in this group, and revealed that endosymbiotic species/strains isolated from *Paramecium bursaria* (green Paramecium) are polyphyletic (Hoshina and Imamura 2008; Pröschold et al. 2011). Similarly, it was shown that endosymbiotic algae isolated from green hydra (*Hydra viridissima*) also have a polyphyletic origin and have many related species that are free-living (Huss et al. 1993/1994; Kovacevic et al. 2010). These phylogenetic studies suggest that auto-/heterotrophic endosymbioses have evolved through algal differentiation from free-living species, and that endosymbiotic species are polyphyletic and coexist with non-endosymbiotic free-living species in the same locality. According to the mechanism suggested by the CET microcosm, during the divergent evolution of *Chlorella*-like species, free-living lineages played an important role in generating a selection pressure for host-benefiting traits in algae, which resulted in the development of endosymbiotic associations. From a viewpoint of ecosystem process, as shown in Fig. 15.1, ecosystem dynamics can downwardly affect the evolution of component species by generating selection pressures on them under a stoichiometric restriction, in which algal-ciliate and algal-bacterial symbiotic associations can emerge and evolve during the mature stage of ecosystem development. The associations in turn change the structure of matter/energy flows in ecosystem.

15.5.4 Horizontal Transmission of Symbiotic Algal Cells: Sudden Acquisition of a New Adaptation?

As described previously, symbiogenesis is not a sudden acquisition of new adaptations, but an outcome of co-evolutionary processes. However, interestingly, some algal clones isolated from ciliate hosts that evolved in the CET microcosm prolonged the longevity of not only the isolated ciliates, but also of the ancestral ciliate (Nakajima et al. 2013). This result suggests that once a cooperative algal genotype evolves in a local population, it can later be transmitted to other individuals of the same or related species, and possibly spread rapidly beyond the local range due to its positive effect on the host fitness. Such transmission suggests the possibility of a sudden acquisition of beneficial autotrophic function by the pre-associated host.

Field researches on the relationship between *Chlorella*-like species and their host species showed that endosymbiotic algae tend to be transmitted between different host species within the same aquatic ecosystem. Summerer et al. (2007, 2008a, b) reported that various zoochlorellae can be harbored by the same ciliate species, depending on their origin, whereas the same algal species can be found in various host species in the same lake. Similarly, Pröschold et al. (2011) reported that *P. bursaria* hosts of different origin have different endosymbiotic algal strains, whereas the endosymbionts isolated from *Climacostomum*, *Coleps* and *Euplotes*, all growing in the same pond, were identical in their SSU and ITS sequences. These findings suggest that the same algal symbionts can spread over a local community through transmission between different ciliate species. In the future, it may be interesting to analyze how algal-ciliate associations can co-evolve through algal transmissions within a multi-species ciliate community. This perspective may help us to develop a new understanding that adaptation by endosymbiosis can evolve through the nexus of multi-species interactions in the local community/ecosystem, involving symbiont transmissions among potential host species.

References

Achuthankutty CT (2006) Environmentally-related seasonal variation in symbiotic associations of heterotrophic dinoflagellates with Cyanobacteria in the western Bay of Bengal. Symbiosis 42:51–58

Allen RD, Fok AK (1983) Non-lysosomal vesicles (acidosomes) are involved in phagosome acidification in *Paramecium*. J Cell Biol 97:566–570

Andrushchyshyn OP, Wilson KP, Williams DD (2007) Ciliate communities in shallow groundwater: seasonal and spatial characteristics. Freshw Biol 52:1745–1761

Azam F, Fenchel T, Field JG, Gray JS, Meyer-Reil LA, Thingstad F (1983) The ecological role of water-column microbes in the sea. Mar Ecol Prog Ser 10:257–263

Bakker C (1967) Massale ontwikkeling van ciliaten met symbiontische algen in het Veerse Meer De Levende. Natuur 70:166–173

Berninger UG, Finlay BJ, Canter HM (1986) The spatial distribution and ecology of zoochlorellae-bearing ciliates in a productive pond. J Protozool 33:557–563

Bull JJ (1994) Virulence. Evolution 48:1423–1437

Bull JJ, Rice WR (1991) Distinguishing mechanisms for the evolution of cooperation. J Theor Biol 149:63–74

Burkholder JM, Glibert PM, Skelton HM (2008) Mixotrophy, a major mode of nutrition for harmful algal species in eutrophic waters. Harmful Algae 8:77–93

Capone DG (2000) The marine nitrogen cycle. In: Kirchman DL (ed) Microbial ecology of the ocean. Wiley-Liss, New York, pp 455–493

Caron DA, Countway PD, Jones AC, Kim DY, Schnetzer A (2012) Marine protistan diversity. Annu Rev Mar Sci 4:467–493

Carvalho WF, Granéli E (2006) Acidotropic probes and flow cytometry: a powerful combination for detecting phagotrophy in mixotrophic and heterotrophic protists. Aquat Microb Ecol 44:85–96

Cernichiari E, Muscatine L, Smith DC (1969) Maltose excretion by the symbiotic algae of *Hydra viridis*. Proc R Soc Lond B 1033:557–576

Christensen V (1995) Ecosystem maturity—towards quantification. Ecol Model 77:3–32

Christopher MH, Patterson DJ (1983) *Coleps hirtus*, a ciliate illustrating facultative symbiosis between protozoa and algae. Ann Stat Biol Besse-en-Chandesse 17:278–297

Cooke GD (1967) The pattern of autotrophic succession in laboratory microcosms. Bioscience 17:717–721

Crawford DW (1989) *Mesodinium rubrum*, a phytoplankter that wasn't. Mar Ecol Prog Ser 58:161–174

Doerder FP, Gates MA, Eberhardt FP, Arslanyolu M (1995) High frequency of sex and equal frequencies of mating types in natural populations of the ciliate *Tetrahymena thermophila*. PNAS 92:8715–8718

Dolan JR (1992) Mixotrophy in ciliates: a review of *Chlorella* symbiosis and chloroplast retention. Mar Microb Food Webs 6:115–132

Dorling M, McAuley Hodge (1997) Effect of pH on growth and carbon metabolism of maltose-releasing *Chlorella* (Chlorophyta). Eur J Phycol 32:19–24

Douglas AE (1994) Symbiotic interactions. Oxford University Press, Oxford

Douglas AE, Smith DC (1984) The green hydra symbiosis. VIII. Mechanisms in symbiont regulation. Proc R Soc Lond B 221:291–319

Dziallas C, Allgaier M, Monaghan MT, Grossart HP (2012) Act together – implications of symbioses in aquatic ciliates. Front Microbio 3:288

Esteban GF, Fenchel T, Finlay BJ (2010) Mixotrophy in ciliates. Protist 161:621–641

Faith BD, Jørgensen SE, Patten BC, Straskrabe M (2004) Ecosystem growth and development. Biosystems 77:213–228

Fenchel T, Finlay BJ (1991) The biology of free-living anaerobic ciliates. Eur J Protistol 26:201–215

Fenchel T, Finlay BJ (1995) Ecology and evolution in anoxic worlds. Oxford University Press, New York

Fenchel T, Finlay BJ (2010) Free-living protozoa with endosymbiotic methanogens. In: Hackstein JHP (ed) (Endo)symbiotic methanogens. Springer, Heidelberg

Ferhman JA (2009) Microbial community structure and its functional implications. Nature 459:193–199

Finlay BJ, Embley TM, Fenchel T (1993) A new polymorphic methanogen, closely related to *Methanocorpusculum parvum*, living in stable symbiosis within the anaerobic ciliate *Trimyema* sp. J Gen Microbiol 139:371–378

Finlay BJ, Esteban G, Clarke KJ, Williams AG, Embley M, Hirt RP (1994) Some rumen ciliates have endosymbiotic methanogens. FEMS Microb Lett 117:157–161

Finlay BJ, Maberly SC, Esteban GF (1996) Spectacular abundance of ciliates in anoxic pondwater: contribution of symbiont photosynthesis to host respiratory oxygen requirements. FEMS Microbiol Ecol 20:229–235

Finlay BJ, Maberly S, Esteban G (2006) Spectacular abundance of ciliates in anoxic pond water: contribution of symbiont photosynthesis to host respiratory oxygen requirements. FEMS Microbiol Ecol 20(4):229–235

Fitt WK, Trench RK (1983) Endocytosis of the symbiotic dinoflagellate *Symbiodinium microadriaticum* Freudenthal by endodermal cells of the scyphistomae of *Cassiopeia xamachana* and resistance of the algae to host digestion. J Cell Sci 64:195–212

Fok AK, Allen RD (1988) The lysosome system. In: Görtz HD (ed) Paramecium. Springer, Berlin, pp 301–324

Gerashchenko NI, Nishijara N, Ohara T, Tosuji H, Kosaka T, Hosoya H (2000) Flow cytometry as a strategy to study the endosymbiosis of algae in Paramecium bursaria. Cytometry 41(3):209–215

Germond A, Hata H, Fujikawa Y, Nakajima T (2013a) The phylogenetic position and phenotypic changes of a *Chlorella*-like alga during a 5-year microcosm culture. Eur J Phycol 48:485–496

Germond A, Kunihiro T, Inouhe M, Nakajima T (2013b) Physiological changes of a green alga (*Micractinium* sp.) involved in an early stage of association with *Tetrahymena thermophila* during 5-year microcosm culture. BioSystems 114:164–171

Graham LE, Graham J (1978) Ultrastructure of endosymbiotic *Chlorella* in a *Vorticella*. J Prot 25:207–210

Gustafson E, Stoecker DK, Johnson MD, Van Heukelem WF, Sneider K (2000) Cryptophyte algae are robbed of their organelles by the marine ciliate *Mesodinium rubrum*. Nature 405:1049–1050

Kawakami H (1984) Ultrastructural study of an endosymbiotic alga and its host ciliate *Stentor niger*. J Prot 31(2):247–253

Hackstein JHP (2010) (Endo)symbiotic Methanogenic Archaea. In: Hackstein JHP (ed) Microbiology monographs 19. Springer, Heidelberg

Hackstein JHP, van Hoek AHAM, Leunissen JAM, Huynen M (2004) Anaerobic ciliates and their methanogenic endosymbionts. In Seckbach J (ed) Cellular origin, life in extreme habitats and astrobiology. Springer, The Netherlands, pp 451–464

Hemprich FW, Ehrenberg CG (1831) Symbolae physicae. Evertebrata I Phytozoa Abh Akad Wiss Berl Kahl A (1930–35) Urtiere oder Protozoa. In Die Tierwelt Deutschlands, Teil 18, 21, 25, 30. Gustav Fisher, Jena

Horiguchi T (2006) Algae and their chloroplasts with particular reference to the dinoflagellates. Paleontol Res 10:299–309

Hoshina R, Imamura N (2008) Multiple origins of the symbioses in *Paramecium bursaria*. Protist 159:53–63

Hoshina R, Kato Y, Kamako S-I, Imamura N (2005) Genetic evidence of 'American' and 'European' type symbiotic algae of *Paramecium bursaria* Ehrenberg. Plant Biol 7:526–532

Howe CJ, Barbrook AC, Nisbet RER, Lockhart PJ, Larkum AWD (2008) The origin of plastids. Phil Trans R Soc B 363:2675–2685

Huss VAR, Holweg C, Seidel B, Reich V, Rathat M, Kessler E (1993/1994). There is an ecological basis for host/symbiont specificity in *Chlorella*/*Hydra* symbioses. Endocyt Cell Res 10:35–46

Ishida K, Green BR (2002) Second- and third-hand chloroplasts in dinoflagellates: Phylogeny of oxygen-evolving enhancer 1 (PsbO) protein reveals replacement of a nuclear-encoded plastid gene by that of a haptophyte tertiary endosymbiont. PNAS 99:9294–9299

Jones RI (1994) Mixotrophy in planktonic protists as a spectrum of nutritional strategies. Mar Microb Food Webs 8:87–96

Jonsson PR (1987) Photosynthetic assimilation of inorganic carbon in marine oligotrich ciliates (Ciliophora, Oligotrichina). Mar Microb Food Webs 2:55–68

Kawakami H, Kawakami N (1985) 7 Int Congr Protozool Nairobi Abstr N°300

Klaveness D (1984) Studies on the morphology, food selection and growth of two planktonic freshwater strains of *Coleps* sp. Protistologica 20:335–349

Kneip C, Lockhart P, Voß C, Maier UG (2007) Nitrogen fixation in eukaryotes—new models for symbiosis. BMC Evol Biol 7:55

Kodama Y, Fujishima M (2005) Symbiotic *Chlorella* sp. of the ciliate *Paramecium* do not prevent acidification and lysosomal fusion of the host digestive vacuoles during infection. Protopl 225:191–203
Kodama Y, Fujishima M (2009) Timing of perialgal vacuole membrane differentiation from digestive vacuole membrane in infection of symbiotic algae *Chlorella vulgaris* of the ciliate *Paramecium bursaria*. Protist 160:65–74
Kovačević G, Kalafatić M, Jelenčić B, Franjević D (2010) Endosymbiotic alga as the stronger evolutionary partner in green hydra symbiosis. J Endocyt Cell Res 20:13–15
Laval-Peuto M, Febvre MM (1986) On plastid symbiosis in Tontonia appendiculariformis. BioSyst 19:137–157
Laval-Peuto M, Rassoulzadegan F (1988) Autofluorescence of marine planktonic oligotrichina and other ciliates. Hydrobiologia 159:99–110
Lee JL, Soldo AT, Reisser W, Lee KM, Jeon KWn Görtz HD (1985) The extent of algal and bacterial endosymbiosis in Protozoa. J Prot 32(3):391–403
Lenci F, Ghetti F, Song PS (2001) Photomovement in ciliates. In: Häder DP, Lebert M (eds) Photomovements. Elsevier Science BV, Amsterdam, The Netherlands, pp 475–503
Lin KL, Wang JT, Fang LS (2000) Participation of glycoproteins on zooxanthellal cell walls in the establishment of a symbiotic relationship with the sea anemone, *Aiptasia pulchella*. Zool Stud 39:172–178
Lobban CS, Schefter M, Simpson AGB, Pochon X, Pawloski J, Foissner W (2002) *Maristentor dinoferus* n. gen., n. sp., a giant heterotrich ciliate (Spirotrichea: Heterotrichida) with zooxanthellae, from coral reefs on Guam, Mariana Islands. Mar Biol 140:411–423; 141: 207–208
Lobban CS, Modea L, Verni F, Rosati G (2005) *Euplotes uncinatus* (Ciliophora, Hypotrichia), a new species with zooxanthellae. Mar Biol 147:1055–1061
Margalef R (1963) On certain unifying principles in ecology. Am Nat 97:357–374
Margulis L (1993) Symbiosis in cell evolution, 2nd edn. Freeman WH, New York
Maynard Smith J (1991) A Darwinian view of symbiosis. In: Margulis L, Fester R (eds) Symbiosis as a source of evolutionary innovation. The MIT Press, Cambridge, pp 26–39
McAuley PJ (1986) The cell cycle of symbiotic Chlorella. III. Numbers of algae in green hydra digestive cells are regulated at digestive cell division. J Cell Sci 85:63–71
McAuley PJ (1988) Uptake of leucine by *Chlorella* symbionts of green *Hydra*. Proceedings of the Royal Society of London. Series B. Biol Sci 234:319–332
McAuley PJ, Dorling M, Hodge H (1996) Effect of maltose release on uptake and assimilation of ammonium by symbiotic *Chlorella* (Chlorophyta). J Phycol 32:839–846
Meints R, Pardy R (1980) Quantitative demonstration of cell surface involvement in a plant-animal symbiosis: lectin inhibition of reassociation. J Cell Sci 43:239–251
Mitra A, Flynn KJ, Burkholder JM, Berge T, Calbet A, Raven JA, Granéli E, Glibert PM, Hansen PJ, Stoecker DK, Thingstad F, Tillmann U, Våge S, Wilken S, Zubkov MV (2014) The role of mixotrophic protists in the biological carbon pump. Biogeosciences 11:995–1005
Montagnes DJS, Barbosa AB, Boenigk J, Davidson K, Jürgens K, Macek M, Parry JD, Roberts EC, Simek K (2008) Selective feeding behaviour of key free-living protists: avenues for continued study. Aquat Microb Ecol 53:83–98
Muscatine L, Pool RR (1979) Regulation of numbers of intracellular algae. Proc R Soc Lond B 204:131–139
Muscatine L, Karakashian SJ, Karakashian MW (1967) Soluble extracellular products of algae symbiotic with a ciliate, a sponge and a mutant *Hydra*. Comp Biochem Physiol 20:1–12
Nakajima T, Sano A, Matsuoka H (2009) Auto-/heterotrophic endosymbiosis evolves in a mature stage of ecosystem development in a microcosm composed of an alga, a bacterium and a ciliate. BioSystems 96:127–135
Nakajima T, Matsubara T, Ohta Y, Miyake D (2013) Exploitation or cooperation? Evolution of a host (ciliate) -benefiting alga in a long-term experimental microcosm culture. BioSystems 113:127–139

Nakajima T, Fujikawa Y, Matsubara T, Karita M, Sano A (2015) Differentiation of a free-living alga into forms with ecto- and endosymbiotic associations with heterotrophic organisms in a 5-year microcosm culture. BioSystems 131:9–21

Nyholm SV, McFall-Ngai M (2004) The winnowing: establishing the squid–vibrio symbiosis. Nat Rev Microbiol 2:632–642

O'Brien TL (1982) Inhibition of vacuolar membrane fusion by intracellular symbiotic algae in *Hydra viridis* (Florida Strain). J Exp Zool 223:211–218

Odum EP (1969) The strategy of ecosystem development. Science 164:262–270

Pierce RW, Turner JT (1992) Ecology of planktonic ciliates in marine food webs. Rev Aquat Sci 6:139–181

Pröschold T, Darienko T, Silva PC, Reisser W, Krienitz L (2011) The systematics of *Zoochlorella* revisited employing an integrative approach. Environ Microbiol 13:350–364

Reisser W (1986) Endosymbiotic associations of freshwater protozoa and algae. In: Corliss JO, Patterson DJ (eds) Progress in protistology, pp 195–214

Reisser W, Kurmeier B (1984) The endosymbiotic unit of *Climacostomum virens* and *Chlorella* sp. Symbiotic features of the association and host-symbiont regulatory mechanisms. Protistologica 20:265–270

Reisser W, Radunz A, Wiessner W (1982) Participation of algal surface structures in the cell recognition process during infection of aposymbiotic *Paramecium bursaria* with symbiotic chlorellae. Cytobios 33(129):39–50

Reisser W, Meier R, Kurmeier B (1983) The regulation of the endosymbiotic algal population size in ciliate-algae associations. An ecological model. In: Schwemmler W, Schenk HEA (eds) Endocytobiology, vol 2. Walter de Gruyter & Co. Berlin, pp 533–543

Rinke C, Schmitz-Esser S, Stoecker K, Nussbaumer AD, Molnár DA, Vanura K, Wagner M, Horn M, Ott JA, Bright M (2006) "Candidatus Thiobios zoothamnicoli", an ectosymbiotic bacterium covering the giant marine ciliate Zoothamnium niveum. Appl Environ Microbiol 72:2014–2021

Roossinck MJ (2005) Symbiosis versus competition in plant virus evolution. Nat Rev Microbiol 3:917–924

Sachs JL, Wilcox TP (2006) A shift to parasitism in the jellyfish symbiont *Symbiodinium* microadriaticum. Proc R Soc B 273:425–429

Sachs JL, Mueller UG, Wilcox TP, Bull JJ (2004) The evolution of cooperation. Quart Rev Biol 79:135–160

Saldarriaga JF, Taylor FJR, Keeling PJ, Cavalier-Smith T (2001) Dinoflagellate nuclear SSU rRNA phylogeny suggests multiple plastid losses and replacements. J Mol Evol 53:204–213

Seckbach J (ed) (1994) Evolutionary pathways and enigmatic algae *Cyanidium caldarium* (Rhodophyta) and related cells. Springer science+Business Media, Dordrecht

Shinzato N, Kamagata Y (2010) The methanogenic and eubacterial endosymbionts of *Trimyema*. In: Hackstein JHP (ed) (Endo)symbiotic methanogens. Springer, Heidelberg

Sommaruga R, WhiteheadK Shick JM, Lobban CS (2006) Mycosporine-like amino acids in the Zooxanthella-ciliate symbiosis *Maristentor dinoferus*. Protist 157:185–191

Stabell T, Andersen T, Klaveness D (2002) Ecological significance of endosymbionts in a mixotrophic ciliate—an experimental test of a simple model of growth coordination between host and symbiont. J Plank Res 9:889–899

Stoecker DK (1999) Mixotrophy among dinoflagellates. J Eukaryot Microbiol 24:397–401

Stoecker DK, Sliver MW, Michaels AE, Davis LH (1988–1989) Enslavement of algal chloroplasts by four *Strombidium* spp. (Ciliophora, Oligotrichida). Mar Microb Food Webs 3:79–100

Stoecker DK, Johnson MD, de Vargas C, Not F (2009) Acquired phototrophy in aquatic protists. Aquat Microb Ecol 57:279–310

Strom S, Wolfe G, Slajer A, Lambert S, Clough J (2003) Chemical defense in the microplankton. II: Inhibition of protist feeding by beta-dimethylsulfoniopropionate (DMSP). Limnol Oceanogr 48:230–237

Strom SL, Wolfe GV, Bright KJ (2007) Responses of marine planktonic protists to amino acids: feeding inhibition and swimming behavior in the ciliate *Favella* sp. Aquat Microb Ecol 47:107–121

Sud GC (1968) Volumetric Relationships of Symbiotic Zoochlorellae to Their Hosts. J Protozool 15:605–607

Summerer M, Sonntag B, Sommaruga R (2007) An experimental test of the symbiosis specificity between the ciliate *Paramecium bursaria* and strains of the unicellular green alga *Chlorella*. Environ Microbiol 9:2117–2122

Summerer M, Sonntag B, Sommaruga R (2008a) Ciliate-symbiont specificity of freshwater endosymbiotic *Chlorella* (Trebouxiophyceae, Chlorophyta). J Phycol 44:77–84

Summerer M, Sonntag B, Sommaruga R (2008b) Ciliate-symbiont specificity of freshwater endosymbiotic *Chlorella* (Trebouxiophyceae, Chlorophyta). J Phycol 44:77–84

Tannock GW (1995) Normal microflora: An Introduction to Microbes Inhabiting the Human Body. Chapman and Hall, London

Trivers RL (1971) The evolution of reciprocal altruism. Quart Rev Biol 46:35–57

van Hoek AHAM, van Alen TA, Sprakel VSI, Leunissen JAM, Brigge T, Vogels GD, Hackstein JHP (2000) Multiple acquisition of methanogenic archaeal symbionts by anaerobic ciliates. Mol Biol Evol 17:251–258

Venn AA, Loram JE, Douglas AE (2008) Photosynthetic symbiosis in animals. J Exp Bot 59:1069–1080

Wagener S, Bardele CF, Pfennig N (1990) Functional integration of *Methanobacterium formicicum* into the anaerobic ciliate *Trimyema compressum*. Arch Microbiol 153:496–501

Watson RA (2006) Compositional evolution: the impact of sex, symbiosis, and modularity on the gradualist framework of evolution. The MIT Press, Cambridge

Weis D (1977) Synchronous development of symbiotic *Chlorella* within *Paramecium bursaria*. Trans Am Microsc Soc 96:82–86

Williams AG, Coleman GS (1991) The rumen protozoa. Springer, Berlin

Wood-Charlson M, Hollingsworth LL, Krupp DA, Weis VM (2006) Lectin/glycan interactions play a role in recognition in a coral/dinoflagellate symbiosis. Cell Microb 8:1985–1993

Wootton EC, Zubkov MV, Jones DH, Martel CM, Thornton CA, Roberts EC (2007) Biochemical prey recognition by planktonic protozoa. Environ Microbiol 9:216–222

Chapter 16
Paramecium as a Model Organism for Studies on Primary and Secondary Endosymbioses

Yuuki Kodama and Masahiro Fujishima

Abstract Endosymbiosis is a driving force in eukaryotic cell evolution. This phenomenon has occurred several times and has yielded a wide diversity of eukaryotic cells. Despite the importance of endosymbiosis, however, molecular mechanisms for its induction between different microorganisms are not so well known. To elucidate these mechanisms, experiments for synchronous induction of the endosymbiosis by symbionts isolated from the symbiont-bearing host cells and the symbiont-free host cells are indispensable. Also, the infection process needs to be easily observable under a microscope. In many endosymbiotic communities, however, both the endosymbionts and the symbiont-free host cells have already lost the ability to survive and grow independently. Consequently, re-induction of the endosymbiosis was difficult. We have developed optimum experimental conditions for the induction of primary and secondary endosymbiosis using the ciliate *Paramecium* and their endosymbionts.

16.1 Introduction

The ciliate *Paramecium* species are valuable cells to study mechanisms for re-establishment of endosymbiosis, in that they frequently bear prokaryotic or eukaryotic (or both) endosymbionts. Most endosymbiotic bacteria of *Paramecium* species cannot grow outside the host cell because of their reduced genome size. Although the endonuclear symbiotic bacteria species *Holospora* are also unable to grow outside the host cell, they can maintain their infectivity to new host cells for a few days at room temperature even after isolation from the host cells (Fujishima

Y. Kodama (✉)
Department of Biological Science, Faculty of Life and Environmental Sciences, Shimane University, Matsue, Japan
e-mail: kodama@life.shimane-u.ac.jp

M. Fujishima
Department of Environmental Science and Engineering, Graduate School of Science and Engineering, Yamaguchi University, Yamaguchi, Japan

G. Witzany and M. Nowacki (eds.), *Biocommunication of Ciliates*,
DOI 10.1007/978-3-319-32211-7_16

et al. 1991). Although the host can acquire various stress resistances by infection of *Holospora*, this symbiont is not necessary for the host's survival. Consequently, re-establishment of endosymbiosis between the *Holospora*-free paramecia and *Holospora* cells isolated from the *Holospora*-bearing paramecia can be induced easily through the host active phagocytosis by mixing them. *P. caudatum* and *Holospora* species are model organisms for researches on the induction of primary symbiosis, because the endosymbiosis can be induced synchronously and whole processes of the re-establishment of endosymbiosis are observable under an ordinal light microscope. Furthermore, macronuclear genomes of *P. tetraurelia* (Aury et al. 2006), *P. caudatum* (McGrath et al. 2014) and draft genomes of three *Holospora* species (Dohra et al. 2013, 2014) were sequenced.

On the other hand, *P. bursaria* and *P. chlorelligerum* (Kreutz et al. 2012) has the ability to keep symbiotic *Chlorella* species in the cytoplasm among *Paramecium* species. Irrespective of the mutual relationship between *P. bursaria* and the symbiotic algae, both cells are still keeping the ability to grow independently, and the endosymbiosis can be re-established synchronously by mixing them. Kodama and Fujishima (2005) found four important cytological events needed for establishing endosymbiosis and their timings in the infection process by pulse-labeling of the alga-free paramecia for 1.5 min with the symbiotic algae isolated from the alga-bearing paramecia and then chasing at known times. *P. bursaria* and the symbiotic *Chlorella* cells also became model organisms for studying the induction of secondary symbiosis. The nuclear genome of the symbiotic *Chlorella variabilis* was sequenced (Blanc et al. 2010), and RNAseq analysis between *P. bursaria* with and without the algae has been done (Kodama et al. 2014). Thus, interactions between *Paramecium* and *Holospora* species and between *P. bursaria* and its symbiotic *C. variabilis* cells provide excellent opportunities to study control mechanisms for establishment of the primary and the secondary symbioses leading to eukaryotic cell evolution. In this chapter, we introduce recent studies on (1) how the symbiont invades the host cytoplasm, (2) how the symbiont can avoid digestion by the host's lysosomal enzymes, (3) how the symbiont can grow synchronously with the host cell, (4) how the host gene expressions are affected by the symbiont, and (5) what benefit the host cell receives which enables it to expand its ecological niche.

16.2 Induction of Re-establishment of Primary Endosymbiosis Between *Paramecium* and *Holospora*

The Gram-negative bacterium *Holospora* species are endonuclear symbionts of the ciliate *Paramecium* species (Fokin and Sabaneyeva 1997; Fokin and Görtz 2009; Fujishima 2009; Gibson et al. 1986; Preer 1969; Ossipov 1973; Ossipov et al. 1975, 1980; Skoblo and Lebedeva 1986) and belong to alfa-proteobacteria (Amann et al. 1991). Phylogenetically most related bacteria with *Holospora* is Rickettsia (Amann et al. 1991; Lang et al. 2005). *Holospora* species are usually found in paramecia

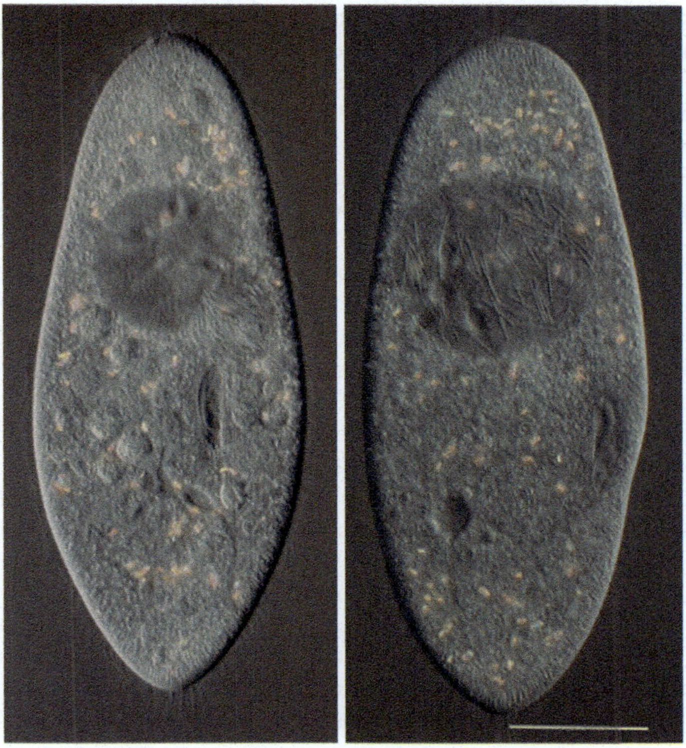

Fig. 16.1 Photomicrographs of *Holospora* in *Paramecium caudatum*. *Left H. undulata* in a micronucleus. *Right H. obtusa* in a macronucleus. *Bar* 50 μm

living in cold areas, such as northern Europe and the Kamchatka Peninsula (Fokin et al. 1996). To date, nine *Holospora* species have been described (Fokin et al. 1996). All show species-specificity and nucleus-specificity in their habitats (Fig. 16.1). They cannot grow outside the host cell with ordinary culture media because of their reduced genome (Dohra et al. 2013, 2014). *Holospora* species show two different forms in their life cycle: a reproductive short form (RF, 1.5–2 μm long) and an infectious long form (IF, 10–15 μm long) (Fokin et al. 1996; Fujishima et al. 1990b; Görtz 1980; Görtz et al. 1989; Gromov and Ossipov 1981). The bacterium exists as a short RF cell and divides by binary fission in the host nucleus when the host is growing. The RF stops dividing and differentiates into a longer IF cell through intermediate forms when the host cell starves (Fujishima et al. 1990a; Görtz 1983), or the host protein synthesis is inhibited (Fujishima, unpublished data). During this differentiation, the bacterium forms a distinctive structure, one-half of which contains the cytoplasm; the other half is a periplasmic lumen with an electron-translucent tip called as invasion tip (Dohra and Fujishima 1999; Fujishima and Hoshide 1988; Görtz 1980; Görtz and Wiemann 1989; Görtz et al. 1989; Iwatani et al. 2005). The IF cells engulfed into the host digestive vacuoles (DVs) escape with the invasion tip ahead and penetrate the target nuclear envelope with this special tip (Fujishima and Fujita 1985; Fujishima and Kawai 2004; Görtz and Wiemann 1989). Under a phase-contrast microscope, the cytoplasmic region looks dark, but the periplasmic region looks as a refractile (Dohra and Fujishima 1999). In the macronucleus-specific *H. obtusa* of *P. caudatum*, the IF cells show clear two nucleoids (Fujishima et al. 1990a; Dohra and Fujishima 1999) stained with 4′, 6-diamidino-2-phenylindole dihydrochloride (DAPI). This bacterium changes the buoyant density, protein composition (Fujishima et al. 1990a), and surface morphology of the outer membrane (Fujishima et al. 1990b) during differentiation. When the host divides again, the IF cells of *H. obtusa* are collected in a connecting piece of the dividing nucleus and they are freed from the nucleus by wrapping with the nuclear membrane. They are eventually expelled from the host cytoproct (Wiemann 1989). On the other hand, the outer membrane of the RF has a stronger affinity to bind the host chromatin than the IF cells, so that the RF cells can remain in the daughter nuclei when the host divides (Ehrsam and Görtz 1999; Fokin et al. 1996; Görtz et al. 1992; Wiemann 1989). When the macronucleus is filled with so many infectious forms, the host cells cannot grow even in sufficient foods in the culture medium and eventually killed by the bacteria; the infectious forms are freed from the cells (Fujishima, unpublished observation). Consequently, the infectious forms appear outside the host cell by these two means and can then infect new host cells. A *Paramecium* cell has a limited life span. Therefore, *Holospora* species need to escape from the host to infect more young cells. For this reason, a different nature of the outer membranes of these two forms is indispensable for *Holospora*'s survival strategy.

The phenomenon of bacterial invasion into a target nucleus is designated as "infection", and stable growth of the infected bacteria in the nucleus is designated as "maintenance" (Fujishima and Fujita 1985). The infection is controlled by (1) engulfment of the IFs into the host DVs (Fujishima and Görtz 1983), (2) escape from the DV before the host's lysosomal fusion to appear in the host cytoplasm (Iwatani et al. 2005), (3) migration to the target nucleus by a help of the host actins (Fujishima 2009; Fujishima et al. 2007; Sabaneyeva et al. 2009), (4) recognition of a target nuclear envelope by a specific binding between *Holospora*'s outer membrane substance and their target nuclear envelope (Fujishima and Kawai 2004) and by a penetration of the host nuclear envelope with the invasion tip (Iwatani et al. 2005). On the other hand, the maintenance is controlled by the host genotypes (Fujishima and Mizobe 1988). Namely, infection and maintenance are independently controlled phenomena. The whole infection process occurs within 10 min (Fujishima and Görtz 1983). To date, the only organism having an ability to distinguish a somatic macronucleus from a germinal micronucleus of the host *Paramecium* species is *Holospora* species. Thus, these bacteria can recognize some differences between the two kinds of the host nuclei originated from a common fertilization nucleus and timing of the nuclear differentiation (Fujishima and Görtz

1983). After infection, *Holospora* alters the host gene expressions (Hori and Fujishima 2003; Hori et al. 2008; Nakamura et al. 2004), and the host acquires various stress resistances (Fujishima et al. 2005; Hori and Fujishima 2003; Hori et al. 2008; Smurov and Fokin 1998).

16.2.1 Genome of Holospora

Draft genome sequences have been determined in three *Holospora* species of *P. caudatum*; a macronucleus-specific *H. obtusa* and micronucleus specific *H. undulate* and *H. elegans* (Dohra et al. 2013, 2014). Among these three *Holospora* genomes, assembly lengths and GC% varied from 1.27 to 1.40 Mbp and 35.2–36.1 %, respectively (Dohra et al. 2014). The FASTORTHO program (http://enews.patricbrc.org/fastortho/) grouped a total of 3553 protein-coding sequences from the three *Holospora* species genomes into 1610 ortholog clusters, of which 572 were identified as single-copy core orthologous genes shared by the three genomes. Of the 572 *Holospora* core genes, 488 (85.3 %) were assigned to at least one of the cluster of orthologous groups (COGs). The 46 genes were assigned to multiple functional categories; for example, type II secretory pathway proteins were assigned to the COG category of cell motility and of intracellular trafficking, secretion, and vesicular transport.

The cytoplasmic endosymbiotic bacterium *Polynucleobacter necessarius* of the ciliate *Euplotes aediculatus* possesses glycolysis/gluconeogenesis, citrate cycle (TCA cycle), and pyruvate metabolism pathways for energy production (Boscaro et al. 2013). However, The *Holospora* genomes lacked many proteins involved in these pathways, indicating that *Holospora* species strongly depend on the host for energy production (Dohra et al. 2014).

16.2.2 How Does Holospora *Invade the Host Cytoplasm and Migrates to the Target Nucleus?*

Life cycle of *Holospora* is shown in Fig. 16.2. When IF long form of *H. obtusa* cells isolated from the symbiotic host *P. caudatum* cells by Percoll density gradient centrifugation are mixed with the aposymbiotic hosts, the bacteria are soon ingested into the host DVs. The DVs of *P. caudatum* can be classified into four different stages according to Fok and Allen (1988) in *P. multimicronucleatum*. The IF cells ingested in DV-I vacuole escape there by destruction of the DV membrane while the DV-I vacuole is acidified by acidosomal fusion and becomes a condensed DV-II vacuole (Fujishima 2009). In the presence of vacuolar-type ATPase (V-ATPase) inhibitors, concanamycin A, both the acidification of the DV and the bacterial escape from the DV are inhibited completely (Fujishima and Kawai 1997). These

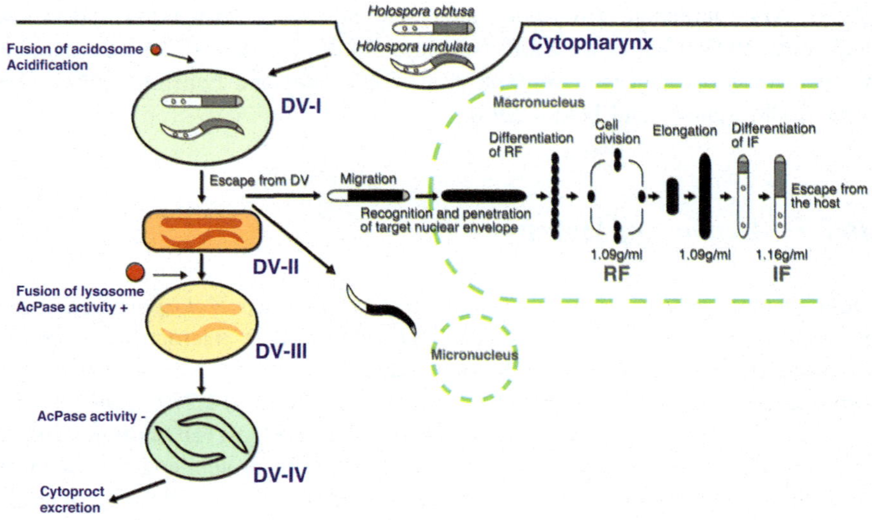

Fig. 16.2 The infection route and life cycle of *Holospora* species. Spherical DV-I vacuole differentiates to condensed and acidified DV-II vacuole by fusion of acidosomes and evagination of the DV membrane to cytoplasm. Then the vacuole differentiates to swollen DV-III vacuole by fusion of lysosomes. Undigested materials remain in the DV-IV vacuole. The DV-IV vacuole fuses to a cytoproct of the host cell and the contents are discharged. Some IF cells of *Holospora* escape from the acidified DV without wrapping with the DV membrane. By acidification of the DV, the bacteria differentiate to the activated forms, and migrate toward the target nucleus with a help of the host actin polymerization. The bacteria distinguish their target nucleus by specific binding between lipopolysaccharides of the outer membrane and the unknown nuclear envelope substance. Then, the bacterium penetrates the target nuclear envelope with an invasion tip leading. After the invasion, bacterial cytoplasmic region increases and large periplasmic region decreases to form constrictions for differentiation to the RFs. During this infection process, the bacterium decreases it buoyant density from 1.16 to 1.09 g/ml. The RF continues to divide by binary fission when the host cell is growing, but the RF halts the binary fission, elongates itself and differentiates to the IF when the host cell starves or the host's protein synthesis is inhibited. During this differentiation, the bacterium increases the buoyant density, forms a large periplasmic region, an invasion tip, and two nucleoids. The infectious forms are freed from the cells (see text). From Fujishima (2009)

results depicted that the acidification of the host DV is an indispensable phenomenon for the bacterial escape from the host DV. Bacteria in the host DVs just before the escape there and those appearing in the host cytoplasm are designated as an activated form (AF) cell. The AF cell looks darker than the IF cell under a phase-contrast microscope (Görtz and Wiemann 1989). The IF cell always escapes from the DV with the invasion tip ahead (Görtz and Dieckmann 1980; Iwatani et al. 2005), and penetrates the target nucleus with the invasion tip leading. We harvested the tips of the IF cells of *H. obtusa* from 3438 bacteria using a laser capture microdissection system (LM 100; Olympus), and loaded to SDS-PAGE. Then, three bands of 89, 76, and 63 kDa were detected by silver staining. Using proteins of 60–90 kDa extracted from the gel as antigens, we developed monoclonal antibodies (mAbs) against the 89 kDa protein in the invasion tip (Iwatani et al. 2005).

Indirect immunofluorescent microscopy and immunoblotting showed that this protein is specific for the invasion tip of *H. obtusa* but not with the RF cells. Subsequently, using partial amino acid sequence of the purified 89-kDa protein, a novel gene encoding the 89 kDa protein was cloned from genomic DNA. The open reading frame of the gene was 2253 nt long with a 32.5 % G + C content. The predicted amino acid sequence of the 89 kDa protein showed two transmembrane signal peptides at N-terminal and two actin-binding motifs near the N-terminal (Iwatani et al. 2005).

Indirect immunofluorescence microscopy with mAbs specific for the 89 kDa protein and the host actin 1–1 showed that the epitopes of the five kinds of mAbs against the 89 kDa protein were present in a lumen of the invasion tip of the IF cell. However, some epitopes translocate outside the bacterial outer membrane of the invasion tip when the IFs were engulfed into the host DVs. Bacterium appeared in the host cytoplasm kept the 89 kDa proteins outside the tip, and the host actins accumulated around the 89 kDa proteins immediately after the bacterial escape from the host DV. When the bacterium penetrated the host macronuclear envelope, a complex of the 89 kDa proteins and the host actins were left behind at the entry point on the nuclear envelope as a cylindrical structure (Iwatani et al. 2005; Fujishima 2009; Fujishima et al. 2007; Fujishima and Kodama 2012). Sabaneyeva et al. (2009) also observed similar actin-based *H. obtusa* motility. These results suggest that the 89 kDa protein and the host actin are responsible for the infection of *Holospora*. However, how the bacteria destruct the host DV membrane by the invasion tip, how the bacterium appeared in the cytoplasm knows a direction to the target nucleus, how the bacterium penetrates the target nuclear envelope, and what is the moving force to push the bacteria from nuclear envelope to the inside the nucleus are unknown.

16.2.3 How Can **Holospora** *Distinguish Their Target Nucleus in Infection Process?*

To know how the *Holospora* recognizes two kinds of the host nuclei, a macronucleus and a micronucleus, mAbs specific for outer membranes of IF cells of *H. obtusa* and *H. undulata* were developed respectively. When the antigens extracted from the SDS-PAGE gels were mixed with freshly isolated nuclei of *P. caudatum,* indirect immunofluorescence with the mAbs showed that the antigens bound with nuclear envelopes of their target nuclei (Fujishima and Kawai 2004). Namely, outer membrane substances of *H. obtusa* bound only to the macronuclear envelope of *P. caudatum,* and those of *H. undulate* bound with the micronuclear envelope. These antigens are resistant against proteinase K and can be stained neither with Coomassie Brilliant Blue R-250 nor by an ordinary silver stain. However, the bands of the antigens were stained with silver for bacterial lipopolysaccharide (LPS). This indicates that their outer membrane substances are

LPSs (Fujishima and Kawai 2004). These results show that the bacterial recognition of their target nuclei is controlled by a specific binding between the LPS of the IF cell and an unknown receptor substance exposed on the target nuclear envelope.

When the IF cells of *H. obtusa* and exconjugants of *P. caudatum* at various stages were mixed, the infectability against *H. obtusa* was acquired by four of the eight post-zygotic nuclei as soon as the four nuclei differentiated morphologically into macronuclear anlagen. Old macronuclear fragments were also infected. These results indicate that the nuclear envelope of the macronuclear anlagen exposes the receptor substance against *H. obtusa* LPS at almost the same time as the first recognizable change in the macronuclear anlagen, and that the receptor substance has been kept on the old macronuclear fragments (Fujishima and Görtz 1983).

The property of the macronucleus, necessary for it to be recognized and infected by *H. obtusa,* is commonly provided by *P. caudatum, P. multimicronucleatum,* and *P. aurelia* species but not by *P. bursaria, P. trichium* (=*P. putrinum*), *P. duboscqui* and *P. woodruffi*, although the bacteria can appear in the cytoplasm through the DVs (Fujishima and Fujita 1985; Fujishima 1986). *P. calkinsi, P. polycaryum* and *P. nephridiatum* also could not be infected by *H. obtusa* (Fujishima, unpublished data). Infectivity of *P. jenningsi* by *H. obtusa* was strain-specific (Fujishima, unpublished data). All strains of *P. caudatum, P. multimicronucleatum,* and *P. aurelia* species examined were infected by *H. obtusa*. However, stable maintenance of the infected *H. obtusa* in the host nucleus was achieved only in specific strains of *P. caudatum.* Species-specific infectivity of *H. obtusa* (Fujishima and Fujita 1985; Fujishima 1986) and phylogenetic tree of *Paramecium* species (Fokin et al. 2004; Hori et al. 2006; Kreutz et al. 2012) show that *H. obtusa* recognizes and invades the macronucleus of closely related species with *P. caudatum* (Fujishima 2009).

16.2.4 How Can Holospora *Avoid Digestion by the Host's Lysosomal Enzymes?*

It is known that intracellular symbionts or parasites use one of three strategies to survive against the host lysosomal digestion; (1) escape from the DVs before lysosomal fusion, (2) prevent fusion of the DVs with lysosomes, and (3) survive in the DVs even after the lysosomal fusion. In case of *Holospora*, they escape from the host DVs after acidosomal fusion but before the host lysosomal fusion (see Fig. 16.2). The IFs hole in the acidified DV membrane by their invasion tip and escape there without wrapping with the DV membrane (Fujishima, unpublished observation). Lysosomal fusion occurs 5–10 min after the DV formation in DV-III. The IFs could not escape from the DV-III to appear in the host cytoplasm. Therefore, only few IFs can escape from the DVs, and most of the IFs in the DVs are partially digested in the DV-III and discharged from a host cytoproct. On the other hand, RFs cannot escape from the host DVs and digested.

16.2.5 How Does *Holospora* Grow in Well Accordance with the Host Growth?

RF cells of *Holospora* continue binary fission in the host nucleus when the host is growing. However, the RF halts the binary fission and differentiates into an IF cell through intermediate forms when the host cell starves or host's protein synthesis is inhibited (Görtz 1983; Fujishima et al. 1990a). This suggests a possibility that RF cell is importing the host nuclear proteins for their growth and for keeping functions and morphology of the RF cell. Actually, 2D-SDS-PAGE showed more than 60 % of proteins of the RF and the IF cells of *H. obtusa* were different (Fujishima et al. 1990a). During this differentiation, *Holospora* changes nature of their outer membrane. The outer membrane of the RF cells has a stronger affinity to bind the host chromatin than the IF cells, so that the RF cells remain in each daughter nucleus when the host cell divides (Ehrsam and Görtz 1999; Fokin et al. 1996; Görtz et al. 1992; Wiemann 1989). On the other hand, the IF cells are collected in a connecting piece of the dividing nucleus. Then, they are freed from the dividing nucleus by wrapping with the nuclear membrane, and eventually expelled from the host cytoproct (Wiemann 1989). Furthermore, when the macronucleus is filled with many IF cells, the host cannot grow and killed by the bacteria. Eventually, the IF cells appeared outside the host cell by these two means, and can infect new host cells. Because the host *Paramecium* cell has a limited life span, *Holospora* must escape from the host to infect young cells. For this reason, different natures of the outer membranes of the RF and the IF cells are indispensable for *Holospora*'s survival strategy.

16.2.6 How Does *Holospora* Alter Host Gene Expressions by Infection?

Differential display and reverse transcribed PCR analysis showed that *H. obtusa* alters multiple gene expression of the host after establishing endosymbiosis (Nakamura et al. 2004) including *hsp60* and *hsp70* gene of the host (Hori and Fujishima 2003). We found that a periplasmic 63 kDa protein of *H. obtusa* might be one of the causes for induction of the host's gene alteration (Abamo et al. 2008). The 63 kDa protein is an IF specific protein and presence in the periplasmic lumen except an invasion tip. Indirect immunofluorescence microscopy showed that not only the pre-existing but also a newly synthesized 63 kDa protein was secreted into the host macronucleus in early infection process (Abamo et al. 2008). A gene encoding the 63 kDa protein was cloned from genomic DNA of *H. obtusa*. This novel gene included 1644 nucleotides encoding a 547-amino acid sequence. Comparison between the deduced amino acid sequence and the N-terminal amino acid sequence of the 63 kDa protein purified from 2D-SDS-PAGE gels revealed that the protein was preceded by a putative signal peptide consisted of 24 amino acids. Therefore, the mature protein comprises 523 amino acids with a predicted molecular mass of 62.151 kDa and a predicted pI of 8.92 (Abamo et al. 2008). Considering the amount of the 63 kDa protein secreted into the macronucleus and the fact that the fluorescence of the 63 kDa protein cannot be observed in the host cytoplasm, we can speculate that this protein might bind to the host DNA or chromatin and changes the host gene expression to the advantage of the bacteria as shown in pathogenic bacterium *Listeria monocytogenes* (Lebreton et al. 2011).

Alteration of the host's gene expression is a general phenomenon for endosymbiosis. Therefore, *Paramecium* and *Holospora* might serve as a good model system to elucidate the mechanism of the pathogen-induced alteration of the host's gene expression.

16.2.7 What Kind of Benefit Does the Host Cell Receive by Infection of Holospora*?*

Three types of *P. caudatum* cells (*H. obtusa*-free cells, reproductive form of *H. obtusa*-bearing cells and predominantly infectious form of *H. obtusa*-bearing cells) cultured at 25 °C were transferred to 4, 10, 25, 35 and 40 °C and their swimming velocities were measured by taking photomicrographs with two-second exposures. The *H. obtusa*-free cells almost ceased swimming at 4 and died soon at 40 °C, while the reproductive form-bearing cells still swam even at these temperatures. Predominantly infectious form of *H. obtusa*-bearing cells also swam though their swimming velocity was statistically slower than that of the reproductive form-bearing *P. caudatum* cells. Thus, *Holospora*-bearing *Paramecium* cells can acquire heat-shock resistance if the host bears RF cells (Fujishima 2009; Fujishima et al. 2005; Hori et al. 2008; Hori and Fujishima 2003). Furthermore, the *Holospora*-bearing paramecia become osmotic shock resistance (Smurov and Fokin 1998). Therefore, *Paramecium* cells become adapted to unsuitable environments for their growth by endosymbiosis with *Holospora* species. Actually, *Holospora*-bearing paramecia can be collected in brackish water.

16.3 Induction of Re-establishment of Secondary Endosymbiosis Between *Chlorella* Species and *P. bursaria*

P. bursaria can maintain several hundred endosymbiotic algae in their cytoplasm (Figs. 16.3a and 16.4a). Each symbiotic alga wrapped with a perialgal vacuole (PV) membrane (Fig. 16.3b), and attaches near the host cell cortex (Fig. 16.3a). The PV membrane has an ability for avoiding host lysosomal fusion (Gu et al. 2002; Kodama and Fujishima 2009b). The association of *P. bursaria* with the

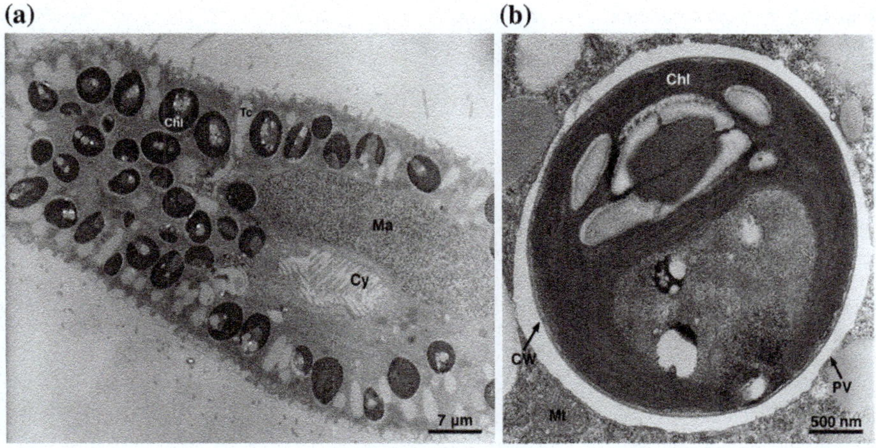

Fig. 16.3 Transmission electron micrographs of *P. bursaria*. **a** Algae-bearing *P. bursaria*. **b** Symbiotic alga near the host cell cortex. Each symbiotic alga wrapped with a PV membrane. *Chl* Chloroplast; *Cy* Cytopharynx; *CW* Cell wall; *PV* Perialgal vacuole; *Ma* Macronucleus; *Mt* Mitochondrion; *Tc* Trichocyst. From Kodama and Fujishima (2010a)

symbiotic *Chlorella* sp. is a mutualism. The host supplies the algae with nitrogen components and CO_2 (Reisser 1976, 1980; Albers and Wiessner 1985), and the host protects algae in the PVs from infection by the *Chlorella* virus (Kawakami and Kawakami 1978; Van Etten et al. 1985; Reisser et al. 1988; Yamada et al. 2006). Also, algal carbon fixation is enhanced in the host (Kamako and Imamura 2006; Kato and Imamura 2009). On the other hand, the symbiotic algae can supply the host with photosynthetic products, mainly maltose (Reisser 1976; Brown and Nielsen 1974; Reisser 1986). The algae in the host show a higher rate of photosynthetic oxygen production than in their isolated state, thereby guaranteeing an oxygen supply for the host respiration (Reisser 1980). Algae-bearing *P. bursaria* can grow better than the algae-free cells (Görtz 1982; Karakashian 1963, 1975); the algae have UV-protective role for the host (Hörtnagl and Sommaruga 2007; Summerer et al. 2009). Because timing of cell divisions of both the algae and the host cells is well coordinated, the symbiotic algae are transferred to the both daughter cells (Kadono et al. 2004; Takahashi et al. 2007).

Irrespective of the mutually beneficial relationships between *P. bursaria* and symbiotic algae, their relationship is facultative mutualism. Algae-free *P. bursaria* as shown in Fig. 16.4b can be easily produced from algae-bearing cells by rapid fission (Jennings 1938), cultivation in darkness (Karakashian 1963; Pado 1965; Weis 1969), X-ray irradiation (Wichterman 1948), treatment with 3-(3,4-dichlorophenyl)-1,1-dimethylurea (DCMU), which is an inhibitor of photosynthesis (Reisser 1976), by treatment with the herbicide paraquat (Hosoya et al. 1995; Tanaka et al. 2002) or by treatment with cycloheximide (Kodama et al. 2007; Weis 1984; Kodama and Fujishima 2008). Furthermore, endosymbiosis between the algae-free *P. bursaria* cells and the symbiotic algae isolated from the algae-bearing

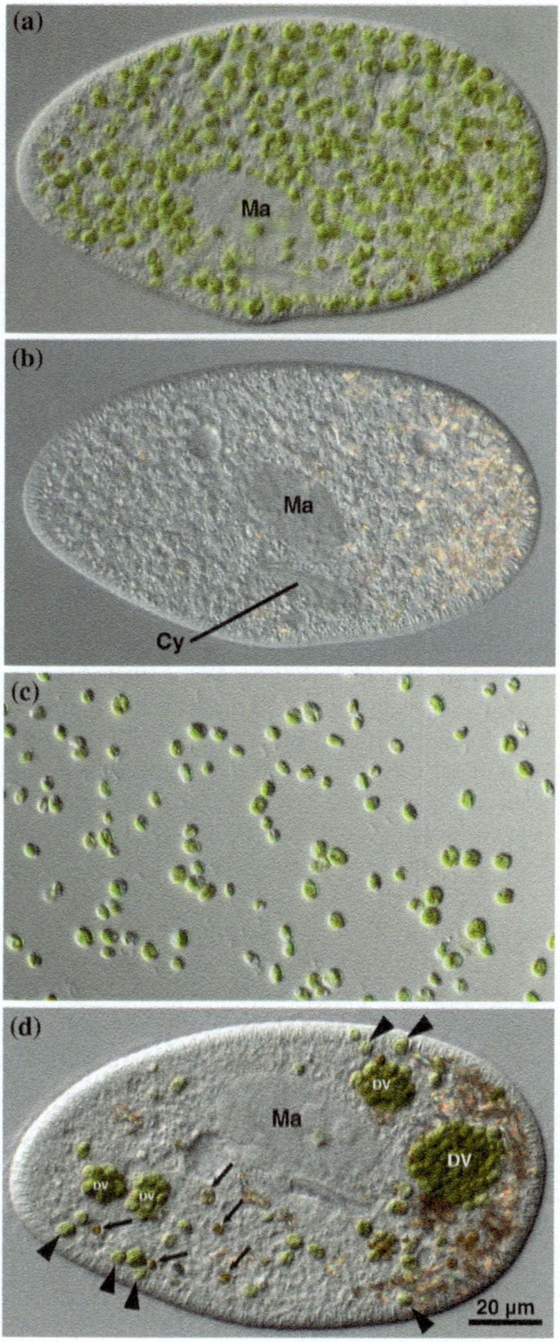

Fig. 16.4 Photomicrographs of *P. bursaria* and isolated symbiotic *Chlorella* sp. **a** Algae-bearing *P. bursaria* cell. **b** Algae-free *P. bursaria* cell. **c** Isolated symbiotic *C. variabilis* cells. **d** Algae-free *P. bursaria* cell, which was mixed with isolated symbiotic algae for 1.5 min, was washed and incubated for 3 h. Arrowhead shows green alga, which establishes endosymbiosis with algae-free cell. The alga localized immediately beneath the paramecium cell cortex. Arrows indicate digested brown alga. *DV* Digestive vacuole. Both green and digested algae appear in the cytoplasm as a result of the budding from the DV. *Ma* Macronucleus; *Cy* Cytopharynx. From Kodama and Fujishima (2012b). (Color figure online)

P. bursaria cells (Fig. 16.4c) is artificially re-established by just mixing them together (Siegel and Karakashian 1959; Karakashian 1975) (Fig. 16.4d). Therefore, the symbiotic associations between these eukaryotic cells are excellent models for studying cell-to-cell interaction and the evolution of eukaryotic cells through secondary endosymbiosis.

16.3.1 Classification of the Host DVs Appearing in Re-Establishment of Endosymbiosis with Chlorella

To understand the re-establishment route of symbiotic *Chlorella* cells, stages of DVs that appear during re-establishment of endosymbiosis were classified and the timing of the appearance of each stage was determined by mixing algae-free paramecia with the isolated symbiotic algae. The cells were mixed at a density of 5×10^3 paramecia per ml with isolated *Chlorella* sp. at 5×10^7 algae per ml in a centrifuge tube (volume, 10 ml) under a fluorescent light (20–30 μ mol photons m^{-2} s^{-1}) for 1.5 min at 25 ± 1 °C. The ciliate-algae mixture was transferred to a centrifuge tube equipped with a 15 μm pore size nylon mesh and filtered. By pouring 30 ml of fresh modified Dryl's solution (MDS, KH_2PO_4 was used instead of $NaH_2PO_4 \cdot 2H_2O$) (Dryl 1959) into this tube, the paramecia were washed and algal cells outside the paramecia were simultaneously removed through the mesh. The paramecia retained on the mesh were harvested and transferred to a centrifuge tube and resuspended in 1 ml of MDS, and then chased for various times under a fluorescent light at 25 ± 1 °C. Aliquot of the cell suspension was fixed by 4 % paraformaldehyde (PFA) at various time points, and the cells were observed under a differential-interference-contrast (DIC) microscope (Kodama and Fujishima 2005).

The DVs observed during the algal infection process were classified into eight different stages on the basis of their morphologies and on changes in algal color and pH in the DVs (Fig. 16.5). The DV-I vacuole has a rounded vacuole membrane containing only green algae. Its membrane is clearly visible under a DIC microscope. DV-II has a reduced size and the vacuole membrane barely visible; the algae are green. In DV-III, the vacuole has increased in size, making its membrane visible; the algae are discolored–either faint yellow or green, or both. The DV-III stage is further classified into three substages: DV-IIIa contains green algae only; DV-IIIb contains both faint yellow and green algae, and DV-IIIc contains faint yellow algae only. In the final stage, DV-IV, the vacuolar size is again reduced, as in DV-II, rendering the vacuole membrane barely visible under a DIC microscope; the algae are green or brown, or both. This vacuole was observed in cells after 20–30 min. DV-IV was also further classified into three sub-stages: DV-IVa contains green algae only; DV-IVb contains both green and brown algae, and DV-IVc contains brown algae only. DVs containing single green *Chlorella* (SGC) were observed in cells fixed 30 min after mixing, but all SGCs present in cells before 30 min after mixing were digested for 30 min (Kodama and Fujishima 2005).

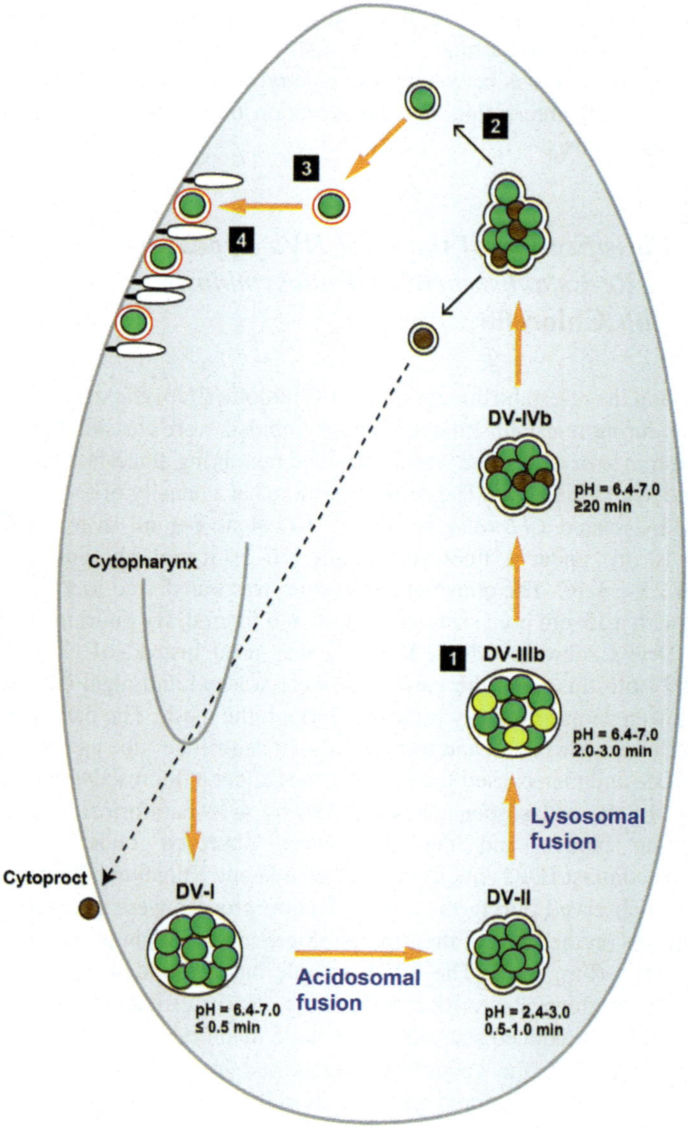

Fig. 16.5 Schematic representation of the algal reinfection process. Using pulse label and chase method, four important cytological events necessary to establish endosymbiosis were clarified. About 3 min after mixing with algae-free *P. bursaria* cells and isolated symbiotic algae, some algae acquire resistance to the host lysosomal enzymes in the DVs. About 30 min after the mixing, the algae start to escape from the DVs as a result of the budding of the membrane into the cytoplasm. About 45 min after the mixing, the DV membrane enclosing SGC differentiates to the PV membrane, which provides protection from lysosomal fusion. Then, the SGC localizes beneath the host cell cortex. About 24 h after the mixing, the SGC starts to increase by cell division and establishes endosymbiosis. Modified from Kodama and Fujishima (2005)

16.3.2 *Four Important Events in Re-establishment of Endosymbiosis*

Four important cytological events needed for establishing endosymbiosis and their timings in the infection process were clarified on the basis of the DV stages as described above (Kodama 2013; Kodama and Fujishima 2005, 2007, 2008, 2009a, b, c, 2010a, b, 2011, 2012a, b, 2014; Kodama et al. 2007, 2011). These four cytological events are described below.

16.3.2.1 Event One

After the lysosomal fusion to the DVs, some algae show temporary resistance to the host's lysosomal enzymes in the DV-IIIb and DV-IVb, even when the digested ones coexist. This phenomenon depends on photosynthetic activity of the isolated algae before mixing with *P. bursaria*. When the isolated algae were kept constantly under dark (DD) conditions for 24 h, almost all algae were digested in the DV. The detailed results were shown in Sect. 16.3.3. Thus, it can be said that the symbiotic algae do not prevent acidification and lysosomal fusion of the host's DV during the re-establishment of endosymbiosis (Kodama and Fujishima 2005).

16.3.2.2 Event Two

Thirty minutes after the mixing algae-free *P. bursaria* and isolated symbiotic algae, the algae start to escape from DV-IVb vacuoles as the result of budding of the membrane into the cytoplasm. Both living and digested algae bud from the DVs of *P. bursaria* (Kodama and Fujishima 2005). *Saccharomyces cerevisiae* cells and polystyrene latex beads of a diameter of 3 μm or greater were able to bud, too (Kodama and Fujishima 2012b). However, this budding is not observed when India ink, 0.81 μm diameter polystyrene latex beads, or food bacteria (*Klebsiella pneumonia*) were ingested into the DVs (Kodama and Fujishima 2005). These results suggest that *P. bursaria* can recognize the diameter of the contents of the DVs, and that those with a diameter of about 3 μm or greater can escape from the DV by the budding of the DV membrane. Because Dynasore, a dynamin inhibitor, greatly inhibited DV budding, dynamin might be involved in this process.

16.3.2.3 Event Three

After the budding from the DV-IVb vacuole, the DV membrane enclosing SGC differentiates into the PV membrane, which provides protection from lysosomal fusion (Kodama and Fujishima 2005, 2009a, b). To understand the timing of differentiation of PV from the host DV, algae-free *P. bursaria* cells were mixed with isolated symbiotic algae for 1.5 min, washed, chased, and fixed at various times after mixing. Then, lysosomal enzyme, acid phosphatase (AcPase) activity in the vacuoles enclosing the algae was detected using Gomori's staining (Gomori 1952). This activity appears in 3 min-old vacuoles; all DVs containing algae demonstrate the activity at 30 min. Algal budding from the DVs begins at 30 min as described above. In the budded membrane, each alga is surrounded by a layer of Gomori's thin positive staining. The vacuoles involving a SGC move quickly and attach immediately beneath the host cell cortex. The first SGC and the first attachment of the SGC beneath the host cell cortex, respectively, occur at 30 and 45 min after mixing. These results suggest that differentiation of the PV membrane occurs within 15 min after the algal budding from the host DV (Kodama and Fujishima 2009c). We have succeeded in developing monoclonal antibodies (mAb)s specific for the DV membrane of *P. bursaria*. These mAbs do not react with the PV membrane, which containing SGC(s). This indicates that both membranes are substantially different (Fujishima and Kodama, unpublished data).

16.3.2.4 Event Four

The SGC(s) wrapped by the PV membrane localize beneath the host cell cortex (Kodama 2013; Kodama and Fujishima 2005, 2011, 2013). Both many trichocysts and mitochondria also localize in this area (Fujishima and Kodama 2012). Gomori's staining showed that the AcPase activity is low in this area (Kodama and Fujishima 2008, 2009b). These observations reflect the possibility that the PV membrane might have no capability for protection from lysosomal fusion, but can avoid lysosomal fusion by binding to the mitochondoria, trichocysts or unknown structures near the host cell cortex to localize at the area of the cell where primary lysosomes are usually missing. To confirm this possibility, preexisting trichocysts beneath the host cell cortex were removed from *P. bursaria* cells through treatment with lysozyme, thereby reducing the AcPase activity-negative area and exposing the PVs to the AcPase activity-positive area, and examined whether the PV's protection from the lysosomal fusion is still achieved or not. The trichocyst-free cell reduced the AcPase activity-negative cortical layer to less than 3 μm depth at the dorsal cortex. However, even if a part of the algal cell had been exposed in the AcPase activity-positive area, the algae were able to attach beneath the host cell cortex and to protect it from lysosomal fusion (Kodama and Fujishima 2009b). This is the first evidence to demonstrate that the PV membrane can give protection from host lysosomal fusion, and that the PV membrane does not require trichocysts for intracellular localization, because the PV membrane could localize the trichocyst-free cell cortex. This result suggests the possibility that the mitochondria anchor the PV membrane near the host cell cortex (Kodama and Fujishima, unpubl. data). Schematic representation of algal reinfection process and four important events in re-establishment of endosymbiosis is summarized in Fig. 16.5.

16.3.3 Algal Resistance to the Host Lysosomal Enzymes

During the algal infection process, the first hurdle for the algae is acquisition of resistance to the host's lysosomal enzymes in the DV as the event one (Kodama and Fujishima 2005). In the event one, some of the algae are not digested in DVs that had been fused with the host lysosomes even in the presence of others that are being digested in a same DV (Figs. 16.6 and 16.7). This differential fate of algae in the same DV is not an inherent property of the algae because this phenomenon occurs

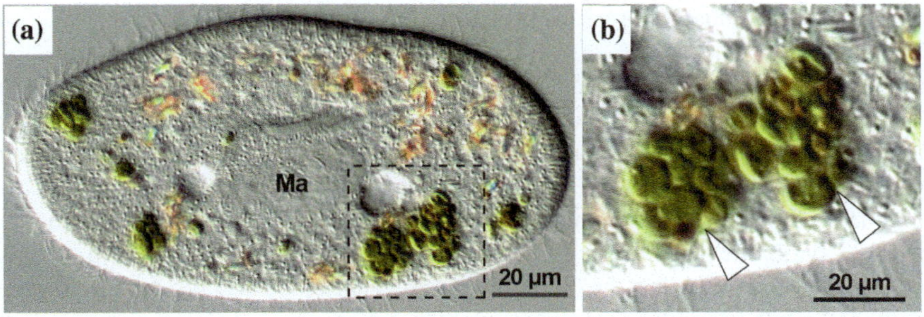

Fig. 16.6 Photomicrograph of algae-free *P. bursaria* 1 h after mixing with isolated symbiotic algae. **b** Shows highly magnified images of the square enclosed area in **a**. As shown by the white arrowhead in **b**, some algae were not digested even if coexisted with the digested brown ones in the same DVs after lysosomal fusion. *Ma* Macronucleus. From Kodama and Fujishima (2010b)

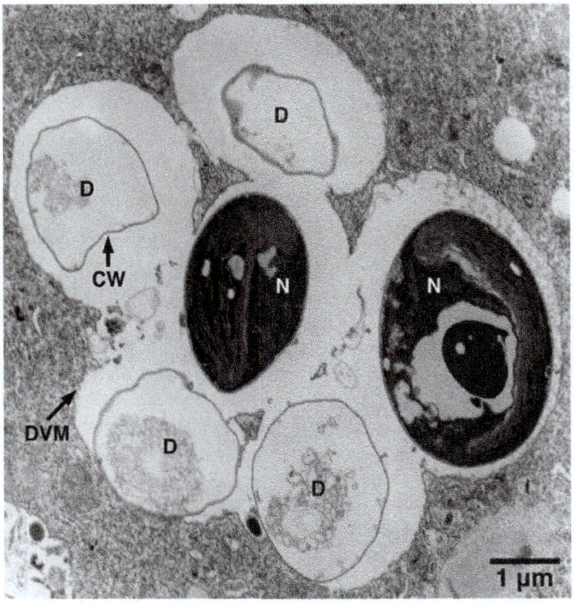

Fig. 16.7 Transmission electron micrograph of a DV-IVb. Three hours after mixing with algae and algae-free *P. bursaria* cells, algae-ingested cells were fixed for transmission electron microscopy. Partially digested (D) and nondigested (N) algae are enclosed in the same DV. The nondigested algae are not separated from the digested algae by a membrane representing a PV membrane. DVM, DV membrane; *CW* Cell wall. From Kodama and Fujishima (2010a)

even with clonal symbiotic algae. Furthermore, this algal fate is independent of the algal cell cycle stage and location of the algae in the DV. Moreover, this resistance to digestion is not related to the algal protein synthesis (Kodama et al. 2007). Gu et al. (2002) showed that degeneration of the symbiotic *Chlorella* under DD conditions is induced by the host lysosomal fusion to PV membranes. This report suggests that the photosynthetic activity and/or related cellular processes of the algae play important functions in protection from the lysosome fusion to the PV membrane (Kodama and Fujishima 2014).

16.3.3.1 Effects of Various Treatments of Isolated Symbiotic *Chlorella Variabilis* Before Mixing with Algae-Free *P. bursaria*

Most of the isolated symbiotic *C. variabilis* incubated under constant light (LL) conditions for 24 h were able to resist digestion in the host DV. The undigested algae then bud from the DVs, and the algae localized beneath the host cell cortex to establish endosymbiosis with algae-free *P. bursaria* cells as shown above and Fig. 16.8a (Kodama and Fujishima 2005, 2012a, b, 2014). However, by incubation of isolated symbiotic algae under DD conditions for 24 h before mixing with the host cells, most of the algae lost the capability of resistance to the host lysosomal enzymes in the DV. Only a few algae are able to avoid digestion and could be localized beneath the host cell cortex after budding from the DVs (Fig. 16.8b, arrowheads). We looked for morphological differences of the vacuole in LL-incubated, LL-incubated with photosynthesis inhibitor DCMU, and DD-incubated algae by staining with LysoSensor Yellow/Blue DND-160 (LysoSensor) (Fig. 16.9). In live cells, LysoSensor accumulates in acidic vacuoles of plant cells (Swanson et al. 1998), and exhibits predominantly yellow fluorescence. As presented in Fig. 16.9b, several

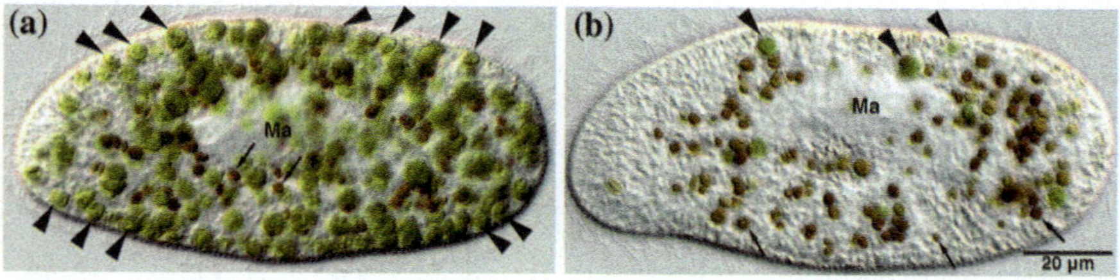

Fig. 16.8 Photomicrographs of algae-free *P. bursaria* cells after mixing with LL- (**a**) or DD- (**b**) incubated algae for 24 h. Both cells were mixed and kept under LL or DD conditions. As shown in (**a**), many LL-incubated algae showed resistance to the host lysosomal enzymes, and the undigested green algae localized beneath the host cell cortex (**a**, *arrowheads*). On the other hand, most of the DD-incubated algae were digested, and the algal color changed from green to brown (**b**, *arrows*). Few algae were able to avoid digestion and establish endosymbiosis (**b**, *arrowheads*). These results show that the algal incubation under LL conditions before ingestion by the alga-free *P. bursaria* cells is necessary to prevent algal digestion. Arrowhead, undigested SGC(s) localized beneath the cortex; arrow, digested brown alga; *Ma* Macronucleus. From Kodama and Fujishima (2014)

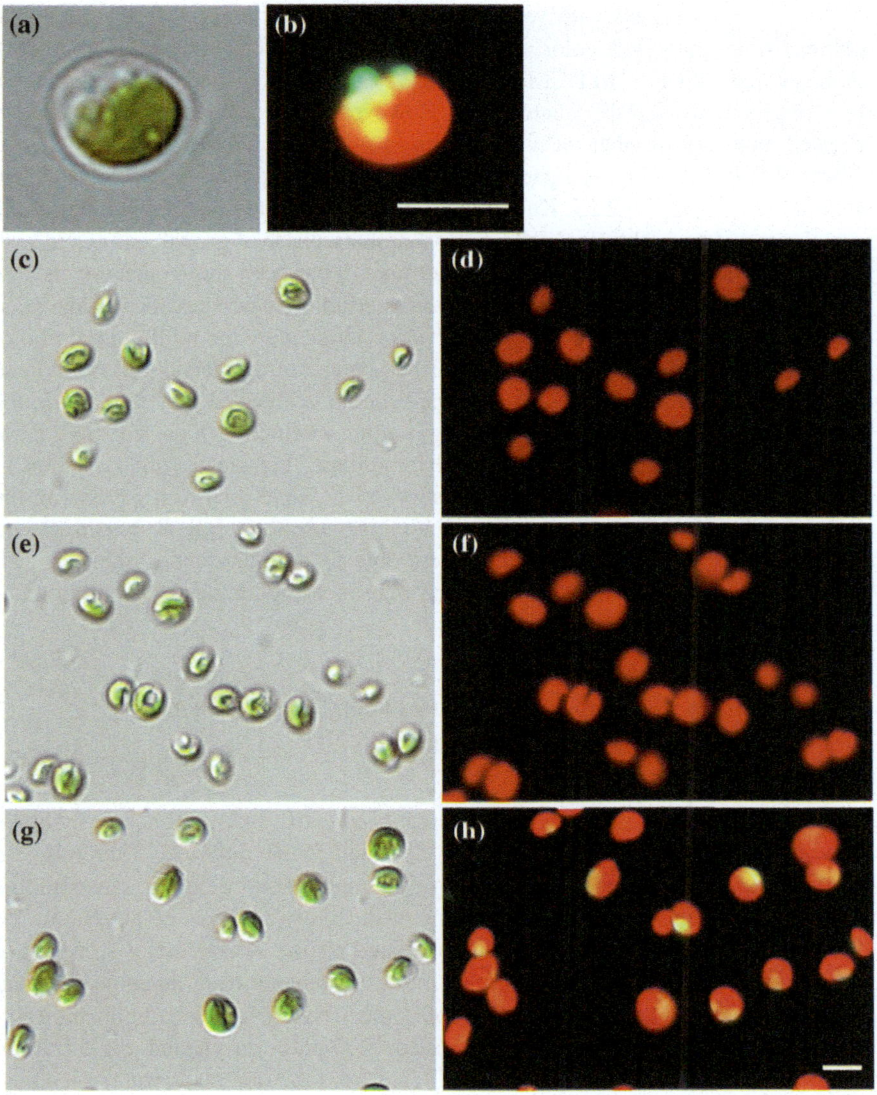

Fig. 16.9 DIC photomicrographs of LL-incubated (**c**), LL-incubated with 10^{-5} M DCMU (**e**), and DD-incubated (**a** and **g**) algae, and fluorescence photomicrographs of LysoSensor-treated LL-incubated (**d**), LL-incubated with 10^{-5} M DCMU (**f**), and DD-incubated (**b** and **h**) isolated symbiotic algae. LysoSensor accumulates in acidic vacuoles, and shows yellow fluorescence (**b**). These figures show that the DD-incubated algae have many yellow fluorescence vacuoles (**h**), more than those incubated under LL conditions with (**f**) or without (**d**) DCMU. The red color shows chlorophyll autofluorescence in the chloroplast. *Scale bars* 5 μm. From Kodama and Fujishima (2014). (Color figure online)

small spherical vacuoles with yellow fluorescence were observed in the algae. No differences in the algal color, shape or volume in LL-incubated (Fig. 16.9c), LL-incubated with 10^{-5} M DCMU (Fig. 16.9e), or DD-incubated (Fig. 16.9g) algae were observed using DIC microscopy. However, fluorescent microscopy clearly revealed that the number of vacuoles in the DD-incubated algae (Fig. 16.9h) increased more than those in algae incubated under LL conditions with (Fig. 16.9f) or without (Fig. 16.9d) DCMU. Kuchitsu et al. (1987) reported that the number of the vacuoles increases in the algal cells at the stationary phase of growth compared with the cells in the log phase of growth. Furthermore, it has been shown that the vacuole volume becomes extremely large after a long period of sugar starvation in the plant cell (Yu 1999). Taken together, algal starvation induced by the inhibition of photosynthesis under the DD conditions might be a cause of the vacuole development. Although the reason why the alga with the vacuole is digested preferentially in the host DV remains unknown, our results suggest that whether the algae are digested or not in the host DVs can de determined by staining the algae with LysoSensor. Figure 16.10 shows schematic representation of the algal digestion patterns in the DVs after the various treatments before mixing with algae-free *P. bursaria.* Our results show that a few of the algae were able to establish endosymbiosis with algae-free *P. bursaria* cells when the algae were incubated under DD conditions.

16.3.4 Transcriptome Analysis Between Algae-Free and -Bearing P. bursaria *Cells*

Despite the importance of *P. bursaria-Chlorella* sp. endosymbiosis as shown above, genomic resources had not been identified for *P. bursaria*. Therefore, we compared gene expressions through RNA-Seq analysis and de novo transcriptome assembly of algae-free and algae-bearing host cells (Kodama et al. 2014). To expedite the process of gene discovery related to the endosymbiosis, we have undertaken Illumina deep sequencing of mRNAs prepared from algae-bearing and algae-free *P. bursaria* cells. We assembled the reads de novo to build the transcriptome. Sequencing using Illumina HiSeq 2000 platform yielded 232.3 million paired-end sequence reads. Clean reads filtered from the raw reads were assembled into 68,175 contig sequences. Of these, 10,557 representative sequences were retained after removing *Chlorella* sequences and lowly expressed sequences. Nearly 90 % of these transcript sequences were annotated by similarity search against protein databases. Hsp70 and glutathione S-transferase (GST) genes were up-regulated and down-regulated as shown by the positive and negative values of logFC, respectively, in algae-bearing cells compared to algae-free cells (Table 16.1). Of the 10,557 unigenes, 8 were annotated as Hsp70 with logFC of −1.3 to 5.6, with a median of 0.92.

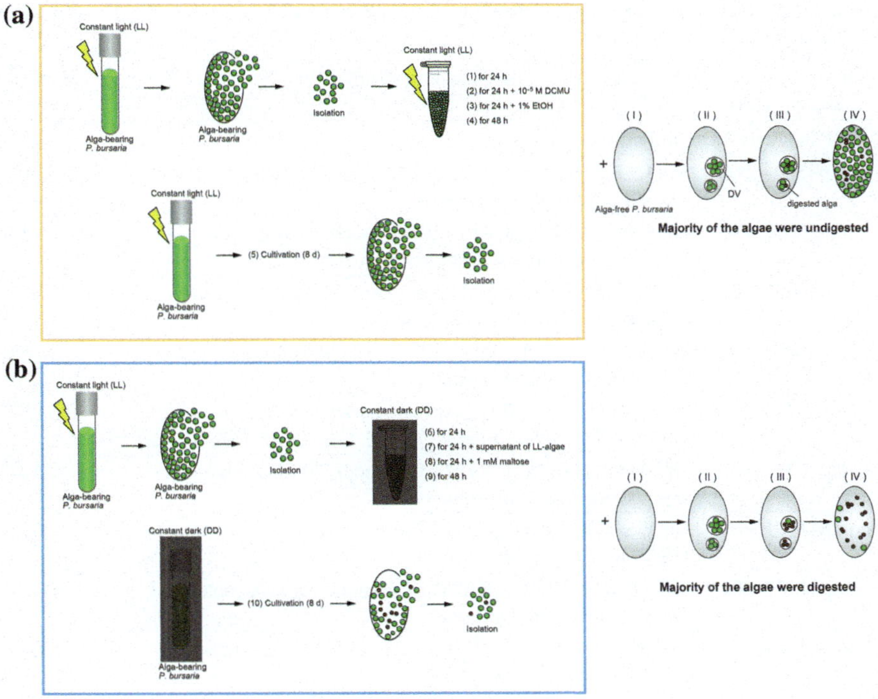

Fig. 16.10 Schematic representation of the algal digestion patterns in the DVs. Under LL conditions (**a**), the isolated symbiotic algae from LL-incubated algae-bearing *P. bursaria* cells were incubated for 24 h (*1*), for 24 h with 10^{-5} M DCMU (*2*), for 24 h with 1 % ethanol (EtOH) (*3*), for 48 h (*4*), and for 8 days (*5*) after (*1–4*) and before (*5*) isolation from algae-bearing *P. bursaria* cells. After mixing with algae-free *P. bursaria* cells (**a** *I*), some algae were ingested from the host cytopharynx and were enclosed in the DVs (**a** *II*). After the lysosomal fusion with the DVs, few algae were digested (*brown alga* in **a** *III*), but most of the algae showed resistance to the host lysosomal enzymes and were not digested (*green algae* in **a** *III*). Finally, most of the algae ingested in the DVs were able to establish endosymbiosis with algae-free *P. bursaria* cells (**a** *IV*). On the other hand, under the DD conditions (**b**), the isolated symbiotic algae from LL-incubated algae-bearing *P. bursaria* cells were incubated for 24 h (*6*), for 24 h with supernatant of LL-incubated algae (*7*), for 24 h with 1 mM maltose (*8*), for 48 h (*9*) and for 8 days (*10*) after (*6–9*) and before (*10*) isolation from algae-bearing *P. bursaria* cells. After mixing with algae-free *P. bursaria* cells (**b** *I*), some algae were enclosed in the DVs as with the LL-incubated algae (**b** *II*) as shown in (**a** *II*). After the lysosomal fusion, most of the algae were digested (*brown alga* in **b** *III*) and a few algae showed resistance to the host lysosomal enzymes (*green alga* in **b** *III*). Most of the algae were digested and excreted from the host cytopharynx (**b** *IV*). From Kodama and Fujishima (2014)

16.3.4.1 Glutathione S-transferase

It is conceivable that photo-oxidative stress is greater in algae-bearing *P. bursaria* cells than in algae-free ones. To determine whether oxidative stress and UV-induced photo-oxidative stress are greater in algae-bearing *P. bursaria* cells than in algae-free ones, Hörtnagl and Sommaruga (2007) examined the level of

Table 16.1 Transcripts encoding glutathione S-transferase and heat shock 70 kDa protein in *P. bursaria*

Trinity transcript name	Annotation from the SwissProt database	logFC
Heat shock 70 kDa protein		
comp43044_c0	sp|P09446|HSP7A_CAEEL|Heat shock 70 kDa protein A OS = *Caenorhabditis elegans*	5.601
comp36402_c4	sp|P09446|HSP7A_CAEEL|Heat shock 70 kDa protein A OS = *Caenorhabditis elegans*	4.183
comp36402_c6	sp|P14834|HSP70_LEIMA|Heat shock 70 kDa protein (Fragment) OS = *Leishmania major*	1.975
comp36402_c1	sp|Q9S9N1|HSP7E_ARATH|Heat shock 70 kDa protein 5 OS = *Arabidopsis thaliana*	1.555
comp37280_c1	sp|P37899|HSP70_PYRSA|Heat shock 70 kDa protein OS = *Pyrenomonas salina*	0.287
comp43771_c0	sp|P09446|HSP7A_CAEEL|Heat shock 70 kDa protein A OS = *Caenorhabditis elegans*	−0.594
comp41901_c0	sp|Q9S7C0|HSP7O_ARATH|Heat shock 70 kDa protein 14 OS = *Arabidopsis thaliana*	−1.076
comp41912_c0	sp|F4JMJ1|HSP7R_ARATH|Heat shock 70 kDa protein 17 OS = *Arabidopsis thaliana*	−1.337
comp43044_c0	sp|P09446|HSP7A_CAEEL|Heat shock 70 kDa protein A OS = *Caenorhabditis elegans*	5.601
Glutathione S-transferase		
comp37410_c0	sp|P78417|GSTO1_HUMAN|Glutathione S-transferase omega-1 OS = *Homo sapiens*	−0.119
comp32377_c0	sp|Q9ZRT5|GSTT1_ARATH|Glutathione S-transferase T1 OS = *Arabidopsis thaliana*	−0.288
comp36943_c0	sp|Q9ZVQ3|GSTZ1_ARATH|Glutathione S-transferase Z1 OS = *Arabidopsis thaliana*	−0.748
comp37841_c0	sp|Q9ZRT5|GSTT1_ARATH|Glutathione S-transferase T1 OS = *Arabidopsis thaliana*	−0.851
comp36483_c0	sp|Q9ZRT5|GSTT1_ARATH|Glutathione S-transferase T1 OS = *Arabidopsis thaliana*	−1.557
comp35816_c1	sp|P78417|GSTO1_HUMAN|Glutathione S-transferase omega-1 OS = *Homo sapiens*	−1.564
comp36242_c0	sp|P16413|GSTMU_CAVPO|Glutathione S-transferase B OS = *Cavia porcellus*	−5.749

oxidative stress by assessing reactive oxygen species with two fluorescent probes (hydroethidine and dihydrorhodamine 123) by flow cytometry. Their results indicated that oxidative stress is higher in algae-free *P. bursaria* cells than in algae-bearing one. Our results showed that expression levels of GST genes in algae-free cells were down-regulated than that in algae-bearing cells (Kodama et al. 2014). This enzyme is related to protect cells from oxidative stress as shown by

McCord and Fridovich (1969), Veal et al. (2002), and our results agreed with the results of Hörtnagl and Sommaruga (2007).

16.3.4.2 Hsp70

Furthermore, it is known that *Paramecium* cell acquires heat-shock resistance by infection of endonucler symbiotic bacteria *Holospora* as shown above (Hori et al. 2008; Hori and Fujishima 2003), and osmotic-shock resistance (Smurov and Fokin 1998). Hori and Fujishima (2003) found that *H. obtusa*-bearing paramecia expressed high levels of *hsp70* mRNA even at 25 °C. Algae-bearing cells show a higher survival ratio against 0.5 mM nickel chloride, high temperatures (42 °C), and 150 mM hydrogen peroxide than the algae-free cells (Kinoshita et al. 2009; Miwa 2009). We found that most of isoforms of the *hsp70* transcripts showed up-regulation by algal infection (Kodama et al. 2014). This up-regulation may be related to the host's tolerance to environmental fluctuations.

16.4 Conclusion

Recently, we succeeded draft genome sequences of three *Holospora* species, *H. obtusa*, *H. undulata*, and *H. elegans* (Dohra et al. 2013, 2014). Furthermore, whole transcriptome analysis between algae-free and algae-bearing *P. bursaria* was succeeded (Kodama et al. 2014). We can expect that these data enable us to understand the molecular mechanisms for establishments of the primary and the secondary symbioses and for the host evolutionary adaptation to global climate change.

Acknowledgements This work was supported by a JSPS KAKENHI Grant-in-Aid for Young Scientists (B) Grant Number 26840119 to Y.K., and also by a JSPS KAKENHI Grant Number 26291073 and a MEXT TOKUBETSUKEIHI to M.F. *Paramecium* strains used in this chapter were provided by the Symbiosis Laboratory, Yamaguchi University, with support in part by the National Bio-Resource Project of the Japan Agency for Medical Research and Development (AMED).

References

Abamo F, Dohra H, Fujishima M (2008) Fate of the 63-kDa periplasmic protein of the infectious form of the endonuclear symbiotic bacterium *Holospora obtusa* during the infection process. FEMS Microbiol Lett 280:21–27

Albers D, Wiessner W (1985) Nitrogen nutrition of endosymbiotic *Chlorella* spec. Endocytobiol Cell Res 2:55–64

Amann R, Springer N, Ludwig W, Görtz H-D, Schleifer K-H (1991) Identification in situ and phylogeny of uncultured bacterial endosymbionts. Nature 351:161–164

Aury JM, Jaillon O, Duret L, Noel B, Jubin C, Porcel BM, Ségurens B, Daubin V, Anthouard V, Aiach N, Arnaiz O, Billaut A, Beisson J, Blanc I, Bouhouche K, Câmara F, Duharcourt S, Guigo R, Gogendeau D, Michael Katinka M, Keller AM, Kissmehl R, Klotz C, Koll F, Mouël AL, Lepère G, Malinsky S, Nowacki M, Nowak JK, Plattner H, Poulain J, Ruiz F, Serrano V, Zagulski M, Dessen P, Bétermier M, Weissenbach J, Scarpelli C, Schächter V, Sperling L, Meyer E, Cohen J, Wincker P (2006) Global trends of whole-genome duplications revealed by the ciliate *Paramecium tetraurelia*. Nature 444:171–178

Blanc G, Duncan G, Agarkove I, Borodovsky M, Gumon J, Kuo A, Lindquist E, Lucas S, Pangilinan J, Polle J, Salamov A, Terry A, Yamada T, Dunigan DD, Grigoriev IV, Claverie JM, Van Ettan JL (2010) The *Chlorella variabilis* NC64A genome reveals adaptation to photosynthesis, coevolution with viruses, and cryptic sex. Plant Cell 22:2943–2955

Boscaro V, Felletti M, Vannini C, Ackerman MS, Chain PSG, Malfatti S, Vergez LM, Shin M, Doak TG, Lynch M, Petroni G (2013) *Polynucleobacter necessarius*, a model for genome reduction in both free-living and symbiotic bacteria. Proc Natl Acad Sci USA 110:18590–18595

Brown JA, Nielsen PJ (1974) Transfer of photosynthetically produced carbohydrate from endosymbiotic *Chlorellae* to *Paramecium bursaria*. J Protozool 21(4):569–570

Dohra H, Fujishima M (1999) Cell structure of the infectious form of *Holospora*, an endonuclear symbiotic bacterium of the ciliate *Paramecium*. Zool Sci 16:93–98

Dohra H, Suzuki H, Suzuki T, Tanaka K, Fujishima M (2013) Draft genome sequence of *Holospora undulata* strain HU1, a micronucleus-specific symbiont of the ciliate *Paramecium caudatum*. Genome Announc 1:e00664–13

Dohra H, Tanaka K, Suzuki T, Fujishima M, Suzuki H (2014) Draft genome sequences of three *Holospora* species (*Holospora obtusa*, *Holospora undulata*, and *Holospora elegans*), endonuclear symbiotic bacteria of the ciliate *Paramecium caudatum*. FEMS Microbiol Lett Genome Announc 359:16–18

Dryl S (1959) Antigenic transformation in *Paramecium aurelia* after homologous antiserum treatment during autogamy and conjugation. J Protozool 6(Suppl):25

Ehrsam E, Görtz H-D (1999) Surface proteins of the Gram-negative bacterium *Holospora obtusa* bind to macronuclear proteins of its host *Paramecium caudatum*. Europ J Protistol 35:304–308

Fok AK, Allen RD (1988) The lysosome system. In: Görtz HD (ed) *Paramecium*. Springer, Berlin, pp 301–324

Fokin SI, Görtz HD (2009) Diversity of *Holospora* bacteria in *Paramecium* and their characterization. In: Fujishima M (ed) Endosymbionts in *Paramecium*, microbiology monographs, vol 12. Springer, pp 161–199

Fokin SI, Sabaneyeva EV (1997) Release of endonucleobiotic bacteria *Holospora bacillata* and *Holospora curvata* from the macronucleus of their host cells *Paramecium woodruffi* and *Paramecium calkinsi*. Endocyt Cell Res 12:49–55

Fokin SI, Brigge T, Brenner J, Görtz HD (1996) *Holospora* species infected the nuclei of *Paramecium* appear to belong into two groups of bacteria. Europ J Protistol 32:19–24

Fokin SI, Przybos E, Chivilev SM, Beier CL, Horn M, Skotarczak B, Wodecka B, Fujishima M (2004) Morphological and molecular investigations of *Paramecium schewiakoffi* sp nov (Ciliophora, Oligohymenophorea) and current status of distribution and taxonomy of *Paramecium* spp. Europ J Protistol 40:225–243

Fujishima M (1986) Further study of the infectivity of *Holospora obtusa*, a macronucleus specific bacterium of the ciliate *Paramecium caudatum*. Acta Protozool 25:345–350

Fujishima M (2009) Infection and maintenance of *Holospora* species in *Paramecium caudatum*. In: Fujishima M (ed) Endosymbionts in *Paramecium*, microbiology monographs, vol 12. Springer, pp 201–225

Fujishima M, Görtz HD (1983) Infection of macronuclear anlagen of *Paramecium caudatum* with the macronucleus-specific symbiont *Holospora obtusa*. J Cell Sci 64:137–146

Fujishima M, Fujita M (1985) Infection and maintenance of *Holospora obtusa*, a macronucleus-specific bacterium of the ciliate *Paramecium caudatum*. J Cell Sci 76:179–187

Fujishima M, Hoshide K (1988) Light and electron microscopic observations of *Holospora obtusa*: A macronucleus-specific bacterium of the ciliate *Paramecium caudatum*. Zool Sci 5:791–799
Fujishima M, Kawai M (1997) Acidification in digestive vacuoles is an early event required for *Holospora* infection of *Paramecium* nucleus. In: Achenk HEA, Herrmann RG, Jeon KW, Müller NE, Schwemmler W (eds) Eukaryotism and symbiosis. Springer, Berlin, pp 367–370
Fujishima M, Mizobe Y (1988) Host genes controlling maintenance of a macronucleus-specific symbiont *Holospora obtusa* of *Paramecium caudatum*. Zool Sci 5:1272
Fujishima M, Kawai M (2004) Endonuclear symbiotic bacteria *Holospora* species distinguish the host nuclear envelopes. Endocyt Cell Res 15:71–76
Fujishima M, Kodama Y (2012) Endosymbionts in *Paramecium*. Eur J Protistol 48:124–137
Fujishima M, Nagahara K, Kojima Y (1990a) Changes in morphology, buoyant density and protein composition in differentiation from the reproductive short form to the infectious long form of *Holospora obtusa*, a macronucleus-specific symbiont of the ciliate *Paramecium caudatum*. Zool Sci 7:849–860
Fujishima M, Sawabe H, Iwatsuki K (1990b) Scanning electron microscopic observations of differentiation from the reproductive short form to the infectious long form of *Holospora obtusa*. J Protozool 37:123–128
Fujishima M, Nagahara K, Kojima Y, Sayama Y (1991) Sensitivity of the infectious long form of the macronuclear endosymbiont *Holospora obtusa* of the ciliate *Paramecium caudatum* against chemical and physical factors. Europ J Protistol 27:119–126
Fujishima M, Kawai M, Yamamoto R (2005) *Paramecium caudatum* acquires heat-shock resistance in ciliary movement by infection with the endonuclear symbiotic bacterium *Holospora obtusa*. FEMS Microbiol Lett 243:101–105
Fujishima M, Iwatani K, Nakamura Y, Kodama Y (2007) Infection of *Holospora* is controlled by 89-lDa proplasmic proteins and the host actin. Protistology 5(Abst):31
Gibson I, Bedingfield G, Horne RW, Bird B (1986) Electron microscope study of the structure of *Holospora caryophila*. Micron Microsc Acta 17:247–257
Gomori G (1952) Microscopic Histochemistry: *Principles and Practice*. University of Chicago Press, Chicago
Görtz HD (1980) Nucleus-specific symbionts in *Paramecium caudatum*. In: Schwemmler W, Schenk HEA (eds) Endocytobiology, endocytobiosis and cell biology I. Walter de Gruyter & Co, Berlin, pp 381–392
Görtz HD (1982) Infections of *Paramecium bursaria* with bacteria and yeasts. J Cell Sci 58:445–453
Görtz HD (1983) Endonuclear symbionts in ciliates. Intern Rev Cytol 14:145–176
Görtz HD, Dieckmann J (1980) Life cycle and infectivity of *Holospora elegans* (Hafkine), a micronucleus-specific symbiont of *Paramecium caudatum* (Ehrenberg). Protistologica 16:591–603
Görtz HD, Wiemann M (1989) Route of infection of the bacteria *Holospora elegans* and *Holospora obtusa* into the nuclei of *Paramecium caudatum*. Eur J Protistol 24:101–109
Görtz HD, Ahlers N, Robenek H (1989) Ultrastructure of the infectious and reproductive forms of *Holospora obtusa*, a bacterium infecting the macronucleus of *Paramecium caudatum*. J Gen Microbiol 135:3079–3085
Görtz HD, Benting J, Ansorge I, Freiburg M (1992) Cell surface proteins of the infectious form of the symbiotic bacterium *Holospora obtusa*. Symbiosis 14:391–397
Gromov BV, Ossipov DV (1981) *Holospora* (ex Hafkine 1980) nom. rev., a genus of bacteria inhabiting the nuclei of paramecia. Int J Syst Bacteriol 31:348–352
Gu F, Chen L, Ni B, Zhang X (2002) A comparative study on the electron microscopic enzymo-cytochemistry of *Paramecium bursaria* from light and dark cultures. Eur J Protistol 38(3):267–278
Hori M, Fujishima M (2003) The endosymbiotic bacterium *Holospora obtusa* enhances heat-shock gene expression of the host *Paramecium caudatum*. J Euk Microbiol 50:293–298
Hori M, Tomikawa I, Przybos E, Fujishima M (2006) Comparison of the evolutionary distances among syngens and sibling species of *Paramecium*. Mol Phyl Evol 38:697–704

Hori M, Fujii K, Fujishima M (2008) Micronucleus-specific bacterium *Holospora elegans* irreversibly enhances stress gene expression of the host *Paramecium caudatum*. Euk Microbiol 55:515–521

Hörtnagl PH, Sommaruga R (2007) Photo-oxidative stress in symbiotic and aposymbiotic strains of the ciliate *Paramecium bursaria*. Photochem Photobiol Sci Official J Eur Photochem Assoc Eur Soc Photobiol 6(8):842–847

Hosoya H, Kimura K, Matsuda S, Kitaura M, Takahashi T (1995) Symbiotic alga-free strains of the green *Paramecium bursaria* produced by herbicide paraquat. Zool Sci 12:807–810

Iwatani K, Dohra H, Lang BF, Burger G, Hori M, Fujishima M (2005) Translocation of an 89-kDa periplasmic protein is associated with *Holospora* infection. Biochem Biophys Res Comm 337:1198–1205

Jennings HS (1938) Sex reaction types and their interrelations in *Paramecium bursaria* I and II. Clones collected from natural habitats. Proc Natl Acad Sci USA 24:112–120

Kadono T, Kawano T, Hosoya H, Kosaka T (2004) Flow cytometric studies of the host-regulated cell cycle in algae symbiotic with green paramecium. Protoplasma 223:133–141

Kamako S-I, Imamura N (2006) Effect of Japanese *Paramecium bursaria* extract on photosynthetic carbon fixation of symbiotic algae. J Euk Microbiol 53:136–141

Karakashian SJ (1963) Growth of *Paramecium bursaria* as influenced by the presence of algal symbionts. Physiol Zool 36:52–68

Karakashian SJ (1975) Symbiosis in *Paramecium bursaria*. Symp Soc Exp Biol 29:145–173

Kato Y, Imamura N (2009) Metabolic control between the symbiotic *Chlorella* and the host *Paramecium*. In: Fujishima M (ed) Endosymbionts in *Paramecium*, microbiology monographs, vol 12. Springer, pp 57–82

Kawakami H, Kawakami N (1978) Behavior of a virus in a symbiotic system, *Paramecium bursaria*—zoochlorella. J Protozool 25:217–225

Kinoshita H, Oomori S, Nozaki M, Miwa I (2009) Timing of establishing symbiosis during the re-infection of *Chlorella* sp in *Paramecium bursaria*. Jpn J Protozool 42:88–89

Kodama Y (2013) Localization of attachment area of the symbiotic *Chlorella variabilis* of the ciliate *Paramecium bursaria* during the algal removal and reinfection. Symbiosis 60(1):25–36

Kodama Y, Fujishima M (2005) Symbiotic *Chlorella* sp. of the ciliate *Paramecium bursaria* do not prevent acidification and lysosomal fusion of host digestive vacuoles during infection. Protoplasma 225:191–203

Kodama Y, Fujishima M (2007) Infectivity of *Chlorella* species for the ciliate *Paramecium bursaria* is not based on sugar residues of their cell wall components, but based on their ability to localize beneath the host cell membrane after escaping from the host digestive vacuole in the early infection process. Protoplasma 231:55–63

Kodama Y, Fujishima M (2008) Cycloheximide induces synchronous swelling of perialgal vacuoles enclosing symbiotic *Chlorella vulgaris* and digestion of the algae in the ciliate *Paramecium bursaria*. Protist 159:483–494

Kodama Y, Fujishima M (2009a) Timing of perialgal vacuole membrane differentiation from digestive vacuole membrane in infection of symbiotic algae *Chlorella vulgaris* of the ciliate *Paramecium bursaria*. Protist 160:65–74

Kodama Y, Fujishima M (2009b) Localization of perialgal vacuoles beneath the host cell surface is not a prerequisite phenomenon for protection from the host's lysosomal fusion in the ciliate *Parameium bursaria*. Protist 160:319–329

Kodama Y, Fujishima M (2009c) Infection of *Paramecium bursaria* with symbiotic *Chlorella* species. In: Fujishima M (ed) Endosymbionts in *Paramecium*, microbiology monographs, vol 12. Springer, pp 31–55

Kodama Y, Fujishima M (2010a) Secondary symbiosis between *Paramecium* and *Chlorella* cells. Int Rev Cell Mol Biol 279:33–77

Kodama Y, Fujishima M (2010b) Elucidation of establishment of secondary endosymbiosis as a driving force for biodiversity. In: Miyamoto A and Fujishima M (eds) Proceedings of infrastructure and environmental management symposium in Yamaguchi 2010, research center for environmental safety (RCES). Yamaguchi University, pp 1–39

Kodama Y, Fujishima M (2011) Endosymbiosis of *Chlorella* species to the ciliate *Paramecium bursaria* alters the distribution of the host's trichocysts beneath the host cell cortex. Protoplasma 248:325–337

Kodama Y, Fujishima M (2012a) Cell division and density of symbiotic *Chlorella variabilis* of the ciliate *Paramecium bursaria* is controlled by the host's nutritional conditions during early infection process. Environ Microbiol 14(10):2800–2811

Kodama Y, Fujishima M (2012b) Characteristics of the digestive vacuole membrane of the alga-bearing ciliate *Paramecium bursaria*. Protist 163(4):658–670

Kodama Y, Fujishima M (2013) Synchronous induction of detachment and reattachment of symbiotic *Chlorella* spp from the cell cortex of the host *Paramecium bursaria*. Protist 164(5):660–672

Kodama Y, Fujishima M (2014) Symbiotic *Chlorella variabilis* incubated under constant dark condition for 24 hours loses ability to avoid digestion by host lysosomal enzymes in digestive vacuoles of host ciliate *Paramecium bursaria*. FEMS Microbiol Ecol 90:946–955

Kodama Y, Nakahara M, Fujishima M (2007) Symbiotic alga *Chlorella vulgaris* of the ciliate *Paramecium bursaria* shows temporary resistance to host lysosomal enzymes during the early infection process. Protoplasma 230:61–67

Kodama Y, Inouye I, Fujishima M (2011) Symbiotic *Chlorella vulgaris* of the ciliate *Paramecium bursaria* plays an important role in maintaining perialgal vacuole membrane functions. Protist 162(2):288–303

KodamaY Suzuki H, Dohra H, Sugii M, Kitazume T, Yamaguchi K, Shigenobu S, Fujishima M (2014) Comparison of gene expression of *Paramecium bursaria* with and without *Chlorella variabilis* symbionts. BMC Genom 15:183

Kreutz M, Stoeck T, Foissner W (2012) Morphological and molecular characterization of *Paramecium* (Viridoparamecium nov. subgen.) *chlorelligerum* Kahl 1935 (Ciliophora). J Eukaryot Microbiol 59:548–563

Kuchitsu K, Oh-hama T, Tsuzuki M, Miyachi S (1987) Detection and characterization of acidic compartments (vacuoles) in *Chlorella vulgaris* 11 h cells by 31P-in vivo NMR spectroscopy and cytochemical techniques. Arc Microbiol 148(2):83–87

Lang BF, Brinkmann H, Koski LB, Fujishima M, Görtz HD, Burger G (2005) On the origin of mitochondria and Rickettsia-related eukaryotic endosymbionts. Jpn J Protozool 38:171–183

Lebreton A, Lakisic G, Job V, Fritsch L, Tham TN, Camejo A, Mattei P-J, Regnault B, Nahori M-A, Cabannes D, Gautreau A, Ait-Si-Ali S, Dessen A, Cossart P, Bierne H (2011) A bacterial protein targets the BAHD1 chromatin complex to stimulate type III interferon response. Science 331:1319–1321

McCord JM, Fridovich I (1969) Superoxide dismutase. An enzymic function for erythrocuprein (hemocuprein). J Biol Chem 244(22):6049–6055

McGrath CL, Gout JF, Doak TG, Yanagi A, Lynch M (2014) Insights into three whole-genome duplications gleaned from the *Paramecium caudatum* genome sequence. Genetics 114

Miwa I (2009) Regulation of circadian rhythms of *Paramecium bursaria* by symbiotic *Chlorella* species. In: Fujishima M (ed) Endosymbionts in *Paramecium*, microbiology monographs, vol 12. Springer, pp 83–110

Nakamura Y, Aki M, Aikawa T, Hori M, Fujishima M (2004) Differences in gene expression of the ciliate *Paramecium caudatum* caused by endonuclear symbiosis with *Holospora obtusa*, revealed using differential display reverse transcribed PCR. FEMS Microbiol Lett 240:209–213

Ossipov DV (1973) Specific infectious specificity of the omega-particles, micronuclear symbiotic bacteria of *Paramecium caudatum*. Cytology (Saint-Petersburg) 15:211–217

Ossipov DV, Skoblo II, Rautian MS (1975) Iota-particles, macronuclear symbiotic bacteria of ciliate *Paramecium caudatum* clone M-115. Acta Protozool 14:263–280

Ossipov DV, Skoblo II, Borschsenius ON, Rautian MS (1980) *Holospora acuminata*—a new species of symbiotic bacterium from the micronucleus of the ciliate *Paramecium bursaria* Focke. Cytology (Saint-Petersburg) 22:922–929

Pado R (1965) Mutual relation of protozoans and symbiotic algae in *Paramaecium bursaria*. I. The influence of light on the growth of symbionts. Folia Biol 13(2):173–182

Preer LB (1969) Alpha, an infectious macronuclear symbiont of *Paramecium aurelia*. J Protozool 16:570–578

Reisser W (1976) The metabolic interactions between *Paramecium bursaria* Ehrbg. and *Chlorella* spec. in the *Paramecium bursaria*—symbiosis. I. The nitrogen and the carbon metabolism. Arc Microbiol 107(3):357–360

Reisser W (1980) The metabolic interactions between *Paramecium bursaria* Ehrbg. and *Chlorella* spec. in the *Paramecium bursaria*-symbiosis. III. The influence of different CO_2-Concentrations and of glucose on the photosynthetic and respiratory capacity of the symbiotic unit. Arc Microbiol 125(3):291–293

Reisser W (1986) Endosymbiotic associations of freshwater protozoa and algae. In: Corliss JO, Patterson DJ (eds) Prog in protistology, vol 1. Biopress Ltd., Bristol, pp 195–214

Reisser W, Klein T, Becker B (1988) Studies of phycoviruses I. On the ecology of viruses attacking *Chlorellae* exsymbiotic from a European strain of *Paramecium bursaria*. Arch Hydrobiol 111:575–583

Sabaneyeva EV, Derlacheva ME, Benken KA, Fokin SI, Vainio S, In Skovorodkin (2009) Actin-based mechanism of *Holospora obtusa* trafficking in *Paramecium caudatum*. Protist 160:205–219

Siegel R, Karakashian S (1959) Dissociation and restoration of endocellular symbiosis in *Paramecium bursaria*. Anat Rec 134:639

Skoblo II, Lebedeva NA (1986) Infection of the nuclear apparatus of *Paramecium bursaria* (Ciliata) by the symbiotic bacterium *Holospora curviuscula*. Cytology (Sankt-Petersburg) 28:367–372

Smurov AO, Fokin SI (1998) Resistance of *Paramecium caudatum* infected with endonuclear bacteria *Holospora* against salinity impact. Proc Zool Inst RAS 276:175–178

Summerer M, Sonntag B, Hortnagl P, Sommaruga R (2009) Symbiotic ciliates receive protection against UV damage from their algae: a test with *Paramecium bursaria* and *Chlorella*. Protist 160(2):233–243

Swanson SJ, Bethke PC, Jones RL (1998) Barley aleurone cells contain two types of vacuoles: Characterization of lytic organelles by use of fluorescent probes. Plant Cell Online 10(5):685–698

Takahashi T, Shirai Y, Kosaka T, Hosoya H (2007) Arrest of cytoplasmic streaming induces algal proliferation in green paramecia. PLoS ONE 2(12):e1352

Tanaka M, Murata-Hori M, Kadono T, Yamada T, Kawano T, Kosaka T, Hosoya H (2002) Complete elimination of endosymbiotic algae from *Paramecium bursaria* and its confirmation by diagnostic PCR. Acta Protozool 41:255–261

Van Etten JL, Burbank DE, Schuster AM, Meints RH (1985) Lytic viruses infecting a *Chlorella*-like alga. Virology 140(1):135–143

Veal EA, Toone WM, Jones N, Morgan BA (2002) Distinct roles for glutathione S-transferases in the oxidative stress response in *Schizosaccharomyces pombe*. J Biol Chem 277(38): 35523–35531

Weis DS (1969) Regulation of host and symbiont population size in *Paramecium bursaria*. Experientia 25(6):664–666

Weis DS (1984) The effect of accumulation time of separate cultivation on the frequency of infection of aposymbiotic ciliates by symbiotic algae in *Paramecium bursaria*. J Protozool 31:14A

Wichterman R (1948) The biological effects of X-rays on mating types and conjugation of *Paramecium bursaria*. Biol Bull 93:201

Wiemann M (1989) The release of *Holospora obtusa* from *Paramecium caudatum* observed with a new device for extended in vivo microscopy. J Protozool 36:176–179

Yamada T, Onimatsu H, Van Etten JL (2006) *Chlorella* viruses. In: Karl M, Aaron JS (eds) Adv Virus Res, vol 66. Academic Press, pp 293–336

Yu SM (1999) Cellular and genetic responses of plants to sugar starvation. Plant Physiol 121(3):687–693

Chapter 17
An Integrated Model of the Biology of the Marine Symbiosis *Maristentor dinoferus*

Christopher S. Lobban and María Schefter

Abstract *Maristentor dinoferus* (Heterotrichida: Maristentoridae) is a symbiosis comprising a very large ciliate with hundreds of endosymbiotic zooxanthellae (*Symbiodinium* sp.). Its large size, large amounts of pigment that make it appear black, tendency to cluster, and preferred substratum of the light-colored blades of the seaweed *Padina* make it visible to the naked eye and observable in the field. Here we review the knowledge of *Maristentor* behavior and ecology through the lens of biocommunication theory and use analogies with other organisms to develop an integrated framework of understanding as a basis for future experimental and observational research. We are particularly interested in the roles and integration of the three most outstanding features of this symbiosis: the zooxanthellae, the densely pigmented cortical granules, and the complex clustering/dispersal behavior of the cells.

17.1 Introduction

At each level in a biological hierarchy there are emergent properties not evident from the properties of the component parts (Mayr 1982); life is not "simply organized organic molecules" as Emmeche (2002: 155) asserted. At each level (including populations up to the biosphere) every system is made up of component parts *and their interactions*. Thus a defining emergent property at all levels in this hierarchy is signal-mediated interactions, which are "the precondition for all cooperation and coordination between at least two biological agents such as organisms, organs, tissues, cells and even subcellular components" (Witzany 2008: 24). Biosemiotics, the study of signs and sign processes in living systems, is not an

C.S. Lobban (✉)
Division of Natural Sciences, University of Guam, Mangilao, GU 96923, USA
e-mail: clobban@guam.net

M. Schefter
UOG Station, PO Box 5126, Mangilao, GU 96923, USA

G. Witzany and M. Nowacki (eds.), *Biocommunication of Ciliates*,
DOI 10.1007/978-3-319-32211-7_17

alternative to the hierarchical approach to understanding biological systems but rather a complementary approach to understanding emergent properties.

Photosymbioses—mutualistic symbioses in which one individual unicellular organism is host to many individuals of a photosynthetic, endosymbiotic unicell (Mordret et al. 2015)—do not fit comfortably into the usual categories in the biological hierarchy because the endosymbionts constitute a population; i.e., within each single ciliate individual there is a population of dinoflagellates. Yet, in consequence, such symbioses depend even more deeply on signal-mediated interactions. Our example to support this assertion is the photosymbiosis *Maristentor dinoferus* (Figs. 17.1 and 17.2) and we will develop a model of its biology through the lens of signal-mediated interactions at various levels of biological organization. Whereas plant cells contain numerous photosynthetic endosymbionts (chloroplasts)

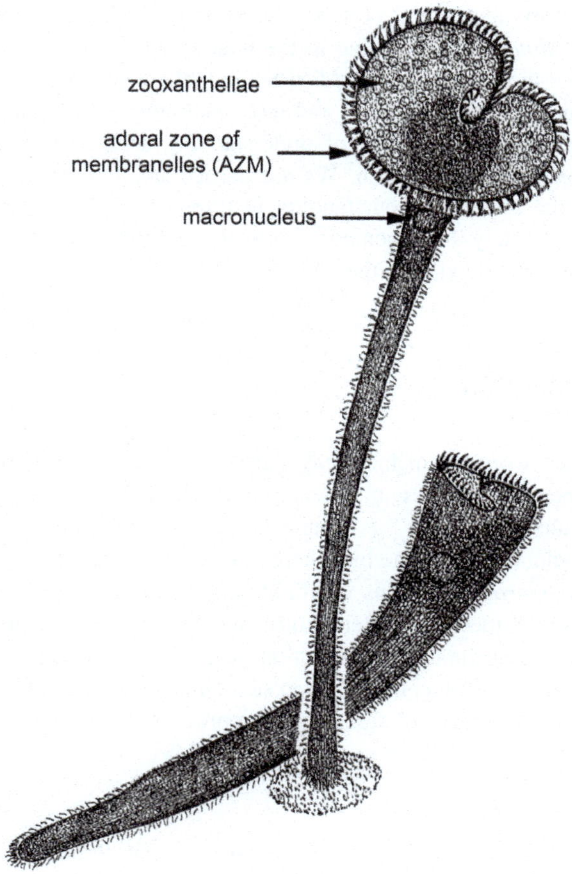

Fig. 17.1 Taxonomic drawing of *Maristentor* in erect and gliding postures. *Scale bar* 200 μm (drawing by Wilhelm Foissner, *Coral Reefs* 21: 332, © 2002 and reprinted with permission of Springer)

Fig. 17.2 *Maristentor* clusters on *Padina*, showing erect cells in clusters, **a** showing of *Maristentor* near the margin and in concentric rings; also showing where algal tissue was removed by fish predation; **b** detail of same image showing the daytime concentration of pigment below the cap (e.g., at *black arrow*) and the location of one of the rings of carbonate deposition (*white arrow*) (*photo* taken at about 10 m depth by Tom Schils)

that have been reduced over eons to organelles, the 500–700 *Symbiodinium* (dinoflagellate) cells within a *Maristentor* cell are fully functional organisms (also called zooxanthellae). They are complete except for cellulose walls and flagella, and fully capable of independent existence [at least this has been shown for the zooxanthellae from corals (Fitt and Trench 1983), which are genetically close to those in *Maristentor* (Lobban et al. 2002)].

Maristentor is rare among ciliates in having zooxanthellae, but many marine organisms in other phyla have zooxanthellae and the entire coral reef ecosystem depends on photosymbiosis (Stanley and Swart 1995; Stat et al. 2006). Along with *Maristentor*, only two other known benthic ciliates are symbiotic with zooxanthellae: *Paraeuplotes tortugensis* and *Euplotes uncinatus* (Lobban et al. 2005). One pelagic zooxanthellate ciliate, a calcified colepid in the genus *Tiarina*, was very recently described from diverse tropical ocean locations (Mordret et al. 2015). In contrast, in freshwaters and cooler marine waters ciliates hosting green algal cells (zoochlorellae) are common and include some that are large-celled, colonial and sessile (Stoecker et al. 2009, Germond and Nakajima *this volume*). These photosymbioses will make useful comparisons with *Maristentor* as we review its signal-mediated interactions. Ciliate symbioses (chiefly with bacteria) were reviewed by Dziallas et al. (2012) and we have adapted their diagram, originally captioned "dynamics," to serve as a model of communication in the context of signal-mediated interactions of *Maristentor* (Fig. 17.3). The bigger picture of biocommunication in coral reefs was discussed by Madl and Witzany (2014).

In this article we will use *Maristentor* to mean *M. dinoferus*, recognizing the possibility that other populations that have been observed without taxonomic analysis may be different species (even on Guam); and that the term *Maristentor* is to be taken to mean the symbiotic association including at least the ciliate cell and the endosymbiotic zooxanthellae. As we refer to literature on other ciliates, zooxanthellae from other symbioses, and even other organisms that can suggest possibilities of what may happen in *Maristentor*, we will make clear the distinction between these analogies and what has been established or hypothesized for *Maristentor* itself.

17.2 Basic Biology of the Ciliate and Dinoflagellate Symbionts

Maristentor is a "majestic" heterotrich ciliate (Class Heterotrichea) (Lynn 2008: 345), i.e., a large (up to 800 μm long) unicellular protozoan with two sets of cilia: the body is covered by longitudinal rows of short cilia, while the oral apparatus at the anterior end is encircled by a set of long cilia organized into sheets (membranelles) to form the adoral zone of membranelles (AZM) (Fig. 17.1). Structurally and molecularly it is distinctive enough to have been placed in its own family, Maristentoridae (Miao et al. 2005), and Foissner (2006:127) calls it a "flagship species" because of its size and distinctiveness. It has several generative micronuclei and a single, globular somatic macronucleus. The cells can adopt four postures: extended and attached by the basal pole, extended and gliding on a substrate, contracted and attached,

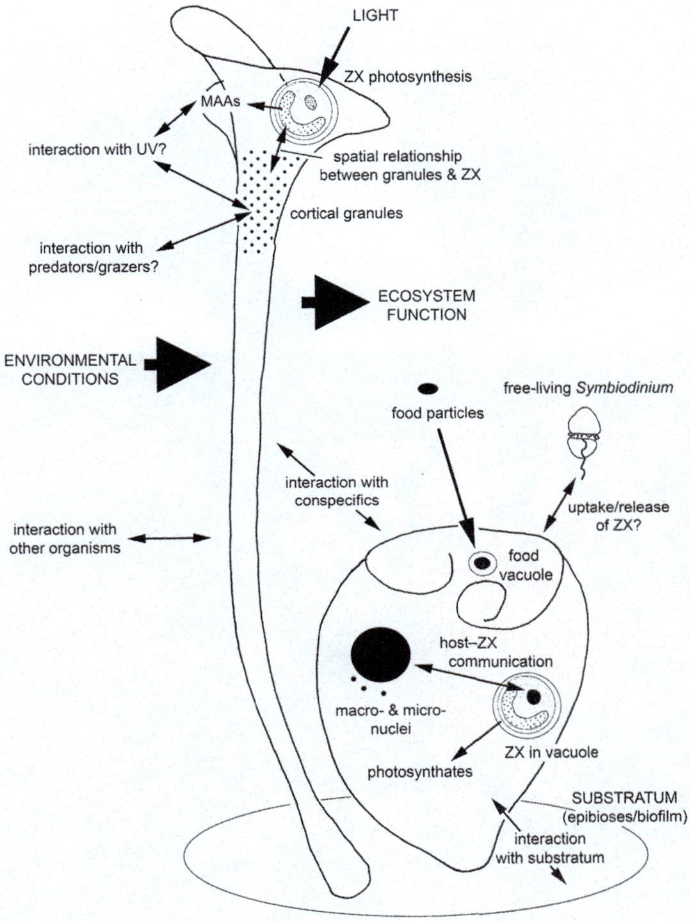

Fig. 17.3 Model of the biology of *Maristentor* in the context of biocommunication, showing known and potential interactions and signaling among symbiont components and environment. Details in the text. *Note* While the outline shapes of the *Maristentor* in erect and contracted postures are realistic, the zooxanthellae (ZX) are not to scale and other physical structures are largely diagrammatic (Adapted from a diagram in Dziallas et al. 2012 and drawings of *Maristentor* by Wilhelm Foissner in Lobban et al. 2002)

contracted and swimming (Figs. 17.1 and 17.3) (Lobban et al. 2002). They form clusters with the basal poles close together (Figs. 17.2b and 17.4c), but cells come and go from the clusters and are not permanently attached to each other or to the substratum. As in *Stentor* and other genera in related families, the complex cell membrane in *Maristentor* has strips between the ciliary rows packed with Golgi-derived cortical granules, or extrusomes, that in some species are known to eject their contents upon stimulation [e.g., by a predator attack (Miyake et al. 2001; Buonanno et al. 2005)]. The pigment granules of *Maristentor* are densely

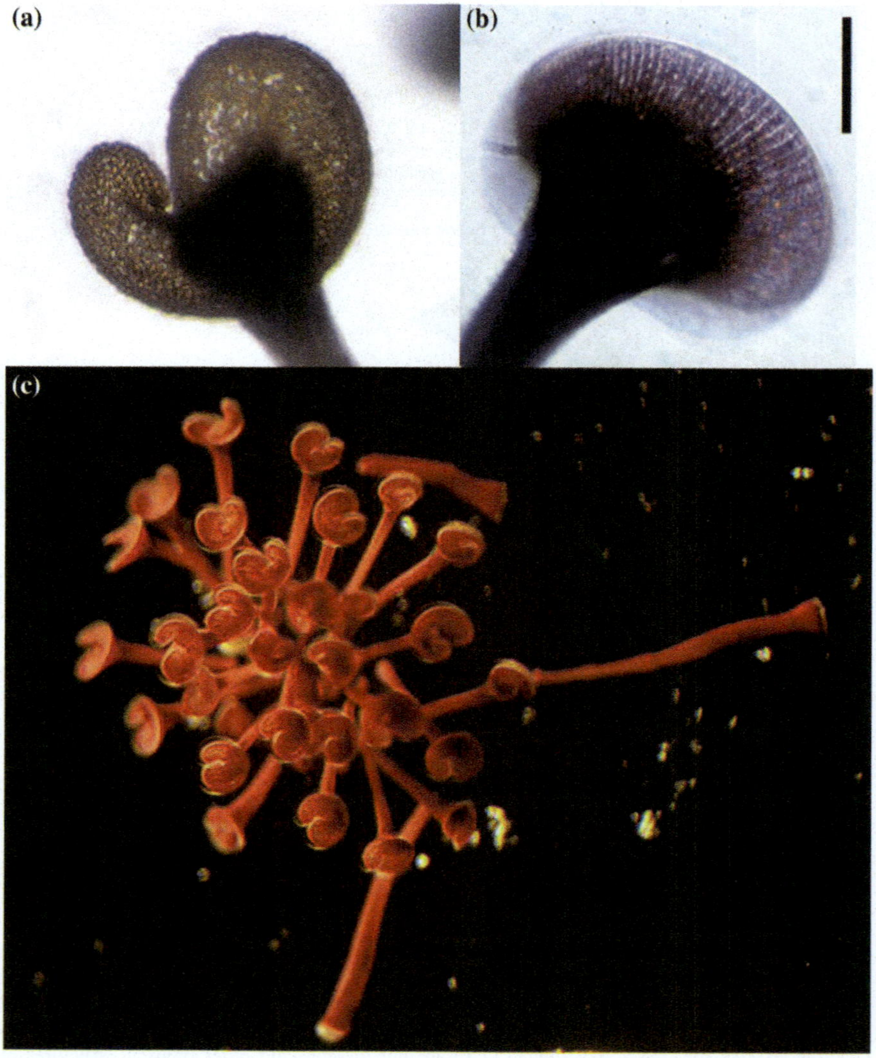

Fig. 17.4 **a**, **b** Differences in distribution of cortical pigment and zooxanthellae in the cap of *Maristentor* by day (**a**) versus by night (**b**); **c** cluster of cells at night imaged with side light, showing uniform fluorescence of cortical pigment. *Scale bar* on (**b**) 100 μm, also applies approximately to (**a**) (**b** reprinted from Lobban et al. 2002, *Marine Biology*, with permission of Springer)

pigmented, making the cells appear black and very evident against the light backgrounds on which it most frequently occurs. The purplish pigment, maristentorin, was structurally characterized by Mukherjee et al. (2006) and is related to hypericin, a compound known for its toxicity and medicinal properties and to other compounds known to act as feeding deterrents (Lobban et al. 2007). While *Maristentor* is easy to observe in the field, it has so far been intractable to laboratory culture and such observations as we have made in vitro have been on populations that were gradually dying.

Cell division has not been observed but is presumed to occur by binary fission, as in other stentorids, i.e., by the formation of a new oral primordium on the posterior ventral surface and subsequent cleavage into the proter (the anterior daughter cell) and the opisthe (the posterior cell with the new oral apparatus), and with division of both micronuclei and the macronucleus (Lynn 2008). However, there are different details in the several families of stentorids (Foissner 1996); in the most extreme case, folliculinids (the sister group to *Maristentor*) produce a non-feeding motile cell from the proter and a sessile feeding cell (like the parent), after the original oral apparatus is resorbed (Mulisch and Patterson 1988). This may be a functional response to the cells living in attached loricas—this constraint does not apply to *Maristentor*—but it does illustrate how weak the analogies can be, even among closely related families. Conjugation has also not been seen in *Maristentor*, but would presumably involve generation of new macronuclei from the micronuclei, characteristic of ciliates (Lynn 2008).

Symbiodinium zooxanthellae are dinoflagellate algae belonging to a complex of nine phylogenetically distinct clades, each with several subclades (Pochon et al. 2014). A number of species have been described using molecular criteria (LaJeunesse et al. 2012; Parkinson et al. 2015). Zooxanthellae lack the characteristic (and taxonomically important) walls and flagella of free-living gymnodinioid dinoflagellates. Eyespots occur in free-living *Symbiodinium* and related species (Hansen and Daugbjerg 2009) but not in zooxanthellae; however, uric acid crystal arrays in zooxanthellae may be involved in light detection (Yamashita et al. 2009). The algal symbiont of *Maristentor* was shown both by its distinctive dinoflagellate nuclei and its partial LSU rDNA gene sequence to belong to *Symbiodinium* clade C, a clade which also includes zooxanthellae of several local corals (Lobban et al. 2002); again, this is based on only a few sequences from one population.

17.3 Genetic Vocabulary: Still a Black Box

Our total molecular knowledge of *Maristentor* and of its zooxanthellae is a few sequences of SSU rDNA from each, obtained for phylogeny (Lobban et al. 2002; Miao et al. 2005). The dearth of knowledge at the genome level precludes our exploration of the signal communications among micronuclei and between micronuclei and macronucleus, and thus the genetic and epigenetic effects on phenotypes. Similarly, it would be useful to have information on the transcriptome, proteome, and metabolome of *Maristentor* under various conditions.

Moreover, the genetic structure of the populations is unknown, as cells from only one collection were used for the phylogenetic analysis (Miao et al. 2005). Indeed, the formal description was based on staining cells from only one collection (Lobban et al. 2002), and it is possible that populations even on Guam include more than one species with differences visible only after the elaborate staining process, or even several syngens, i.e., reproductively isolated but morphologically indistinguishable taxa, as is seen in some *Paramecium* species and other ciliates (Hori et al. 2006; Zagata et al. 2015). Equally, it may be that there are different *Symbiodinium* taxa in *Maristentor* populations, as has been found in the zoochlorellae in *Paramecium* (Summerer et al. 2007, 2008). One of the disadvantages of any very distinctive "flagship" species is that it is easy to overlook small differences and assume that everything with the distinctive character(s) is the same species (Lobban 2015).

17.4 Interpretation of Abiotic Influences

The environment in which *Maristentor* has been observed is the marine benthos (Lobban et al. 2002, 2014; Lobban and Schefter 2012). We hypothesized that it can also exist in the hyponeuston (suspended from the water surface) but this has not been observed in the field (Lobban et al. 2014). In the benthic habitat *Maristentor* is usually attached by the basal pole, standing erect with the cap spread out, but it moves by gliding on its side (Figs. 17.1 and 17.2). It is a competent swimmer but swimming has been observed only in disturbed populations (Lobban et al. 2002, 2014). Water movement (hydrodynamic forces such as surge; nutrient fluxes across boundary layers) no doubt influence *Maristentor*, as they do other organisms (Hurd et al. 2014), and *Maristentor* responds to sudden vibration of the substrate by contracting (Lobban and Schefter 1996), but key environmental factors for *Maristentor* are likely to be (1) suitable substrate for attachment and (2) light for photosynthesis. We will explore each of these in turn.

The substrate for *Maristentor* is most commonly chalky brown algae in the genus *Padina*, which form curved, fan-shaped blades (Fig. 17.2a), but it also occurs on several other substrates including other seaweeds, cyanobacterial mats, rock, and sand (Lobban and Schefter 2012; Lobban et al. 2014). It has even been photographed on a live juvenile scorpionfish (Michael Martin, unpublished image posted on Facebook)! During 3 years when *Maristentor* populations exploded, we found that they would spread onto limestone marker tiles placed adjacent to populations on *Padina* (Lobban and Schefter 2012). When we moved these tiles we sometimes observed that standing clusters would disperse, all cells adopting the gliding position, but not swimming. Images in Fig. 17.5 (see Sect. 17.6) also suggest that net movement during the morning diaspora was oriented and not random (perhaps towards the rising sun?). However, when populations on *Padina* were collected in plastic bags and even when plastic bottles were placed over otherwise undisturbed populations, cells swam off the substrate and headed toward the brightest place. In dishes in the laboratory, this resulted in their gathering near the surface on the brightest side of the dish; in the field they accumulated on the part of the bottle in line with the sun (Lobban et al. 2014, fig. 6). These observations suggest that cells can detect the substrate and respond to it (see Sect. 17.7), clinging to it or swimming

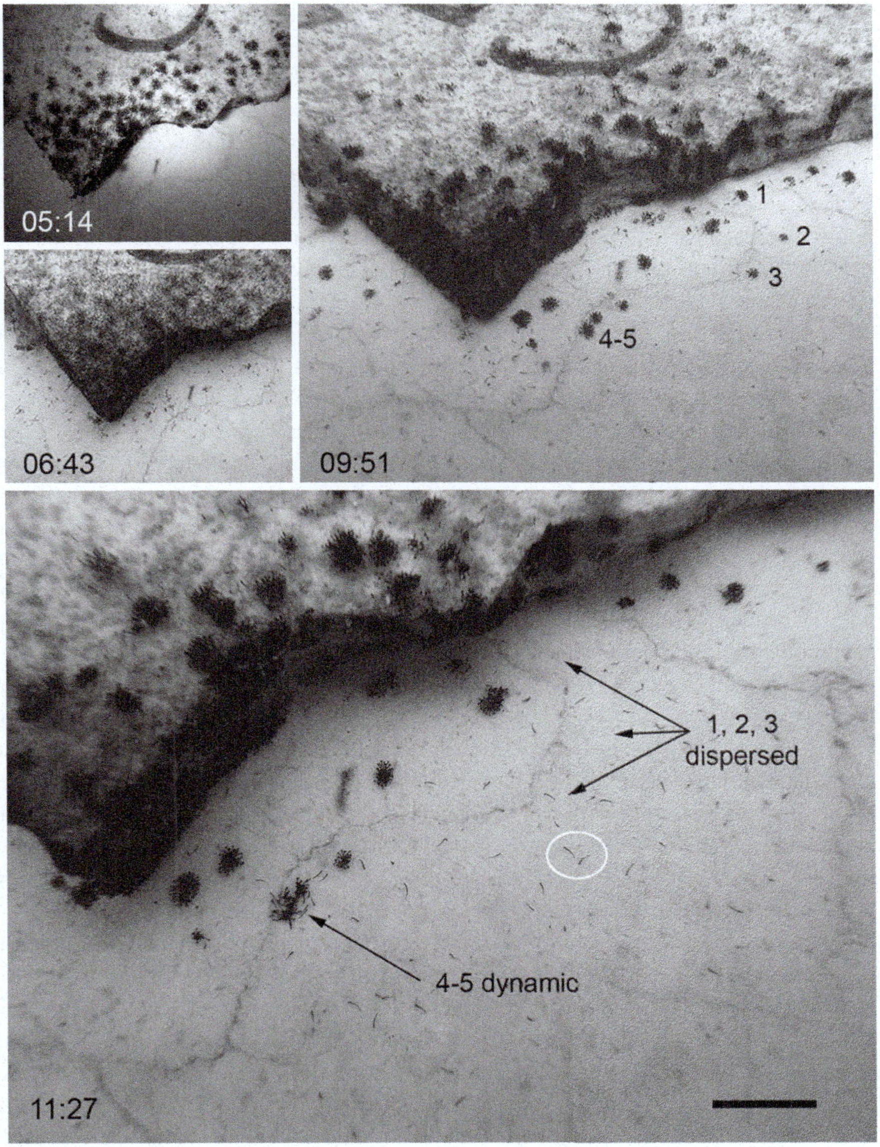

Fig. 17.5 Dispersal and dispersion in *Maristentor*: series of images in situ of experiment in which a piece of marble tile with a population of *Maristentor* (*old tile*) was moved a short distance and placed on a clean marble tile (*new tile*) (development of the population on *Padina* and this marker tile over the preceding 3 weeks shown in Lobban and Schefter 2012, figs. 7–10). (05:14) Population before dawn (dawn at 06:00), shortly after set up; no shock dispersal occurred. Besides the obvious clusters on the face of the old tile there are dense clusters on the vertical edge. (06:43) Soon after dawn the clusters had dispersed and some cells can already be seen on the new tile. (09:51) Re-clustering more or less complete, several clusters on the new tile up to 12 mm from the old tile. (11:27) Cocktail-party dynamics is underway: three clusters indicated at 09:51 (# 1, 2 and 3) have dispersed and two clusters (# 4, 5) are in a dynamic state compared to their appearance at 09:51. Many gliding cells can still be seen across the new tile as far as 50 mm from the old tile (two marked by the white oval). *Scale bar* for this image = ca. 10 mm

away from it under different circumstances. They also show that cells can detect light gradients and respond to them (phototaxis), maintaining a benthic distribution in spite of the positive phototactic capacity, except when disturbed in certain ways. This benthic attraction is not necessarily negative phototaxis, because the cells do not seem to avoid light or seek shade; being photosynthetic symbioses, they need light. The environmental cues that switch these responses have not been investigated.

Responses to light are typically mediated by photoreceptor pigments, which could be present in the ciliate, the zooxanthellae, or both. For example, some strains of *Symbiodinium* in the motile stage are positively phototactic to green light (Hollingsworth et al. 2005), which could come from green fluorescence by coral polyps and/or from the asymmetrical underwater light field in coastal waters. *Symbiodinium* also exhibited chemosensory responses (Fitt 1985). Although these responses were seen in free-living, motile cells, one must wonder whether they could also be active in symbio. However, since corals and many other zoothanthellate symbioses are nonmotile, they are weak analogies to help understand *Maristentor* behavior. Turning to ciliates, the responses of heterotrich ciliates to light have been explored largely in *Stentor coeruleus* and *Blepharisma japonica* (Song et al. 1991; Lenci et al. 2001). Both have hypericin-like compounds in cortical pigment granules, as does *Maristentor*, but both are harmed by the phototoxic properties of their pigments and are negatively phototatic, seeking deep shade or darkness. Neither has photosynthetic endosymbionts, of course, but several species of *Stentor* do have zoochlorellae and various cortical pigments (or none). Only *Stentor araucanus* has both a blue pigment, possibly similar to that of *S. coeruleus*, and zoochlorellae (Lobban et al. 2007); it forms dense planktonic aggregations and is evidently not harmed by any phototoxic effect (Wölfl and Geller 2002). In spite of these obvious pigments in the granules, positive phototaxis in heterotrich ciliates such as *Fabrea salina* is probably due to rhodopsin dispersed invisibly over the cell surface (Podestà et al. 1994) and probably also present in other heterotrichs (detailed review in Lobban et al. 2007).

There has been much effort to understand the trigger and signal transduction mechanism in *S. coeruleus* and *B. japonica*, in which there is a predictable photophobic response, and speculative models have been proposed (e.g., Lenci et al. 2001; Wood 2001; Sobierajska et al. 2006). In these species, an increase in light is detected by the cortical pigment (signal perception) and from there signal transduction, amplification and transmission leads to a change in the beating of the cilia (mechanotransduction), perhaps the result of changes in transmembrane electrical potential. Such a response is not expected in *Maristentor*, because it is positively phototactic when it swims. Nevertheless, there is some stimulus that triggers *Maristentor* cells to abandon the substrate. The stimulus has been induced by enclosure, but we have to assume that this ability to leave the substrate en masse and swim toward light can also occur under natural conditions; this was part of the evidence that led us to hypothesize hyponeustonic *Maristentor* populations (Lobban et al. 2014).

17.5 Intraorganismic Communication Between the Photosymbiotic Partners

The behavior of *Maristentor* must meet the needs of both partners in the photosymbiosis. If we think of the ciliate as a motor home for a large family of zooxanthellae, then we can more easily see that there are certain requirements for the vehicle, such as roads and gasoline, and other requirements for the family, such as food. In describing a symbiosis with endosymbionts it is more useful to classify the interaction between the containing partner ("host") and contained partner (endosymbiont) as intraorganismic rather than transorganismic, and to reserve transorganismic to refer to interactions between the symbiosis and other species in its surroundings. In this section we discuss acquisition and regulation of zooxanthellae and the distribution of zooxanthellae and cortical granules. Before speculating about *Maristentor*, we describe some potential analogies: organelles and other situations involving zooxanthellae.

For organelles derived originally by endosymbiosis, i.e., plastids and mitochondria, which are now tightly integrated into the host cells, control of the system firmly belongs to the host cell and the organelles divide or are partitioned when the host cell divides (vertical transmission). At the other extreme, most corals must re-establish the photosymbiosis soon after metamorphosis (horizontal transmission); this involves both the selection of the coral by certain *Symbiodinium* strains and the acquisition of the algae as zooxanthellae (Yamashita et al. 2014). Two-way cell recognition is implied, and evidence for a lectin/glycan system has been found in establishment of initial symbiosis by larvae of the coral *Fungia scutaria* (Wood-Charlson et al. 2006). Once acquired by corals, zooxanthellae divide and are partitioned between the daughter cells; supernumerary cells are continually expelled into the coelenteron and may be picked up by other cells in the colony (either recently divided cells or those that have lost zooxanthellae due to bleaching). Coral cells can also continue to acquire algal cells from the environment. Densities of zooxanthellae per unit area of coral surface are fairly consistent and apparently regulated by space constraints rather than by the host restricting their nutrient supply (Jones and Yellowlees 1997), and perhaps also by digestion and extrusion of zooxanthellae (Titlyanov et al. 1996). Germond and Nakajima (*this volume*) discuss further the steps required for cells ingested by ciliates to avoid digestion and establish symbiosis, and the subsequent regulation of endosymbiont populations. Perhaps the closest analogy to *Maristentor* is *Euplotes uncinatus*, a ciliate but in a different Class (Spirotrichea). There Lobban et al. (2005) counted 56 ± 24 zooxanthellae packed into the whole cell except for the cytostome ("mouth") and associated digestive vacuoles. *E. uncinatus* cells are dorsoventrally flattened averaging 93 × 57 μm and they scurry about on a series of cirri that function as "legs." The number of zooxanthellae fell to 12 ± 7 after 3 days in darkness and there appeared to be spaces where zooxanthellae had been. There appeared to be many intact but not swimming zooxanthellae free on the floor of the culture well, strongly suggesting that the zooxanthellae had been expelled. Alongside these two extremes of mutualism are several well-studied examples of kleptoplastidy, where some ciliates and metazoa retain plastids "stolen" from their food (Stoecker et al. 1988; McManus et al. 2012; Dziallas et al. 2012). This is not mutualism, since it does not benefit the original owner of the plastids; it has been described as "predation with farming of the prey organelles" (Nowack and Melkonian 2010: 708). Nevertheless, the intercellular communication requirements are probably parallel to those in symbioses. Much more work remains to be done on the intracellular signals that prevent digestion of endosymbiotic algae or plastids in mixotrophic ciliates (McManus et al. 2012).

The range of symbiotic relationships means that analogies from one to another are often weak. In the case of *Maristentor* neither cell division nor loss of zooxanthellae has been observed, and we do not know whether zooxanthellae are acquired from the environment periodically/continually or are all descendants of a relationship established long ago. Nevertheless, the analogies allow us to hypothesize that in *Maristentor* (1) the relationship between these ciliates and their zooxanthellae is stable and depends at least on the continual division of the zooxanthellae and partition between daughter cells when the ciliate divides; (2) the number of zooxanthellae per cell is regulated in some way. There is as yet no basis for proposing loss of zooxanthellae and/or occasional acquisition, or not.

In *Maristentor* the distribution of pigment granules and zooxanthellae changes from day to night in a cyclic rhythm (Lobban et al. 2002) (Fig. 17.4a, b). It is not known whether these are endogenous (circadian) rhythms or direct responses to environmental cues such as the light:dark cycle. During the day, pigment granules are concentrated below the cap (Figs. 17.2b, black arrow and 17.4a), where they are out of the way of the zooxanthellae spread out in the cap and stalk. At night the granules disperse over the entire surface and the zooxanthellae occur deeper in the cytoplasm (Fig. 17.4b, c). The daytime pattern has been observed in all in situ situations, the nighttime pattern chiefly in vitro.

The reasons for these patterns are not clear; a priori explanations involve either UV protection or photosynthesis. Maristentorin absorbs strongly in the UV and could perhaps provide protection for the nucleus when the granules concentrate just below the cap (Lobban et al. 2007). However, *Maristentor* has UV-blocking mycosporine-like amino acids (MAAs), almost certainly produced by the zooxanthellae (Sommaruga et al. 2006), which could replace or complement maristentorin, implying yet another path of intracellular communication. As for photosynthesis: during the day, of course, the zooxanthellae need to be well situated for photosynthesis, but optimal conditions for photosynthesis do not necessarily mean maximum exposure to light. The pigment granules potentially interfere with light absorption by zooxanthellae, but maristentorin fluoresces in the red region (Mukherjee et al. 2006) (Fig. 17.4c), where chlorophylls also have an important absorption peak. Indeed, in deep-water corals, fluorescent pigments may occur in a layer below the zooxanthellae, trapping and re-emitting light that passed the zooxanthellae (Schlichter et al. 1994). One could imagine that *Maristentor* cells could act as funnels to trap light in deep-water situations, but the consistent daytime pattern of granule distribution, even at depth, suggests that this is not the case. Nevertheless, *Maristentor* may need to adjust the position of the zooxanthellae in response to irradiance, since it lives in a range of irradiance environments from about 2 m below sea level to at least 20 m deep, in turbid harbor waters (where most of our studies were done) to clearer coastal waters. Movement of plastids in plant/algal cells is common but not ubiquitous and involves microtubules and microfibrils of the cytoskeleton. For example, Chen and Li (1991) compared two diatoms, one of which showed plastid movements on a light:dark cycle and in response to high irradiance, the other did not. In the latter there were no microtubules (tubulin) or microfibrils (actin) associated with the plastids. In *Maristentor*, zooxanthellae exist inside host vacuoles but no connections between these vacuoles and cytoskeleton were evident to Lobban et al. (2002). But these analogies, all to do with light, do nothing to suggest reasons for the different distribution of granules and zooxanthellae at night.

The purpose of the intracellular movements in *Maristentor* may be a mystery, but their existence implies communication within the cell: perception of environmental stimuli, translation and transmission of those stimuli into responses by actin and other molecules at appropriate places.

17.6 Interorganismic Communication Among *Maristentor* Individuals

At the level of interaction of individuals within populations, *Maristentor* is better known than most ciliates because of two particular characters: (1) it is large enough and dark enough to see and photograph against a light background in the field (Figs. 17.2 and 17.5); (2) it is gregarious and forms attached clusters of commonly 25–100 individuals in populations that can reach 1×10^6 cells m^{-2}. The clusters are dynamic during the day, dispersing and re-forming and also, during the afternoon, becoming larger but fewer as a result of "cocktail party" dynamics (Fig. 17.5) (Lobban and Schefter 2012; Lobban et al. 2002, 2014). Individual cells seem to have opposing tendencies to cluster and to disperse. In this section we examine the clustering behavior of *Maristentor* through the lens of signal mediation.

In situ, the dispersal–re-clustering cycle in *Maristentor* begins at dawn when tight clusters that appear to have been stable overnight completely disperse within 2 h (Fig. 17.5: 05:14 vs. 06:43). Cells glide around on the surface, sometimes occupying new territory in the process, and then re-cluster, with the majority of the population being back in clusters about 4 h after dawn (Fig. 17.5: 09:51). The clusters do not form in the same places as before, and we found that on a tile with heterogeneity in the biofilm caused by fish grazing, the clusters tended to re-group on thinner biofilm (Lobban et al. 2014). By taking series of close-up photographs of populations over time we were able to identify clusters and estimate the numbers of cells in each. Many of the clusters increased or decreased in size over the rest of the day (Fig. 17.5: 09:51 vs. 11:27), with the end result being fewer, larger clusters than initially. In some 10-minute interval series we were able to observe the sudden break up of clusters, but most of the dynamics involved individuals coming and going, hence the name "cocktail-party dynamics." In vitro, where cells congregated on the lighted side of the dish, dispersal still occurred at dawn but re-clustering did not take place until 2–4 p.m. (Lobban and Schefter, unpublished). This suggests that there are signals in the field that alter the timing of a circadian rhythm, but preliminary attempts to manipulate the dispersal phase with early and delayed dawn were inconclusive. Observations that cells did not swim off *Padina* when collected late in the day, and did not disperse immediately on tile moved before dawn, hinted that there may also be a rhythm in dispersability (Lobban et al. 2014). The factors affecting individual choices to stay in a group or to leave or join a group are completely unknown, but the end result suggests communication among members of a cluster and between cluster members and gliding individuals. In the following paragraphs we explore analogies that could generate hypotheses about how clusters form and why, beccause typically, organisms do not cluster or aggregate unless there is a value to that behavior.

Aggregation by independent but gregarious individuals is common but mostly involves very mobile individuals, such as schools of fish and flocks of birds (Camazine et al. 2001). Other aggregations include simple feeding responses such as the formation of blooms of the photosymbiotic ciliate *Mesodinium rubrum* via chemosensory detection of prey (Wilkerson and Grunseich 1990; Fenchel and Blackburn 1999), or are highly structured events such as nest building by social insects (Camazine et al. 2001). Nevertheless, some principles of self-organized groups may be helpful in understanding *Maristentor* behavior. In particular, members of a self-organized group often rely on a few simple behavioral rules, which, applied collectively, result in emergent properties such as a cluster (Camazine et al. 2001). For example, larvae of a bark beetle, *Dendroctonus micans* feed in clusters. They emit an attractant hormone and when larvae were spaced randomly in an experimental arena they spontaneously formed a cluster if the density of larvae was high enough (Deneubourg et al. 1990). In this case the simple rule guiding the behavior would be "feed where another is feeding" and the signal mediating it is the hormone (Camazine et al. 2001). Pheromones are also involved in the well-known example of social amoebae (*Dictyostelium discoideum*) aggregation (Gilbert 2013) and in aggregations of bacteria (Greenberg 2003; Mansfield and Turner 2002). However, there are other insects in which thigmotropism has been invoked as a cohesion mechanism (Demoulin 1962), while fish can use both vision and the lateral line to sense neighbors.

Organisms can typically detect only the local surroundings and cannot see the pattern, so that behavioral rules relate to what the nearby others are doing (Camazine et al. 2001). In *Maristentor*, the first two behavioral rules are probably (1) "stand erect and optimize photosynthesis by zooxanthellae" and (2) "stand where others stand." The first of these rules provides maximum benefit for the photosymbiosis by spreading out the majority of the zooxanthellae in a monolayer in the cap, the second produces the clusters. But what is the signal by which cells recognize where others are standing?

Clustering behavior implies that *Maristentor* individuals can sense others with which they are not in contact, and recognize them as conspecifics as opposed to potential predators. In the insect analogies above, hormones were involved. Although there is no evidence of the release of chemicals by *Maristentor*, release of molecules from cortical granules is conceivable, given that granules can act as extrusomes. In other ciliates, cortical granules can contain a variety of contents, and may often be mixtures, as reported for the toxisomes of *Coleps hirtus* (Buonanno et al. 2014) (see Sect. 17.7). It is not likely that all cortical granules in a cell have the same contents, and in a highly pigmented species like *Maristentor* it will be harder to detect colorless granules, or those with less pigment but more of other compounds. One can even imagine that maristentorin, unique to this organism, could be used as the clustering signal and to recognize conspecifics. Alternatively, hydromechanical signals could be detected by the cilia. Both motile micropredators and prey (e.g., copepods and ciliates) generate hydromechanical signals and can detect them (Kiørboe and Visser 1999; Visser 2001), and hydromechanical signals could be produced by stationary *Maristentor* clusters. *Stentor* has mechanoreceptors distributed on its cell surface that are thought to enable response to predator contact (Wood 1989); if such receptors are present in *Maristentor* they may play a role in recognition of conspecifics. Finally, one may ask whether it is even possible, given *Maristentor*'s phototactic ability, that cells could sense dark spots (clusters) and glide toward them? (Going back to the motor home analogy, this would be like a rule to "drive toward a patch of trees.")

A signal that brings organisms closer together is likely to be limited by some negative feedback to prevent contact, not necessarily itself a signal. For *Maristentor*, closeness might be physically limited by the width and activity of the caps and by the need for space for them to snap into the contracted position when startled, which makes them wider than the base in the extended posture (outline drawings in Fig. 17.3). Clusters vary greatly in size and packing density (compare clusters in Figs. 17.2b, 17.4c and 17.5), the latter so far only subjectively estimated and not yet quantified through the afternoon period of cocktail-party dynamics or day versus night. At very high densities cells can even form a continuous lawn (Lobban and Schefter 2012, Fig. 6).

The feature that makes *Maristentor* different from the many studies and models of aggregation in insects (Deneubourg et al. 1990) is that there is not only a clustering force but also a dispersive force. Additional rules are needed to explain why cells leave clusters in the process of cocktail-party dynamics and why they disperse as a population at dawn, and here it may be necessary to invoke collective decision making (Sumpter and Pratt 2009), or quorum responses, wherein the organisms' probability of exhibiting a particular behavior increases as the number of others already doing it increases (Ward et al. 2008). This could explain why small clusters tend to get smaller and large clusters larger. For example, in a study of cockroaches choosing between two equal shelters, the rule leading to consensus was that an individual's probability of leaving a shelter decreased as the shelter's population increased (Ame et al. 2006).

By whatever means they are formed, aggregations are expected to provide some benefit to the organisms. They often improve individual survival, allow better exploitation of food resources and provide better protection against enemies, by reinforcing the effects of chemical defense and aposematic signals (Deneubourg et al. 1990 and references therein). Some of these benefits may also accrue to *Maristentor*. The question of defense will be taken up in (see Sect. 17.7). As for exploitation of food resources, there are potential benefits of feeding in groups, as can be seen by analogy with peritrich ciliate colonies (Hartmann et al. 2007), inasmuch as *Maristentor* has a substantial feeding apparatus and is presumed to ingest bacteria/particles as well as benefitting from photosynthate from the zooxanthellae. The colonial peritrich *Opercularia asymmetrica*, has a prominent apical ring of cilia, like *Maristentor*, and adjacent zoids have a fixed physical relationship within the stalked colony. Hartmann et al. (2007) found adjacent zoids alternately beat their cilia, resulting in a non-stationary flow field and improved flux of food particles. Our observations of *Maristentor* showed only continuous beating of the AZM, even when cells were gliding, but hydrodynamics over a cluster of *Maristentor* may be more advantageous to individual members than the flow generated by single cells. However, we do not know whether *Maristentor* is feeding constantly even though the AZM is continuously active. It is conceivable that effective feeding occurs while the cells are gliding across the biofilm and that few particles are obtained from the water column while cells are erect. Logically, that could lead to a hypothesis that standing erect optimizes photosynthesis and gliding optimizes particle capture. Yet, from this hypothesis one would predict that clustering would predominate by day and gliding would predominate at night, and this is clearly not the case. While gliding may be more effective for feeding, or for acquiring certain kinds of food, there appears to be the least amount of gliding during the night, when photosynthesis is impossible.

17.7 Interspecies Communication: Interactions with Biofilm and Fish

The coral reef habitat of *Maristentor* includes several seaweeds, especially *Padina* spp.; microbial communities on the substrata on which *Maristentor* populations develop; an unknown community of benthic ciliates and other eukaryotic microorganisms; and the communities of fish and invertebrates, especially those that consume seaweeds and biofilm. There are different kinds of microbial communities on living and non-living surfaces. Living surfaces (basibionts) of all kinds in water become colonized by epibionts, but the assemblages of epibionts (the epibioses) vary because of interaction with the living substrate, with age of the underlying tissue, and with many other factors, in contrast to the biofilm of fouling organisms on nonliving surfaces (Wahl 2008). No organism is presently known to directly prey on *Maristentor*, but it is likely that predatory ciliates exist there, comparable to *Dileptus* in freshwater habitats, and it is possible that small invertebrates with diets that include ciliates would target *Maristentor*. In this section we review and interpret our in situ observations of responses of *Maristentor* to biofilm and to grazing fish that suggest a coordination of defense. Note that most observations were made during bloom years when *Maristentor* populations were very high and would move onto pieces of limestone tile placed next to populations on *Padina* (Lobban and Schefter 2012).

During the morning dispersal, cells glide around and can migrate as much as several centimeters, as seen from observations of cells migrating from a populated tile to a new, clean tile (Fig. 17.5). On tiles with established biofilm, grazing by the fish *Ctenochaetus striatus* produced heterogeneity by removing discs of the biofilm down to apparently bare tile; these patches were filled in by development of new biofilm. Thus by dawn the next day there were patches of thinner biofilm (12–16 h development since the bites were taken), among the older, thicker biofilm. It was clear from new bite marks that *C. striatus* grazing on heterogeneous biofilm preferred thicker biofilm. *Maristentor*, on the other hand, preferred to cluster on thinner biofilm, statistically moving onto previous bite marks and the edges of the tiles (Lobban et al. 2014, Fig. 9). This behavior was likely mediated by the cells sensing different chemical or physical (structural) conditions of the biofilm, if they encountered it, rather than detecting the thickness of the film per se. Cells that remained on thick biofilm presumably did so because they did not encounter thinner biofilm during their movements. The ability of small organisms to detect the character of the substrate has been particularly addressed in metamorphosis and settlement of coral larvae, which prefer crustose coralline algae (CCA). Larvae will settle on polystyrene in the presence of CCA extract. Actually, the morphogenetic and settlement signals appear to originate from bacteria that may produce the signals only when they are attached to suitable substrate (Heyward and Negri 1999; Negri et al. 2001).

The net effect of *Maristentor* location preferences and *Ctenochaetus* feeding preferences was to reduce the likelihood of the ciliate being caught in a bite but the pigment and clustering seemed to contribute directly to feeding deterrence. Maristentorin is related to a number of compounds known to be used as feeding deterrents in organisms from flowering plants to crinoids (marine invertebrates) (Lobban et al. 2007), and including particularly responses of several heterotrich ciliates to predatory ciliates, which they deter by extruding the contents of cortical granules (e.g., Miyake et al. 2001, 2003). Maristentorin has not yet been tested directly as a feeding deterrent to *C. striatus* but evidence of the location of bite marks relative to clusters suggested that *C. striatus* made some effort to avoid large clusters (Lobban et al. 2014). We (Lobban et al. 2007, 2014) have argued that the high concentration of toxic pigment in the cortical granules is likely to serve as a feeding deterrent, and that clustering into dark spots, especially on the most common substratum *Padina*, could serve as a visual warning to fish feeding on the biofilm. Further, we have proposed that their selection of thin biofilms and edges provides a measure of protection against grazing fish. Although *C. striatus* did incidentally consume *Maristentor* on biofilm, it cannot be considered a predator of *Maristentor*. As with the defensive effect of biofilm preference, the protective effect of maristentorin and clustering against *C. striatus* may be an indirect benefit of other functions. *Maristentor* is not commonly exposed to grazing by *C. striatus* and the toxin and behaviors are likely to have evolved in response to other factors and to serve other purposes instead or in addition, and just happened to be effective in the situations we observed.

Natural biofilm will provide a complex set of chemical signals, and it is likely that successionally less-developed biofilm has a different chemical "fingerprint" from more-developed biofilm, and that *Maristentor* can detect this. The fact that it moved out onto tile with no biofilm (Fig. 17.5) (an unnatural situation) suggests that biofilm has a negative effect on site selection, and potentially explains in part the apparent preference of *Maristentor* for *Padina* (i.e., for the epibioses thereon). We have not studied the distribution of *Maristentor* on *Padina*, especially during normal (non-bloom) seasons, but our impression is that in uncrowded conditions populations tend to gather in particular places on the frond, including the outer, youngest part of the fronds, which will have the least epibiota (Fig. 17.2a). In *Padina* there are concentric variations in the seaweed tissue, e.g., in calcium carbonate deposition (Fig. 17.2b) and the presence of hairs and sometimes a concentric pattern in *Maristentor* clusters (Fig. 17.2a). This pattern is not consistent, however, and there is much scope for study of its surface preferences.

17.8 Conclusions: Three-Leveled Biosemiotics

Biocommunication theory posits that "organisms interact by using signals as signs according to combinatorial rules (syntax), contextual rules (pragmatic) and content-specific rules (semantic) which cannot be reduced to one another but are crucial for successful communication processes" (Witzany 2008: 24). Syntax describes how signs may be combined, like words in a sentence; pragmatics are the rules that determine the interactional context; and semantics is the meaning—function—of the message. The knowledge we have about *Maristentor*, and the speculation we can make based on some weak and some strong analogies, amount to only a framework of understanding of how this integrated photosymbiosis is itself integrated into its natural environment. Nevertheless, biocommunication theory is a useful lens with which to look for ways to improve our understanding. For example, one could study the changes in the combinatorial rules and/or contextual rules that result at one time in individual decisions to leave a cluster and at another to the morning dispersal and swimming away from the substrate. First it will be necessary to determine the signs in use, but with our lack of knowledge of the genome we can scarcely guess what these may be.

All the hypotheses presented here on the basis of analogy need experimental or observational data to support or refute them, and we have tried to remain clear about what is knowledge and what is speculation. Hopefully, the biocommunication lens is also useful in showing how *Maristentor* can contribute to understanding of communicative processes in other organisms. If nothing else, this review serves to bring together the many outstanding questions about the biology of *Maristentor dinoferus*. Among the most intriguing are: How do cells divide asexually and do they have sexual reproduction? Are populations clonal or of mixed genotypes? Is there turnover of the endosymbiont population? What are the pragmatic rules and semantic contexts that mediate the complex clustering and dispersal behavior? What are the ranges of function of the pigment maristentorin and the clustering behavior in the context of the photosymbiosis? Addressing many of these questions will not be easy until or unless someone can successfully culture *Maristentor*.

Acknowledgements Many people have contributed to discussions of *Maristentor* over two decades. Here we wish to thank G. Curt Fiedler for reviewing a draft manuscript and Toshiyuki Nakajima for sharing a draft of his chapter for this book.

References

Ame JM, Halloy J, Rivault C, Detrain C, Deneubourg JL (2006) Collegial decision making based on social amplification leads to optimal group formation. Proc Natl Acad Sci USA 103:5835–5840

Buonanno F, Saltalamacchia P, Miyake A (2005) Defence function of pigmentocysts in the karyorelictid ciliate *Loxodes striatus*. Eur J Protistol 41:151–158

Buonanno F, Anesi A, Guella G, Kumar S, Bharti D, La Terza A, Quassinti L, Bramucci M, Ortenzi C (2014) Chemical offense by means of toxicysts in the freshwater ciliate, *Coleps hirtus*. J Eukaryot Microbiol 61:293–304

Camazine S, Deneubourg J-L, Franks NR, Sneyd J, Theraulaz G, Bonaneau E (2001) Self-organization in biological systems. Princeton University Press, Princeton

Chen S-T, Li C-W (1991) Relationships between the movements of chloroplasts and cytoskeletons in diatoms. Bot Mar 34:505–511

Demoulin G (1962) Comportement des chenilles de *Thaumetopoea pityocampa* Schiff. au cours des processions de nymphose. Comptes Rendues de l'Academie des Sciences de Paris 254:733–734

Deneubourg J-L, Grégoire J-C, Le Fort E (1990) Kinetics of larval gregarious behavior in the bark beetle *Dendroctonus micans* (Coleoptera: Scolytidae). J Insect Behav 3:169–182

Dziallas C, Allgaier M, Monaghan MT, Grossart H-P (2012) Act together—implications of symbioses in aquatic ciliates. Frontiers Microbiol 3(288):17

Emmeche C (2002) Taking the semiotic turn, or how significant philosophy of biology should be done. Nord J Philos 3:155–162

Fenchel T, Blackburn N (1999) Motile chemosensory behaviour of phagotrophic protists: mechanisms for and efficiency in congregating at food patches. Protist 150:325–336

Fitt WK (1985) Chemosensory responses of the symbiotic dinoflagellates *Symbiodinium microadriaticum* (Dinophyceae). J Phycol 21:62–67

Fitt WK, Trench RK (1983) The relation of diel patterns of cell division to diel pattern of motility in the symbiotic dinoflagellate *Symbiodinium microadriaticum* Freudenthal in culture. New Phytol 94:421–432

Foissner W (1996) Ontogenesis in ciliated protozoa with emphasis on stomatogenesis. In: Hausmann K, Bradbury PC (eds) Ciliates. Cells as organisms. Gustav Fischer Verlag, Stuttgart, pp 95–177

Foissner W (2006) Biogeography and dispersal of micro-organisms: a review emphasizing protists. Acta Protozoologica 45:111–136

Germond A, Nakajima T (*this volume*) Symbiotic associations in ciliates: ecological and evolutionary perspectives. In: Witzany G, Nowacki M (eds) Biocommunication of ciliates. Springer

Gilbert SF (2013) Developmental Biology, 10th edn. Sinauer, Sunderland, USA 719 pp

Greenberg PE (2003) Bacterial communication: tiny teamwork. Nature 424:134

Hansen G, Daugbjerg N (2009) *Symbiodinium natans* sp. nov.: a "free-living" dinoflagellate from Tenerife (Northeast-Atlantic Ocean). J Phycol 21:251–263

Hartmann C, Özmutlu Ö, Petermeier H, Fried J, Delgado A (2007) Analysis of the flow field induced by the sessile peritrichous ciliate *Opercularia asymmetrica*. J Biomech 40:137–148

Heyward AJ, Negri AP (1999) Natural inducers for coral larval metamorphosis. Coral Reefs 18:273–279

Hollingsworth LL, Kinzie RA III, Lewis TD, Krupp DA, Leong JAC (2005) Phototaxis of motile zooxanthellae to green light may facilitate symbiont capture by coral larvae. Coral Reefs 24:523

Hori M, Tomikawa I, Przybos E, Fuishima M (2006) Comparison of the evolutionary distances among syngens and sibling species of *Paramecium*. Mol Phylogenet Evol 38:697–704

Hurd CL, Harrison PJ, Bischof K, Lobban CS (2014) Seaweed ecology and physiology, 2nd edn. Cambridge University Press, Cambridge 562 pp

Jones R, Yellowlees D (1997) Regulation and control of intracellular algae (=zooxanthellae) in hard corals. Philos Trans R Soc Lond B 352:457–468

Kiørboe T, Visser AW (1999) Predator and prey perception in copepods due to hydromechanical signals. Mar Ecol Prog Ser 179:81–95

LaJeunesse TC, Parkinson JE, Reimer JD (2012) A genetics-based description of *Symbiodinium minutum* sp. nov. and *S. psygmophilum* sp. Nov. (Dinophyceae), two dinoflagellates symbiotic with cnidaria. J Phycol 48:1380–1391

Lenci F, Ghetti F, Song P-S (2001) Photomovement in ciliates. In: Häder D-P, Lebert M (eds) Photomovement. Elsevier Science B.V, Amsterdam, The Netherlands, pp 475–503

Lobban CS (2015) A second species of *Microtabella* (Grammatophoraceae, Bacillariophyta) from Guam. Mar Biodivers Rec 8:e151, 5 pp

Lobban CS, Schefter M (1996) An abundance of marine *Stentor* (Ciliophora: Spirotrichea) epiphytic on *Padina* (Phaeophyta). Micronesica 29:99–100

Lobban CS, Schefter M (2012) Blooms of a benthic ciliate, *Maristentor dinoferus* (Heterotrichida: Maristentoridae), on coral reefs of Guam, Mariana Islands. Micronesica 43:114–127

Lobban CS, Schefter M, Simpson AGB, Pochon X, Pawlowski J, Foissner W (2002) *Maristentor dinoferus* n. gen., n. sp., a giant heterotrich ciliate (Spirotrichea: Heterotrichida) with zooxanthellae, from coral reefs on Guam, Mariana Islands. Mar Biol 140:411–423+141: 207–208

Lobban CS, Modeo L, Verni F, Rosati G (2005) *Euplotes uncinatus* (Ciliophora, Hypotrichia), a new species with zooxanthellae. Mar Biol 147:1055–1061

Lobban CS, Hallam SJ, Mukherjee P, Petrich JW (2007) Photophysics and multifunctionality of hypericin-like pigments in heterotrich ciliates: a phylogenetic perspective. Photochem Photobiol 83:1074–1094

Lobban CS, Schefter M, Donaldson TJ (2014) Cluster dynamics in *Maristentor dinoferus*, a gregarious benthic ciliate with zooxanthellae and a hypericin-like pigment, in relation to biofilm grazing by the fish *Ctenochaetus striatus*. Symbiosis 63:137–147

Lynn DH (2008) The Ciliated Protozoa. Characterization, classification, and guide to the literature. 3rd edn. Springer, New York, 605 pp

Madl P, Witzany G (2014) How corals coordinate and organize: an ecosystemic analysis based on biocommunication and fractal properties. In: Witzany G (ed) Biocommunication of animals, Springer Science+Business Media, Dordrecht, Germany, pp 351–382

Mansfield M, Turner SL (2002) Quorum sensing in context: out of molecular biology and into microbial ecology. Microbiology 148:3762–3764

Mayr E (1982) The Growth of Biological Thought: diversity, evolution and inheritance. Belknap Press, Cambridge 974 pp

McManus GB, Schoener DM, Haberlandt K (2012) Chloroplast symbionts in a marine ciliate: ecophysiology and the risks and rewards of hosting foreign organelles. Frontiers Microbiol 3:321. doi:10.3389/fmicb.2012.00321

Miao W, Simpson AGB, Fu C, Lobban CS (2005) The giant zooxanthellae-bearing ciliate *Maristentor dinoferus* is closely related to Folliculinidae. J Eukaryot Microbiol 52:11–16

Miyake A, Harumoto, Iio H (2001) Defense function of pigment granules in *Stentor coeruleus*. Eur J Protistol 37:77–88

Miyake AF, Buonanno P, Saltalamacchia P, Masaki ME, Iio H (2003) Chemical defense by means of extrusive cortical granules in the heterotrich ciliate *Climacostomum virens*. Eur J Protistol 39:25–36

Mordret S, Romac S, Henry N, Colin S, Carmichael M, Berney C, Audic S, Richter DJ, Pochon X, de Vargas C, Decelle J (2015) The symbiotic life of *Symbiodinium* in the open ocean within a new species of calcifying ciliate (*Tiarina* sp). ISME J (in press)

Mukherjee P, Fulton DB, Halder M, Han X, Armstrong DW, Petrich J, Lobban CS (2006) Maristentorin, a novel pigment from the positively phototactic marine ciliate *Maristentor dinoferus,* is structurally related to hypericin and stentorin. J Phys Chem B 110:6359–6364

Mulisch M, Patterson DJ (1988) Stomatogenesis during cell division in the loricate ciliate *Eufolliculina uhligi*: a scanning electron microscope study. Eur J Protistol 23:193–201

Negri AP, Webster NS, Hill RT, Heyward AJ (2001) Metamorphosis of broadcast spawning corals in response to bacteria isolated from crustose algae. Mar Ecol Prog Ser 223:121–131

Nowack ECM, Melkonian M (2010) Endosymbiotic associations within protists. Philos Trans R Soc Lond B Biol Sci 365:699–712

Parkinson JE, Coffroth MA, LaJeunesse TC (2015) New species of Clade B *Symbiodinium* (Dinophyceae) from the greater Caribbean belong to different functional guilds: *S. aenigmaticum* sp. nov., *S. antillogorgium* sp. nov., *S. endomadracis* sp. nov., and *S. pseudominutum* sp. nov. J Phycol 51:850–858

Pochon X, Putnam HM, Gates RD (2014) Multi-gene analysis of *Symbiodinium* dinoflagellates: a perspective on rarity, symbiosis, and evolution. PeerJ 2:e394. https://dx.doi.org/10.7717/peerj.394

Podestà A, Marangoni R, Villani C, Colombetti G (1994) A rhodopsin-like molecule on the plasma membrane of *Fabrea salina*. J Eukaryot Microbiol 41:565–569

Schlichter D, Meier U, Fricke HW (1994) Improvement of photosynthesis in zooxanthellate corals by autofluorescent chromatophores. Oecologia 99:124–131

Sobierajska K, Fabczak H, Fabczak S (2006) Photosensory transduction in unicellular eukaryotes: a comparison between related ciliates *Blepharisma japonicum* and *Stentor coeruleus* and photoreceptor cells of higher organisms. J Photochem Photobiol B Biol 83:163–171

Sommaruga R, Whitehead K, Shick JM, Lobban CS (2006) Mycosporine-like amino acids in the zooxanthella-ciliate symbiosis *Maristentor dinoferus*. Protist 157:185–191

Song P-S, Kim I-H, Rhee JS, Huh J, Florell S, Faure B, Lee KW, Kahsai T, Tamai N, Yamazaki T, Yamazaki I (1991) Photoreception and photomovements in *Stentor coeruleus*. In: Lenci F, Ghetti F, Colombett G, Häder D-P, Song P-S (eds) Biophysics of photoreceptors and photomovements in microorganisms. Plenum Press, New York, pp 267–279

Stanley GD, Swart PK (1995) Evolution of the coral-zooxanthella symbiosis during the Triassic: a geochemical approach. Paleobiology 21:179–199

Stat M, Carter D, Hoegh-Guldberg O (2006) The evolutionary history of *Symbiodinium* and scleractinian hosts—symbiosis, diversity, and the effect of climate change. Perspect Plant Ecol Evol Syst 8:23–43

Stoecker DK, Silver MW, Michaels AE, Davis LH (1988) Enslavement of algal chloroplasts by four *Strombidium* spp (Ciliophora, Oligotrichida). Mar Microb Food Webs 3:79–100

Stoecker DK, Johnson MD, de Vargas C, Not F (2009) Acquired phototrophy in aquatic protists. Aquat Microb Ecol 57:279–310

Summerer M, Sonntag B, Sommaruga R (2007) An experimental test of the symbiosis specificity between the ciliate *Paramecium bursaria* and strains of the unicellular green alga *Chlorella*. Environ Microbiol 9:2117–2122

Summerer M, Sonntag B, Sommaruga R (2008) Ciliate-symbiont specificity of freshwater endosymbiotic *Chlorella* (Trebouxiophyceae, Chlorophyta). J Phycol 44:77–84

Sumpter DJT, Pratt SC (2009) Quorum responses and consensus decision making. Philos Trans R Soc B 364:743–753

Titlyanov EA, Titlyanova TV, Leletkin VA, Tsukahara J, van Woesik R, Yamazato K (1996) Degradation of zooxanthellae and regulation of their density in hermatyopic corals. Mar Ecol Prog Ser 139:167–178

Visser AW (2001) Hydromechanical signals in the plankton. Mar Ecol Prog Ser 222:1–24

Wahl M (2008) Ecological lever and interface ecology: epibiosis modulates the interface between host and environment. Biofouling 24:427–438

Ward AJW, Sumpter DJT, Couzin ID, Hart PJB, Krause J (2008) Quorum decision-making facilitates information transfer in fish shoals. Proc Natl Acad Sci USA 105:6948–6953

Wilkerson FP, Grunseich G (1990) Formation of blooms by the symbiotic ciliate *Mesodinium rubrum*: the significance of nitrogen uptake. J Plankton Res 12:973–989

Witzany G (2008) Biocommunication of unicellular and multicellular organisms. The biosemiotic categorization of rule-governed sign-mediated interactions within and between bacteria, fungi and plants. TripleC (Cognition, Communication, Co-operation) 6:24–53

Wölfl S, Geller W (2002) *Chlorella*-bearing ciliates dominate in an oligotrophic North Patagonian lake (Lake Pirehueico, Chile): abundance, biomass and symbiotic photosynthesis. Freshw Biol 47:231–242

Wood DC (1989) Localization of mechanoreceptors in the protozoan, *Stentor coeruleus*. J Comp Physiol A 165:229–235

Wood DC (2001) Electrophysiology and light responses in *Stentor* and *Blepharisma*. In: Häder D-P, Lebert M (eds) Photomovement. Elsevier Science BV, Amsterdam, pp 505–518

Wood-Charlson EM, Hollingsworth LL, Krupp DA, Weis VM (2006) Lectin/glycan interactions play a role in recognition in a coral/dinoflagellate symbiosis. Cell Microbiol 8:1985–1993

Yamashita H, Kobiyama A, Koike K (2009) Do uric acid deposits in zooxanthellae function as eye-spots? PLoS One 4(7):e6303. doi:10.1371/journal.pone.0006303

Yamashita H, Suzuki G, Kai S, Hayashibara T, Koike K (2014) Establishment of coral–algal symbiosis requires attraction and selection. PLoS One 9(5):e97003. doi:10.1371/journal.pone.0097003

Zagata P, Greczek-Stachura M, Tarcz S, Rautian M (2015) Molecular identification of *Paramecium bursaria* syngens and studies on geographic distribution using mitochondrial cytochrome c oxidase subunit I (COI). Folia Biol (Krakow) 63:77–83

Chapter 18
Interactions Between Parasitic Ciliates and Their Hosts: *Ichthyophthirius multifiliis* and *Cryptocaryon irritans* as Examples

Kassandra E. Zaila, Deanna Cho and Wei-Jen Chang

Abstract Among all known parasitic ciliates, *Ichthyophthirius multifiliis* and *Cryptocaryon irritans* are probably the two best studied "model parasites," which cause the white spot diseases in freshwater and marine fish, respectively. These two parasites are not phylogenetically closely related, but share very similar life cycle and pathological manifestations. Both parasites invade and feed in the epithelial layer of the skin and gill, and trigger both innate and adaptive immune responses in fish hosts, locally and systemically. The parasite-host interactions not only help add to our knowledge of fish immunity, but also shed light on research regarding other less well-studied ciliate parasites. Future research directions are also discussed in this chapter.

18.1 Parasites

Ciliates comprise a group of morphologically diverse, unicellular organisms unified by both their nuclear dimorphism and infraciliary system. The great morphological diversities reflect how ciliates evolve and adapt to various ecological environments. Taking size as an example, the smallest ciliates are around 10 micrometers in diameter, and on the other end of the spectrum the ciliate *Stentor*, can grow to 2 mm in length. Currently there are more than 10,000 described ciliate species, and a good portion of them are parasites of mammals (Schuster and Visvesvara 2004; Schuster and Ramirez-Avila 2008), amphibians (Densmore and Green 2007), fish (Scholz 1999; Valladao et al. 2013, 2015), insects (Grasse and de Boissezon 1929; Egerter et al. 1986), and other invertebrates (Morado and Small 1995). However, most of these ciliate parasites have not been well studied except for the two fish parasites

K.E. Zaila · D. Cho · W.-J. Chang
Department of Biology, Hamilton College, Clinton, NY 13323, USA

W.-J. Chang (✉)
Institute of Molecular Biology, Academia Sinica, Taipei 11529, Taiwan
e-mail: wchang@hamilton.edu

G. Witzany and M. Nowacki (eds.), *Biocommunication of Ciliates*,
DOI 10.1007/978-3-319-32211-7_18

Ichthyophthirius multifiliis and *Cryptocaryon irritans*, which cause "white spot diseases" in freshwater and marine fish, respectively. In the past 20 years much attention has been given to understanding interactions between these two fish parasites and their hosts, primarily due to their severe impacts on both aquaculture and ornamental fish markets.

18.1.1 Life Cycles

I. multifiliis and *C. irritans* possess striking similarities in their life cycles. They are ectoparasites with no known intermediate hosts, and both exhibit broad host ranges and high fatality rates in hosts caused by their infections. Both *I. multifiliis* and *C. irritans* consist of three morphologically and physiologically distinct life stages: an infectious theront stage, a parasitizing trophont stage, and an asexually reproducing tomont stage. The spindle-shaped theront cells are free-swimming, and are looking for host fish. Once theronts establish infections in the epithelial layer of host fish, they quickly transform into round-shaped trophont cells. Trophonts feed on host tissues and can grow up to several hundred micrometers, thus the visible "white spots" on fish skin. After 7–10 days of infection, depending on water temperature, mature trophonts will disassociate from the host and enter the reproductive tomont stage by attaching themselves to a flat surface. There, tomont cells are encysted and begin rounds of cellular divisions, which can give rise to up to hundreds if not thousands of theront cells per tomont (MacLennan 1937; Hoffman 1967).

Despite the similarities in their infectious cycles, *I. multifiliis* and *C. irritans*, however, belong to two different classes in the phylum of Ciliophora. *I. multifiliis* is a member of Oligohymenophorea and is closely related to the free-living *Tetrahymena thermophila* in this class. Another free-living model organism *Paramecium tetraurelia* is also a member of Oligohymenophorea, but is further related to *I. multifiliis* (Wright and Lynn 1995; Gentekaki et al. 2014). *C. irritans* was first assigned to the family Ichthyophthiriidae, class Oligohymenophorea, because of its pathological resemblance to that of *I. multifiliis* (Corliss 1979). However, cytological evidence does not support such classification (Cheung et al. 1981; Lynn and Corliss 1991; Colorni and Diamant 1993; Diggles 1997), and later results derived from molecular phylogenetic analyses using SSU rDNA sequences reassign *C. irritans* to the class Prostomatea (Wright and Colorni 2002). The phylogenetic relationship between classes Oligohymenophorea and Prostomatea have not been fully resolved (Gao and Katz 2014; Zhang et al. 2014), but it is clear that the common features and parallel life cycles shared by *I. multifiliis* and *C. irritans* are likely to be a result of convergent evolution.

18.1.2 Fish Hosts and Parasite Strains

I. multifiliis and *C. irritans* both show low host specificities, and they can infect a wide range of freshwater and marine fish, respectively. However, due to economic interests most research has been done using common or high value aquaculture/ornamental fish species. Furthermore, both *I. multifiliis* and *C. irritans* are obligate parasites, and would thus require continuous supplies of fish hosts in order to study the different stages of these parasites. Different labs may use different fish species depending on how convenient it is to obtain and maintain fish cultures.

For *I. multifiliis* the most commonly used fish species are rainbow trout (*Oncorhynchus mykiss*), common carp (*Cyprinus carpio*), grass carp (*Ctenopharyngodon idella*), channel catfish (*Ictalurus punctatus*), Nile tilapia (*Oreochromis niloticus*), and European chub (*Squalius cephalus*). While these species cover a broad range of ray-finned fishes, they represent only four out of approximately 42 orders in the class Actinopterygii. Comparatively, researchers studying *C. irritans* have frequently used orange-spotted grouper (*Epinephelus coioides*), red sea bream (*Pagrus major*), Asian seabass/barramundi (*Lates calcarifer*), thick-lipped mullet (*Chelon labrosus*), and large yellow croaker (*Larimichthys crocea*), as their host organisms. Except for thick-lipped mullet, the other four species belong to the large, and highly diverse order Perciformes.

Without using a standardized host species, it may be sensible to keep in mind that different fish hosts may respond differently when being challenged by the same parasite species. In fact, different fish species showed different susceptibilities to *I. multifiliis* (Clayton and Price 1994) or *C. irritans* infections (Luo et al. 2008). Furthermore, *C. irritans* isolated from different fish hosts vary in their pathogenicity (Burgess and Matthews 1995b). In other words, different isolates (or strains) (Swennes et al. 2006), or even different passages of the same parasite species may also differ in their pathogenicity/virulences (Xu and Klesius 2004). Correlations between parasite isolates/strains and their levels of pathogenicity/virulences remain unclear, however, despite the fact that molecular characterization using ITS sequences has been available in *C. irritans* (Diggles and Adlard 1997; Yambot et al. 2003), and recently in *I. multifiliis* by using mitochondrial sequences (MacColl et al. 2015).

18.1.3 Chemotaxis: Transmission of Theronts to Fish Hosts

Free-living theronts have a short-lifespan and their host-finding mechanism is crucial to its survival and progression in the life cycle. Evidences suggest that the initial response of *I. multifiliis* theronts is mediated by environmental factors, specifically light rather than an attraction to a fish host (Wahli et al. 1991). The theronts positive phototaxis influences its upward movement towards the surface, which then provides an opportunistic region for host finding. Once in close proximity to the fish host, theronts change behavioral patterns in response to different chemical stimuli unique to teleosts (Haas et al. 1999). Buchmann and Nielsen later designed a novel bioassay and demonstrated that both serum factors and mucus, rather than low molecular weight carbohydrates, amino acids, fatty acids, and urea, have a high chemoattractant effect on theronts (Buchmann and Nielsen 1999). Interestingly, *I. multifiliis* is also attracted to sera from marine fish that it can not infect (hake (*Merluccius merluccius*) and dab (*Limanda limanda*)). The low specificity against fish sera could probably explain the broad host-range of *I. multifiliis* (Buchmann and Nielsen 1999). Similar chemoattraction assays were conducted in *C. irritans*, which was also attracted to both sera and mucus collected from fish (Luo et al. 2008).

18.1.4 The Invasion of I. multifiliis

After being released from tomont cysts, theronts of *I. multifiliis* have a window of 24 h before they lose the ability to infect fish at 20 °C (McCallum 1982). It is unclear what changes on *I. multifiliis* occur after those 24 h, but theronts in *in vitro* culture generally can live for longer than 2 days.

How theronts invade fish's gill epithelium has been well documented in *I. multifiliis* by extensive microscopic observations (Ewing et al. 1985). Theronts first penetrate through margins of epithelial cells and quickly (within 5 min) move the whole cell into the epithelial layer before they stop and transform into the trophont stage by the basement membrane. The penetration also cause disruptions of host cells, whose debris seem to be absorbed by *I. multifiliis* through its cytostome (Ewing et al. 1985). While the whole invasion process seem efficient and effective, only approximately 50 % of theronts gain entry into the epithelial layer within the first 5 min of infection. Furthermore, less than 50 % of theronts which successfully invade the epithelial layer survive and continue developing into full trophonts 10 min post infection. In other words, only about 20 % of theronts successfully infect fish and continue to grow as trophonts (Ewing et al. 1986). It is not clear what caused the huge decline of *I. multifiliis* population in fish's gill between 5 and 10 min post infection.

18.2 Host Responses

Similar to other vertebrates, fish also have both innate and adaptive (acquired) immune systems. These two systems use different types of receptors to recognize pathogens (Medzhitov 2007; Alvarez-Pellitero 2008). The innate immune response is mediated by pattern recognition receptors (PRRs), which are germline encoded and have broad specificity for conserved regions known as pathogen-associated molecular patterns (PAMPs) (Janeway and Medzhitov 2002; Medzhitov and Janeway 2002). In contrast, adaptive immunity is mediated by antigen receptors that have random, but narrow specificities. Antigen receptors are distributed on B and T lymphocytes and each cell has receptors of a single specificity, meaning a binding site for a particular epitope. When a pathogen is first presented, only populations of lymphocytes that can recognize epitopes from this pathogen will elicit a response. Some of these activated lymphocytes will become memory B, or T cells, and can elicit a faster and stronger response when being challenged by the same pathogen again, which constitutes the basis for immunological memory.

While innate and adaptive immune systems interact with pathogens in fundamentally different ways and initiate distinctive host responses, the two systems do cross-talk and coordinate with each other via complicated networking. Fish, in particular, have a relatively primitive adaptive immune system compared to that in vertebrates (Du Pasquier 1982; Ellis 2001). Innate immunity thus plays the key role in combating pathogens before the much slower adaptive immunity rolls in (Magnadottir 2006). Moreover, the innate system is needed to activate the adaptive system, and research has shown that innate responses may help determine the nature of adaptive responses (Fearon and Locksley 1996; Bendelac and Fearon 1997; Fearon 1997; Carroll and Prodeus 1998; Vilmos and Kurucz 1998). A general picture of how and when these two immune systems are activated against *I. multifiliis* and *C. irritans* infections in fish can now be drawn based on findings in the past few decades (Dickerson and Findly 2014). We now know more about responses from individual (sub)systems, but interconnections between these (sub) systems remain largely unknown.

18.2.1 Innate Immune System

The innate immune system is the collaboration of several subsystems that carry out different functions to elicit a host defense mechanism. Some of these subsystems and their functions in the host defense are described below.

18.2.1.1 Initiation of the Immune Response via Pathogen Recognition Receptors (PRR)

Both *I. multifiliis* and *C. irritans* invasions activate fish's innate or nonspecific immune responses. This first line of fish defense is triggered by the recognition of PAMPs by PRRs. In teleost fish, there are four types of PRRs including, Toll-like receptors, C-type lectin receptors, NOD-like receptors, and RIG-I-like receptors (Alvarez-Pellitero 2008). The first and best described PRR is Toll-like receptors (TLRs), which are type 1 transmembrane proteins with two distinct domains: an extracellular leucine-rich repeat and an intracellular Toll/IL-1 receptor (Medzhitov 2001). To date, at least 20 TLRs have been identified in numerous fish species (reviews in (Alvarez-Pellitero 2008; Buchmann 2014)). Members of the TLR family directly or indirectly bind to PAMPs and induce signal-transduction cascades.

Expression levels of several TLR genes in orange-spotted grouper significantly changed after fish were infected by *C. irritans*. The changes post infection (could be up- or down-regulation) varied by time point and tissue type, suggesting that the primary response to *C. irritans* infection is tightly controlled and/or there is a possible involvement of other TLR genes that have not yet been discovered (Li et al. 2011, 2012). Similar expression changes were also found in genes on the Toll-like receptor pathway in yellow croaker infected by *C. irritans* (Wang et al. 2015; Zhang et al. 2015). Furthermore, genes connected to the Toll-like receptor signal transduction pathway, such as *MyD88* (Li et al. 2011), tumor necrosis factor-associated factor 6 (*TRAF6*) (Li et al. 2014a), interleukin 1 receptor-associated kinase 4 (*IRAK-4*) (Li et al. 2014b), and transforming growth factor-beta-activated kinase 1 (*Tak1*) (Li et al. 2015), also showed changes in expression levels following *C. irritans*, or bacterial infections (Wei et al. 2011). These observations strongly support the view that TLR-mediated host defense mechanisms are activated soon after *C. irritans* (and/or bacterial) infections, and the signals are passed down to elicit inflammatory responses (see below).

Similar to what were observed in *C. irritans* infection, changes at the expression levels of six TLR genes in channel catfish (Zhao et al. 2013a), *TRAF6* and *Tak1* in grass carp (Zhao et al. 2013b), were also detected following *I. multifiliis* infection. Moreover, transcripts of Tak1-binding proteins (TAB1 and TAB2) that regulate the activation of Tak1 (Landstrom 2010), were also significantly upregulated post infection (Zhao et al. 2014). Taken together, although much more details of activation mechanisms remain unclear, fish initiate proinflammatory and many other responses following *C. irritans* or *I. multifiliis* infection (Gonzalez et al. 2007a, b, c; Alvarez-Pellitero 2008), likely through TLR signaling pathways.

18.2.1.2 Local Cellular Response: Inflammation

Activation of TLRs leads to a cascade of intracellular pathways that result in upregulation of proinflammatory and antimicrobial genes (Neves et al. 2015). Specifically, TLRs activate resident macrophages, which trigger the production of proinflammatory cytokines: tumor necrosis factor (TNF)-α and interleukin (IL)-1β (Medzhitov 2007). TNF-α, IL-1β, and chemokines are key signal modulators of local and systemic inflammatory responses in fish (Secombes et al. 2001). These protein-signaling modulators direct the migration of leukocytes to the site of infection by increasing vascular permeability and/or by the stimulation of chemotaxis (Medzhitov 2007).

Proinflammatory Molecules: IL-1β, TNF-α, and Chemokines

The cytokine IL-1β is considered to be the most important mediator of the inflammatory response in teleost fish (Engelsma et al. 2002). After *I. multifiliis* infection, IL-1β mRNA levels increased as early as three hours post infection in rainbow trout larvae (Heinecke and Buchmann 2013), and peaked at 36 h post infection in the skin of both rainbow trout (Sigh et al. 2004a) and common carp (Gonzalez et al. 2007a). The rapid peak in expression levels of IL-1β in the fish larvae blood followed by a peak at 36 h in the skin represents the activation and subsequent migration of immune-related cells to the site of infection. A similar IL-1β expression hike was also observed in *C. irritans* infected orange-spotted grouper (Li et al. 2011).

Expressions of other proinflammatory genes, such as TNF-α and different classes of chemokines, were also upregulated following *I. multifiliis* infection. TNF-α has been described as one of key modulators in fish innate immunity specifically by its ability to activate macrophages and to attract neutrophils (Hardie et al. 1994). After *I. multifiliis* infection, TNF-α gene expression was upregulated in rainbow trout (Sigh et al. 2004a) and common carp (Gonzalez et al. 2007a). Expression levels of CXC chemokines, chemotactic cytokines involved in leukocyte migration to the site of infection, and CXC receptors (CXCRs) also increased (Gonzalez et al. 2007a). Additionally, changes in expression levels of the CC chemokine subfamily in response to both *I. multifiliis* and *C. irritans* infections have been observed. After infection of *C. irritans* in barramundi, significant upregulation of CC chemokine transcripts were detected in spleen, liver, gill, and kidney tissues (Mohd-Shaharuddin et al. 2013). On the contrary, CC chemokine in common carp after exposure to *I. multifiliis* showed significant upregulation only in the liver, but not in the skin (Gonzalez et al. 2007d).

Leukocyte Migration After Infection of I. multifiliis

Within 24 h of exposure to *I. multifiliis* the cellular response of the fish is dominated by the migration of neutrophils to the site of infection in both primary exposed and immunized fish (Hines and Spira 1973; Ventura and Paperna 1985; Cross and Matthews 1993; Dickerson and Clark 1996). In the epidermis of carp at 2–3 days (early phase) after exposure parasites were surrounded by neutrophils and by 5–6 days (late phase) the cellular response became extensive with major infiltration of granulocytes including eosinophils, basophils and neutrophils (Cross and Matthews 1993). It was also noted that during the course of infection, a layer of necrotic host tissue debris surrounding the feeding parasite gradually formed and the migrating leukocytes were found in this layer. However, cytological studies showed that the leukocytes did not come into direct membrane-to-membrane contact with the parasite (Cross 1994). The leukocytes instead appeared to be degraded within the necrotic tissue layer presumably by the secretion of lytic enzymes and metabolites from the feeding parasites (Cross 1994; Dickerson and Clark 1996). It is thus hypothesized that the localized cellular response to *I. multifiliis* is geared more towards mediating pathogenesis than protecting the host (Cross 1994).

Inducible Nitric Oxide Synthase (INOS) and Nitric Oxide (NO)

Nitric oxide synthases (NOS) catalyze the production of nitric oxide (NO) from the amino acid L-arginine (Hibbs et al. 1987). Inducible nitric oxide synthase (iNOS) is an isoform of NOS and its expression can be induced by proinflammatory cytokines (i.e. TNF-α and IL-1β). The functions of NO are diverse, ranging from being secondary messengers, immunoregulators, to toxic agents against microbes (Coleman 2001). It has been shown that iNOS genes were upregulated in fish following exposures to microbial and cytokine stimuli, and some host protections were linked to the production of NO (Hodgkinson et al. 2015). However, when common carps were infected by kinetoplastid parasites *Trypanosoma borreli* and *Trypanosoma carassii*, the former one induced NO production but not the latter (Saeij et al. 2003). Furthermore, carps expressing NO following *T. borreli* infection had a higher mortality rate than those who did not (by using the iNOS inhibitor aminoguanidine), suggesting that NO might be toxic to fish hosts under some circumstances (Saeij et al. 2003).

After infection with *I. multifiliis,* increased expression of the iNOS gene has been observed in the skin of both common carp (Gonzalez et al. 2007a) and rainbow trout (Sigh et al. 2004a), despite differences in magnitude. Increased expression of both iNOS and proinflammatory genes (TNF-α and IL-1β) post *I. multifiliis* infection supports that the NO pathway is activated. The effects of NO production during *I. multifiliis* infection, however, remain unknown.

18.2.1.3 Systemic Innate Immunity

The Complement System

The complement cascade plays essential roles in both the innate and adaptive immune responses. Activation of the complement system in fish, as in higher vertebrates, is dependent on three pathways: the alternative pathway, the lectin pathway, and the classical pathway (Boshra et al. 2006). The alternative pathway is spontaneously activated by complement protein C3 covalently binding to various microbial surfaces. The lectin pathway requires the interaction between carbohydrates ligands found on the surface of pathogens to lectins such as mannose-binding lectin (MBL) and ficolin. When MBL binds to carbohydrate moieties on microbial pathogens, proteases associated with MBL activate and cleave complement proteins to active forms. The classical pathway is triggered by antibodies derived from the adaptive immune response binding to the C1 complex. Once activated, the complement system has numerous functions including, opsonization of the pathogens, activation and chemotaxis of leukocytes, disposing of immune complexes and the products of inflammatory injury, and the ability to create a membranolytic complex (Gasque 2004).

After infection with *I. multifiliis* upregulation of C3 gene expression has been detected in the gills, skin, head kidney, and spleen of rainbow trout (Sigh et al. 2004a;

von Gersdorff Jorgensen et al. 2008; Olsen et al. 2011), but not in the skin or blood of common carp (Gonzalez et al. 2007b). Rather, a moderate increase in the expression of C3 was observed in the liver 24 h post infection in common carp. Additionally, transcripts of complement factors C2 and C4 were upregulated as early as 24 h post infection in the skin and liver of common carp (Gonzalez et al. 2007b).

Interestingly, in barramundi infected with *C. irritans* no significant changes in the expression levels of the C3 gene were detected in all tested tissues (spleen, liver, kidney, and gills). Expression of the C4 gene in the liver, in contrast, showed a 58-fold increase on day 3 post-infection (Mohd-Shaharuddin et al. 2013). In orange-spotted grouper serum alternative complement activities steadily and significantly rose from one to four weeks after immunization with *C. irritans* (Dan et al. 2013). Taken together, these observations indicate that after infection of *I. multifiliis* or *C. irritans* different host fish may respond with activating different complement factors. Locations, magnitudes, and timing of expression of the factors may also vary.

Antimicrobial Peptides

Antimicrobial peptides (AMPs) are a major humoral component in the innate defense of teleost fish. AMPs are small peptides, mostly cationic or amphipathic, that play a role in lytic activity by directly interacting with pathogens and disrupting the permeability of the microorganism's membrane (Valero et al. 2013). AMPs act against a broad spectrum of pathogens, including viruses, bacteria, fungi, and protozoan parasites (Rajanbabu and Chen 2011). To date, approximately 62 different AMPs, have been identified in several teleost fish species and their activities tested against various common fish pathogens (Rajanbabu and Chen 2011; Valero et al. 2013).

In teleost fish, piscidin-2 of the piscidin family is one of the most potent broad-spectrum host-produced AMPs (Zahran and Noga 2010). Piscidin-2 is lethal to the infective stage (theront) of ectoparasites that infect marine (the ciliates *C. irritans,* and *Trichodina* sp., and the dinoflagellate *Amyloodinium ocellatum*), and freshwater fish (the ciliate *I. multifiliis*) (Colorni et al. 2008). Moreover, the efficacy of piscidin-2 depends on the salt environment and at lower cationic concentrations piscidin-2 is more effective. Research showed that *I. multifiliis* theronts could all be killed at a piscidin-2 concentration of 6.3 ug mL^{-1}. To achieve the same effect against *C. irritans* theronts in seawater it would require a concentration of 25 ug mL^{-1} piscidin-2, a nearly four folds of increase (Colorni et al. 2008).

Another antiparasitic AMP HbβP-1 which belongs to the β-hemoglobin peptide family has been characterized in the epithelium of channel catfish infected with *I. multifiliis* (Ullal et al. 2008; Ullal and Noga 2010). HbβP-1 is stage specific, lethal to trophonts and inactive against theronts or tomonts, and draws lytic effects on trophont cells. Compared to piscidin-1, which could also lyse *I. multifiliis* trophonts, HbβP-1 showed no detectable hemolytic effect on erythrocytes of channel catfish whereas piscidin-1 did (Ullal et al. 2008). During the course of a 2 week infection, expression of HbβP-1 and HbβP-1 associated activities gradually increased in the skin and gill epithelium of infected fish (Ullal et al. 2008). Because HbβP-1 could also lyse trophonts of dinoflagellate *A. ocellatum*, a major ectoparasite for warmwater marine fish, this AMP might be providing broad spectrum protection for both marine and freshwater fish (Ullal and Noga 2010).

Finally, in large yellow croakers, Niu and colleagues reported a novel piscidin-like antimicrobial peptide, known as Pc-pis, against *C. irritans* (Niu et al. 2013). Pc-pis showed not only lytic effects against the trophonts of *C. irritans*, but also broad antibacterial and antifungal activities (Niu et al. 2013).

Acute Phase Response and the Role of Serum Amyloid A, an Acute Phase Protein

The later phase of the innate immune system is driven by cytokines and chemokines (IL-1β and TNF-α), which induce the acute phase response (APR). APR is a systemic defense mechanism in response to infection and inflammation, and involves the production of proteins known as acute phase proteins (APPs) (Gonzalez et al. 2007c). Serum amyloid A (SAA) is one of the major APPs in mammals and is synthesized by monocytes and macrophages. In fish, it has been suggested that SAA acts as a proinflammatory cytokine involved in lymphocyte migration to the site of infection (Heinecke and Buchmann 2013). Following infection with *I. multifiliis,* SAA transcripts in the skin, liver, and blood of common carp were elevated (Gonzalez et al. 2007c). Similar increase was also detected in rainbow trout after exposure to *I. multifiliis* (Sigh et al. 2004a; Olsen et al. 2011). These results were consistent with detected expression levels of SAA following immunization with a live vaccine of *I. multifiliis* (von Gersdorff Jorgensen et al. 2008). Immunohistochemical stainings also confirmed the presence of SAA at sites of parasite location (Heinecke and Buchmann 2013). Finally, upregulation of SAA gene expression was observed in the spleen, liver, and kidney of *C. irritans* infected barramundi (Picón-Camacho et al. 2012). In sum, these observations suggest that fish SAA is commonly involved in defense against *I. multifiliis* or *C. irritans* infections, possibly through APR activation.

Other Immune Related Genes

Two other cytokines are also involved in the inflammatory response, and they act as bridges connecting the innate and adaptive immune responses: interleukin (IL)-12 and interferon (IFN)-γ. IL-12, produced by macrophages or dendritic cells, promotes the maturation of T helper 1 cells (in humans) (Trinchieri 2003). Interferons are hallmarks of antiviral defense in vertebrates and can be classified into three subfamilies, type I, II and III (Zou and Secombes 2011). The fish type II IFN includes the molecule IFN-γ, which plays a role in enhancing the phagocytic and nitric oxide activities, and modulating expressions of proinflammatory cytokines such as IL-12, IL-1β, IL-6, and TNF-α (Zou and Secombes 2011). Trout macrophages treated with recombinant fish IFN-γ expressed more MHCI and MHCII molecules, indicating that IFN-γ may help promote antigen presentation to the adaptive immune system (Zou et al. 2005; Martin et al. 2007a, b, c). Upregulation of these cytokine genes (IFNγ, TNF-α, and IL-12) has been recently reported in the spleens of immunized orange-spotted grouper (Josepriya et al. 2015), and genes involved in the adaptive immune response were simultaneously activated (discussed below), providing the link between both the innate and adaptive immune responses against protozoan pathogens in fish.

18.2.2 Adaptive Immune System

Unlike the innate immune system, which is driven by various subsystems, the adaptive immune system is a single entity primarily mediated by B- and T-lymphocytes. The primary function of B cells is to produce antibodies, also known as immunoglobulins, in response to antigen binding to the B-cell receptor (BCR). To date, there have been three classes of immunoglobulin isotypes identified in teleost fish including, IgM, IgD, and IgT/IgZ (Hansen et al. 2005; Zhang et al. 2010, 2011), with IgM being the most abundantly distributed (Mutoloki et al. 2014). T cells play a central role in cell-mediated immunity and can be further characterized into two functional groups including, cytotoxic T cells and helper T cells. Cytotoxic T cells directly kill infected cells and helper T cells function by modulating other immune cells through cytokine production (Mutoloki et al. 2014). Antigens bound to major histocompatibility complex (MHC) proteins on the surface of macrophages or dendritic cells of the innate system are presented to T cell receptors (TCRs) on T cells. The antigen presentation, and the presence of T cell co-receptors (CD4 on helper T cells and CD8 on cytotoxic T cells), activate corresponding T cells (Castro and Tafalla 2015).

18.2.2.1 Adaptive Mucosal Response

Both *I. multifiliis* and *C. irritans* infections elicit a local adaptive immune response in both the epithelia of the gills and skin tissues of naive and immunized teleost fish species, evidenced by the presence of adaptive immune relevant proteins (IgM, IgT, and MHC II), as well as T cells.

Adaptive Responses in the Gills

Immunohistochemical stainings show that in rainbow trout IgT positive cells densely populate the epithelial lining of the gills, whereas IgM positive cells are identified in the gill capillaries (Olsen et al. 2011). This is in agreement with early suggestion that IgT plays a specialized role in mucosal immunity (Zhang et al. 2010), and supports the notion that there may be two separate antibody production compartments (St Louis-Cormier et al. 1984; Lin et al. 1996; Clark and Dickerson 1997; Sigh et al. 2004b; Zhang et al. 2010).

After rainbow trout being immunized with *I. multifiliis* theronts, IgT and IgM antibodies were shown to bind in situ to the surface of early trophonts in the gills two hours after new infections (Von et al. 2011). Naive rainbow trout showed significantly less IgM, and no IgT, binding to trophonts (Von et al. 2011). The expression of IgM and IgT genes in rainbow trout gills were upregulated in both immunized and challenged fish (Olsen et al. 2011). These findings indicate the presence of immunological memory against *I. multifiliis* infections, and support the hypothesis that cross-binding of *I. multifiliis* surface antigens by fish antibodies acts as a stimulus which elicits the exit of the parasite (Clark and Dickerson 1997; Buchmann et al. 2001).

In addition to IgT and IgM positive cells, MHCII+ and CD8+ cells were also detected in the gills of rainbow trout. MHCII+ cells carried different morphologies, whereas CD8+ cells had a lymphocyte-like morphology and were primarily located at the base of the primary gill filaments and interbranchial septum (Olsen et al. 2011). Numbers and distribution of MHCII+ cells did not change significantly in immunized and naive fish, but the abundance of MHCII+ cells decreased significantly in the gills of immuno-suppressant hydrocortisone treated fish (Olsen et al. 2011).

Adaptive Responses in the Skin

Teleost skin differs from the mammalian one because it lacks a keratinized epithelial layer and is instead coated by a mucosal layer. Teleost fish have been shown to contain a skin-associated lymphoid tissue (SALT) that structurally resembles the teleost gut- associated lymphoid tissue (GALT). Interestingly, it has been shown that the B cell and antibody responses in the skin of rainbow trout are similar to those in the gut (Xu et al. 2013). IgT, antibody predominantly located in the gut mucosa, was the dominant type against *I. multifiliis* infection in the skin mucosal layer in rainbow trout (Roca et al. 2008; Xu et al. 2013; Mutoloki et al. 2014). IgT protein concentrations in the trout's skin mucus increased by more than 6 folds in infected fish, but IgM concentrations and IgM+ B cell numbers remained unchanged after *I. multifiliis* infection (Xu et al. 2013).

The anomaly with respect to the types of antibodies produced against epidermal pathogen infections in teleost fish may be channel catfish, which lacks the antibody IgT (Bengten et al. 2002, 2006; Edholm et al. 2010). It can thus be speculated that this species uses IgM for both mucosal and systemic immune responses. In fact, research shows that IgM against *I. multifiliis*, particularly to the surface immobilization antigens (I-antigens), was present in both cutaneous mucus and sera of infected as well as immunized fish (Clark et al. 1996; Dickerson and Clark 1998; Xu and Klesius 2002;

Xu et al. 2002; Maki and Dickerson 2003). However, it was also shown that only monoclonal IgG, but not the larger tetrameric IgM antibodies, against I-antigens conferred protection against lethal challenges of *I. multifiliis* in naive channel catfish when antibodies were injected intraperitoneally (Lin et al. 1996). These findings suggest that protective IgM antibodies found on cutaneous mucus were not diffused from serum, instead they were produced locally from the epidermis layer. Maki and Dickerson later showed that while both *I. multifiliis* infected (surface exposure) and I-antigen immunized (intraperitoneal injection) channel catfish produced comparable titers of antibodies in sera, a higher titer of antibodies was found in the cutaneous mucus of infected fish, but not in the I-antigen immunized fish (Maki and Dickerson 2003). Skin explants from *I. multifiliis* immunized channel catfish were also capable of secreting *I. multifiliis* specific antibodies in vitro, further supporting a parallelism between the mucosal and systemic immune response (Xu et al. 2002). Recently, Zhao and colleagues identified specific IgM antibody-secreting cells (ASC) in the skin of channel catfish after *I. multifiliis* infection (Zhao et al. 2008). Upon reinfection *I. multifiliis* immunized catfish showed increased numbers of ASC in the skin, further demonstrating that these ASC cells were associated with immune memory and augmented responses (Zhao et al. 2008).

Antibodies specifically against *C. irritans* after surface exposure or intraperitoneal vaccination were also detected in both sera and/or mucus of grouper (Yambot and Song 2006; Luo et al. 2007), barramundi (Bryant et al. 1999), thick-lipped mullet (Burgess and Matthews 1995a), and mummichog (Yoshinaga and Nakazoe 1997). Vaccinations also conferred immunity against *C. irritans* infection (Burgess and Matthews 1995a; Yoshinaga and Nakazoe 1997; Yambot and Song 2006; Luo et al. 2007).

The T cell responses to *I. multifiliis* infection have not been examined as extensively as those of B cells, but T cells were recently implicated to be present in the skin and respond to infections (Findly et al. 2013). In channel catfish, lymphocytes have been detected using H&E staining, excised using laser dissection microscopy, and shown by RT-PCR to contain cells expressing T cell receptor beta chain (TCRβ) gene (Findly et al. 2013). The detection of TCRβ transcripts indicates that T cells reside in the epidermis of infected fish at the site of infection. Additionally, in grouper immunized with *C. irritans* DNA vaccine increased number of CD8+ leukocytes were detected in the skin (Josepriya et al. 2015).

18.2.2.2 Adaptive Systemic Responses

I. multifiliis infection induces both adaptive mucosal and systemic immune responses. Antibodies specifically against *I. multifiliis* were detected circulating in the blood and were derived from ASC in the spleen and head kidney (Bromage et al. 2004; Zhao et al. 2008). Immunized channel catfish developed serum antibody responses at 5 weeks that continued to increase until 14 weeks post immunization (Maki and Dickerson 2003). As stated above, despite the increase in titers of serum antibodies, I-antigen specific IgM molecules probably did not reach surface epithelia and confer protection (Borji et al. 2012). The larger size of IgM molecules (~750 kDa) presumably could not diffuse through the vasculature to surface epithelia (Lin et al. 1996; Dickerson and Findly 2014). Protection against *I. multifiliis* was likely conferred by ASC in the skin. While it seems that the adaptive systemic and local responses are two independent systems, It is, however, possible that during local inflammation capillaries become permeable and serum antibodies reach the parasite in the skin and gills when blood enters tissues. Phagocytes may also enter mucosal tissues, and probably take up, process, and present specific antigens released by the parasite to B and T cells in the skin and lymphoid tissues in the spleen and head kidney (Dickerson 2012).

Furthermore, Findly and Dickerson conducted spectratype analysis to follow the repertoires of the TCRβ CDR3 in channel catfish after *I. multifiliis* infection (unpublished results described in (Dickerson and Findly 2014)). A systemic expansion of T cells expressing a similar subset of TCRβ variable genes in the tissues including, head kidney, spleen, circulation, and in skin, was detected, and it was concluded that T cells were probably activated through a similar pathway upon *I. multifiliis* infection (Dickerson and Findly 2014).

Adaptive systemic responses against *C. irritans* in orange-spotted grouper have been well demonstrated in a recent DNA-vaccine study (Josepriya et al. 2015). Expression vectors of *C. irritans* I-antigens and/or heat shock protein 70 (Hsp70C) encapsulated in chitosan nanoparticles were administered into grouper fingerlings through oral intubation. Titers and immobilization ability of mucosal antibodies from immune fish against I-antigens were significantly higher and better. In spleens, both adaptive cellular and humoral immune responses were also activated that expression of MHCII, MHCI, CD4+, CD8α+, interferon (IFN)-γ, TNF-α, IL-12, IgM heavy chain, and IgT heavy chain genes were significantly upregulated. Immune fish conferred immunity against subsequent lethal dose challenges for at least 7 weeks post immunization, and increased numbers of CD8α+ cells in the skin, but not in PBL, were observed when being challenged by *C. irritans* (Josepriya et al. 2015). Interestingly, cytotoxic effects of CD8α+ cells against *C. irritans* theronts showed no significant difference between those derived from naive and immune fish (Josepriya et al. 2015). Noteworthy, in the spleen of barramundi infected by *C. irritans* expression of MHCIIα and MHCIIβ genes were also significantly elevated (Mohd-Shaharuddin et al. 2013), which is in agreement with observations made by Josepriya et al. (2015).

18.2.2.3 Long-Term Protective Immunity

The successes of vaccinations against *I. multifiliis* and *C. irritans* infections clearly demonstrate that while being primitive, the fish immune system has the ability to memorize and to counter protozoan pathogens with elevated responses. In channel fish such immunological memory could last for three years against *I. multifiliis* infection (Findly et al. 2013). Channel catfish were first immunized by exposure to a low dose of *I. multifiliis*, and it was found that titers of serum antibodies against

I. multifiliis I-antigens peaked at 14 weeks (Maki and Dickerson 2003), and then dropped to an undetectable level one year post infection (Findly et al. 2013). When being challenged at higher doses of *I. multifiliis* theronts three years after the initial immunization, immune fish responded faster in I-antigen antibody production, and survived longer than control fish (Findly et al. 2013), suggesting the existence of a long-term immunity against the parasite. However, the location from where these IgM memory B cells were induced remains unknown.

18.3 Conclusions

Both *I. multifiliis* and *C. irritans* are ideal organisms to study multiple aspects of parasitism. Compared to many other protozoan parasites, such as the causative agent of the sleeping sickness *Trypanosoma brucei* which has six morphologically distinct stages, *I. multifiliis* and *C. irritans* each has only three. The genome of *I. multifiliis* has been published (Coyne et al. 2011), as well as genomes of two closely related free-living model organisms *T. thermophila* and *P. tetraurelia* (Aury et al. 2006; Eisen et al. 2006). More extensive genome/transcriptome comparisons should help elucidate how a parasite like *I. multifiliis* arose from once a free-living organism. When *C. irritans* genome is available, comparisons among these four organisms can then be made from the evolutionary perspective and help us understand adaptations and transitions at the genome level toward a parasitic life cycle.

In the past few years, there have been smaller scale transcriptome studies attempting to identify stage-specific transcripts in *I. multifiliis* (Abernathy et al. 2007; Cassidy-Hanley et al. 2011), and in *C. irritans* (Lokanathan et al. 2010; Khoo et al. 2012), and with the costs of next-generation sequencing continuing to drop, more larger scale analyses can be expected in the next few years. In fact, in 2015 there have already been several new transcriptome/proteome studies published that systematically analyze fish responses against parasite infections (Mai et al. 2015; Tong et al. 2015; Wang et al. 2015). Genomes of common carp (Xu et al. 2014), channel catfish (Jiang et al. 2013), rainbow trout (Berthelot et al. 2014), yellow croaker (Wu et al. 2014), and many other fish species have also been made available in the past few years, in addition to numerous transcriptome projects associated with each of them. The newly available, as well as forthcoming, sequencing data should help complement findings derived from real-time quantitative PCR by providing a more comprehensive picture on how fish hosts interact with parasites.

One other major area that has not been addressed is how genetic variations in these parasites relate to pathogenicity and virulence. Taking I-antigen, the favorite candidate for vaccine development, as an example, it has been shown that different isolates of *C. irritans* may express different types of I-antigens. Antibodies recognizing one type might not be able to bind to another type, thus losing the neutralization ability (Hatanaka et al. 2007, 2008). This is also the basis of serotyping used in distinguishing different strains/isolates of *I. multifiliis* (Dickerson et al. 1993). Moreover, the I-antigen genes, at least in *I. multifiliis*, are products of multiple gene and/or cluster duplications (Coyne et al. 2011), and how a locus was chosen to be expressed and whether the expression could be shifted to another locus remain poorly understood (Lin et al. 2002; Wang et al. 2002; Clark and Forney 2003). Further genomic, cytological, and genetic studies are needed to address these fundamental questions. The findings can also help study other less popular ciliate parasites.

Acknowledgements The authors would like to thank Dr. Hui Gong for his critical comments, and Glynis Asu for her research and instructional support.

References

Abernathy JW, Xu P, Li P, Xu DH, Kucuktas H, Klesius P, Arias C, Liu Z (2007) Generation and analysis of expressed sequence tags from the ciliate protozoan parasite Ichthyophthirius multifiliis. BMC Genom 8:176. doi:10.1186/1471-2164-8-176

Alvarez-Pellitero P (2008) Fish immunity and parasite infections: from innate immunity to immunoprophylactic prospects. Vet Immunol Immunopathol 126:171–198. doi:10.1016/j.vetimm.2008.07.013

Aury JM, Jaillon O, Duret L, Noel B, Jubin C, Porcel BM, Segurens B, Daubin V, Anthouard V, Aiach N, Arnaiz O, Billaut A, Beisson J, Blanc I, Bouhouche K, Camara F, Duharcourt S, Guigo R, Gogendeau D, Katinka M, Keller AM, Kissmehl R, Klotz C, Koll F, Le Mouel A, Lepere G, Malinsky S, Nowacki M, Nowak JK, Plattner H, Poulain J, Ruiz F, Serrano V, Zagulski M, Dessen P, Betermier M, Weissenbach J, Scarpelli C, Schachter V, Sperling L, Meyer E, Cohen J, Wincker P (2006) Global trends of whole-genome duplications revealed by the ciliate Paramecium tetraurelia. Nature 444:171–178

Bendelac A, Fearon DT (1997) Innate pathways that control acquired immunity. Curr Opin Immunol 9:1–3

Bengten E, Quiniou SM, Stuge TB, Katagiri T, Miller NW, Clem LW, Warr GW, Wilson M (2002) The IgH locus of the channel catfish, Ictalurus punctatus, contains multiple constant region gene sequences: different genes encode heavy chains of membrane and secreted IgD. J Immunol 169:2488–2497

Bengten E, Quiniou S, Hikima J, Waldbieser G, Warr GW, Miller NW, Wilson M (2006) Structure of the catfish IGH locus: analysis of the region including the single functional IGHM gene. Immunogenetics 58:831–844. doi:10.1007/s00251-006-0139-9

Berthelot C, Brunet F, Chalopin D, Juanchich A, Bernard M, Noel B, Bento P, Da Silva C, Labadie K, Alberti A, Aury JM, Louis A, Dehais P, Bardou P, Montfort J, Klopp C, Cabau C, Gaspin C, Thorgaard GH, Boussaha M, Quillet E, Guyomard R, Galiana D, Bobe J, Volff JN, Genet C, Wincker P, Jaillon O, Roest Crollius H, Guiguen Y (2014) The rainbow trout genome provides novel insights into evolution after whole-genome duplication in vertebrates. Nat Commun 5:3657. doi:10.1038/ncomms4657

Borji H, Naghibi A, Nasiri MR, Ahmadi A (2012) Identification of Dactylogyrus spp. and other parasites of common carp in northeast of Iran. J Parasitic Dis 36:234–238. doi:10.1007/s12639-012-0115-2

Boshra H, Li J, Sunyer JO (2006) Recent advances on the complement system of teleost fish. Fish Shellfish Immunol 20:239–262 S1050-4648(05)00077-X [pii]

Bromage ES, Kaattari IM, Zwollo P, Kaattari SL (2004) Plasmablast and plasma cell production and distribution in trout immune tissues. J Immunol 173:7317

Bryant MS, Lee RP, Lester RJG, Whittington RJ (1999) Anti-immunoglobulin antisera used in an ELISA to detect antibodies in barramundi Lates calcarifer to Cryptocaryon irritans. Dis Aquat Org 36:21–28

Buchmann K (2014) Evolution of innate immunity: clues from invertebrates via fish to mammals. Front Immunol 5:459. doi:10.3389/fimmu.2014.00459

Buchmann K, Nielsen ME (1999) Chemoattraction of Ichthyophthirius multifiliis (Ciliophora) theronts to host molecules. Int J Parasitol 29:1415–1423. doi:10.1016/S0020-7519(99)00091-0

Buchmann K, Sigh J, Nielsen CV, Dalgaard M (2001) Host responses against the fish parasitizing ciliate Ichthyophthirius multifiliis. Vet Parasitol 100:105–116. doi:10.1016/S0304-4017(01)00487-3

Burgess PJ, Matthews RA (1995a) Cryptocaryon irritans(Ciliophora): acquired protective immunity in the thick-lipped mullet, Chelon labrosus. Fish Shellfish Immunol 5:459–468. doi:10.1006/fsim.1995.0043

Burgess PJ, Matthews RA (1995b) Fish host range of seven isolates of Cryptocaryon irritans (Ciliophora). J Fish Biol 46:727–729. doi:10.1006/jfbi.1995.0066

Carroll MC, Prodeus AP (1998) Linkages of innate and adaptive immunity. Curr Opin Immunol 10:36–40

Cassidy-Hanley D, Cordonnier-Pratt M, Pratt LH, Devine C, Mozammal Hossain M, Dickerson HW, Clark TG (2011) Transcriptional profiling of stage specific gene expression in the parasitic ciliate Ichthyophthirius multifiliis. Mol Biochem Parasitol 178:29–39. doi:10.1016/j.molbiopara.2011.04.004

Castro R, Tafalla C (2015) 2—Overview of fish immunity. In: Peatman BHB (ed) Mucosal Health in AquacultureAcademic Press, San Diego, pp 3–54

Cheung PF, Nigrelli RF, Ruggieri GD (1981) Scanning electron microscopy on *Cryptocaryon irritans* Brown, 1951, a parasitic ciliate in marine fish. J Aquaricult Aquat Sci 2:70–72

Clark TG, Dickerson HW (1997) Antibody-mediated effects on parasite behavior: evidence of a novel mechanism of immunity against a parasitic protist. Parasitol Today 13:477–480. doi:10.1016/S0169-4758(97)01152-6

Clark TG, Forney JD (2003) 17—Free-living and Parasitic Ciliates. In: Scherf AC (ed) Antigenic VariationAcademic Press, London, pp 375–402

Clark TG, Lin TL, Dickerson HW (1996) Surface antigen cross-linking triggers forced exit of a protozoan parasite from its host. Proc Natl Acad Sci USA 93:6825–6829

Clayton GM, Price DJ (1994) Heterosis in resistance to Ichthyophthirius multifiliis infections in poeciliid fish. J Fish Biol 44:59–66. doi:10.1006/jfbi.1994.1006

Coleman JW (2001) Nitric oxide in immunity and inflammation. Int Immunopharmacol 1:1397–1406 S1567-5769(01)00086-8 [pii]

Colorni A, Diamant A (1993) Ultrastructural features of Cryptocaryon irritans, a ciliate parasite of marine fish. Eur J Protistol 29:425–434

Colorni A, Ullal A, Heinisch G, Noga EJ (2008) Activity of the antimicrobial polypeptide piscidin 2 against fish ectoparasites. J Fish Dis 31:423–432. doi:10.1111/j.1365-2761.2008.00922.x

Corliss JO (1979) Chapter 20—The Ciliate Taxa, including Families and their Genera. In: Corliss JO (ed) The Ciliated Protozoa (Second Edition) Pergamon, pp 206–326

Coyne RS, Hannick L, Shanmugam D, Hostetler JB, Brami D, Joardar VS, Johnson J, Radune D, Singh I, Badger JH, Kumar U, Saier M, Wang Y, Cai H, Gu J, Mather MW, Vaidya AB, Wilkes DE, Rajagopalan V, Asai DJ, Pearson CG, Findly RC, Dickerson HW, Wu M, Martens C, Van de Peer Y, Roos DS, Cassidy-Hanley DM, Clark TG (2011) Comparative genomics of the pathogenic ciliate Ichthyophthirius multifiliis, its free-living relatives and a host species provide insights into adoption of a parasitic lifestyle and prospects for disease control. Genome Biol 12:R100. doi:10.1186/gb-2011-12-10-r100

Cross ML (1994) Localized cellular responses to Ichthyophthirius multifiliis: Protection or pathogenesis? Parasitol Today 10:364–368. doi:10.1016/0169-4758(94)90253-4

Cross ML, Matthews RA (1993) Localized leucocyte response to Ichthyophthirius multifiliis establishment in immune carp Cyprinus carpio L. Vet Immunol Immunopathol 38:341–358. doi:10.1016/0165-2427(93)90092-I

Dan XM, Zhang TW, Li YW, Li AX (2013) Immune responses and immune-related gene expression profile in orange-spotted grouper after immunization with Cryptocaryon irritans vaccine. Fish Shellfish Immunol 34:885–891. doi:10.1016/j.fsi.2012.12.011

Densmore CL, Green DE (2007) Diseases of amphibians. ILAR J 48:235–254

Dickerson HW (2012) Ichthyophthirius multifiliis. Fish Parasites: Pathobiology and Prot. 55–72

Dickerson HW, Clark TG (1996) Immune response of fishes to ciliates. Annu Rev Fish Dis 6:107–120. doi:10.1016/S0959-8030(97)00007-0

Dickerson H, Clark T (1998) Ichthyophthirius multifiliis: a model of cutaneous infection and immunity in fishes. Immunol Rev 166:377–384. doi:10.1111/j.1600-065X.1998.tb01277.x

Dickerson HW, Findly RC (2014) Immunity to Ichthyophthirius infections in fish: a synopsis. Dev Comp Immunol 43:290–299. doi:10.1016/j.dci.2013.06.004

Dickerson HW, Clark TG, Leff AA (1993) Serotypic variation among isolates of Ichthyophthirius multifiliis based on immobilization. J Eukaryot Microbiol 40:816–820

Diggles BK (1997) Some information on the morphology of Cryptocaryon irritans from South-East Queensland, Australia. Eur J Protistol 33:200–210

Diggles BK, Adlard RD (1997) Intraspecific variation in Cryptocaryon irritans. J Eukaryot Microbiol 44:25–32

Du Pasquier L (1982) Antibody diversity in lower vertebrates—why is it so restricted? Nature 296:311–313

Edholm ES, Bengten E, Stafford JL, Sahoo M, Taylor EB, Miller NW, Wilson M (2010) Identification of two IgD+ B cell populations in channel catfish, Ictalurus punctatus. J Immunol 185:4082–4094. doi:10.4049/jimmunol.1000631

Egerter DE, Anderson JR, Washburn JO (1986) Dispersal of the parasitic ciliate Lambornella clarki: implications for ciliates in the biological control of mosquitoes. Proc Natl Acad Sci USA 83:7335–7339

Eisen JA, Coyne RS, Wu M, Wu D, Thiagarajan M, Wortman JR, Badger JH, Ren Q, Amedeo P, Jones KM, Tallon LJ, Delcher AL, Salzberg SL, Silva JC, Haas BJ, Majoros WH, Farzad M, Carlton JM, Smith RK Jr, Garg J, Pearlman RE, Karrer KM, Sun L, Manning G, Elde NC, Turkewitz AP, Asai DJ, Wilkes DE, Wang Y, Cai H, Collins K, Stewart BA, Lee SR, Wilamowska K, Weinberg Z, Ruzzo WL, Wloga D, Gaertig J, Frankel J, Tsao CC, Gorovsky MA, Keeling PJ, Waller RF, Patron NJ, Cherry JM, Stover NA, Krieger CJ, del Toro C, Ryder HF, Williamson SC, Barbeau RA, Hamilton EP, Orias E (2006) Macronuclear genome sequence of the ciliate Tetrahymena thermophila, a model eukaryote. PLoS Biol 4: e286

Ellis AE (2001) Innate host defense mechanisms of fish against viruses and bacteria. Dev Comp Immunol 25:827–839 S0145305X01000386 [pii]

Engelsma MY, Huising MO, van Muiswinkel WB, Flik G, Kwang J, Savelkoul HF, Verburg-van Kemenade BM (2002) Neuroendocrine-immune interactions in fish: a role for interleukin-1. Vet Immunol Immunopathol 87:467–479 S0165242702000776 [pii]

Ewing MS, Kocan KM, Ewing SA (1985) Ichthyophthirius multifiliis (Ciliophora) invasion of Gill Epithelium1. J Protozool 32:305–310. doi:10.1111/j.1550-7408.1985.tb03055.x

Ewing MS, Lynn ME, Ewing SA (1986) Critical periods in development of Ichthyophthirius multifiliis (Ciliophora) Populations1. J Protozool 33:388–391. doi:10.1111/j.1550-7408.1986.tb05628.x

Fearon DT (1997) Seeking wisdom in innate immunity. Nature 388:323–324. doi:10.1038/40967

Fearon DT, Locksley RM (1996) The instructive role of innate immunity in the acquired immune response. Science 272:50–53

Findly RC, Zhao X, Noe J, Camus AC, Dickerson HW (2013) B cell memory following infection and challenge of channel catfish with Ichthyophthirius multifiliis. Dev Comp Immunol 39:302–311. doi:10.1016/j.dci.2012.08.007

Gao F, Katz LA (2014) Phylogenomic analyses support the bifurcation of ciliates into two major clades that differ in properties of nuclear division. Mol Phylogenet Evol 70:240–243. doi:10.1016/j.ympev.2013.10.001

Gasque P (2004) Complement: a unique innate immune sensor for danger signals. Mol Immunol 41:1089–1098 S0161-5890(04)00218-4 [pii]

Gentekaki E, Kolisko M, Boscaro V, Bright KJ, Dini F, Di Giuseppe G, Gong Y, Miceli C, Modeo L, Molestina RE, Petroni G, Pucciarelli S, Roger AJ, Strom SL, Lynn DH (2014) Large-scale phylogenomic analysis reveals the phylogenetic position of the problematic taxon Protocruzia and unravels the deep phylogenetic affinities of the ciliate lineages. Mol Phylogenet Evol 78:36–42. doi:10.1016/j.ympev.2014.04.020

Gonzalez SF, Buchmann K, Nielsen ME (2007a) Real-time gene expression analysis in carp (Cyprinus carpio L.) skin: inflammatory responses caused by the ectoparasite Ichthyophthirius multifiliis. Fish Shellfish Immunol 22:641–650. doi:10.1016/j.fsi.2006.08.011

Gonzalez SF, Buchmann K, Nielsen ME (2007b) Complement expression in common carp (Cyprinus carpio L.) during infection with Ichthyophthirius multifiliis. Dev Comp Immunol 31:576–586. doi:10.1016/j.dci.2006.08.010

Gonzalez SF, Buchmann K, Nielsen ME (2007c) Ichthyophthirius multifiliis infection induces massive up-regulation of serum amyloid A in carp (Cyprinus carpio). Vet Immunol Immunopathol 115:172–178. doi:10.1016/j.vetimm.2006.09.007

Gonzalez SF, Chatziandreou N, Nielsen ME, Li W, Rogers J, Taylor R, Santos Y, Cossins A (2007d) Cutaneous immune responses in the common carp detected using transcript analysis. Mol Immunol 44:1664–1679. doi:10.1016/j.molimm.2006.08.016

Grasse P, de Boissezon P (1929) Turchiniella culicis n. g., n. sp., infusoire parasite de l'hemocoele d'un Culex adults. Bull Soc Zool Fr 54:187–191

Haas W, Haberl B, Hofmann M, Kerschensteiner S, Ketzer U (1999) Ichthyophthirius multifiliis invasive stages find their fish hosts with complex behavior patterns and in response to different chemical signals. Eur J Protistol 35:129–135

Hansen JD, Landis ED, Phillips RB (2005) Discovery of a unique Ig heavy-chain isotype (IgT) in rainbow trout: implications for a distinctive B cell developmental pathway in teleost fish. Proc Natl Acad Sci USA 102:6919–6924 0500027102 [pii]

Hardie LJ, Chappell LH, Secombes CJ (1994) Human tumor necrosis factor alpha influences rainbow trout Oncorhynchus mykiss leucocyte responses. Vet Immunol Immunopathol 40:73–84

Hatanaka A, Umeda N, Yamashita S, Hirazawa N (2007) Identification and characterization of a putative agglutination/immobilization antigen on the surface of Cryptocaryon irritans. Parasitology 134:1163–1174 S003118200700265X [pii]

Hatanaka A, Umeda N, Hirazawa N (2008) Molecular cloning of a putative agglutination/immobilization antigen located on the surface of a novel agglutination/immobilization serotype of Cryptocaryon irritans. Parasitology 135:1043–1052. doi:10.1017/S0031182008004617

Heinecke RD, Buchmann K (2013) Inflammatory response of rainbow trout Oncorhynchus mykiss (Walbaum, 1792) larvae against Ichthyophthirius multifiliis. Fish Shellfish Immunol 34:521–528. doi:10.1016/j.fsi.2012.11.036

Hibbs J, Taintor R, Vavrin Z (1987) Macrophage cytotoxicity: Role for L-arginine deiminase and imino nitrogen oxidation to nitrite. Science 235:473–476. doi:10.1126/science.2432665

Hines RS, Spira DT (1973) Ichthyophthiriasis in the mirror carp. J Fish Biol 5:527–534. doi:10.1111/j.1095-8649.1973.tb04484.x

Hodgkinson JW, Grayfer L, Belosevic M (2015) Biology of Bony Fish Macrophages. Biology (Basel) 4:881–906 biology4040881 [pii]

Hoffman GL (1967) Parasites of North American freshwater fishes. University of California Press, Berkeley

Janeway CA Jr, Medzhitov R (2002) Innate immune recognition. Annu Rev Immunol 20:197–216. doi:10.1146/annurev.immunol.20.083001.084359

Jiang Y, Gao X, Liu S, Zhang Y, Liu H, Sun F, Bao L, Waldbieser G, Liu Z (2013) Whole genome comparative analysis of channel catfish (Ictalurus punctatus) with four model fish species. BMC Genom 14:780-2164-14-780, doi:10.1186/1471-2164-14-780

Josepriya TA, Chien KH, Lin HY, Huang HN, Wu CJ, Song Y-L (2015) Immobilization antigen vaccine adjuvanted by parasitic heat shock protein 70C confers high protection in fish against cryptocaryonosis. Fish Shellfish Immunol 45:517–527. doi:10.1016/j.fsi.2015.04.036

Khoo C-K, Abdul-Murad AM, Kua B-C, Mohd-Adnan A (2012) Cryptocaryon irritans infection induces the acute phase response in Lates calcarifer: a transcriptomic perspective. Fish Shellfish Immunol 33:788–794. doi:10.1016/j.fsi.2012.07.005

Landstrom M (2010) The TAK1-TRAF6 signalling pathway. Int J Biochem Cell Biol 42:585–589. doi:10.1016/j.biocel.2009.12.023

Li Y-W, Luo X-C, Dan X-M, Huang XZ, Qiao W, Zhong ZP, Li A-X (2011) Orange-spotted grouper (Epinephelus coioides) TLR2, MyD88 and IL-1beta involved in anti-Cryptocaryon irritans response. Fish Shellfish Immunol 30:1230–1240. doi:10.1016/j.fsi.2011.04.012

Li Y-W, Luo X-C, Dan X-M, Qiao W, Huang XZ, Li A-X (2012) Molecular cloning of orange-spotted grouper (Epinephelus coioides) TLR21 and expression analysis post Cryptocaryon irritans infection. Fish Shellfish Immunol 32:476–481. doi:10.1016/j.fsi.2011.11.021

Li Y-W, Li X, Xiao X-X, Zhao F, Luo X-C, Dan X-M, Li A-X (2014a) Molecular characterization and functional analysis of TRAF6 in orange-spotted grouper (Epinephelus coioides). Dev Comp Immunol 44:217–225. doi:10.1016/j.dci.2013.12.011

Li Y-W, Mo X-B, Zhou L, Li X, Dan X-M, Luo X-C, Li A-X (2014b) Identification of IRAK-4 in grouper (Epinephelus coioides) that impairs MyD88-dependent NF-Î°B activation. Dev Comp Immunol 45:190–197. doi:10.1016/j.dci.2014.02.019

Li Y-W, Li X, Wang Z, Mo ZQ, Dan X-M, Luo X-C, Li A-X (2015) Orange-spotted grouper Epinephelus coioides Tak1: molecular identification, expression analysis and functional study. J Fish Biol 86:417–430. doi:10.1111/jfb.12550

Lin TL, Clark TG, Dickerson H (1996) Passive immunization of channel catfish (Ictalurus punctatus) against the ciliated protozoan parasite Ichthyophthirius multifiliis by use of murine monoclonal antibodies. Infect Immun 64:4085–4090

Lin Y, Lin TL, Wang C-C, Wang X, Stieger K, Klopfleisch R, Clark TG (2002) Variation in primary sequence and tandem repeat copy number among i-antigens of Ichthyophthirius multifiliis. Mol Biochem Parasitol 120:93–106

Lokanathan Y, Mohd-Adnan A, Wan KL, Nathan S (2010) Transcriptome analysis of the Cryptocaryon irritans tomont stage identifies potential genes for the detection and control of cryptocaryonosis. BMC Genom 11, doi:10.1186/1471-2164-11-76

Luo X-C, Xie M-Q, Zhu X-Q, Li A-X (2007) Protective immunity in grouper (Epinephelus coioides) following exposure to or injection with Cryptocaryon irritans. Fish Shellfish Immunol 22:427–432. doi:10.1016/j.fsi.2006.04.011

Luo XC, Xie MQ, Zhu XQ, Li AX (2008) Some characteristics of host-parasite relationship for Cryptocaryon irritans isolated from South China. Parasitol Res 102:1269–1275. doi:10.1007/s00436-008-0904-9

Lynn DH, Corliss JO (1991) Ciliophora. In: Corliss JO, Harrison FW (eds) Microscopic anatomy of invertebrates. Wiley-Liss, New York, pp 333–467

MacColl E, Therkelsen MD, Sherpa T, Ellerbrock H, Johnston LA, Jariwala RH, Chang W, Gurtowski J, Schatz MC, Mozammal Hossain M, Cassidy-Hanley DM, Clark TG, Chang W-J (2015) Molecular genetic diversity and characterization of conjugation genes in the fish parasite Ichthyophthirius multifiliis. Mol Phylogenet Evol 86:1–7. doi:10.1016/j.ympev.2015.02.017

MacLennan RF (1937) Growth in the ciliate Ichthyophthirius I. Maturity and encystment. J Exp Zool 76:423–440. doi:10.1002/jez.1400760306

Magnadottir B (2006) Innate immunity of fish (overview). Fish Shellfish Immunol 20:137–151 S1050-4648(05)00078-1 [pii]

Mai YZ, Li Y-W, Li RJ, Li W, Huang XZ, Mo ZQ, Li A-X (2015) Proteomic analysis of differentially expressed proteins in the marine fish parasitic ciliate Cryptocaryon irritans. Vet Parasitol 211:1–11. doi:10.1016/j.vetpar.2015.05.004

Maki JL, Dickerson HW (2003) Systemic and cutaneous mucus antibody responses of channel catfish immunized against the protozoan parasite Ichthyophthirius multifiliis. Clin Diagn Lab Immunol 10:876–881. doi:10.1128/CDLI.10.5.876-881.2003

Martin SA, Mohanty BP, Cash P, Houlihan DF, Secombes CJ (2007a) Proteome analysis of the Atlantic salmon (Salmo salar) cell line SHK-1 following recombinant IFN-gamma stimulation. Proteomics 7:2275–2286. doi:10.1002/pmic.200700020

Martin SA, Taggart JB, Seear P, Bron JE, Talbot R, Teale AJ, Sweeney GE, Hoyheim B, Houlihan DF, Tocher DR, Zou J, Secombes CJ (2007b) Interferon type I and type II responses in an Atlantic salmon (Salmo salar) SHK-1 cell line by the salmon TRAITS/SGP microarray. Physiol Genomics 32:33–44 00064.2007 [pii]

Martin SA, Zou J, Houlihan DF, Secombes CJ (2007c) Directional responses following recombinant cytokine stimulation of rainbow trout (Oncorhynchus mykiss) RTS-11 macrophage cells as revealed by transcriptome profiling. BMC Genom 8:150 1471-2164-8-150 [pii]

McCallum HI (1982) Infection dynamics of Ichthyophthirius multifiliis (Poecilia latipinna). Parasitology 85:475–488

Medzhitov R (2001) Toll-like receptors and innate immunity. Nat Rev Immunol 1:135–145. doi:10.1038/35100529

Medzhitov R (2007) Recognition of microorganisms and activation of the immune response. Nature 449:819–826 nature06246 [pii]

Medzhitov R, Janeway CA Jr (2002) Decoding the patterns of self and nonself by the innate immune system. Science 296:298–300. doi:10.1126/science.1068883

Mohd-Shaharuddin N, Mohd-Adnan A, Kua BC, Nathan S (2013) Expression profile of immune-related genes in Lates calcarifer infected by Cryptocaryon irritans. Fish Shellfish Immunol 34:762–769. doi:10.1016/j.fsi.2012.11.052

Morado JF, Small EB (1995) Ciliate parasites and related diseases of crustacea: a review. Rev Fish Sci 3:275–354. doi:10.1080/10641269509388575

Mutoloki S, Jørgensen JB, Evensen Ø (2014) The adaptive immune response in fish. In: Gudding R, Lillehaug A, Evensen Ø (ed) Fish Vaccination. John Wiley and Sons, Ltd, Chichester, UK, pp 104–115

Neves LR, Braga ECR, Tavares-Dias M (2015) Diversity of parasites in Curimata incompta (Curimatidae), a host from Amazon river system in Brazil. J Parasitic Dis. doi:10.1007/s12639-015-0674-0

Niu S-F, Jin Y, Xu X, Qiao Y, Wu Y, Mao Y, Su Y-Q, Wang J (2013) Characterization of a novel piscidin-like antimicrobial peptide from Pseudosciaena crocea and its immune response to Cryptocaryon irritans. Fish Shellfish Immunol 35:513–524. doi:10.1016/j.fsi.2013.05.007

Olsen MM, Kania PW, Heinecke RD, Skjoedt K, Rasmussen KJ, Buchmann K (2011) Cellular and humoral factors involved in the response of rainbow trout gills to Ichthyophthirius multifiliis infections: molecular and immunohistochemical studies. Fish Shellfish Immunol 30:859–869. doi:10.1016/j.fsi.2011.01.010

Picón-Camacho SM, Taylor NGH, Bron JE, Guo FC, Shinn AP (2012) Effects of long duration, low dose bronopol exposure on the control of Ichthyophthirius multifiliis (Ciliophora), parasitising rainbow trout (Oncorhynchus mykiss Walbaum). Vet Parasitol 186:237–244, DOI:10.1016/j.vetpar.2011.11.022

Rajanbabu V, Chen JY (2011) Applications of antimicrobial peptides from fish and perspectives for the future. Peptides 32:415–420. doi:10.1016/j.peptides.2010.11.005

Roca FJ, Mulero I, Lopez-Munoz A, Sepulcre MP, Renshaw SA, Meseguer J, Mulero V (2008) Evolution of the inflammatory response in vertebrates: fish TNF-alpha is a powerful activator of endothelial cells but hardly activates phagocytes. J Immunol 181:5071–5081 181/7/5071 [pii]

Saeij JP, de Vries BJ, Wiegertjes GF (2003) The immune response of carp to Trypanoplasma borreli: kinetics of immune gene expression and polyclonal lymphocyte activation. Dev Comp Immunol 27:859–874 S0145305X03000831 [pii]

Scholz T (1999) Parasites in cultured and feral fish. Vet Parasitol 84:317–335 S0304-4017(99)00039-4 [pii]

Schuster FL, Ramirez-Avila L (2008) Current world status of Balantidium coli. Clin Microbiol Rev 21:626–638. doi:10.1128/CMR.00021-08

Schuster FL, Visvesvara GS (2004) Amebae and ciliated protozoa as causal agents of waterborne zoonotic disease. Vet Parasitol 126:91–120 S0304-4017(04)00401-7 [pii]

Secombes CJ, Wang T, Hong S, Peddie S, Crampe M, Laing KJ, Cunningham C, Zou J (2001) Cytokines and innate immunity of fish. Dev Comp Immunol 25:713–723 S0145305X01000325 [pii]

Sigh J, Lindenstrom T, Buchmann K (2004a) Expression of pro-inflammatory cytokines in rainbow trout (Oncorhynchus mykiss) during an infection with Ichthyophthirius multifiliis. Fish and Shellfish Immunol 17:75–86, doi:10.1016/j.fsi.2003.12.005

Sigh J, Lindenstrom T, Buchmann K (2004b) The parasitic ciliate Ichthyophthirius multifiliis induces expression of immune relevant genes in rainbow trout, Oncorhynchus mykiss (Walbaum). J Fish Dis 27:409–417. doi:10.1111/j.1365-2761.2004.00558.x

St Louis-Cormier EA, Osterland CK, Anderson PD (1984) Evidence for a cutaneous secretory immune system in rainbow trout (Salmo gairdneri). Dev Comp Immunol 8:71–80

Swennes AG, Noe JG, Findly RC, Dickerson HW (2006) Differences in virulence between two serotypes of Ichthyophthirius multifiliis. Dis Aquat Org 69:227–232

Tong C, Lin Y, Zhang C, Shi J, Qi H, Zhao K (2015) Transcriptome-wide identification, molecular evolution and expression analysis of Toll-like receptor family in a Tibet fish, Gymnocypris przewalskii. Fish Shellfish Immunol 46:334–345. doi:10.1016/j.fsi.2015.06.023

Trinchieri G (2003) Interleukin-12 and the regulation of innate resistance and adaptive immunity. Nat Rev Immunol 3:133–146. doi:10.1038/nri1001

Ullal AJ, Noga EJ (2010) Antiparasitic activity of the antimicrobial peptide HbβP-1, a member of the β-haemoglobin peptide family. J Fish Dis 33:657–664. doi:10.1111/j.1365-2761.2010.01172.x

Ullal AJ, Wayne Litaker R, Noga EJ (2008) Antimicrobial peptides derived from hemoglobin are expressed in epithelium of channel catfish (Ictalurus punctatus, Rafinesque). Dev Comp Immunol 32:1301–1312. doi:10.1016/j.dci.2008.04.005

Valero Y, Chaves-Pozo E, Meseguer J, Esteban MA, Cuesta A (2013) Biological role of fish antimicrobial peptides. In: Seong M (ed) Antimicrobial peptides: Properties, functions and role in immune response. Nova Biomedical, New York, pp 31–60

Valladao GM, Padua SB, Gallani SU, Menezes-Filho RN, Dias-Neto J, Martins ML, Ishikawa MM, Pilarski F (2013) Paratrichodina africana (Ciliophora): a pathogenic gill parasite in farmed Nile tilapia. Vet Parasitol 197:705–710. doi:10.1016/j.vetpar.2013.04.043

Valladao GM, Giannecchini LG, Martins ML, de Padua SB (2015) Trichodina modesta: an exotic ciliate in the Neotropical region parasitizing an unusual host. Rev Bras Parasitol Vet 24:162–167. doi:10.1590/S1984-29612015024

Ventura MT, Paperna I (1985) Histopathology of Ichthyophthirius multifiliis infections in fishes. J Fish Biol 27:185–203

Vilmos P, Kurucz E (1998) Insect immunity: evolutionary roots of the mammalian innate immune system. Immunol Lett 62:59–66 S0165-2478(98)00023-6 [pii]

von Gersdorff Jorgensen L, Nemli E, Heinecke RD, Raida MK, Buchmann K (2008) Immune-relevant genes expressed in rainbow trout following immunisation with a live vaccine against Ichthyophthirius multifiliis. Dis Aquat Org 80:189–197. doi:10.3354/dao01935

Von GJ, Heinecke RD, Skjødt K, Rasmussen KJ, Buchmann K (2011) Experimental evidence for direct in situ binding of IgM and IgT to early trophonts of Ichthyophthirius multifiliis (Fouquet) in the gills of rainbow trout, Oncorhynchus mykiss (Walbaum). J Fish Dis 34:749–755, doi:10.1111/j.1365-2761.2011.01291.x

Wahli T, Meier W, Schmitt M (1991) Affinity of Ichthyophthirius multifiliis theronts to light and/or fish. J Appl Ichthyol 7:244–248. doi:10.1111/j.1439-0426.1991.tb00603.x
Wang X, Clark TG, Noe J, Dickerson HW (2002) Immunisation of channel catfish, Ictalurus punctatus, with Ichthyophthirius multifiliis immobilisation antigens elicits serotype-specific protection. Fish Shellfish Immunol 13:337–350. doi:10.1006/fsim.2001.0410
Wang P, Wang J, Su YQ, Mao Y, Zhang JS, Wu CW, Ke QZ, Han KH, Zheng WQ, Xu ND (2015) Transcriptome analysis of the Larimichthys crocea liver in response to Cryptocaryon irritans. Fish Shellfish Immunol, S1050-4648(15)30230-8 [pii]
Wei YC, Pan TS, Chang MX, Huang B, Xu Z, Luo TR, Nie P (2011) Cloning and expression of Toll-like receptors 1 and 2 from a teleost fish, the orange-spotted grouper Epinephelus coioides. Vet Immunol Immunopathol 141:173–182. doi:10.1016/j.vetimm.2011.02.016
Wright AG, Colorni A (2002) Taxonomic re-assignment of Cryptocaryon irritans, a marine fish parasite. Eur J Protistol 37:375–378. doi:10.1078/0932-4739-00858
Wright AD, Lynn DH (1995) Phylogeny of the fish parasite Ichthyophthirius and its relatives Ophryoglena and Tetrahymena (Ciliophora, Hymenostomatia) inferred from 18S ribosomal RNA sequences. Mol Biol Evol 12:285–290
Wu C, Zhang D, Kan M, Lv Z, Zhu A, Su Y, Zhou D, Zhang J, Zhang Z, Xu M, Jiang L, Guo B, Wang T, Chi C, Mao Y, Zhou J, Yu X, Wang H, Weng X, Jin JG, Ye J, He L, Liu Y (2014) The draft genome of the large yellow croaker reveals well-developed innate immunity. Nat Commun 5:5227. doi:10.1038/ncomms6227
Xu D-H, Klesius PH (2002) Antibody mediated immune response against Ichthyophthirius multifiliis using excised skin from channel catfish, Ictalurus punctatus (Rafinesque), immune to Ichthyophthirius. J Fish Dis 25:299–306. doi:10.1046/j.1365-2761.2002.00377.x
Xu D-H, Klesius PH (2004) Two year study on the infectivity of Ichthyophthirius multifiliis in channel catfish Ictalurus punctatus. Dis Aquat Org 59:131–134
Xu D-H, Klesius PH, Shelby RA (2002) Cutaneous antibodies in excised skin from channel catfish, Ictalurus punctatus Rafinesque, immune to Ichthyophthirius multifiliis. J Fish Dis 25:45–52. doi:10.1046/j.1365-2761.2002.00339.x
Xu Z, Parra D, Gómez D, Salinas I, Zhang YA, Von GJ, Heinecke RD, Buchmann K, LaPatra S, Oriol Sunyer J (2013) Teleost skin, an ancient mucosal surface that elicits gut-like immune responses. Proc Natl Acad Sci USA 110:13097–13102, doi:10.1073/pnas.1304319110
Xu P, Zhang X, Wang X, Li J, Liu G, Kuang Y, Xu J, Zheng X, Ren L, Wang G, Zhang Y, Huo L, Zhao Z, Cao D, Lu C, Li C, Zhou Y, Liu Z, Fan Z, Shan G, Li X, Wu S, Song L, Hou G, Jiang Y, Jeney Z, Yu D, Wang L, Shao C, Song L, Sun J, Ji P, Wang J, Li Q, Xu L, Sun F, Feng J, Wang C, Wang S, Wang B, Li Y, Zhu Y, Xue W, Zhao L, Wang J, Gu Y, Lv W, Wu K, Xiao J, Wu J, Zhang Z, Yu J, Sun X (2014) Genome sequence and genetic diversity of the common carp, Cyprinus carpio. Nat Genet 46:1212–1219. doi:10.1038/ng.3098
Yambot AV, Song Y-L (2006) Immunization of grouper, Epinephelus coioides, confers protection against a protozoan parasite, Cryptocaryon irritans. Aquaculture 260:1–9. doi:10.1016/j.aquaculture.2006.05.055
Yambot AV, Song Y-L, Sung H-H (2003) Characterization of Cryptocaryon irritans, a parasite isolated from marine fishes in Taiwan. Dis Aquat Org 54:147–156
Yoshinaga T, Nakazoe J-I (1997) Acquired protection and production of immobilization antibody against Cryptocaryon irritans (Ciliophora, Hymenostomatida) in Mummichog (Fundulus heteroclitus). Fish Pathol 32:229–230
Zahran E, Noga EJ (2010) Evidence for synergism of the antimicrobial peptide piscidin 2 with antiparasitic and antioomycete drugs. J Fish Dis 33:995–1003. doi:10.1111/j.1365-2761.2010.01205.x
Zhang YA, Salinas I, Li J, Parra D, Bjork S, Xu Z, LaPatra SE, Bartholomew J, Sunyer JO (2010) IgT, a primitive immunoglobulin class specialized in mucosal immunity. Nat Immunol 11:827–835. doi:10.1038/ni.1913
Zhang YA, Salinas I, Oriol Sunyer J (2011) Recent findings on the structure and function of teleost IgT. Fish Shellfish Immunol 31:627–634. doi:10.1016/j.fsi.2011.03.021

Zhang Q, Yi Z, Fan X, Warren A, Gong J, Song W (2014) Further insights into the phylogeny of two ciliate classes Nassophorea and Prostomatea (Protista, Ciliophora). Mol Phylogenet Evol 70:162–170. doi:10.1016/j.ympev.2013.09.015

Zhang DL, Yu DH, Chen J, Chen C, Wang ZY (2015) Co-expression of march5b and tlr7 in large yellow croaker Larimichthys crocea in response to Cryptocaryon irritans infection. J Fish Biol 87:360–370. doi:10.1111/jfb.12726

Zhao X, Findly RC, Dickerson HW (2008) Cutaneous antibody-secreting cells and B cells in a teleost fish. Dev Comp Immunol 32:500–508. doi:10.1016/j.dci.2007.08.009

Zhao F, Li Y-W, Pan H-J, Shi C-B, Luo X-C, Li A-X, Wu S-Q (2013a) Expression profiles of toll-like receptors in channel catfish (Ictalurus punctatus) after infection with Ichthyophthirius multifiliis. Fish Shellfish Immunol 35:993–997. doi:10.1016/j.fsi.2013.05.023

Zhao F, Li Y-W, Pan H-J, Wu S-Q, Shi C-B, Luo X-C, Li A-X (2013b) Grass carp (Ctenopharyngodon idella) TRAF6 and TAK1: molecular cloning andexpression analysis after Ichthyophthirius multifiliis infection. Fish Shellfish Immunol 34:1514–1523. doi:10.1016/j.fsi.2013.03.003

Zhao F, Li Y-W, Pan H-J, Shi C-B, Luo X-C, Li A-X, Wu S-Q (2014) TAK1-binding proteins (TAB1 and TAB2) in grass carp (Ctenopharyngodon idella): identification, characterization, and expression analysis after infection with Ichthyophthirius multifiliis. Fish Shellfish Immunol 38:389–399. doi:10.1016/j.fsi.2014.04.008

Zou J, Secombes CJ (2011) Teleost fish interferons and their role in immunity. Dev Comp Immunol 35:1376–1387. doi:10.1016/j.dci.2011.07.001

Zou J, Carrington A, Collet B, Dijkstra JM, Yoshiura Y, Bols N, Secombes C (2005) Identification and bioactivities of IFN-gamma in rainbow trout Oncorhynchus mykiss: the first Th1-type cytokine characterized functionally in fish. J Immunol 175:2484–2494 175/4/2484 [pii]

Chapter 19
Ciliates in Planktonic Food Webs: Communication and Adaptive Response

Thomas Weisse and Bettina Sonntag

Abstract Ciliates are key elements of aquatic food webs, acting as predators of bacteria, algae, other protists and even some metazoans. Planktonic ciliates are important food for zooplankton, and mixotrophic and functionally autotrophic species may significantly contribute to primary production in lakes and in the ocean. Ciliates are linked to other plankton organisms and to abiotic parameters by various direct (e.g., predation and parasitism) and indirect (e.g., nutrient release via excretion, competition for food) interactions. Communication is involved in many of those interactions, but direct evidence from the field is scarce. This is mainly because aquatic ciliates live in a dilute environment and most species occur in low cell numbers. Key processes of chemical and hydromechanical communication among ciliates, between ciliates and their predators and prey, and between ciliates and their endosymbionts have been studied in some detail in vitro. Results from microcosm experiments suggest that ciliates also cooperate with each other and use information mediated by different other organisms. Extrapolating those results obtained in small-scale laboratory experiments to the ocean level is a major challenge for future research. To this end, more communication and cooperation is needed between cell biologists, ecologists, and evolutionary biologists. In contrast to communication, adaptations of natural ciliate populations to their abiotic and biotic environment have been well documented in selected freshwater and marine species. The effect of individual environmental variables and, less often, the interactive effect of several variables on growth and survival rates of ciliates have been studied with a number of ciliate species in the laboratory and inferred from their seasonal occurrence in the field. However, more information is needed on the autecology of free-living ciliate species and their role in the planktonic community. In particular, the significance of indirect ('lateral') effects in the food web received little attention and awaits future research.

T. Weisse · B. Sonntag (✉)
University of Innsbruck, Research Institute for Limnology,
Mondseestrasse 9, 5310 Mondsee, Austria
e-mail: bettina.sonntag@uibk.ac.at

G. Witzany and M. Nowacki (eds.), *Biocommunication of Ciliates*,
DOI 10.1007/978-3-319-32211-7_19

19.1 Introduction

Ciliates are structural elements of virtually any aquatic food web, acting as predators of bacteria, algae, other protists and even some metazoans, and serving as food for zooplankton (mainly microcrustacea such as copepods and daphnids) (Azam et al. 1983; Porter et al. 1985; Weisse 2003, 2006). Mixotrophic and functionally photoautotrophic ciliates such as *Mesodinium rubrum* may at times significantly contribute to primary production in lakes and in the ocean (reviewed by Esteban et al. 2010). Endosymbiotic ciliates are important in certain environments such as the rumen of ruminants and other herbivores but are less common among free-living species (Görtz 1996). Parasitic ciliates, namely in the class Spirotrichea, attack as ectoparasites and endoparasites a variety of aquatic vertebrates and invertebrates (summarised by Lynn 2008). Species-specific ciliate-ciliate parasitism among spirotrichs and suctorians is also important because the parasitic partner may kill the host (summarised by Berger 1999). Free-living ciliates are major players in benthic and pelagic food webs (Fig. 19.1) and linked to the other organisms in the community by various direct and indirect interactions. Direct interactions result mainly from predation (with ciliates acting actively as predators, respectively passively as prey), symbiosis/mutualism, commensalism and parasitism (with ciliates acting as parasites or hosts). Further, many ciliates use other organisms such as benthic cnidarians and planktonic copepods as substrates and (in the latter case) as vectors of passive transport (Fig. 19.2a, b). Conversely, peritrich ciliates may actively relocate planktonic algal colonies to which they attach (Fig. 19.2c). All these direct interactions are interspecific, i.e. the ciliates interact with different other species. However, for the evolutionary success of a (sexual) species, one of the most important interactions is that with its congener. Ciliates have to find a partner for mating; even if ciliates divide asexually, which is in ciliates the only process of reproduction (i.e., leading to an increase in cell numbers), the mother and daughter cell are closely interacting. Examples for indirect interactions are, e.g., competition with other predators for the same food or nutrient release via excretion, thereby supporting growth of algal or bacterial prey; similarly, ciliate faecal pellets may serve as substrate for bacterial decomposition or food for other protists (Buck and Newton 1995; Stoecker 1984). A major conceptual difference between direct and indirect interactions is that only the former require communication, i.e. the active (mutual?) exchange of information. The various modes of ciliate communication with their internal (intracellular) and external environment have been discussed in detail in the other chapters of this book. In most cases, ciliate communication has been studied with a few model organisms such as *Tetrahymena, Paramecium, Euplotes* and *Blepharisma* in the laboratory (e.g., Chaine et al. 2010; Jacob et al. 2015; Kuhlmann et al. 1999; Nanney 1980). Much less is known about communication of ciliates in their natural realm.

Ciliates in aquatic ecosystems live in a dilute environment. In the pelagic, the density of plankton correlates with the trophic conditions of the water body. Under eutrophic (=nutrient-rich) conditions plankton density in a volume of water is much

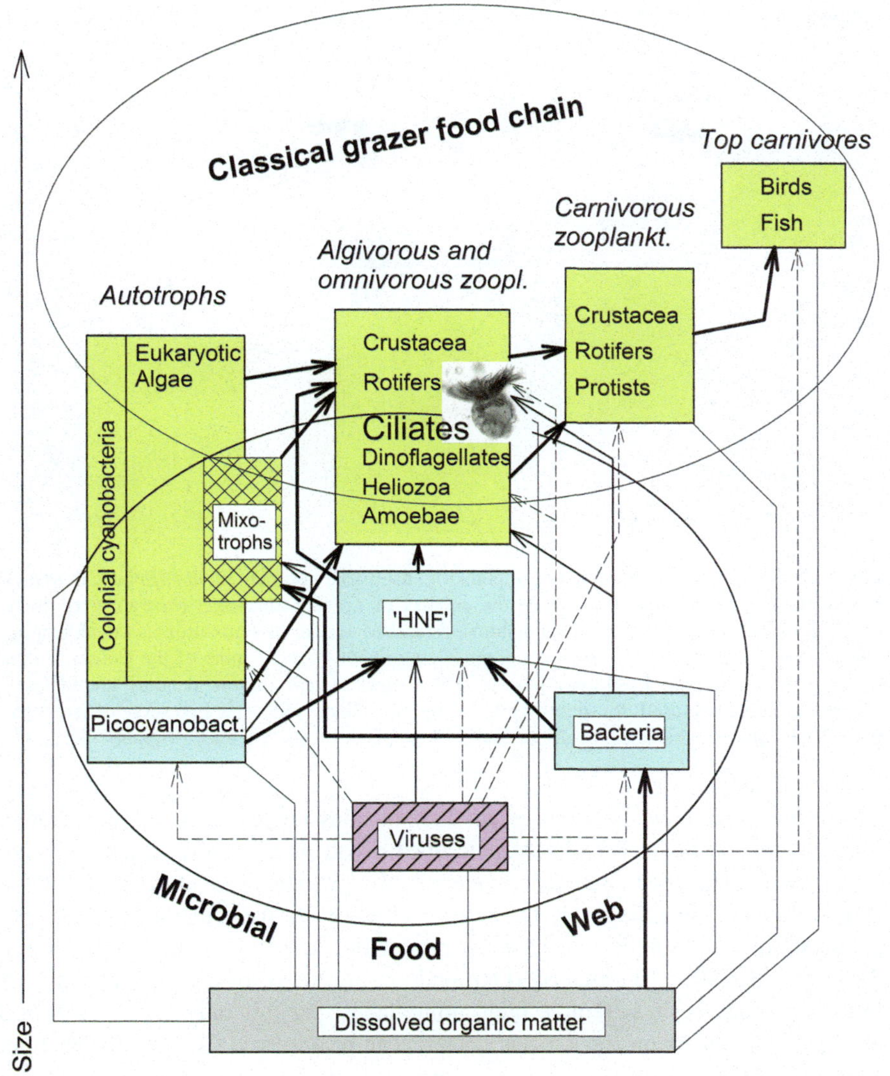

Fig. 19.1 The planktonic food web in lakes. Feeding interactions and bacterial substrate uptake are indicated by *solid lines* and *arrow heads*. Viruses attack several components of the planktonic food web (*broken lines* and *arrows*). There is no strict separation between the classical grazer food chain and the microbial food web, and ciliates and some other protists are at the interface of the two aspects of the pelagic food web; 'HNF' = heterotrophic nanoflagellates (modified from Weisse 2003, 2006)

higher than under oligotrophic (=nutrient-low) conditions and contact rates among planktonic organisms may vary accordingly (Rothschild and Osborn 1988). If we assume a relatively high abundance of one ciliate mL^{-1} of a given species in a lake and that this species is medium-sized (50 μm in length), the distance to its closest

Fig. 19.2 Representative ciliates in lake plankton. **a** Epizoic peritrich ciliates (*Epistylis*; *arrows*) on zooplankton; **b** *Ophryoglena* inside the carapax of a dead crustacean (*arrow*); **c** epiphytic *Vorticella* species attached to colonial diatoms; **d** crushed armour of a mixotrophic (algal-bearing) *Coleps*, *arrow* denotes single algal symbiont; **e** conspicuous caudal spine of the *Coleps* armour (*arrow*); **f** mixotrophic kleptoplast-bearing *Pelagostrombidium* (*arrow* denotes kleptoplast = 'stolen' chloroplasts used for photosynthesis by the ciliate after the digestion of an alga); **g** *Urotricha* species characteristically present in lake plankton (B. Sonntag, original)

conspecific corresponds to, on average, 200x its body length (in linear dimensions). Most ciliate species in the oligotrophic open ocean or in remote alpine lakes are even smaller and their abundances are <1 mL^{-1} (Sonntag et al. 2011a; Weisse 2014), further increasing the theoretical distance to the next neighbour. For example, in one millilitre (mL) in a productive system, we find 550 ciliates, 7 rotifers, and 90 × 10^6 bacteria (Ong'ondo et al. 2013). In contrast, in nutrient-limited systems such as remote high alpine lakes, per mL only 5 ciliates, 1,500 algae, 0.02 zooplankton and 4 × 10^5 bacteria can be observed (Kammerlander et al. subm). It follows that ciliate-ciliate communication is not an easy task in many aquatic environments.

Another challenge of communication is that the high dilution of any chemical substance released into the aquatic environment by an organism implies a major analytical challenge. Even with advanced analytical techniques, there is always the risk of contamination whenever an observer manipulates ciliates in their natural realm. It is, therefore, not surprising that very little is known on the significance of ciliate communication in terms of chemical cues in aquatic ecosystems. The following chapter summarises our knowledge on ciliate communication in aquatic ecosystems with emphasis on planktonic food webs. Since this research discipline is still in its infancy, the major goal of this article is to identify open questions for future research. Accordingly, we will deal with conceptual issues rather than providing solid facts in the following.

19.2 Ciliates in Planktonic Food Webs

The key role of planktonic ciliates in lakes and oceans is undoubted (Azam et al. 1983; Weisse 2003). Though their general roles as consumers of phytoplankton and bacteria and link to higher trophic levels are known, less studies on the autecology of ciliate species are reported. Commonly, seasonal temporal patterns and succession are driven by various environmental factors such as food availability, predation, parasitism and abiotic factors (e.g., temperature, pH). Elucidating particular interactions among ciliates, their potential food organisms and predators are still a major challenge in aquatic ecology (e.g., Sommer et al. 2012). The species differ in their timing of appearance and influence community patterns (Müller 1989; Sonntag et al. 2006). Temporal and spatial variation in the abundance of species can increase or decrease the impact of interspecific interactions on community composition (Morin 2011). Spatial variation can occur at different scales, ranging from the clumping of organisms within small patches of a habitat to large-scale patterns. Micropatches (cm-scale) may provide increased food levels for ciliates (Montagnes et al. 1996). Understanding the complex interactions at the basis of aquatic food webs is therefore a cornerstone for elucidating ecosystem functioning. Inference of such interactions, including co-occurrences and seasonal dynamics is still a major topic in aquatic ecology (Steele et al. 2011; Posch et al. 2015). Recently, the existence of networks has been demonstrated for the ubiquitous marine SAR11 bacteria cluster; Steele et al. (2011) detected multiple instances where bacteria and eukaryotes (including ciliates) were connected. Weisse (2014) postulated that rare ciliates may directly or indirectly require other organisms to survive. For instance, rare species may take refuge from parasites in the presence of an abundant, closely related species. Undoubtedly, the maintenance of networks requires inter- and intraspecific communication and cooperation.

Depending on the lake's trophic status, ciliate assemblages vary considerably in terms of species diversity and abundances, and by far more than 100 ciliate species can be detected in the plankton of a single lake (e.g., Müller et al. 1991; Weisse and Müller 1998; Foissner et al. 1999; Sonntag et al. 2006; Esteban et al. 2012). Predictable recurrent patterns such as the so-called 'spring bloom' are known phenomena in temperate lakes. At this time of the year, ciliate species of the genera *Balanion* and *Urotricha* co-occur with the phytoplankton bloom that is accelerated by higher temperature and increased photosynthetic activity (Müller et al. 1991; Weisse et al. 1990, Fig. 19.2g). Commonly, ciliates of the genera *Rimostrombidium*, *Urotricha* and *Balanion* are key species in temperate lakes as they may consume >50 % of the algal standing stock during the phytoplankton spring bloom (for species-specific information see Foissner et al. 1999 and references therein). In the hypolimnion of deep lakes and highly productive small water bodies, oxygen depleted water layers are a preferred habitat of obligate and facultative anaerobic ciliates (Fenchel and Finlay 1990; Fenchel et al. 1989; Finlay and Fenchel 1986). Other predictable recurrent patterns include a mixotrophic ciliate assemblage typically observed in the epilimnion in late summer and autumn (Sonntag et al. 2006, 2011b). Mixotrophy can be considered as an advantage at times of low food supply and in temperate lakes these ciliates are commonly found in the surface water layers or around the oxycline (=zone of oxygen depletion due to degradation processes; Berninger et al. 1986; Esteban et al. 2010; Sonntag et al. 2011b). Endosymbioses in planktonic ciliates typically include a green algal partner (Pröschold et al. 2011, Fig. 19.2d). In such mutualistic relationships, the ciliate actively ingests food (=heterotrophic nutrition mode) and gets photosynthetic products from its algal partner (=autotrophic). Both live tightly together and algae are passed to the daughter ciliate cell during division. Another form of mixotrophy is that the ciliate accumulates 'kleptoplasts' (=stolen chloroplasts) from ingested algae that are utilised by the host (Fig. 19.2f).

19.3 The Conceptual Background—A Historical Perspective

The state of the art that we have portrayed in the previous section has developed over the past 50 years. In the 1960 and 1970s eutrophication reached its peak in many industrialised countries of the Northern Hemisphere. Resulting from concerted international research programs such as the International Biological Program (IBP) and the OECD eutrophication study (e.g., Schindler 2006; Vollenweider 1968), dissolved reactive phosphorus and nitrogen (as nitrate, nitrite, and ammonium) were identified as the major macronutrients limiting primary production in lakes (mostly P) and in inshore marine systems (mostly N). Improved water management and concomitant legal actions (e.g., banning of phosphorus from detergents) led to a re-oligotrophication of many lakes and coastal areas in the following decades.

With the advent of novel techniques for measuring microbial abundance and production in the 1970s and early 1980s, it became obvious that heterotrophic microbes (bacteria and protists) are quantitatively important in oceanic planktonic food webs as producers and consumers of organic matter (Azam et al. 1983; Pomeroy 1974; Sieburth et al. 1978; Williams 1981); little later this change in the food web paradigm was also accepted by limnologists (reviewed by Weisse 2003, Fig. 19.1). In the final two decades of the last century, a debate followed mainly in the freshwater literature about the primary mechanisms controlling microbial biomass and production in situ. It soon became obvious that planktonic microbes are both controlled 'bottom-up' by resource limitation and 'from above' by consumer regulation via grazing pressure of larger organisms (Billen et al. 1990; McQueen et al. 1989; Sommer et al. 1986; Weisse 1991); the mutual significance of these driving forces changes across different aquatic ecosystems and seasonally within a given habitat. Advocates of the 'top-down' control were inspired by the (partial!) success of biomanipulation. Biomanipulation is a management technique to enhance water quality by controlling phytoplankton biomass via increasing the rate at which algae are removed by zooplankton (Benndorf 1995; Shapiro et al. 1975); to achieve the latter, a lake is stocked with piscivorous fish that reduces the populations of (zoo)planktivorous fish. Biomanipulation was mainly applied at the height of the eutrophication but is still being used occasionally (e.g., Ha et al. 2013; Peretyatko et al. 2012). This method was supported by the finding that predator control at a higher trophic level may 'cascade down' to the lower levels of the food web (Carpenter et al. 1985; Dinasquet et al. 2012).

In our context, the above historical background is important because it is not only reflected by but also coined the conceptual understanding of at least one generation of biological oceanographers and limnologists. The common denominator of all these issues is that since Lindeman's 'trophic pyramid' (Lindeman 1942) food webs are thought to be hierarchically organised, with production at the lower level determining production at the next higher level. This 'vertical structure' largely ignores 'lateral' organism interactions occurring at the same level. Even in Tilman's 'mechanistic resource competition theory' (Tilman 1981, 1982) the 'lateral' effects (inter- and intraspecific competition) were mediated by 'bottom-up' forces (i.e., nutrient supply). This approach is also inherent in functional ecology that relates predator growth, respectively ingestion rates to prey concentration (numerical response and functional response, reviewed by Weisse et al. 2016). Only recently it has become obvious that not only prey but also grazer population density may affect the functional response; using the ciliate predator-prey system of *Didinium-Paramecium* DeLong and Vasseur (2013) indicated that increased predator abundance can reduce *per capita* ingestion rate via interference competition. These authors state that interference competition reduces the amount of available resource that can be acquired due to the presence of competitors. This definition includes conspecifics and thus differs from the usual definition of interference competition as direct negative interactions between two species by aggressively attempting to exclude one another from particular habitats (Fradkin 1995; Lampert and Sommer 2007). Irrespective of how this 'lateral' effect is mediated, interference competition requires communication between sympatric individuals. Another example that lateral processes, which are not obvious in Fig. 19.1, may be significant in planktonic food webs is the nutritional history (i.e., past-prey availability) of the predators, which can also affect both their numerical and functional responses (Li et al. 2013). In other words, the same amount and quality of food may have different effects on the consumers. Furthermore, if a ciliate predator uses chemical communication for aggregation and collective feeding (see next section), the predation pressure may change at short time scales, although the prey density remained unaltered.

19.4 Communication in Natural Ciliate Populations

Although measurements of communication in natural ciliate populations are virtually non-existing, based upon indirect observations in the field and extrapolation from laboratory experiments it appears safe to assume that ciliates communicate both with each other as well as with their prey and also with their predators. The most obvious example of ciliate-ciliate communication originates from sexual recombination. Since Sonneborn's seminal studies with *Paramecium* (Sonneborn 1936, 1937), mating in ciliates has been extensively studied (reviewed by Dini and Nyberg 1993; Luporini et al. 2015; see Chapter 'Pheromone Signalling' by Luporini). In outbreeding ciliates, sexual recombination can be identified relatively easily by observing conjugating specimens (Fig. 19.3).

Ciliate communication via diffusible signalling pheromones (gamones) during mating comprises a suite of processes that are discussed elsewhere in this book (see Chapters 'Communication in *Tetrahymena* Reproduction' by Ashlock and Pearlman, 'Pheromone Signalling' by Luporini, and 'Mating, Sex, and Reproduction in *Tetrahymena*' by Zufall) and shall not be repeated here. Irrespective of minor specific differences, there is no reason to assume that the basic principles of information exchange during conjugation deciphered with *Tetrahymena*, *Euplotes*, *Blepharisma* and a few other model species (Dini and Nyberg 1993; Hausmann et al. 1996; Luporini et al. 2015; Nanney 1980) in the laboratory do not apply to natural populations.

The pheromone peptides and proteins belong to one of the three classes of secondary metabolites synthesised by ciliates (reviewed by Guella et al. 2010; Luporini et al. 2015; see Chapter 'Pheromone Signalling' by Luporini). The second class of products comprises compounds of mixed biosynthetic origin broadly defined as pigments. The third class, polycyclic terpenoids, has been intensively studied over the past two decades with marine interstitial ciliates belonging to the genus *Euplotes* (Guella et al. 2010). Terpenoids are highly bioactive compounds, inhibiting an array of cell structures and processes of potential competitors and predators of various *Euplotes* species. For instance, euplotin C may affect cell cycle,

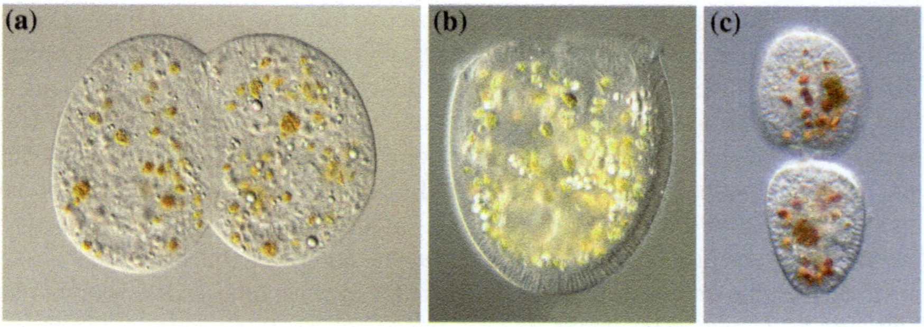

Fig. 19.3 The freshwater ciliate *Bursaridium* sp.; **a** conjugating pair; **b** vegetative cell; **c** vegetative cells near completing cell division (B. Sonntag, original)

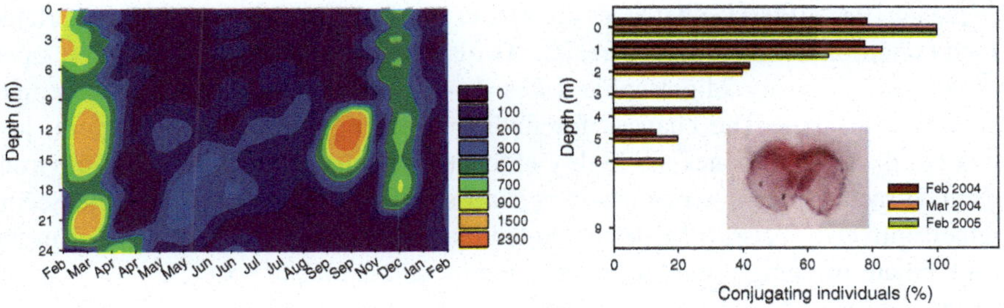

Fig. 19.4 Occurrence of *Rimostrombidium humile* in Lake Piburg, Austria (*left*, cells L^{-1}), and percentage of conjugating pairs (see *insert*) detected in February and March 2004 and 2005 (*right*). During late winter—early spring, up to 100 % of *R. humile* were found in conjugation directly under the ice-cover (lasting from December through April). In 2011, this pattern was again observed (B. Sonntag, original data)

ciliary motility, cell shape, food vacuole formation, and endocytosis in other ciliates (Guella et al. 1994; Ramoino et al. 2007).

With respect to planktonic food webs discussed in this chapter, the external factors triggering or promoting mating deserve special attention. Stress in general and starvation, in particular, induce a shift from asexual to sexual reproduction in various organisms (Hadany and Otto 2009). Accordingly, some authors assume that sex in ciliates (conjugation or autogamy) is generally associated with a decline in (food) resources available for population growth (Dini and Nyberg 1993; Lynn 2008). A likely explanation is the high costs of recombination relative to asexually propagating ciliates; for instance, in *Paramecium tetraurelia*, in the time needed for conjugation and nuclear reorganisation of the two exconjugants, a vigorously growing cell can divide three times to give eight descendant cells (Nanney 1980). Our own previous findings over several years of observation (Fig. 19.4) showed high rates of sex at the beginning of the growing season in some species, while others conjugate more frequently later in the season (B. Sonntag, unpubl. data). Most likely, the combination of low temperature and low food supply will lead to long generation times in late winter. Conjugation in late winter may therefore be associated with lower costs of sex, relative to recombination taking place at more favourable growth conditions in summer.

In conclusion, it is at present an open question if food depletion generally favours sexual recombination in natural ciliate populations. If this is the case, we may assume that ciliates would first increase their swimming speed to (1) search for a patch with higher food supply and (2) enhance the chances of encounter with a conspecific ready to mate. It has been observed, indeed, that ciliates increase their swimming speed at low prey levels (Crawford 1992; Fenchel 1992) but this was not seen in relation to mating. We may, however, speculate that increasing the searching rate at low food levels serves both of the above purposes. If mating is the primary cause and food acquisition is less important, a ciliate may reduce its swimming speed when sensing a higher concentration of gamone (Sugiura et al. 2010). However, aquatic protists typically lead a 'feast and famine' existence

(Fenchel 1987, 1992; Calbet et al. 2013), i.e. they have to cope with highly variable food conditions in their natural realm. As outlined above, in vast parts of the open ocean and in many oligotrophic lakes food is permanently scarce and ciliate abundance is low. The chances for finding a mating partner are, therefore, low unless ciliate pheromones are highly efficient and relatively stable in the environment, attracting potential partners over surprisingly long distances. Since this seems rather unlikely, Weisse (2014) concluded that conjugation is less important in the 'rare ciliate biosphere' than in more common and abundant species.

In more general terms, sexual recombination may provide a significant fitness gain for ciliates, i.e. an adaptive advantage over asexual reproduction under (rapidly) changing environmental conditions as it has been demonstrated recently for facultative sexual rotifers (Becks and Agrawal 2010, 2012). If this assumption holds for ciliates, it may be important in the course of the ongoing, climate-induced environmental changes.

Conjugation is sporadically reported from many ciliate species (Bell 1988; Lynn 2008; Siegel 1961). However, the vast majority of those observations were made with laboratory cultures. We are left with the fact that in spite of its enormous theoretical and practical significance, the occurrence and actual frequency of sex is unknown in >99 % of extant ciliate species in their natural habitats (Weisse 2006, 2008). A detailed, seven-year investigation from marine sandy shores revealed that among 28,890 individual ciliates counted only 20 conjugating pairs (0.07 % of overall ciliates) were observed (Lucchesi and Santangelo 2004). Conjugation was found in less than 7 % of the 121 species encountered in this investigation. These authors concluded that conjugation is an erratic sexual phenomenon among ciliates in the field. Support for this empirical finding is provided by evolutionary theory predicting that sexual recombination has little effect in very large populations; this is because if population size exceeds 10^{12}, as is the case in many freshwater protist species, recombination is not needed to escape Muller's ratchet (Bell 1988). By contrast, other studies have used indirect evidence from mating type experiments and genomic signatures of sex to suggest that sex may be more common in, at least, some natural populations (Doerder et al. 1995; Dunthorn and Katz 2010; Kusch et al. 2000; Schurko et al. 2009, see Chapter 'Mating, Sex, and Reproduction in *Tetrahymena*' by Zufall). Local inbreeding may reduce the effective population size; however, the latter is virtually unknown in ciliates and any other free-living protist species.

Communication between protist predators is known for some species such as the giant heterotrophic dinoflagellate *Noctiluca scintillans. Noctiluca* shows a cooperative feeding strategy by forming large mucus nets that act as efficient particle filters in the marine water column (Omori and Hamner 1982). Since ciliates and dinoflagellates are closely related (within the clade Alveolata), we may assume that similar cooperative feeding strategies have evolved among ciliates. Cooperation affecting growth, aggregation, recruitment and dispersal rates has been demonstrated for the model ciliate *Tetrahymena thermophila* in experimental microcosms (Chaine et al. 2010; Jacob et al. 2015; Pennekamp et al. 2014; Schtickzelle et al. 2009; see Chapter 'Social Information in Cooperation and Dispersal by *Tetrahymena*' by Jacob et al.).

However, the experimental design of these microscosm studies considered only active processes that can be assessed at the 'local' scale (i.e., the organisms' 'home range', Weisse 2008) in a relatively undisturbed environment. The home range can be roughly equated with the 'local' dimension typical for active organisms of different sizes. Weisse (2008) noted that the local dimensions may be worlds apart for passively dispersed microorganisms such as most aquatic ciliates. It is, therefore, an open question to what extent the laboratory results can be extrapolated to the turbulent aquatic environment typically encountered in lakes and in the ocean.

There is indirect evidence that some ciliates communicate when attacking their prey. This is most likely for histophagous species such as the hymenostomes *Ophryoglena* and *Tetrahymena* (reviewed by Lynn 2008, Fig. 19.2b). Similarly, the nutritionally extremely versatile freshwater ciliates of the prostome genus *Coleps* (summarised by Foissner et al. 1999) are also histophagous and may be regarded as scavengers (Auer et al. 2004; Buonanno et al. 2014, Fig. 19.2d, e). We repeatedly observed that *Coleps spetai* cells are attracted by dying or dead other organisms. It seems plausible that this behaviour is mediated by organic substances released from the carcasses; whether or not this involves active communication between the ciliates remains unknown at present. *Coleps* species may also serve as prey for zooplankton. For the planktonic *Coleps hirtus*, Wickham and Gugenberger (2008) performed experiments under the assumption that kairomones released by the predator induced morphological alterations of the armour of the ciliate to reduce its predation rates. The study showed that the conspicuous caudal spine that was supposed to be involved in the defense, was obviously not altered but the length: width ratio of the ciliate was (Fig. 19.2e). Another case study reported adverse effects of the prostome ciliates *Urotricha furcata* and *Urotricha farcta* on their competitor (for small algal food) and predator, the rotifer *Keratella quadrata* (Weisse and Frahm 2001). The ciliates increased significantly rotifer mortality rates in small laboratory containers. Similar to *Coleps*, *Urotricha* spp. were attracted by dying rotifers (T. Weisse, pers. obs.). When the first ciliate arrived in the immediate vicinity of the dying rotifer, several other cells followed suit. Notably, *Urotricha* did not show this behaviour with the closely related, slightly smaller rotifer species *Keratella cochlearis*. In contrast to *K. quadrata*, the presence of the ciliates increased the growth rates of *K. cochlearis* because the latter preyed upon the former. To make the ciliate-rotifer interactions even more complex, the same laboratory experiments also revealed that *K. cochlearis* did not ingest a similar sized other prostome ciliate, *Balanion planctonicum*, while *K. quadrata* drastically reduced the *Balanion* population by grazing and/or mechanical interference (Weisse and Frahm 2002). These laboratory experiments elucidated that interactions between small rotifers and co-occurring prostome ciliates are highly specific and, in part, chemically mediated. Weisse and Frahm (2001) concluded that among *Urotricha* and other small ciliates, chemical defence mechanisms against co-occurring rotifers may have evolved. We will discuss in the next section that many ciliates seem to sense their predators and at least some respond by behavioural changes.

19.5 Adaptation of Planktonic Ciliates

When considering chemical and physical communication between a predator and its prey, several principal phenomena emerge: (1) predator-induced morphologic responses of a (ciliate) species, such as the formation of extended wings and projections, (2) ciliate-predator-induced morphologic responses of prey organisms such as bacteria or algae by colony formation or size changes of single cells, (3) behavioural (escape) responses to avoid the proximity of a predator by, e.g., performing rapid jumps known from several planktonic ciliate species and, (4) active defense or prey capture by the extrusion of cortical granules or extrusomes containing (lethal) toxins. These four responses will be discussed in the next section. The major processes by which ciliates may communicate with each other and their prey and predators are schematically summarised in Fig. 19.5.

In virtually every aquatic ecosystem, ciliates may fall victim to a variety of predators and, therefore, have developed a range of chemical, morphological, and behavioural defense strategies. When a potential predator, no matter if rotifer, crustacean, turbellarian, insect larvae, amoeba or another ciliate is co-occurring with a ciliate (e.g., *Euplotes, Colpidium, Aspidisca*), morphological alterations based on a reorganization of the cytoskeleton can be observed (Kuhlmann et al. 1999). For example, the ciliate-ciliate communication between the predatory *Lembadion bullinum* and *Colpidium kleini* results in shorter and spherical morphotypes in the latter and the induced *Colpidium*-phenotype was attacked less successfully (Fyda 1998). On the other hand, among clonal cultures, different cell

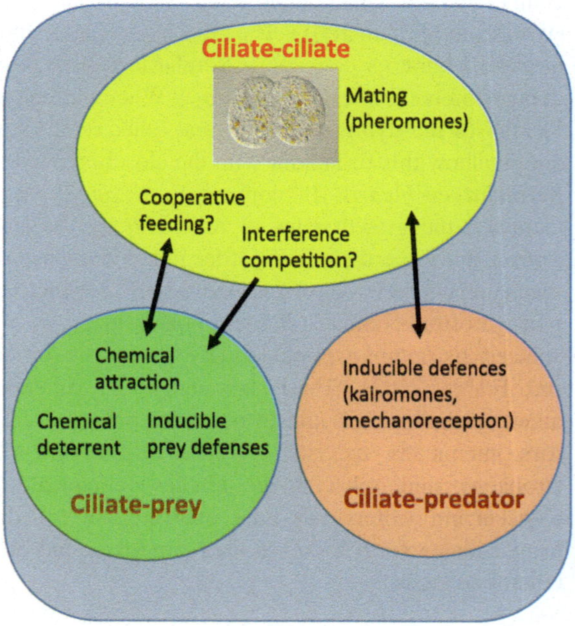

Fig. 19.5 Major processes by which ciliates may communicate with each other and their prey and predators. *Arrows* denote processes that may occur at various levels

types including gigantism are found (Kuhlmann 1993). These large forms are interpreted as individuals that were prepared against cannibalism and that occasionally occur in starving cultures. In all cases, the transformation is a reversible phenomenon. Chemically induced cues (=kairomones) may trigger such morphological alterations. Known signal substances released by predators are, for instance, the L-factor (*Lembadion bullinum*) or the S-factor (released by the turbellarian *Stenostomum sphagnetorum*). For both a protein/peptide nature has been described (see review of Kuhlmann et al. 1999).

Chemical cues may have been involved also when in the presence of a predatory *Cyclidium glaucoma* the co-cultivated bacterial assemblage changed their morphology and became uneatable for the ciliate (Posch et al. 2001). Interestingly, not the length but the width of the bacterial cells significantly increased and the ciliates were not able to ingest their former prey any more. Similarly, Long et al. (2007) reported that chemical cues induced consumer-specific defense during predation on the bloom-forming colonial marine algae *Phaeocystis*. When grazed by ciliates, the algae formed colonies of up to 30,000 µm in diameter that could not be ingested by the ciliates but preferably by copepods. When grazed by copepods, *Phaeocystis* remained as single cells (4–6 µm, Weisse and Scheffel-Möser 1990) which were not in the preferred food size range of the crustaceans. Ciliates and heterotrophic dinoflagellates were identified as the major potential consumers of single cells at the beginning of a dense *Phaeocystis* bloom (Weisse and Scheffel-Möser 1991).

Many pelagic ciliates have jumping responses that reduce their vulnerability to rotifer or copepod nauplii predators and the response is not induced by kairomones but of hydromechanical nature. Common planktonic ciliates of the genera *Urotricha, Halteria, Askenasia, Rimostrombidium, Mesodinium* (*Myrionecta*) or *Balanion* show typical intermittent jumps or bursts of very rapid swimming (for an overview on characteristic movements see species descriptions in Foissner et al. 1999). So far, many ciliates were tested for their individual jumping response which appears to be an effective strategy to escape zooplankton predation (e.g., Gilbert 1994). These escape jumps were not only recognised after direct contact between a zooplankton predator and a ciliate (or an approaching micropipette) but even before both touched each other (e.g., Tamar 1968, 1979; Taylor et al. 1971; Sonntag et al. 2011a). Activated mechanoreceptors in the ciliate cell membrane are likely responsible to perceive the deformation or depolarization of the cell membrane after mechanic contact or hydraulic pressure waves (Naitoh and Eckert 1969). Besides, a non-jumping behaviour as seen in *Paramecium* might be less successful and the species easier preyed (Gilbert 1994). Copepod nauplii are commonly found in the pelagial of lakes and oceans and are known as feeding on protists. Nauplii perform attack jumps against moving prey that touches their sensory setae. Interestingly, these jumps are not directed towards but past a prey and the nauplius sets up a strong intermittent feeding current pulling the prey to the mouth (Bruno et al. 2012). Consequently, a prey ciliate has to be moving to be detected. When movement is a trigger for a predator for catching a prey, it becomes evident that some planktonic ciliates of the genera *Histiobalantium, Askenasia, Halteria, Pelagohalteria* do not only perform (escape) jumps but they also remain motionless floating in the water.

This floating status can be maintained by special ciliature including some few long bristles (described as 'jumping bristles' in halteriids or 'tactile cilia' in *Histiobalantium*). Similar to nauplii, the diffusion feeder *Histiobalantium bodamicum* must rely on prey motility and was therefore able to ingest cryptophytes but not the centric diatom *Stephanodiscus hantzschii* in feeding experiments (Müller and Weisse 1994; Müller and Schlegel 1999). Jumping may therefore not always be an advantage as potential predators are incited by the movements (Jakobsen 2001). Moreover, swimming at high velocities as observed during jumping is energetically very costly (Crawford 1992).

An efficient direct defense mechanism in ciliates is the possession of numerous so-called toxicysts. These are specific cell organelles (=extrusomes filled with a toxic agent) that can be immediately ejected and paralyze or kill an aggressor or prey (e.g., Rosati and Modeo 2003). *Coleps hirtus* has extrusomes located in the oral apparatus and during predation, the extrusomes are extruded into prey ciliates immobilizing the victim and ingesting it (Buonanno et al. 2014). Even young zebrafish larvae have been attacked and killed by assault of hundreds of *Coleps* (Mazanec and Trevarrow 1998). Other efficient lethal toxins are found in coloured and colourless cortical granules of heterotrich ciliates, e.g., climacostol (*Climacostomum virens*), stentorin (*Stentor coeruleus*) and blepharismin (*Blepharisma japonicum*) (Miyake et al. 2003).

In spite of the enormous significance of predation on ciliates, food supply has been identified as the single most important factor affecting growth rates of planktonic ciliates in seasonally varying environments (Weisse 2006). As described above, many free-living ciliate species have to cope with permanent or occasional food shortage in the field. The most obvious adaptations to food depletion are to reduce the cellular metabolism and/or form resting stages (cysts). The sensitivity to low food concentrations can be deduced from a predator's numerical response, i.e. the specific growth rate of a (ciliate) predator as a function of its prey abundance or biomass (Fig. 19.6). The steeper the initial slope (α) and the lower the x-axis intercept (threshold food concentration, V') of a given predator are, the higher are its prey affinity and ability to survive low food levels. Higher maximum growth rates (r_{max}) imply an advantage over a potential competitor at permanently high food levels. In analogy to the mechanistic resource competition theory developed for nutrient uptake of algae (Sommer 1989; Tilman 1982; Tilman et al. 1982), the competitive ability of two co-occurring predator species can be inferred from their respective numerical response curves (Lampert and Sommer 2007; Rothhaupt 1988). In the example shown in Fig. 19.6, species B reaches unusually high maximum growth rates, i.e. it would be the superior competitor relative to species A at high food concentrations; however, species B would be outcompeted by species A at low food abundance and become extinct unless this is prevented by behavioural adaptations. Such a scenario was recently illustrated with two ciliate species from the reservoir of tank bromeliads: the colpodid ciliate *Bromeliothrix metopoides* needed high food concentration to thrive and was able to survive in the presence of the bromeliophryid *Glaucomides bromelicola* at low food levels only because, in contrast to the latter, it can encyst (Weisse et al. 2013a).

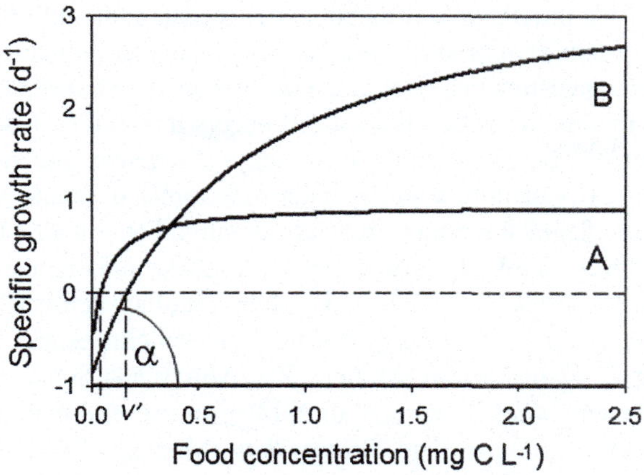

Fig. 19.6 Numerical response of two algivorous planktonic freshwater ciliates, the prostomes *Balanion planctonicum* (*species A*) and *Urotricha farcta* (*species B*). Species A needs a lower threshold food concentration (*V'*, indicated by the *vertical lines*) than species B to survive. If food levels exceed 0.5 mg C L^{-1}, species B reaches higher growth rates and becomes the superior competitor for algal food (data from Weisse et al. 2001, 2002)

Another remarkable example of behavioural adaptation to low food conditions originates from comparing the already mentioned prostome ciliates *B. planctonicum*, *U. furcata* and *U. farcta*. Both *Urotricha* species have significantly higher threshold food concentration than *Balanion*, i.e. they are inferior competitors at low food conditions (Weisse et al. 2001, Fig. 19.6). However, *Balanion* is highly sensitive to starvation; if the food abundance falls below the critical threshold, its growth rate becomes negative and the population goes extinct within a few days, as predicted by its numerical response (Müller and Schlegel 1999; Weisse et al. 2001). The two *Urotricha* species seem to violate the numerical response concept: when food falls below the threshold, their populations decline but do survive at low levels, provided that resources (light, nutrients) are available for the algal food (Weisse et al. 2001). As a result, the prey population may recover, and typical predator-prey cycles emerge. Weisse et al. (2001) reported that they have maintained *U. furcata* cultures fed with the small cryptophyte *Cryptomonas* sp. without exchanging the medium or adding food for several months; there was no indication that the ciliates encysted. An intriguing question is, what happens in *Urotricha* cultures if the food abundance falls below the critical threshold? Does the large majority of the ciliates in the population starve to death to allow the few remaining cells to survive with the scarce food, or do those cells survive in the population that are particularly good at sustaining starvation by reducing their metabolism? These possibilities are not mutually exclusive and it is important to note that the numerical response is averaged over a population, i.e. that phenotypic plasticity in response to food supply is likely. It is obvious that ciliates sense their prey (case I); more exciting is the question if ciliates also communicate with their conspecifics, i.e. exchange information about their food (case II, discussed above). This issue is complicated by the fact that predators may interact with each other via interference competition. A recent study with the model ciliate predator-prey system *Didinium-Paramecium* suggests that predator abundance can reduce *per capita* ingestion rate (DeLong and Vasseur 2013). Although the evidence is indirect, it supports the notion that case II occurs in natural ciliate populations.

A recent review on the functional ecology of aquatic phagotrophic protists (Weisse et al. 2016) identified some major taxonomic differences. For instance, heterotrophic dinoflagellates tend to sustain starvation better than ciliates, surviving longer at food levels below the critical threshold; these authors also suggested that most planktonic ciliates are *r*-selected, while dinoflagellates tend towards *K*-selection. Typical examples for fast growing *r*-selected ciliates are 'boom and bust' species such as *B. planctonicum* (Müller 1991; Müller and Schlegel 1999; Weisse et al. 2001), *B. metopoides* (Weisse et al. 2013b), and several marine species of the oligotrich genera *Strobilidium* and *Strombidium* (Montagnes 1996). Examples for *K*-selected planktonic ciliates are rare; the slow-growing freshwater species *H. bodamicum* may become a superior competitor in periods when algal concentrations are too low to support peaks of rapidly growing opportunistic (*r*-selected) species (Müller and Weisse 1994).

In addition to ecophysiological differences apparent across major taxonomic groups, there are pronounced intraspecific differences (Boenigk 2008; Lowe et al. 2005; Weisse 2002; Weisse and Rammer 2006; Zhang et al. 2014), pointing to large phenotypic flexibility of many protist species. Intraspecific differences were not only reported for growth and grazing rates of planktonic ciliates (Montagnes et al. 1996; Pérez-Uz 1995; Weisse 2006), but they were also found in response to abiotic factors such as temperature and pH (Gächter and Weisse 2006; Weisse et al. 2007).

19.6 Conclusions

Adaptations of natural ciliate populations to their abiotic and biotic environment have been well documented in selected freshwater and marine species. The effect of individual environmental variables and, less often, the interactive effect of several variables on growth and survival rates of ciliates have been studied with a number of species in laboratory experiments. However, more information is needed on the autecology of free-living ciliate species and their role in the planktonic community. In particular, indirect ('lateral') effects received little attention; for instance, per capita ingestion and growth rates of a ciliate predator may change not only as a function of prey density, but may also be affected by the density and nutritional history of the predator.

Some adaptations, e.g., morphological and behavioural defense strategies of ciliates against predators and parasites, and symbiotic relationships between ciliates and algae or bacteria require chemical or hydromechanical communication between ciliates and other organisms. Communication is also undoubtedly involved in conjugation, and the chemical cues triggering ciliate mating have been identified in laboratory cultures with several model species. Much less is known about other forms of communication such as cooperation in the natural realm. Recent evidence gained in microcosm experiments suggests that cooperation in aggregation and active dispersal may be wide spread but the chemical nature of cooperation remains at present obscure.

Ciliate communication over relatively short distances (μm to mm spatial scale) mediated by direct contact as well as by chemical and hydromechanical cues is a fact, but the question is if, and if so, how ciliates communicate over distances beyond their local scale ('home range'). The benefits of cooperation, which are obvious within the home range, may be offset by costs associated with communication at larger spatial scales. To our knowledge, the trade-off between costs and benefits has not yet been systematically investigated for planktonic ciliates. In a first step, chemical and hydromechanical communication should be scaled relative to the size, motility, and generation times of aquatic ciliates.

Communication in ciliates is obvious at various levels, ranging from intracellular communication (cell signalling) to the population and ecosystem level. Both intra- and interspecific communication occur at all levels. However, the first level is the domain of cell biologists, while evolutionary biologists and ecologists primarily address the population and ecosystem level. To advance our understanding of communication in aquatic ciliates, more communication between the various biological disciplines is needed.

References

Auer B, Elzer U, Arndt H (2004) Comparison of pelagic food webs in lakes along a trophic gradient and with seasonal aspects: influence of resource and predation. J Plankton Res 26:697–709

Azam F, Fenchel T, Field J, Gray J, Meyer-Reil L, Thingstad F (1983) The ecological role of water column microbes in the sea. Mar Ecol Prog Ser 10:257–263

Becks L, Agrawal AF (2010) Higher rates of sex evolve in spatially heterogeneous environments. Nature 468:89–92

Becks L, Agrawal AF (2012) The evolution of sex is favoured during adaptation to new environments. PLoS Biol 10:e1001317

Bell G (1988) Sex and death in Protozoa. Cambridge University Press, Cambridge

Benndorf J (1995) Possibilities and limits for controlling eutrophication by biomanipulation. Int Rev Hydrobiol 80:519–534

Berger H (1999) Monograph of the Oxytrichidae (Ciliophora, Hypotrichia). Springer, Dordrecht

Berninger U-G, Finlay BJ, Canter HM (1986) The spatial distribution and ecology of zoochlorellae-bearing ciliates in a productive pond. J Protozool 33:557–563

Billen G, Servais P, Becquevort S (1990) Dynamics of bacterioplankton in oligotrophic and eutrophic aquatic environments: bottom-up or top-down control? Hydrobiologia 207:37–42

Boenigk J (2008) Nanoflagellates: Functional groups and intraspecific variation. Denisia 23:331–335

Bruno E, Borg CMA, Kiørboe T (2012) Prey detection and prey capture in copepod nauplii. PLoS ONE 7:e47906

Buck KR, Newton J (1995) Fecal pellet flux in Dabob Bay during a diatom bloom: contribution of microzooplankton. Limnol Oceanogr 40:306–315

Buonanno F, Anesi A, Guella G, Kumar S, Bharti D, La Terza A, Quassinti L, Bramucci M, Ortenzi C (2014) Chemical Offense by Means of Toxicysts in the Freshwater Ciliate, *Coleps hirtus*. J Eukaryot Microbiol 61:293–304

Calbet A, Isari S, Martínez RA, Saiz E, Garrido S, Peters J, Borrat RM, Alcaraz M (2013) Adaptations to feast and famine in different strains of the marine heterotrophic dinoflagellates *Gyrodinium dominans* and *Oxyrrhis marina*. Mar Ecol Prog Ser 483:67–84

Carpenter SR, Kitchell JF, Hodgson JR (1985) Cascading trophic interactions and lake productivity. Bioscience 35:634–639

Chaine AS, Schtickzelle N, Polard T, Huet M, Clobert J (2010) Kin-based recognition and social aggregation in a ciliate. Evolution 64:1290–1300

Crawford DW (1992) Metabolic cost of motility in planktonic protists: theoretical considerations on size scaling and swimming speed. Microb Ecol 24:1–10

DeLong JP, Vasseur DA (2013) Linked exploitation and interference competition drives the variable behavior of a classic predator-prey system. Oikos 122:1393–1400

Dinasquet J, Titelman J, Møller LF, Setälä O, Granhag L, Andersen T, Båmstedt U, Haraldsson M, Hosia A, Katajisto T, Kragh T, Kuparinen J, Schrøter ML, Søndergaard M, Tiselius P, Riemann L (2012) Cascading effects of the ctenophore *Mnemiopsis leidyi* on the planktonic food web in a nutrient-limited estuarine system. Mar Ecol Prog Ser 460:49–61

Dini F, Nyberg D (1993) Sex in ciliates. In: Jones JG (ed) Advances in Microbial Ecology. Plenum Press, New York, pp 129–144

Doerder FP, Gates MA, Eberhardt FP, Arslanyolu M (1995) High frequency of sex and equal frequencies of mating types in natural populations of the ciliate *Tetrahymena thermophila*. Proc Natl Acad Sci USA 92:8715–8718

Dunthorn M, Katz LA (2010) Secretive ciliates and putative asexuality in microbial eukaryotes. Trends Microbiol 18:183–188

Esteban GF, Fenchel T, Finlay BF (2010) Mixotrophy in ciliates. Protist 161:621–641

Esteban GF, Finlay BJ, Clarke KJ (2012) Priest Pot in the English Lake District: a showcase of microbial diversity. Freshwater Biol 57:321–330

Fenchel T (1987) Ecology of protozoa: The ecology of free-living phagotrophic protists. Science Tech Publishers & Springer-Verlag, Madison, Berlin

Fenchel T, Finlay BJ, Gianni A (1989) Microaerophily in ciliates: responses of an *Euplotes* species (Hypotrichida) to oxygen tension. Arch Protistenkd 137:317–330

Fenchel T, Finlay BJ (1990) Anaerobic free-living protozoa: growth efficiencies and the structure of anaerobic communities. FEMS Microbiol Ecol 74:269–276

Fenchel T (1992) What can ecologists learn from microbes: life beneath a square centimetre of sediment surface. Funct Ecol 6:499–507

Finlay BJ, Fenchel T (1986) Spatial distribution of microaerophilic ciliates in lakes: the interacting roles of oxygen tension, blue light and gravity. In: Megušar F, Gantar M (eds) Fourth International Symposium on Microbial Ecology. Slovene Society for Microbiology, Ljubljana, 343–348

Foissner W, Berger H, Schaumburg J (1999) Identification and Ecology of Limnetic Plankton Ciliates. Bayerisches Landesamt für Wasserwirtschaft, Munich

Fradkin S (1995) Effects of interference and exploitative competition from large-bodied cladocerans on rotifer community structure. Hydrobiologia 313–314:387–393

Fyda J (1998) Predator induced morphological changes in the ciliate *Colpidium* (Protozoa, Ciliophora). Eur J Protistol 34:111–117

Gächter E, Weisse T (2006) Local adaptation among geographically distant clones of the cosmopolitan freshwater ciliate *Meseres corlissi*. I. Temperature response. Aquat Microb Ecol 45:291–300

Gilbert JJ (1994) Jumping behavior in the oligotrich ciliates *Strobilidium velox* and *Halteria grandinella* and its significance as a defense against rotifers. Microb Ecol 27:189–200

Görtz H-D (1996) Symbiosis in ciliates. In: Hausmann K, Bradbury PC, Fischer G (eds) Ciliates: Cells as Organisms. Stuttgart, pp 441–462

Guella G, Dini F, Tomei A, Pietra F (1994) Preuplotin, a putative biogenetic precursor of the euplotins, bioactive sesquiterpenoids of the marine ciliated protist *Euplotes crassus*. J Chem Soc, Perkin Trans I 2:161–166

Guella G, Skropeta D, Di Giuseppe G, Dini F (2010) Structures, biological activities and phylogenetic relationships of terpenoids from marine ciliates of the genus *Euplotes*. Mar Drugs 8:2080–2116

Ha J-Y, Saneyoshi M, Park H-D, Toda H, Kitano S, Homma T, Shiina T, Moriyama Y, Chang K-H, Hanazato T (2013) Lake restoration by biomanipulation using piscivore and *Daphnia* stocking; results of the biomanipulation in Japan. Limnology 14:19–30

Hadany L, Otto SP (2009) Condition-dependent sex and the rate of adaptation. Am Nat 174S71–S78

Hausmann K, Bradbury PC, Fischer G (eds) (1996) Ciliates: Cells as organisms. Stuttgart

Jacob S, Chaine AS, Schtickzelle N, Huet M, Clobert J (2015) Social information from immigrants: multiple immigrant-based sources of information for dispersal decisions in a ciliate. J Animal Ecol 84:1373–1383

Jakobsen HH (2001) Escape response of planktonic protists to fluid mechanical signals. Mar Ecol Progr Ser 214:67–78

Kuhlmann H-W (1993) Giants in *Lembadion bullinum* (Ciliophora, Hymenostomata)—general morphology and inducing conditions. Arch Protistenkd 143:325–336

Kuhlmann H-W, Kusch J, Heckmann K (1999) Predator-induced defenses in ciliated protozoa. In: Tollrian R, Harvell CD (eds) The ecology and evolution of inducible defenses. Princeton University Press. Princeton, New Jersey, pp 142–159

Kusch J, Welter H, Stremmel M, Schmidt HJ (2000) Genetic diversity in populations of a freshwater ciliate. Hydrobiologia 431:185–192

Lampert W, Sommer U (2007) Limnoecology. Oxford University Press, Oxford

Li J, Fenton A, Kettley L, Roberts P, Montagnes DJS (2013) Reconsidering the importance of the past in predator–prey models: both numerical and functional responses depend on delayed prey densities. Proc R Soc B 280

Lindeman RL (1942) The trophic-dynamic aspect of ecology. Ecology 23:399–418

Long JD, Smalley GW, Barsby T, Anderson JT, Hay ME (2007) Chemical cues induce consumer-specific defenses in a bloom-forming marine phytoplankton. Proc Natl Acad Sci USA 104:10512–10517

Lowe CD, Day A, Kemp SJ, Montagnes DJS (2005) There are high levels of functional and genetic diversity in *Oxyrrhis marina*. J Eukaryot Microbiol 52:250–257

Lucchesi P, Santangelo G (2004) How often does conjugation in ciliates occur? Clues from a seven-year study on marine sandy shores. Aquat Microb Ecol 36:195–200

Luporini P, Alimenti C, Vallesi A (2015) Ciliate pheromone structures and activity: a review. Ital J Zool 82:3–14

Lynn DH (2008) The ciliated protozoa—characterization, classification, and guide to the literature. Springer, Dordrecht

Mazanec AI, Trevarrow B (1998) *Coleps*, scourge of the baby zebrafish. Zebrafish Sci Monit 5:1

McQueen DL, Johannes MRS, Post JR, Stewart TJ, Lean DRS (1989) Bottom-up and top-down impacts on freshwater pelagic community structure. Ecol Monogr 59:289–309

Miyake A, Buonanno F, Saltalamacchia P, Masaki ME, Iio H (2003) Chemical defense by means of extrusive cortical granules in the heterotrich ciliate *Climacostomum virens*. Eur J Protistol 39:25–36

Montagnes DJS (1996) Growth responses of planktonic ciliates in the genera *Strobilidium* and *Strombidium*. Mar Ecol Prog Ser 130:241–254

Montagnes DJS, Berger JD, Taylor FJR (1996) Growth rate of the marine planktonic ciliate *Strombidinopsis cheshiri* Snyder and Ohman as a function of food concentration and interclonal variability. J Exp Mar Biol Ecol 206:121–132

Morin PJ (2011) Community Ecology, 2nd edn. John Wiley & Sons Ltd, Chichester, UK

Müller H (1989) The relative importance of different ciliate taxa in the pelagic food web of Lake Constance. Microb Ecol 18:261–273

Müller H (1991) *Pseudobalanion planctonicum* (Ciliophora, Prostomatida): ecological significance of an algivorous nanociliate in a deep meso-eutrophic lake. J Plankton Res 13:247–262
Müller H, Schöne A, Pinto-Coelho RM, Schweizer A, Weisse T (1991) Seasonal succession of ciliates in Lake Constance. Microb Ecol 21:119–138
Müller H, Weisse T (1994) Laboratory and field observations on the scuticociliate *Histiobalantium* from the pelagic zone of Lake Constance, FRG. J Plankton Res 16:391–401
Müller H, Schlegel A (1999) Responses of three freshwater planktonic ciliates with different feeding modes to cryptophyte and diatom prey. Aquat Microb Ecol 17:49–60
Naitoh Y, Eckert R (1969) Ironic mechanisms controlling behavioral responses of *Paramecium* to mechanical stimulation. Science 164:963–965
Nanney DL (1980) Experimental ciliatology. An introduction to genetic and developmental analysis in ciliates. Wiley, Somerset, New Jersey
Omori M, Hamner WM (1982) Patchy distribution of zooplankton: behavior, population assessment and sampling problems. Mar Biol 72:193–200
Ong'ondo GO, Yasindi AW, Oduor SO, Jost S, Schagerl M, Sonntag B, Boenigk J (2013) Ecology and community structure of ciliated protists in two alkaline–saline Rift Valley lakes in Kenya with special emphasis on *Frontonia*. J Plankton Res 35:759–771
Pennekamp F, Mitchell KA, Chaine A, Schtickzelle N (2014) Dispersal propensity in *Tetrahymena thermophila* ciliates–a reaction norm perspective. Evolution 68:2319–2330
Peretyatko A, Teissier S, De Backer S, Triest L (2012) Biomanipulation of hypereutrophic ponds: when it works and why it fails. Environ Monit Assess 184:1517–1531
Pérez-Uz B (1995) Growth rate variability in geographically divers clones of *Uronema* (Ciliophora: Scuticociliatida). FEMS Microb Ecol 16:193–204
Pomeroy LR (1974) The ocean's food web: a changing paradigm. Bioscience 24:499–504
Porter KG, Sherr EB, Sherr BF, Pace M, Sanders RW (1985) Protozoa in planktonic food webs. J Protozool 32:409–415
Posch T, Jezbera J, Vrba J, Šimek K, Pernthaler J, Andreatta S, Sonntag B (2001) Size selective feeding in *Cyclidium glaucoma* (Ciliophora, Scuticociliatida) and its effects on bacterial community structure: a study from a continuous cultivation system. Microb Ecol 42:217–227
Posch T, Eugster B, Pomati F, Pernthaler J, Pitsch G, Eckert EM (2015) Network of interactions between ciliates and phytoplankton during spring. Front Microbiol 6
Pröschold T, Darienko T, Silva DC, Reisser W, Krienitz L (2011) The systematics of *Zoochlorella* revisited employing an integrative approach. Environ Microbiol 13:350–364
Ramoino P, Usai C, Maccione S, Beltrame F, Diaspro A, Fato M, Guella G, Dini F (2007) Effect of the bioactive metabolite euplotin C on phagocytosis and fluid-phase endocytosis in the single-celled eukaryote *Paramecium*. Aquat Toxicol 85:67–75
Rosati G, Modeo L (2003) Extrusomes in ciliates: diversification, distribution, and phylogenetic implications. J Eukaryot Microbiol 30:383–402
Rothhaupt KO (1988) Mechanistic resource competition theory applied to laboratory experiments with zooplankton. Nature 333:660–662
Rothschild BJ, Osborn TR (1988) Small-scale turbulence and plankton contact rates. J Plankton Res 3:465–474
Schindler DW (2006) Recent advances in the understanding and management of eutrophication. Limnol Oceanogr 51:356–363
Schtickzelle N, Fjerdingstad EJ, Chaine A, Clobert J (2009) Cooperative social clusters are not destroyed by dispersal in a ciliate. BMC Evol Biol 9
Schurko AM, Neiman M, Logsdon JM Jr (2009) Signs of sex: what we know and how we know it. Trends Ecol Evol 24:208–217
Shapiro J, Lamarra V, Lynch M (1975) Biomanipulation: an ecosystem approach to lake restoration. In: Brezonik PL, Fox JL (eds) Water Quality Management through Biological Control. University of Florida. Gainesville, Fl pp 85–96
Sieburth JM, Smetacek V, Lenz J (1978) Pelagic ecosystem structure: heterotrophic compartments of the plankton and their relationship to plankton size fractions. Limnol Oceanogr 23:1256–1263

Siegel RW (1961) Direct and indirect evidence that free-living ciliates conjugate in nature. J Eukaryot Microbiol 8:27–29

Sommer U, Gliwicz ZM, Lampert W, Duncan A (1986) The PEG model of a seasonal succession of planktonic events in fresh waters. Arch Hydrobiol 106:433–471

Sommer U (ed) (1989) The role of competition for resources in phytoplankton succession. In: Plankton ecology. Springer Verlag. Berlin, pp 57–106

Sommer U, Adrian R, De Senerpont Domis L, Elser JJ, Gaedke U, Ibelings B, Jeppesen E, Lurling M, Molinero JC, Mooij WM, van Donk E, Winder M (2012) Beyond the Plankton Ecology Group (PEG) Model: mechanisms driving plankton succession. Annu Rev Ecol Evol Syst 43:429–448

Sonneborn TM (1936) Factors determining conjugation in *Paramecium aurelia*. I. The cyclical factor: The recency of nuclear reorganization. Genetics 21:503–514

Sonneborn TM (1937) Sex, sex inheritance and sex determination in *Paramecium aurelia*. Proc Natl Acad Sci USA 23:378–385

Sonntag B, Posch T, Klammer S, Teubner K, Psenner R (2006) Phagotrophic ciliates and flagellates in an oligotrophic deep alpine lake: contrasting variability with seasons and depths. Aquat Microb Ecol 43:193–207

Sonntag B, Summerer M, Sommaruga R (2011a) Factors involved in the distribution pattern of ciliates in the water column of a transparent alpine lake. J Plankton Res 33:541–546

Sonntag B, Summerer M, Sommaruga R (2011b) Are freshwater mixotrophic ciliates less sensitive to solar ultraviolet radiation than heterotrophic ones? J Eukaryot Microbiol 58:196–202

Steele JA, Countway PD, Xia L, Vigil PD, Beman JM, Kim DY, Chow C-ET, Sachdeva R, Jones AC, Schwalbach MS, Rose JM, Hewson I, Patel A, Sun F, Caron DA, Fuhrman JA (2011) Marine bacterial, archaeal and protistan association networks reveal ecological linkages. ISME J 5:1414–1425

Stoecker DK (1984) Particle production by planktonic ciliates. Limnol Oceanogr 29:930–940

Sugiura M, Shiotani H, Suzaki T, Harumoto T (2010) Behavioural changes induced by the conjugation-inducing pheromones, gamone 1 and 2, in the ciliate *Blepharisma japonicum*. Europ J Protistol 46:143–149

Tamar H (1968) Observations on *Halteria bifurcata* sp. n. and *Halteria grandinella*. Acta Protozool 6:175–183

Tamar H (1979) The movements of jumping ciliates. Arch Protistenk 122:290–327

Taylor FJR, Blackbourn DJ, Blackbourn J (1971) The red-water ciliate *Mesodinium rubrum* and its "incomplete symbionts": a review including new ultrastructural observations. J Fish Res Board Can 28:391–407

Tilman D (1981) Test of resource competition theory using four species of Lake Michigan algae. Ecology 62:802–815

Tilman D (1982) Resource competition and community structure. Princeton University Press, Princeton, New Jersey

Tilman D, Kilham SS, Kilham P (1982) Phytoplankton community ecology: the role of limiting nutrients. Annu Rev Ecol Syst 13:349–372

Vollenweider RA (1968) Scientific fundamentals of the eutrophication of lakes and flowing waters, with particular reference to nitrogen and phosphorus as factors in eutrophication Technical Report. OECD

Weisse T, Müller H, Pinto-Coelho RM, Schweizer A, Springmann D, Baldringer G (1990) Response of the microbial loop to the phytoplankton spring bloom in a large prealpine lake. Limnol Oceanogr 35:781–794

Weisse T, Scheffel-Möser U (1990) Morphometric characteristics and carbon content of *Phaeocystis* cf. *pouchetii*. Bot Mar 33:197–203

Weisse T (1991) The annual cycle of heterotrophic freshwater nanoflagellates: role of bottom-up versus top-down control. J Plankton Res 13:167–185

Weisse T, Scheffel-Möser U (1991) Uncoupling the microbial loop: growth and grazing loss rates of bacteria and heterotrophic nanoflagellates in the North Atlantic. Mar Ecol Progr Ser 71:195–205

Weisse T, Müller H (1998) Planktonic protozoa and the microbial food web in Lake Constance. Arch Hydrobiol Spec Iss Adv Limnol 53:223–254
Weisse T, Frahm A (2001) Species-specific interactions between small planktonic ciliates (*Urotricha* spp.) and rotifers (*Keratella* spp.). J Plankton Res 23:1329–1338
Weisse T, Karstens N, Meyer VCM, Janke L, Lettner S, Teichgräber K (2001) Niche separation in common prostome freshwater ciliates: the effect of food and temperature. Aquat Microb Ecol 26:167–179
Weisse T (2002) The significance of inter- and intraspecific variation in bacterivorous and herbivorous protists. Antonie Van Leeuwenhoek 81:327–341
Weisse T, Frahm A (2002) Direct and indirect impact of two common rotifer species (*Keratella* spp.) on two abundant ciliate species (*Urotricha furcata, Balanion planctonicum*). Freshwater Biol 47:53–64
Weisse T, Stadler P, Lindström ES, Kimmance SA, Montagnes DJS (2002) Interactive effect of temperature and food concentration on growth rate: a test case using the small freshwater ciliate *Urotricha farcta*. Limnol Oceanogr 47:1447–1455
Weisse T (2003) Pelagic microbes—Protozoa and the microbial food web. In: O'Sullivan P, Reynolds CS (eds) The lakes handbook. Blackwell Science Ltd., Oxford, pp 417–460
Weisse T (2006) Freshwater ciliates as ecophysiological model organisms—lessons from *Daphnia*, major achievements, and future perspectives. Arch Hydrobiol 167:371–402
Weisse T, Rammer S (2006) Pronounced ecophysiological clonal differences of two common freshwater ciliates, *Coleps spetai* (Prostomatida) and *Rimostrombidium lacustris* (Oligotrichida), challenge the morphospecies concept. J Plankton Res 27:55–63
Weisse T, Scheffel U, Stadler P, Foissner W (2007) Local adaptation among geographically distant clones of the cosmopolitan freshwater ciliate *Meseres corlissi*. II. Response to pH. Aquat Microb Ecol 47:289–297
Weisse T (2008) Distribution and diversity of aquatic protists: an evolutionary and ecological perspective. Biodivers Conserv 17:243–259
Weisse T, Scheffel U, Stadler P, Foissner W (2013a) Functional ecology of the ciliate *Glaucomides bromelicola*, and comparison with the sympatric species *Bromeliothrix metopoides* J Eukaryot Microb 60:578–587
Weisse T, Scheffel U, Stadler P, Foissner W (2013b) *Bromeliothrix metopoides*, a boom and bust ciliate (Ciliophora, Colpodea) from tank bromeliads. Eur J Protistol 49:406–419
Weisse T (2014) Ciliates and the rare biosphere—community ecology and population dynamics. J Eukaryot Microb 61:419–433
Weisse T, Anderson R, Arndt H, Calbet A, Hansen PJ, Montagnes DJS (2016) Functional Ecology of aquatic phagotrophic protists—concepts, limitations, and perspectives. Eur J Protistol (in press)
Wickham SA, Gugenberger E (2008) Evaluating inducible morphological defences in the common freshwater ciliate, *Coleps hirtus*. J Plankton Res 30:1315–1321
Williams PJLeB (1981) Incorporation of microheterotrophic processes into the classical paradigm of the planktonic food web. Kieler Meeresforsch Sonderh 1:1–28
Zhang Y, Klapper R, Lohbeck KT, Bach LT, Schulz KG, Reusch TBH, Riebesell U (2014) Between- and within-population variations in thermal reaction norms of the coccolithophore *Emiliania huxleyi*. Limnol Oceanogr 59:1570–1580

The manufacturer's authorised representative in the EU is Springer Nature Customer Service Centre GmbH, Europaplatz 3, 69115 Heidelberg, Germany. If you have any concerns regarding our products, please contact ProductSafety@springernature.com

Printed and bound by CPI Group (UK) Ltd, Croydon, CR0 4YY
30/09/2026
02225501-0009